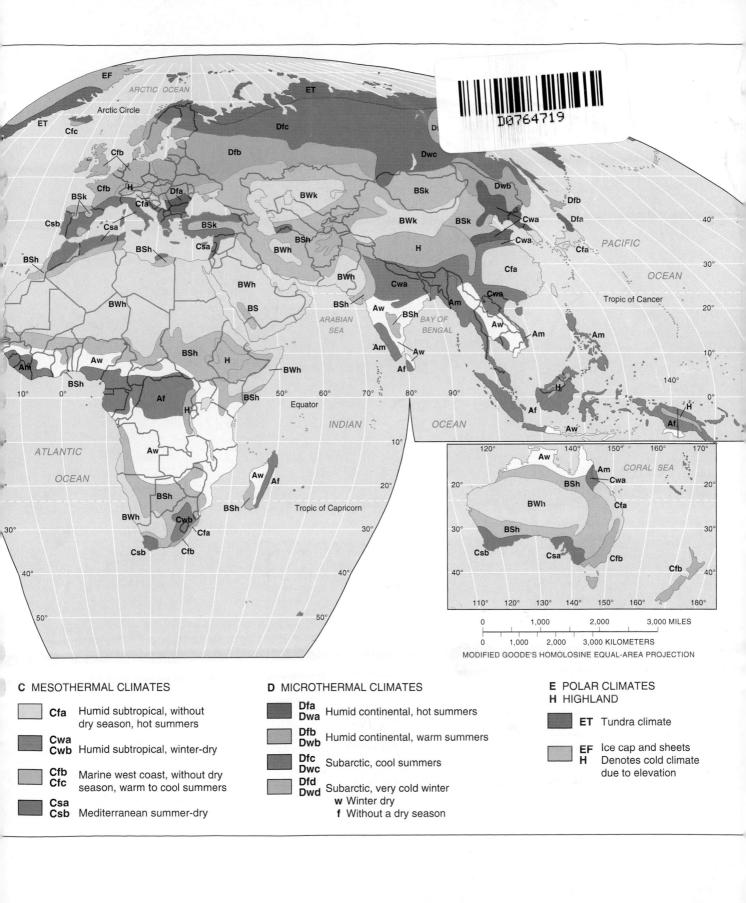

C MESOTHERMAL CLIMATES

	Cfa	Humid subtropical, without dry season, hot summers
	Cwa **Cwb**	Humid subtropical, winter-dry
	Cfb **Cfc**	Marine west coast, without dry season, warm to cool summers
	Csa **Csb**	Mediterranean summer-dry

D MICROTHERMAL CLIMATES

	Dfa **Dwa**	Humid continental, hot summers
	Dfb **Dwb**	Humid continental, warm summers
	Dfc **Dwc**	Subarctic, cool summers
	Dfd **Dwd**	Subarctic, very cold winter
	w	Winter dry
	f	Without a dry season

E POLAR CLIMATES
H HIGHLAND

	ET	Tundra climate
	EF **H**	Ice cap and sheets Denotes cold climate due to elevation

MODIFIED GOODE'S HOMOLOSINE EQUAL-AREA PROJECTION

Dawn at Maligne Lake, Jasper National Park, Alberta, Canada. [Photo by Raymond Gehman.]

ELEMENTAL GEOSYSTEMS

A FOUNDATION IN PHYSICAL GEOGRAPHY

Robert W. Christopherson

PRENTICE HALL
Englewood Cliffs, New Jersey 07632

To all the students and teachers of Earth,
our home planet,
and its sustainable future.

LIBRARY OF CONGRESS CATALOGING-IN-PUBLICATION DATA
Christopherson, Robert W.
    Elemental geosystems : a foundation in physical geography / Robert
W. Christopherson.
        p.   cm.
    Includes bibliographical references and index.
    ISBN 0-02-322461-4
    1. Physical geography. I. Title.
GB54.5 C47   1995
910'.02—dc20   94-38541
                                                       CIP

Front Cover Photo: Maroon Peak, central peak of the Maroon Bells (4271 m, 14,014 ft).
Maroon Lake in the foreground is at 2987 m (9800 ft) elevation. The scene is in the
Maroon Bells-Snowmass Wilderness Area, southwest of Aspen, Colorado (39° N, 107° W).
Photo by © Galen Rowell.
Back Cover Photo: Full Earth photo by Apollo 17 astronauts, December 1972, from NASA.
Frontispiece: Dawn at Maligne Lake, Spirit Island at center, in the Canadian Rockies
(52.5° N, 117° W), Jasper National Park, Alberta, Canada. Photo by Raymond Gehman.

Editor-in-Chief, ESM: Paul F. Corey
Geography Editor: Ray Henderson
Production Assistant: Bobbé Christopherson
Photo Editor: Chris Migdol
Editorial Production Service: Electronic Publishing Services Inc.
Illustrations: Maryland CartoGraphics Inc., Earth Imaging, Inc., Precision Graphics, and Tasa
Graphic Arts Inc.

This book was set in ITC Garamond Light and Helvetica by Electronic Publishing Services
Inc. and was printed and bound by R.R. Donnelley & Sons, Inc. The cover was printed by
Lehigh Press.

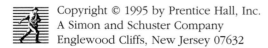

PRINTING 1 2 3 4 5 6 7 8 year 5 6 7 8 9 00 1 2 3

Brief Contents

1   Foundations of Geography   1

PART 1   The Energy-Atmosphere System   34

2   Solar Energy, Seasons, and the Atmosphere   37
3   Atmospheric Energy and Global Temperatures   75
4   Atmospheric and Oceanic Circulation   107

PART 2   Water, Weather, and Climate   138

5   Atmospheric Water and Weather   141
6   Water Resources   185
7   Earth's Climates   213

PART 3   Earth's Changing Landscapes   248

8   The Dynamic Planet   251
9   Earthquakes and Volcanoes   281
10   Weathering, Karst Landscapes, and Mass Movement   315
11   Rivers and Related Landforms   341
12   Wind Processes and Desert Landscapes   371
13   Coastal Processes and Landforms   395
14   Glacial and Periglacial Landscapes   421

PART 4   Biogeography   450

15   The Geography of Soils   453
16   Ecosystems and Biomes   481
17   Earth, Humans, and the New Millennium   529

APPENDIX **A**   Information Sources, Organizations, and Agencies   A.1
APPENDIX **B**   Topographic Map Symbols—USGS   B.1
Suggested Readings   S.1
Glossary   G.1
Index   I.1
Topographic Maps   T.1

CONTENTS

1 Foundations of Geography —————————— 1

THE SCIENCE OF GEOGRAPHY   2
   The Geographic Continuum   4

EARTH SYSTEMS CONCEPTS   5
   Systems Theory   5
   Earth as a System   8
   Earth's Four Spheres   9

A SPHERICAL PLANET   10

LOCATION ON EARTH   12
   Latitude   13
   Longitude   14
   Great Circles and Small Circles   16

PRIME MERIDIAN AND STANDARD TIME   17
   International Date Line   18
   Coordinated Universal Time   18
   Daylight Saving Time   19

MAPS, SCALES, AND PROJECTIONS   19
   Map Scales   20
   Map Projections   21

MAPPING AND TOPOGRAPHIC MAPS   25

REMOTE SENSING AND GIS   27
   Remote Sensing   27
   Geographic Information System (GIS)   30

SUMMARY   32
   Key Terms   32   Review Questions   33

FYI REPORT **1-1**   *Measuring Earth in 247 B.C.*   11
FYI REPORT **1-2**   *The Search for Longitude*   15
NEWS REPORT **#1**   *"GPS: A Personal Locator"*   17
NEWS REPORT **#2**   *"A New Clock for UTC"*   19
NEWS REPORT **#3**   *"Careers in GIS"*   30

PART ONE
The Energy-Atmosphere System   34

2 Solar Energy, Seasons, and the Atmosphere ————— 37

THE SOLAR SYSTEM, SUN, AND EARTH   38
   Dimensions and Distances   38
   Earth's Orbit   38

SOLAR ENERGY: FROM SUN TO EARTH   39
   Solar Wind   41
   Electromagnetic Spectrum of Radiant Energy   42

ENERGY AT THE TOP OF THE ATMOSPHERE   44
   Intercepted Energy   44

THE SEASONS   46
   Seasonality   47
   Reasons for Seasons   48
   Annual March of the Seasons   50

EARTH'S ATMOSPHERE   53
   Heterosphere   54
   Homosphere   56

VARIABLE ATMOSPHERIC COMPONENTS   62
   Natural Sources   62
   Natural Factors That Affect Air Pollution   63
   Anthropogenic Pollution   64

SUMMARY   70
   Key Terms   71   Review Questions   72

FYI REPORT **2-1**   *The Scientific Method*   *39*
FYI REPORT **2-2**   *Stratospheric Ozone Losses: A Worldwide Health Hazard*   *59*
NEWS REPORT **#1**   *"Scientists Discover the Atmosphere's Origin"*   *41*
NEWS REPORT **#2**   *"Does Solar Wind Affect Earth's Weather Patterns?"*   *44*
NEWS REPORT **#3**   *"Earth's Atmosphere Unique Among Planets"*   *54*
NEWS REPORT **#4**   *"New UV Index Announced to Help Save Your Skin"*   *58*
NEWS REPORT **#5**   *"Air Pollution Abatement—Costs vs. Benefits"*   *70*

3 Atmospheric Energy and Global Temperatures —— 75

ENERGY BALANCE IN THE TROPOSPHERE 76
　　Energy in the Atmosphere: Some Basics 76
　　Earth-Atmosphere Radiation Balance 81
ENERGY AT THE SURFACE 81
　　Daily Radiation Curves 82
　　Simplified Surface Energy Balance 82
PRINCIPAL TEMPERATURE CONTROLS 83
　　Latitude 86
　　Altitude 86
　　Cloud Cover 89
　　Land-Water Heating Differences 90
EARTH'S TEMPERATURE PATTERNS 93
　　January Temperature Map 94
　　July Temperature Map 96
　　Annual Range of Temperatures 97
　　Air Temperature and the Human Body 98
　　The Urban Environment 100
SUMMARY 102
　　Key Terms 103　　Review Questions 103
FYI REPORT **3-1**　　*Solar Energy Collection and Concentration* 84
FYI REPORT **3-2**　　*Temperature Concepts, Terms, and Measurements* 87

4 Atmospheric and Oceanic Circulation —— 107

WIND ESSENTIALS 108
　　Wind: Description and Measurement 110
　　Global Winds 110
　　Air Pressure 111
DRIVING FORCES WITHIN THE ATMOSPHERE 115
　　Pressure Gradient Force 115
　　Coriolis Force 116
　　Friction Force 117
ATMOSPHERIC PATTERNS OF MOTION 118
　　Primary High-Pressure and Low-Pressure Areas 118
　　Upper Atmospheric Circulation 125
　　Local Winds 128
OCEANIC CURRENTS 134
　　Surface Currents 134
　　Deep Currents 135
SUMMARY 136
　　Key Terms 136　　Review Questions 137
FYI REPORT **4-1**　　*Wind Power: An Energy Resource* 131
NEWS REPORT **#1**　　*"Coriolis, A Forceful Effect on Drains?"* 121
NEWS REPORT **#2**　　*"Winds Can Be 'Foolish' and 'Uncertain'"* 125
NEWS REPORT **#3**　　*"Jet Streams Affect Flight Times"* 128

PART TWO

Water, Weather, and Climate   138

5 Atmospheric Water and Weather ———————— 141

WATER ON EARTH   142
  Quantity Equilibrium   142
  Distribution of Earth's Water   143

UNIQUE PROPERTIES OF WATER   144
  Heat Properties   144

HUMIDITY   148
  Relative Humidity   148
  Expressions of Relative Humidity   150

ATMOSPHERIC STABILITY   151
  Adiabatic Processes   151
  Stable and Unstable Atmospheric Conditions   152

CLOUDS AND FOG   154
  Cloud Types and Identification   154
  Fog   159

AIR MASSES   161

ATMOSPHERIC LIFTING MECHANISMS   163
  Convectional Lifting   163
  Orographic Lifting   164
  Frontal Lifting   165

MIDLATITUDE CYCLONIC SYSTEMS   168
  Life Cycle of a Midlatitude Cyclone   168
  Daily Weather Map and the Midlatitude Cyclone   170

VIOLENT WEATHER   172
  Thunderstorms   172
  Tornadoes   172
  Tropical Cyclones   175

SUMMARY   179
  Key Terms   182   Review Questions   183
  FYI REPORT **5-1**   *1992—The Year of Hurricanes*   *180*
  NEWS REPORT **#1**   *"Breaking Roads and Pipes and Sinking Ships"*   *148*
  NEWS REPORT **#2**   *"Harvesting Fog"*   *161*
  NEWS REPORT **#3**   *"Mountains Set Precipitation Records"*   *166*
  NEWS REPORT **#4**   *"Recent Tornado Records Are a Call for Action"*   *176*
  NEWS REPORT **#5**   *"Benefits from Hurricanes?"*   *179*

6   Water Resources ———————————— 185

THE HYDROLOGIC CYCLE   186
    A Hydrologic Cycle Model   186

THE WATER-BALANCE CONCEPT   187
    The Water-Balance Equation   187
    Three Examples of Water Balances   193
    Water Balance and Water Resources   193

GROUNDWATER RESOURCES   196
    Groundwater Description   196
    Groundwater Utilization   198
    Pollution of the Groundwater Resource   198

DISTRIBUTION OF STREAMS   200

OUR WATER SUPPLY   203
    Daily Water Budget   204
    World Water Economy   206

GLOBAL OCEANS AND SEAS   207
    Salinity and Composition   208

SUMMARY   210
    Key Terms   210   Review Questions   211
    FYI REPORT **6-1**   *Ogallala Aquifer Overdraft 199*
    FYI REPORT **6-2**   *Acid Deposition: A Blight on the Landscape 205*
    NEWS REPORT **#1**   *"Water in the Middle East: Running on Empty" 201*
    NEWS REPORT **#2**   *"Personal Water Use" 207*

7   Earth's Climates ———————————— 213

CLIMATE SYSTEM COMPONENTS   214
    Temperature and Precipitation   217

CLASSIFICATION OF CLIMATIC REGIONS   218
    The Köppen Classification System   219
    Tropical A Climates   222
    Mesothermal C Climates   225
    Microthermal D Climates   229
    Polar E Climates   234
    Dry Arid and Semiarid B Climates   235

FUTURE CLIMATE CHANGE   237
    Developing Climate Models   237
    Global Warming   239
    Warming Indications and the Future   241
    Consequences of Climatic Warming   241
    Global Cooling   244

SUMMARY   245
    Key Terms   245   Review Questions   246
    FYI REPORT **7-1**   *The El Niño Phenomenon   215*
    NEWS REPORT **#1**   *"Cwa Climate Region Sets Precipitation Records"   228*

PART THREE

Earth's Changing Landscapes   248

8   The Dynamic Planet ———————————— 251

THE PACE OF CHANGE   252

EARTH'S STRUCTURE AND INTERNAL ENERGY   252
Earth in Cross Section   254

GEOLOGIC CYCLE   259
Rock Cycle   260

CONTINENTAL DRIFT AND PLATE TECTONICS   267
A Brief History   267
Sea-floor Spreading and Production of New Crust   269
Subduction of Crust   270
The Formation and Breakup of Pangaea   270

PLATE TECTONICS   271
Plate Boundaries   271
Earthquakes and Volcanoes   275
Hot Spots   275

SUMMARY   277
Key Terms   278   Review Questions   279
NEWS REPORT **#1**   *"Drilling the Crust to Record Depths"*   259
NEWS REPORT **#2**   *"Canyon Rocks Harder than Steel"*   267
NEWS REPORT **#3**   *"Yellowstone on the Move"*   277

9   Earthquakes and Volcanoes ———————— 281

THE OCEAN FLOOR   282

EARTH'S SURFACE RELIEF FEATURES   282
Crustal Orders of Relief   283
Earth's Topographic Regions   284

CRUSTAL FORMATION PROCESSES   285
Continental Shields   285
Building Continental Crust   286

CRUSTAL DEFORMATION PROCESSES   287
Folding   287
Faulting   290

OROGENESIS (MOUNTAIN BUILDING)   291
Types of Orogenies   292
The Appalachian Mountains   293
World Structural Regions   293

EARTHQUAKES   295
Earthquake Essentials   297
The Nature of Faulting   298
Earthquake Forecasting, Preparedness, and Planning   302

VOLCANISM   303
Locations of Volcanic Activity   303
Types of Volcanic Activity   304

SUMMARY   311
Key Terms   312   Review Questions   313
FYI REPORT **9-1**   *The 1980 Eruption of Mount Saint Helens   409*
NEWS REPORT **#1**   *"James Michener On Terranes and Tectonics in Alaska"   286*
NEWS REPORT **#2**   *"Damage Strikes at Distance from Epicenters" 299*

10 Weathering, Karst Landscapes, and Mass Movement ⎯⎯⎯ 315

LANDMASS DENUDATION   316
Base Level of Streams   316
Dynamic Equilibrium Approach to Landforms   318

WEATHERING PROCESSES   320
Physical Weathering Processes   321
Chemical Weathering Processes   323

KARST TOPOGRAPHY AND LANDSCAPES   326
Caves and Caverns   328

MASS MOVEMENT PROCESSES   330
Mass Movement Mechanics   331
Classes of Mass Movements   333
Human-Induced Mass Movements   336

SUMMARY   337
Key Terms   338   Review Questions   338
NEWS REPORT **#1**   *"Alabama Hills—A Popular Movie Location"   326*
NEWS REPORT **#2**   *"Amateurs Make Cave Discoveries"   331*
NEWS REPORT **#3**   *"Debris Avalanche Destroys Colombian Town"   335*

11 Rivers and Related Landforms ————————— 341

FLUVIAL PROCESSES AND LANDSCAPES   342
   The Drainage Basin System   342
STREAMFLOW CHARACTERISTICS   347
   Stream Gradient   353
   Stream Deposition   354
FLOODS AND RIVER MANAGEMENT   361
   Streamflow Measurement   361
SUMMARY   366
   Key Terms   367   Review Questions   368
FYI REPORT **11-1**   *Floodplain Management Strategies*   *362*
NEWS REPORT **#1**   *"Climate Change and a River Basin"*   *345*
NEWS REPORT **#2**   *"Niagara Falls Closed for Inspection"*   *356*
NEWS REPORT **#3**   *"The 1993 Midwest Floods"*   *358*

12 Wind Processes and Desert Landscapes ————— 371

THE WORK OF WIND   372
   Eolian Erosion   373
   Eolian Transportation   374
   Eolian Depositional Landforms   375
   Loess Deposits   379
OVERVIEW OF DESERT LANDSCAPES   382
   Desert Climates   382
   Desert Fluvial Processes   382
   Desert Landscapes   384
SUMMARY   390
   Key Terms   393   Review Questions   393
FYI REPORT **12-1**   *The Colorado River: A System Out of Balance*   *386*
NEWS REPORT **#1**   *"War Tears up the Pavement"*   *373*
NEWS REPORT **#2**   *"Yardangs from Mars"*   *374*
NEWS REPORT **#3**   *"The Dust Bowl"*   *382*

13 Coastal Processes and Landforms ————————— 395

COASTAL SYSTEM COMPONENTS   396
   The Coastal Environment and Sea Level   396
COASTAL SYSTEM ACTIONS   397
   Tides   398
   Waves   400
   Sea Level Changes   404

COASTAL SYSTEM OUTPUTS   404
   Erosional Coastal Processes and Landforms   404
   Depositional Coastal Processes and Landforms   405
   Emergent and Submergent Coastlines   409
   Organic Processes: Coral Formations   410
   Salt Marshes and Mangrove Swamps   412

HUMAN IMPACT ON COASTAL ENVIRONMENTS   413

SUMMARY   414
   Key Terms   419   Review Questions   419
   FYI REPORT **13-1**   *An Environmental Approach to Shoreline Planning*   416
   NEWS REPORT **#1**   *"Sea Level Varies Along the U.S. Coastline"*   398
   NEWS REPORT **#2**   *"Killer Waves Strike Beachcombers"*   402
   NEWS REPORT **#3**   *"Warmer Water Raises Sea Level"*   405
   NEWS REPORT **#4**   *"Engineers Nourish A Beach"*   410
   NEWS REPORT **#5**   *"Hurricane Hugo Devastates the Beachfront"*   412
   NEWS REPORT **#6**   *"Worldwide Coral Bleaching Worsens"*   414

14   Glacial and Periglacial Landscapes ———————— 421

RESERVOIR OF ICE   422
   Types of Glaciers   422

GLACIAL PROCESSES   424
   Formation of Glacial Ice   424
   Glacial Mass Balance   427
   Glacial Movement   427

GLACIAL LANDFORMS   429
   Erosional Landforms Created by Alpine Glaciation   429
   Depositional Landforms Created by Alpine Glaciation   432
   Erosional and Depositional Features of Continental Glaciation   433

PERIGLACIAL LANDSCAPES   437
   Permafrost   437
   Frozen Ground Phenomena   438
   Humans and Periglacial Landscapes   438

THE PLEISTOCENE ICE AGE EPOCH   441
   Pluvial Periods and Paleolakes   445

ARCTIC AND ANTARCTIC REGIONS   445
   The Antarctic Ice Sheet   446

SUMMARY   446
   Key Terms   448   Review Questions   449
   FYI REPORT **14-1**   *Deciphering Past Climates: Paleoclimatology*   443
   NEWS REPORT **#1**   *"South Cascade Glacier Loses Mass"*   428
   NEWS REPORT **#2**   *"Glacial Erratic Marks Famous Grave"*   434
   NEWS REPORT **#3**   *"The Great Salt Lake Floods"*   446

PART FOUR
Biogeography   450

15   The Geography of Soils ———————— 453

SOIL CHARACTERISTICS   454
  Soil Profiles   454
  Soil Horizons   454
  Soil Properties   456
  Soil Chemistry   457
  Soil Formation Factors and Management   459

SOIL CLASSIFICATION   460
  A Brief History of Soil Classification   460
  Diagnostic Soil Horizons   460
  The Eleven Soil Orders of Soil Taxonomy   461

SUMMARY   478
  Key Terms   478   Review Questions   479
  NEWS REPORT **#1**   *"Soil Losses Estimated"*   466
  NEWS REPORT **#2**   *"Death of the Kesterson National Wildlife Refuge"*   470

16   Ecosystems and Biomes ———————— 481

ECOSYSTEM COMPONENTS AND CYCLES   482
  Communities   482
  Plants: The Essential Biotic Component   485
  Abiotic Ecosystem Components   486
  Biotic Ecosystem Operations   493

STABILITY AND SUCCESSION   496
  Ecosystem Stability and Diversity   496
  Ecological Succession   498

EARTH'S ECOSYSTEMS   502
  Terrestrial Ecosystems   503

EARTH'S MAJOR TERRESTRIAL BIOMES   505
  Equatorial and Tropical Rain Forest   505
  Tropical Seasonal Forest and Scrub   511
  Tropical Savanna   515
  Midlatitude Broadleaf and Mixed Forest   517
  Northern Needleleaf Forest and Montane Forest   518
  Temperate Rain Forest   518

Mediterranean Shrubland   519
Midlatitude Grasslands   520
Desert Biomes   521
Arctic and Alpine Tundra   523

SUMMARY   525
Key Terms   526   Review Questions   526
FYI REPORT **16-1**   *Biodiversity and Biosphere Reserves*   *513*
NEWS REPORT **#1**   *"Humans Dump Carbon into the Atmosphere"*   *493*
NEWS REPORT **#2**   *"Yellowstone Fire Assessment and Fire Ecology"*   *502*
NEWS REPORT **#3**   *"Large Marine Ecosystems: A Management Tool"*   *503*
NEWS REPORT **#4**   *"Annual Losses to the Rain Forest Estimated"*   *516*

17   Earth, Humans, and the New Millennium —————— 529

AN OILY BIRD   531
The Larger Picture   532

GAIA HYPOTHESIS   532

THE NEED FOR INTERNATIONAL COOPERATION   534
The Environmental Cost of Conflict   538

WHO SPEAKS FOR EARTH?   539
Review Questions   540
FYI REPORT **17-1**   *Earth Summit 1992*   *536*
NEWS REPORT **#1**   *"Huge Oil Spills"*   *534*
NEWS REPORT **#2**   *"ANWR Faces Threats"*   *535*

APPENDIX **A**   Information Sources, Organizations, and Agencies ——————————— A.1

APPENDIX **B**   Topographic Map Symbols—USGS ——— B.1

SUGGESTED READINGS ———————————————— S.1

GLOSSARY ———————————————————— G.1

INDEX ——————————————————————— I.1

TOPOGRAPHIC MAPS ———————————————— T.1
Map #1: Floodplain and fluvial processes (Philip, MS)   T.1
Map #2: Oxbow lake, political boundary (Carter Lake, NE)   T.2
Map #3: Folded mountains, water gap (Cumberland, MD)   T.3
Map #4: Alpine glaciers (Mount Rainier, WA)   T.4
Map #5: Features of continental glaciation (Jackson, MI)   T.5
Map #6: Coastal processes (Point Reyes, CA)   T.6

PREFACE

Welcome to *Elemental Geosystems–A Foundation in Physical Geography*. A study of Earth's physical geography is essential as we approach the new millennium and the environmental challenges that await us in the new century. As evidence of this need, the largest gathering of nations occurred at the Earth Summit held in Rio de Janeiro in 1992. Nations acted on an agenda critical to the environment and to many aspects of physical geography.

Recently nature delivered the three most costly natural disasters in history: Hurricane Andrew in Florida and Louisiana (1992), the Midwest floods (1993), and the Northridge (Reseda) earthquake in Los Angeles (1994)—combined damage exceeded $80 billion! Awareness that we should better understand Earth's dynamic systems, such as these events, is at the heart of this text.

In keeping with the timeliness of its companion, *Geosystems–An Introduction to Physical Geography*, second edition, relevant environmental issues, ongoing change in Earth's systems, and related human actions are explored within the core material presented in *Elemental Geosystems*. A dramatic survey of contemporary physical geography introduces Chapter 1.

The goal of physical geography is to explain the *spatial* dimension of Earth's natural systems—its energy, air, water, weather, climates, landforms, soils, plants, and animals. Earth is a place of great physical and cultural diversity, yet people generally know little of it. We must work to reverse this trend. An informed citizenry requires relevant education about the life-sustaining environment. The intent of *Elemental Geosystems* is to assist this learning process.

Elemental Geosystems

From across the United States, Canada, Australia, and New Zealand, we received many positive responses to *Geosystems*. The success of the companion text is helping to define physical geography in this era of Earth system science. This establishes a need for a shorter essential's version that maintains the qualities of the parent book—a contemporary physical geography for general education courses and the nonmajor science student.

Elemental Geosystems has a reduced number of chapters and shortened length. This is accomplished by cutting unnecessary coverage and a thorough rewriting of many sections. Some chapters were combined and subjects logically woven together for improved comprehension using our systems approach. Attention is directed to increased numbers of headings in the outline sequence, simplified terminology, redesigned art, and an attention to readability and vocabulary. To make this text both informative and enjoyable, all figures are carefully integrated with the text. This sensitivity to your needs makes *Elemental Geosystems* an accessible science text.

Systems Organization

This book maintains the careful organization of its companion text through a logical order of topics. As an example of this flowing content the global implications of the Mount Pinatubo eruption in June 1991 are woven throughout six chapters, not just in the section on volcanoes. In this way the text analyzes the worldwide impact of this event, bringing together various physical factors to form a complete picture.

Elemental Geosystems is structured in four parts, each containing related chapters according to the flow of individual systems, or consistent with time and the flow of events. Chapter 1 presents the foundations of physical geography, including a discussion of geography, systems analysis, latitude, longitude, time, the science of mapmaking (cartography), and the modern tools of remote sensing and geographic information systems (GIS). With these foundation tools, each of the text's parts then can be covered, either in their presented order or in any convenient sequence.

Part One exemplifies the systems organization of the text, beginning with the origin of the Solar System and the Sun. Solar energy passes across space to Earth's atmosphere varying in intensity by latitude and seasonal variations of Sun angle and daylength. Energy is traced through the atmosphere to Earth's surface. Earth's atmospheric and surface energy balances are generated by this cascade of solar energy (Chapter 2), producing patterns of world temperatures (Chapter 3), and general and local atmospheric and oceanic circulations (Chapter 4).

Part Two presents hydrology, meteorology and weather, oceanography, and climate, in a flowing sequence through Chapters 5 to 7. Physical geography is linked to Earth sciences, an influence seen in Chapters 8 through 14 of Part Three where we discuss the physical planet and related processes that affect the crust. Earth's surface is a place of an enormous ongoing struggle between the processes that build, warp, and fault the landscape and those that wear it down through the action of rivers, wind, waves, and ice.

Finally, Part Four brings the content of the first three parts together in biogeography, including soils, plants, animals, and Earth's major terrestrial biomes (Chapters 15 and 16). The text culminates with Chapter 17, "Earth, Humans, and the New Millennium," a unique capstone chapter that summarizes human-environment interactions, physical geography, and the coming century by answering the question, "Who speaks for Earth?" This chapter is sure to stimulate further thought and discussion, dealing as it does with the most profound issue of our time, *Earth stewardship*.

Important Learning Features and Tools

Elemental Geosystems contains many features to assist students and teachers:

- Each chapter includes a heading outline, review questions, key terms list, and detailed summaries with highlighted key terms.
- Key chapters (climate, soils, biomes) present large, integrative tables to help you synthesize content. Global climates are featured on a convenient world map located inside the front cover of this text.
- The Glossary provides basic definitions for 550 key terms and concepts that are printed in **boldface** where they are defined within the text; entries include chapter location. Other important terms and ideas in the text are placed in *italics* for emphasis.
- *Elemental Geosystems* features 200 color photographs, many by the author; 70 remote sensing images; 120 maps; 270 art drawings; and numerous tables; all selected and prepared to enhance the learning experience.
- The cartography program is updated to reflect the rapid pace of change in political boundaries and even the 1992 time-zone changes in Eastern Europe. All the maps in this edition reflect the latest placement of international boundaries.
- The text and all figures use metric/English measurement equivalencies appropriate to this transition period in the United States and for science courses. A complete set of measurement conversions is presented in an easy-to-use arrangement inside the back cover.
- Fourteen "**FYI Reports**" provide additional depth for key topics as diverse as the scientific method, the stratospheric ozone predicament, the Mount Saint Helens eruption, the status of the over budgeted Colorado River, Hurricane Andrew, biosphere reserves and biodiversity, and the 1992 UNCED Earth Summit.
- Fifty "**NEWS Reports**" through-out the text present late-breaking environmental issues, new applications in geography, and fascinating items that high-

light content. A sample of titles includes: "GPS: A Personal Locator," "Careers in GIS," "New UV Index Announced to Help Save Your Skin," "Harvesting Fog," "Water in the Middle East: Running on Empty," "Niagara Falls Closed for Inspection," "Engineers Nourish A Beach," "Worldwide Coral Bleaching Worsens," "Humans Dump Carbon into the Atmosphere," and "Huge Oil Spills."

- Updated treatment of global climate change and related potential effects are woven through the text. Sections are developed from current research including the Intergovernmental Panel on Climate Change (IPCC) 1990 reports and the 1992 update and 1994 reports.

- A brief sample of remote sensing includes: images from the Earth Radiation Budget experiments (ERBE), AVHRR images of Hurricane Andrew from *NOAA 10* and *11*, *Nimbus-7* measurements of the Mount Pinatubo eruption, images from the Galileo spacecraft during its fly-by of Earth, X-ray and visible light images of the Sun from the *Yohkoh* satellite, the latest *Nimbus-7* images of stratospheric ozone depletion, and several comparative sets of *Landsat* images of environmental changes.

- "Appendix A" provides addresses for important geographic and environmental organizations, agencies, and general reference works.

- A list of "Suggested Readings" is conveniently grouped by chapter and included at the end of the text. (The *Instructor's Resource Manual* includes a more comprehensive list of readings.)

- Our planet—Earth—is consistently treated in *Elemental Geosystems* as a proper noun. Earth is capitalized! This is consistent with the treatment of other planets and is an effective cue to think of Earth as our significant home planet.

Supplements and Ancillaries

The most important learning aid in the classroom is the teacher! To assist instruction a variety of author-prepared supplements are available with the text.

- A *Student Workbook* is available to provide additional learning tools, objectives, glossary, and crit-

ical thinking exercises, examples, and self-tests.

- A comprehensive *Instructor's Resource Manual* includes extensive support material for the textbook.

- A separate *Test Item File*, in print copy and on disk, is available. The exclusive Prentice Hall Test Manager 2.0, for IBM and Macintosh, eases your test preparation task. The Test Manager allows you to add your own questions to the bank, conveniently rearrange selections, and quickly export to Microsoft Word and WordPerfect.

- *Applied Physical Geography: Geosystems in the Laboratory* is a new laboratory manual that includes 12 comprehensive chapters that follow the text. The lab exercises are divided into a total of 77 convenient steps.

- The *Macmillan Geodisc*. This exciting new videodisc features 1500 still images, over 55 minutes of motion video, and multiple full-color animations. A separate instructor's manual facilitates classroom use with bar code listings.

- The overhead Transparencies set includes 100 full-color overheads of key figures from the text plus 4 of the principal tables. Also, 100 35-mm slides of important figures from the text.

- *ABC News-Prentice Hall Video Library* features twenty-five 3 to 10 minute video segments highlighting recent environmental topics and geographic issues in the news.

- *Themes of the Times* is a periodic newspaper supplement featuring a contemporary view of selected subjects from the pages of *The New York Times*. Prentice Hall and *The New York Times* bring these time-sensitive articles to the classroom directly from one of the world's distinguished newspapers.

Acknowledgments

All too often families go unmentioned as professional ties are cited. However, I could not continue work without the loving support of an extended family; always present when I was absent and immersed in the project. My family has earned such recognition, Mom and Dad, my sister and brothers Lynne, Randy, and Marty, and children Keri, Matt,

Reneé, and Steve, and the next generation—Chavon, Bryce, and Payton. Despite the missed time together they were never far from thought.

I acknowledge the courage and dedication of the faculty and staff of California State University–Northridge. Their campus suffered $350 million damage from the January 1994 earthquake and presently operates out of 450 trailers. The geography department, without access to classrooms, offices, equipment, or maps, taught outdoors for a time. And yet, they graduated majors and masters only five months after the devastating earthquake. Their performance of duty continued despite the destruction of homes, apartments, cars, and computers suffered by members of the department. The faculty, staff, and students, even insisted on hosting a scheduled spring regional geography conference. This represents an incredible dedication to geographic education and provides a positive example to us all.

My gratitude goes to all the students over these past 26 years at American River College for defining the importance of Earth's future, for their questions and enthusiasm. Thanks to the authors and scientists who published the research, articles, and books I used in refining this text. I owe much to the following individuals who consulted with me through the editions of *Geosystems*, reviewed portions of the manuscript, and made suggestions both formal and informal; particularly Dr. Dorothy Sack who completed a comprehensive review of the parent text. I thank the many geographers who spoke to me at our annual national and regional geography meetings and who have written and telephoned, providing valuable guidance and analysis. These most helpful geographers and their affiliations are:

Ted J. Alsop, Utah State University

Ward Barrett, University of Minnesota

David R. Butler, University of North Carolina

Ian A. Campbell, University of Alberta–Edmonton

Armando M. da Silva, Towson State University

Dirk H. de Boer, University of Saskatchewan

Mario P. Delisio, Boise State University

Joseph R. Desloges, University of Toronto

Lee R. Dexter, North Arizona University

John Dixon, University of Arkansas

Don W. Duckson, Jr., Frostburg State University

Christopher H. Exline, University of Nevada–Reno

Michael M. Folsom, Eastern Washington University

Glen Fredlund, University of Wisconsin–Milwaukee

David E. Greenland, University of Oregon

John W. Hall, Louisiana State University–Shreveport

Gail Hobbs, Pierce College

David A. Howarth, University of Louisville

Patricia G. Humbertson, Youngstown State University

David W. Icenogle, Auburn University

Philip L. Jackson, Oregon State University

J. Peter Johnson, Jr., University of Ottawa

Guy King, California State University–Chico

Peter W. Knightes, Central Texas College

Richard Kurzhals, Grand Rapids Junior College

Robert D. Larsen, Southwest Texas State University

William G. Loy, University of Oregon

Joyce Lundberg, Carleton University

W. Andrew Marcus, Montana State University

Elliot G. McIntire, California State University–Northridge

Norman Meek, California State University–San Bernardino

Lawrence C. Nkemdirim, University of Calgary

John E. Oliver, Indiana State University

Bradley M. Opdyke, National University of Australia–Canbarra

Robin J. Rapai, University of North Dakota

Philip D. Renner, American River College

William C. Rense, Shippensburg University

Wolf Roder, University of Cincinnati

Dorothy Sack, University of Wisconsin–Madison

Glenn R. Sebastian, University of South Alabama

Daniel A. Selwa, U.S.C. Coastal Carolina College

Thomas W. Small, Frostburg State University

Daniel J. Smith, University of Victoria

Stephen J. Stadler, Oklahoma State University

Susanna T. Y. Tong, University of Cincinnati

David Weide, University of Nevada–Las Vegas

Gerald R. White, American River College

Brenton M. Yarnal, Pennsylvania State University

My thanks to the staff of Prentice Hall. Their dedication to the craft of modern textbook publishing is greatly appreciated by this author. A special thanks to Paul Corey, Editor-in-Chief, for his editorial skill, guidance, and abiding friendship through the life of the *Geosystems* books. This time of transition to Prentice Hall is made easier through the work and encouragement of Ray Henderson, Geography Editor, and Will Ethridge, President ESM Division, and the enthusiasm of Leslie Cavaliere, Geography Marketing Manager, and the guidance of Ray Mullaney, Editor-in-Chief for Book Development.

My gratitude to Eileen Mitchell, Project Manager, and the staff of Electronic Publishing Services Inc., New York, for their expertise, care, and handling of this text as it made its way from disk to paper through the latest in book publishing technology.

Also, thanks go to Chris Migdol for assisting with photo research. A personal commendation to Fred Schroyer, a wonderful developmental editor, for his skill and dedication to the printed word and his sensitivity to the students who will read this book.

Finally, there is one person with a unique sensitivity to Earth and all living things who is an inspiration. *Geosystems* is possible only with the partnership of my very special production assistant, Bobbé Christopherson. There are not words nor space to express my appreciation and feelings for her tireless work. She invested many months preparing a detailed figure log, obtaining permissions, proofing art, copyediting drafts, page proofs, and ancillaries, and disk-entering revisions and type-marking notations. She is truly a warrior by my side as we attack these projects.

Physical geography teaches us about the intricate supporting web that is Earth's environment. Dramatic changes that demand our attention and understanding are occurring in many human-Earth relationships as we approach the new millennium. All things considered, this is an important time to be enrolled in a relevant geography course! The best to you in all your studies.

Robert W. Christopherson
Folsom, California

ELEMENTAL GEOSYSTEMS

A FOUNDATION IN PHYSICAL GEOGRAPHY

An astronomer uses a simple device to sight the Sun's angle above the horizon and determine latitude. [*From Jacques de Vault,* Cosmographia, *1583.* Bibliotheque National, Paris]

1

FOUNDATIONS OF GEOGRAPHY

THE SCIENCE OF GEOGRAPHY
    The Geographic Continuum
EARTH SYSTEMS CONCEPTS
    Systems Theory   Earth's Four Spheres
    Earth as a System
A SPHERICAL PLANET
LOCATION ON EARTH
    Latitude   Longitude
    Great Circles and Small Circles
PRIME MERIDIAN AND STANDARD
    TIME
    International Date Line
    Coordinated Universal Time
    Daylight Saving Time
MAPS, SCALES, AND PROJECTIONS
    Map Scales   Map Projections
MAPPING AND TOPOGRAPHIC MAPS
REMOTE SENSING AND GIS
    Remote Sensing
    Geographic Information System (GIS)
SUMMARY
FYI REPORT 1-1 MEASURING EARTH IN 247 B.C.
FYI REPORT 1-2 THE SEARCH FOR LONGITUDE

Welcome to a foundation in physical geography! Consider the following events and conditions:

- The three most costly natural disasters in history—Hurricane Andrew in Florida and Louisiana (1992), the Midwest floods (1993), and the Northridge (Reseda) earthquake in Los Angeles (1994)—exceeded $80 billion in public and private damage. Such events turn the public to physical geography for understanding and answers.

- Weather patterns, both expected and unusual, constantly occur. Record numbers of tornadoes struck the United States from 1991 through 1994. Lake Eyre in central Australia filled to unusual depths; Lake Chad in western Africa continues to recede; and the Great Salt Lake in Utah rises to record levels—all in the last ten years. And, record floods inundated Georgia in 1994.

- Rocks found in northwestern Canada are 3.96 billion years old, predating all others on our 4.6 billion-year-old planet. The Appalachian mountains in North America and the Atlas mountains of North Africa are now thousands of kilometers apart, but once were joined. They drifted apart as vast plates of continental crust slowly move.

- Earthquakes struck Ontario, Canada, southern California, Turkey, and Toez, Colombia. A tsunami (earthquake-generated sea wave) devastated Indonesia in June 1994. Volcanoes erupted in Chile, Japan, and Alaska. Mount Pinatubo explodes in the Philippines (1991): colorful afterglows dominate twilight worldwide, global temperatures drop slightly, and resulting sulfuric acid mists further contribute to declining ozone levels in the stratosphere.

- Lush rain forests straddle the equator, stark deserts bake in the subtropics, and cool moist climates dominate northwestern coastlines. Changes in global climates are producing gradual shifts in these expected natural patterns.

- Natural ecosystems show the scars of civilization: only three dozen white rhinos remain in the northern portion of their African habitat, record numbers of species face extinction in the tropics, agricultural soils lose productivity,

ocean fishery yields continue to fall, and forests decline due to logging and acid rain.

- Dynamic patterns of energy, air, water, weather, climates, landforms, soils, plants, and animals vary from place to place and through the seasons, and evolve over time.

Where are these places? Why are they there? Why do physical conditions differ from equator to midlatitudes, and between deserts and polar regions? How does solar energy influence the distribution of trees, soils, climates, and lifestyles? How does it produce the patterns of wind, weather, and ocean currents? How do aspects of nature affect human populations? What impact do humans have on natural systems? Let us explore these questions, and others, through geography's unique perspective—welcome to a foundation in physical geography!

A knowledge of physical geography is important to answer the where, the why, and the how questions about Earth's physical systems and their interaction with living things. Thus, in this chapter, our study of geosystems—Earth systems—begins with a look at the science of physical geography and the geographic tools we will need.

Physical geographers use systems analysis to study the environment. Therefore, we introduce Earth's interrelated natural and human systems. We then consider location, a key theme of geographic inquiry—the latitude, longitude, and time coordinates that delimit Earth's surface. The pursuit of longitude and a world time system provide us with interesting insights into geography. We study maps as critical tools that geographers create to portray physical and cultural information.

Finally, we examine computer-based information systems and remote-sensing capabilities which are adding a new and exciting dimension to physical geography research. Chapter 1 concludes with a brief overview of these rapidly expanding techniques.

The Science of Geography

Geography (from *geo,* "Earth," and *graphein,* "to write") is the science that studies the relationships among geographic areas, natural systems, society, cultural activities, and the interdependence of all these *over space*. The term *spatial* refers to the nature and character of physical space; to measurements, locations, and the distribution of things. Human beings are spatial actors, both affecting and being affected by Earth. We influence vast areas because of our technology, mobility, and numbers. For

FIGURE 1-1

Five fundamental themes in geography defined. *Location*: geographic freeway sign; *Place*: Uluru (Ayers) Rock, Northern Territory, Australia; *Movement*: global and deep-space communications through modern satellites; *Regions*: intensive use of land for rice paddies in China; and *Human-Earth relationships*: thatched houses on stilts along the Amazon River, Brazil. [Regions photo by Frank Balthis; Place photo by Michael Dunning/TSW; Human-Earth Relationships photo by Gael Summer-Hebden; Movement and Location photos by author.]

example, think of your own route to the classroom or library today, your knowledge of street patterns, traffic trouble spots, parking spaces, or bike rack locations used to minimize walking distances.

The Association of American Geographers (AAG) and the National Council for Geographic Education (NCGE) describe modern geographic education using five key themes: **location, place, movement, region,** and **human-Earth relationships** (Figure 1-1). *Elemental Geosystems* draws on each theme.

Geography synthesizes knowledge from many fields, integrating information to form a coherent picture of Earth. Geography is governed by a *method* rather than by a specific body of knowledge and this method is **spatial analysis**. Geographers view phenomena as occurring in spaces and areas that have distinctive characteristics. The language of geography reflects this: space, territory, zone, pattern, distribution, place, location, region, sphere, province, and distance.

Geographers analyze the differences and similarities between places and locations. **Process** is central to geographic synthesis—that is, analyzing a set of actions or mechanisms that operate in some special order. As an example, there are numerous processes involved in the study of Earth's vast water system. Spatial analysis, then, is a process-oriented integrative approach central to geographic inquiry.

Physical geography, therefore, centers on the spatial analysis of all the physical elements and processes that make up the environment: energy, air, water, weather, climate, landforms, soils, ani-

mals, plants, and Earth itself. We add to this the oldest theme in the geographic tradition, that of human activity, impact, and the shared human-Earth relationship.

The Geographic Continuum

Geography is eclectic, integrating a wide range of subject matter from diverse fields; virtually any subject can be examined geographically. Figure 1-2 shows a continuous distribution—a continuum—along which the content of geography is arranged. Disciplines in the physical and life sciences are listed on the continuum's left; those in the cultural/human sciences are listed on the right. Specialties within geography that draw from these disciplines are listed at either end of the continuum.

This continuum reflects a basic duality within geography—*physical* geography versus *human/cultural* geography. This duality is paralleled in society by the tendency of those who live in developed countries to distance themselves from their supportive life-sustaining environment and to think of themselves as exempt from the physical functions of Earth. In contrast, many people in Third World countries live closer to nature and are acutely aware of its importance in their daily lives. Regardless of our philosophies toward Earth, we all depend on Earth's systems to provide oxygen, water, nutrients, energy, and materials.

Our modern world requires that we shift our study of geographic processes, and perhaps our philoso-

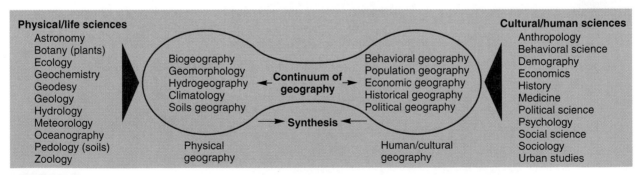

FIGURE 1-2

Distribution of geographic content along a continuum. Our focus is physical geography, but with an integration of the human component. Synthesis of Earth topics and human topics is suggested by movement toward the middle of the continuum.

phies, toward the center of the continuum to attain a more *holistic*, or complete, perspective. Modern environmental problems call for geographers to examine relations among Earth, society, and cultures throughout space and time. Critical environmental concerns that are integral to physical geography include:

1. *Global ozone depletion* in the upper atmosphere that allows increasing amounts of ultraviolet radiation to reach Earth's surface.

2. *Possible global warming* through human-caused increases of carbon dioxide in the atmosphere.

3. *Worsening air pollution,* particularly in metropolitan areas.

4. *Identification of natural hazards* that threaten society, such as hurricanes, earthquakes, tsunamis, landslides, droughts, and floods.

5. *Deliberate destruction of Earth's forests.*

6. *Increasing losses of plant and animal diversity* as habitats disappear.

7. *Spatial analysis of disasters* such as the 1986 Chornobyl' nuclear power station explosion and the 1989 *Exxon Valdez* oil tanker catastrophe.

8. *Accounting for natural resources* as environmental assets on national economic balance sheets.

Geography is in a unique position to synthesize the environmental, spatial, and human aspects of these concerns.

The *United Nations Conference on Environment and Development,* held at Rio de Janeiro in 1992, brought tens of thousands of delegates from over 160 countries and 1000 nongovernmental organizations to the largest world summit ever conducted. On the agenda were global climate change, biodiversity, sustainable forestry, and an "Earth Charter"— themes at the heart of geographic studies (see Chapter 17, FYI Report 17-1).

The spatial patterns of change we face today are unique, for we are taxing Earth's systems in new ways. Some past civilizations adapted to crises, whereas others failed. Perhaps this ability to adapt is the key. If so,

understanding of our relationship to Earth's physical geography is of great importance to human survival, for the innumerable physical processes in the environment operate as interacting life-support systems.

Earth Systems Concepts

You no doubt are familiar with the word system: "check the cooling system"; "how does the grading system work?"; "there is a weather system approaching." Systems surround us. *Systems analysis* began with studies of energy and temperature in the 19th century, and was developed further during World War II. Systems analysis has moved to the forefront as a method for understanding the workings of many sciences. Geographers use systems methodology as an analytical tool at varying levels of complexity.

Systems Theory

Simply stated, a **system** is any ordered, interrelated set of things, and their attributes, linked by flows of energy and matter, as distinct from their surrounding environment outside the system. The elements within a system may be arranged in a series or interwoven with one another. A system can comprise any number of subsystems. Within Earth's systems both matter and energy are stored and retrieved, and energy is transformed from one type to another.

A natural system generally is not self-contained: inputs of energy and matter flow into the system, and outputs flow from the system. Such a system is called an **open system**. Earth is an open system *in terms of energy,* as are most natural systems; solar energy enters freely. A system that is shut off from the surrounding environment so that it is self-contained is a **closed system**. Earth is essentially a closed system *in terms of physical matter and resources* (matter seldom is added to Earth from space, or removed). Figure 1-3 illustrates open and closed systems.

Example. Figure 1-4 illustrates a simple open system, using plant photosynthesis as an example. *Photosynthesis* is the growth process by which plants transform inputs of solar energy, water, nutrients, and carbon dioxide into stored chemical energy (car-

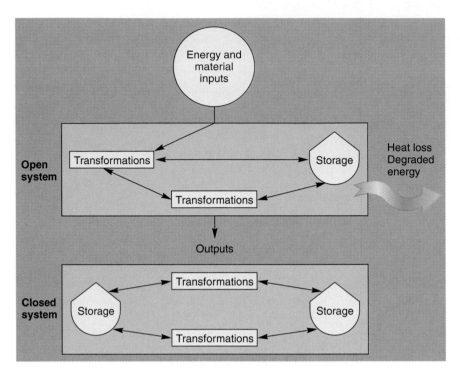

FIGURE 1-3
An open system and a closed system. In either type, inputs are transformed and stored as the system operates.

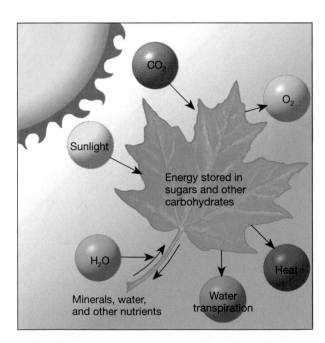

FIGURE 1-4
Simple open system of a plant leaf. Plants take light, carbon dioxide (CO_2), water (H_2O), and nutrients as system inputs to produce outputs of oxygen (O_2) and carbohydrates (sugars) through the photosynthesis process. Plant respiration is simply a reverse of this process.

bohydrates), releasing an output of oxygen as a by-product. Plants derive energy for their operations from a reverse of photosynthesis, a process called *respiration*. In respiration, the plant consumes inputs of chemical energy (carbohydrates) and releases outputs of carbon dioxide, water, and heat into the environment. Thus, a plant acts as an open system. (Photosynthesis and respiration processes are discussed in Chapter 16.)

Feedback. As a system operates, outputs that can influence continuing system operations are generated. This "information" is returned to various points in the system via pathways called **feedback loops**. Feedback can guide further system operations. If the "information" amplifies or encourages response in the system, it is called **positive feedback**. If the "information" slows or discourages response in the system, it is called **negative feedback**. Such negative feedback causes self-regulation in a natural system, maintaining the system within its performance range, and in a dynamic steady state. In the photosynthetic system (Figure 1-4), any increase or decrease in daylength (photoperiod), carbon dioxide, or water availability will produce feedback information that affects specific responses in the plant.

In Earth's systems, negative feedback is far more common than positive feedback. Unchecked positive feedback in a system can create a runaway ("snowballing") condition until a critical limit is reached, leading to instability, disruption, or death.

Equilibrium. Most systems maintain structure and character over time. An energy and material system that remains balanced over time is at **equilibrium**. When the rates of inputs and outputs in the system are equal and the amounts of energy and matter in storage within the system are constant (or more realistically, as they fluctuate around a stable average), the system is in **steady-state equilibrium**.

However, a system fluctuating around an average value may demonstrate a trend over time and represent a changing *dynamic-equilibrium* state through time. A landscape, such as a hillside, that adjusts after a landslide to a new equilibrium among slope, materials, and energy, is a system in dynamic equilibrium. Such is the status of Earth's climate,

long-term climatic changes, and the present global warming.

The interaction of volcanoes and the atmosphere is an example of a system in dynamic equilibrium. In June 1991, Mount Pinatubo erupted violently in the Philippines, injecting 15 to 20 million tons of ash and sulfuric acid mist into the upper atmosphere (Figure 1-5). These materials increased sunlight reflection from the atmosphere and reduced the solar energy passing through the atmosphere, thus lowering global temperatures an average of 0.5C° (0.9F°) or more. (The Mount Pinatubo story is woven throughout this text. In Chapter 4, Figure 4-1 displays the eruption and satellite images of the debris. Volcanic processes are covered in Chapters 8 and 9, past climate effects in Chapter 14, and the relationship of this eruption to the "nuclear winter" hypothesis in Chapter 17.)

Models of Systems. A **model** is a simplified, idealized representation of part of the real world. Models are designed with varying degrees of abstraction. Physical geographers construct simple system models to demonstrate complex associations in

FIGURE 1-5
Debris lofted by the 1991 Mount Pinatubo eruption affected the Earth-atmosphere system. This eruption was one of the largest this century. The atmosphere is a vast system, cooled by the effects of volcanic ash and warmed by increasing levels of heat-absorbing gases produced by human activities. Modern technology makes it possible for geographers to study the atmosphere's dynamic equilibrium. [Photo by Reuters/Bettmann.]

the environment, such as the hydrologic cycle that models Earth's entire water system. The simplicity of a model makes a system easier to comprehend (example: Figure 1-4). It also allows predictions, but such predictions are only as good as the assumptions and accuracy built into the model. Imposing a model too rigidly on a natural setting can lead to misinterpretation, so it is best to view a model for what it is—a simplification.

We discuss many system models in this text, including the hydrologic cycle, water balance, surface energy budgets, earthquakes and faulting, glacier mass budgets, soil profiles, and various ecosystems. Computer-based models are in use to study most natural systems—from climate change to Earth's interior. A general circulation model of the atmosphere operated by the Goddard Institute accurately predicted the effect of Mount Pinatubo's ash on the atmosphere and a lowering of global temperatures.

Earth as a System

Because it receives energy from an outside source—the Sun—and reradiates energy into space, Earth operates as an open system. In terms of matter, however, Earth is essentially a closed system: closed to any large amounts of incoming matter.

Earth's Energy Equilibrium: An Open System. Most Earth systems are dynamic because of the tremendous infusion of radiant energy from ther-

monuclear reactions deep within the Sun. This energy penetrates the outermost edge of Earth's atmosphere and cascades through the terrestrial systems, transforming along the way into various other forms of energy, such as kinetic energy, mechanical energy, or potential energy. Eventually, Earth radiates this energy back to the cold vacuum of space in an amount essentially equal to that which entered the system.

Researchers are examining the dynamics of Earth's energy equilibrium in more detail than ever before to distinguish natural changes from those produced by human intervention. Tremendous breakthroughs in understanding and prediction should occur as general circulation models of Earth's energy-atmosphere-water system become even more accurate. By the late 1990s, at least four polar-orbiting satellites—part of an Earth Observation System, or EOS—are expected to introduce a new era in the monitoring of Earth's open energy system.

Earth's Physical Matter: A Closed System. In terms of physical matter—air, water, and material resources—Earth is nearly a closed system. The only exceptions to this are frequent but tiny meteors, cosmic and meteoric dust, extremely rare comets and asteroids, and outgassing of water from deep within Earth's crust. Since the initial formation of the planet, no significant quantities of matter have entered the system. Just as importantly, no significant quantities have left the system, either. This is it! Earth's physical materials are finite. No matter how numerous and

FIGURE 1-6
Typical urban landfill that essentially is an "urban ore mine." [Photo by author.]

FIGURE 1-7
Earth's four spheres. Each sphere is a model for the four "parts" used to organize this book.

       Part 1 - atmosphere
       Part 2 - hydrosphere
       Part 3 - lithosphere
       Part 4 - biosphere

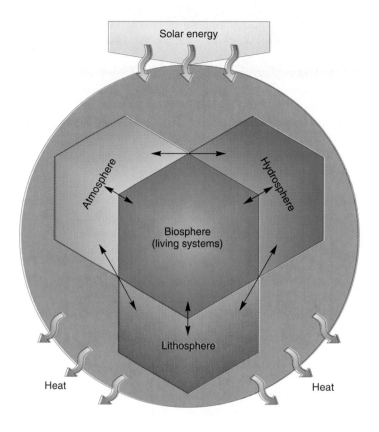

daring the technological reorganizations of matter become, the physical base is fixed.

Society loses track of its resources in what is essentially a once-through resource stream from virgin materials to the landfill (Figure 1-6). As Earth's people come to understand the fact of finite resources, a serious effort to recycle and to make efficient use of energy and materials can begin. Recycling costs for some materials may be prohibitive, but about half the contents of an average landfill are extractable economically—glass, newspaper, aluminum, and some other metals. The fact that Earth is a closed material system makes such recycling efforts inevitable.

Earth's Four Spheres

Earth's surface is where four immense open systems interface and interact. Figure 1-7 shows three **abiotic** (nonliving) systems overlapping to form the realm of the **biotic** (living) system. The abiotic

spheres are the **atmosphere**, **hydrosphere**, and **lithosphere**. The biotic sphere is called the **biosphere**. Because these four models are not independent units in nature, their boundaries must be understood as transition zones rather than sharp delimitations.

Atmosphere. The atmosphere is a thin, gaseous veil surrounding Earth, held to the planet by the force of gravity. Formed by gases arising from within Earth's crust and interior, and the exhalations of all life over time, the lower atmosphere is unique in the Solar System. It is a combination of nitrogen, oxygen, argon, carbon dioxide, water vapor, and small amounts of trace gases.

Hydrosphere. Earth's waters exist in the atmosphere, on the surface, and in the crust near the surface, in liquid, solid, and gaseous states. Water occurs in two forms, fresh and saline (salty), and exhibits important heat properties as well as playing its extraordinary role as a solvent. Among the planets in the Solar System, only Earth possesses water in any quantity.

Lithosphere. Earth's crust and a portion of the upper mantle directly below the crust form the lithosphere. The crust is quite brittle compared to the layers beneath it, which are in motion in response to an uneven distribution of heat and pressure. (In a broad sense, the term lithosphere sometimes refers to the entire solid planet.)

Biosphere. The intricate, interconnected web that links all organisms with their physical environment is the biosphere. Sometimes called the **ecosphere**, the biosphere is the area in which physical and chemical factors form the context of life. The biosphere exists in the overlap among the abiotic spheres, extending from the seafloor to about 8 km (5 mi) into the atmosphere. Life is sustainable within these natural limits. In turn, life processes have powerfully shaped the other three spheres through various interactive processes. The biosphere has evolved, reorganized itself at times, faced extinction, gained new vitality, and managed to flourish overall. Earth's biosphere is the only one known in the Solar System; thus, life as we know it is unique to Earth.

A Spherical Planet

Information about Earth's shape and size is fundamental to physical geography. The roundness or sphericity of Earth is not as modern a concept as many believe. For instance, the Greek mathematician and philosopher Pythagoras (ca. 580–500 B.C.) determined through observation that Earth is round. We don't know what Pythagoras observed to conclude that Earth is a sphere, but we can guess. He might have noticed ships sailing beyond the horizon and apparently sinking below the surface, only to arrive back at port with dry decks. Perhaps he noticed Earth's curved shadow cast on the lunar surface during an eclipse of the Moon. He might have deduced that the Sun and Moon are not just disks but are spherical, and that Earth must be a sphere as well.

Earth's sphericity was generally accepted by the educated populace as early as the first century A.D. (See FYI Report 1-1). Christopher Columbus, for example, knew he was sailing around a sphere in 1492; that is why he expected to reach the East Indies.

In 1687 Sir Isaac Newton postulated that the round Earth, along with the other planets, could not be perfectly spherical. Until that time the spherical-perfection model was a basic assumption of **geodesy**, the science that attempts to determine Earth's shape and size by surveys and mathematical calculations. Newton reasoned that Earth is slightly misshapen by its spinning, making it bulge through the equator and flatten at the poles (Figure 1-8).

Earth's equatorial bulge and its polar oblateness are universally accepted and confirmed by satellite observations. The modern era of Earth measurement is one of tremendous precision and is called the "geoidal epoch" because Earth is considered a **geoid**,

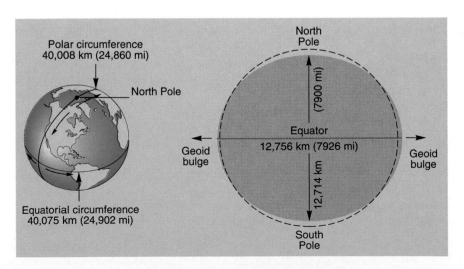

FIGURE 1-8
Earth's diameter and circumference. The dashed line is a perfect circle for reference.

Measuring Earth in 247 B.C.

Eratosthenes, foremost among early geographers, served as the librarian of Alexandria in Egypt during the third century B.C. He was in a position of scientific leadership, for Alexandria's library was the greatest in the ancient world. Among his achievements was calculation of Earth's circumference to a high level of accuracy, quite a feat for 247 B.C.

Travelers told Eratosthenes that on June 21 they had seen the Sun's rays shine directly to the bottom of a well at Syene, the location of present-day Aswan, Egypt (Figure 1). Eratosthenes knew from his own observations that the Sun's rays never were directly overhead at Alexandria, even at noon on June 21, the longest day of the year and the day on which the Sun is at its most northward position in the sky. Unlike Syene, objects always cast a daytime shadow in Alexandria. Using the considerable geometric knowledge of the era, Eratosthenes conducted an experiment.

In Alexandria at noon on June 21, Eratosthenes measured the angle of a shadow cast by an obelisk, a perpendicular column used for telling time by the Sun. Knowing the height of the obelisk and measuring the length of the shadow from its base, he solved the triangle for the angle of the Sun's rays, which he determined to be 7.2°. However, at Syene on the same day, the angle of the Sun's rays was 0° from a perpendicular—that is, the Sun was directly overhead. Geometric principles told Eratosthenes that the distance on the ground between Alexandria and Syene formed an arc of Earth's circumference equal to the angle of the Sun's rays at Alexandria. Since 7.2° is roughly 1/50 of the 360° in Earth's total circumference, the distance between Alexandria and Syene must represent approximately 1/50 of Earth's total circumference.

Camel caravans took 50 days to trek from Syene to Alexandria, covering about 100 stadia each day—

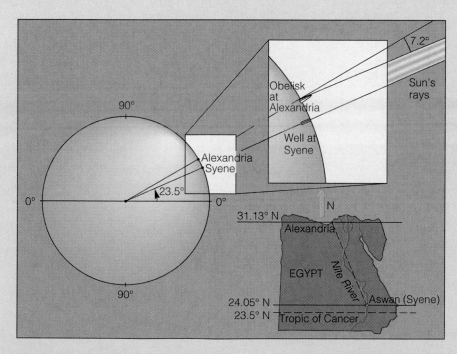

FIGURE 1
Eratosthenes's determination of Earth's circumference. This remarkably accurate measurement was based on a precise geometry and approximate camel-caravan speed.

5000 stadia one way. Eratosthenes determined the surface distance between the two cities as 5000 stadia, a Greek unit of measure. He then multiplied 5000 stadia by 50 to determine that Earth's circumference is about 250,000 stadia. Assuming that a stadium equals approximately 185 m (607 ft), Eratosthenes's calculations convert to roughly 46,250 km (28,738 mi), which is remarkably close to the correct value of 40,075 km (24,902 mi) for Earth's equatorial circumference.

Eratosthenes's work teaches the value of observing carefully and integrating all observations with previous learning. Measuring Earth's circumference required application of his knowledge of Earth-Sun relationships, geometry, and geography to his keen observations to prove his perception about Earth's size.

meaning literally that "the shape of Earth is uniquely Earth-shaped," not a perfect sphere.

Figure 1-8 gives Earth's measurements. Its circumference was first measured over 2200 years ago by the Greek geographer, astronomer, and librarian Eratosthenes (ca. 276–195 B.C.). The ingenious reasoning by which he arrived at his calculation is presented in FYI Report 1-1.

Location on Earth

To know specifically where something is located on Earth's surface, a coordinated grid system is needed, one that is agreed to by all peoples. The terms latitude and longitude were in use on maps as early as the first century A.D., with the concepts themselves dating back to Eratosthenes and others.

The geographer, astronomer, and mathematician Ptolemy (ca. A.D. 90–168) contributed greatly to modern maps, and many of his terms and configurations are still used today. On his map of the known world (Figure 1-9), Ptolemy described 8000 places according to their north-south and east-west coordinates. He also placed north at the top and east at the right, orienting maps in a manner familiar to us.

Ptolemy divided the circle into 360 *degrees* (360°), with each degree comprising 60 *minutes* (60'), and each minute including 60 *seconds* (60"), in a manner adapted from the Babylonians. He located places

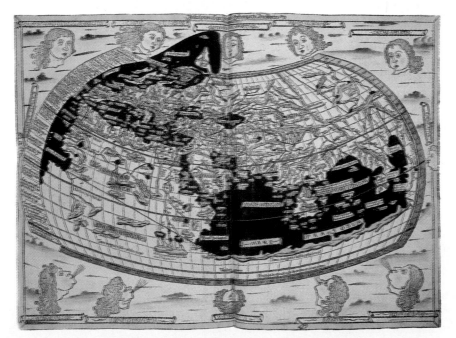

FIGURE 1-9
Ptolemy published an early version of this map of the known world—Europe, Asia, and North Africa—in A.D. 140. Shown is a Renaissance interpretation. [© National Maritime Museum, Greenwich, England.]

using these degrees, minutes, and seconds, although the precise length of a degree of latitude and a degree of longitude on Earth's surface remained unresolved for the next 17 centuries.

Latitude

Latitude *is an angular distance north or south of the equator,* measured from the center of Earth (Figure 1-10a). On a map or globe, the lines designating these angles of latitude run east and west, parallel to the equator. Because Earth's equator divides the distance between the North Pole and the South Pole exactly in half, it is assigned the value of 0° latitude. The angle of latitude increases toward the North Pole at 90° north latitude, and increases toward the South Pole at 90° south latitude.

A line connecting all points along the same latitudinal angle is called a **parallel**. Thus, in Figure 1-10b, *latitude* is the name of the angle (49° north latitude), *parallel* names the line (49th parallel), and both indicate distance north of the equator. In the figure, an angle of 49° north latitude is measured and, by connecting all points at this latitude, the 49th parallel is designated. The 49th parallel is a significant one in the Western Hemisphere, for it forms the boundary between Canada and the United States from Minnesota to the Pacific.

Latitude is readily determined by reference to *fixed celestial objects* like the Sun or the stars, a method dating to ancient times. During daylight hours the angle of the Sun above the horizon indicates the observer's latitude, after adjustment is made for the seasonal tilt of Earth and time of day. Because Polaris, the North Star, is almost directly overhead at the North Pole, persons anywhere in the Northern Hemisphere can determine their latitude at night simply by sighting Polaris and measuring its angle above the local horizon. In the Southern Hemisphere, Polaris cannot be seen (it is below the horizon), so observers sight on the Southern Cross (Crux Australis) constellation.

Latitudinal Geographic Zones. Natural environments differ in appearance from the warm and humid equator to the frigid poles. These differences result from the amount of solar energy received, which varies by latitude and season of the year. As

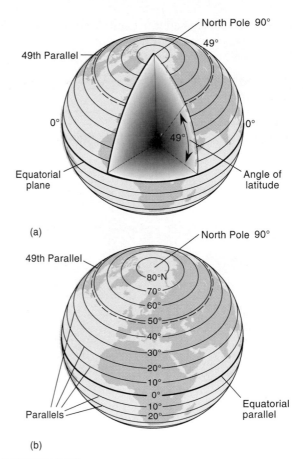

FIGURE 1-10
Angles of latitude measured north or south of the equator (a) determine parallels (b).

a convenience, geographers identify *latitudinal geographic zones* as regions with fairly consistent qualities. Figure 1-11 portrays these zones, their locations, and their names: *equatorial, tropical, subtropical, midlatitude, subarctic* or *subantarctic,* and *arctic* or *antarctic.* "Lower latitudes" are those nearer the equator, whereas "higher latitudes" refer to those nearer the poles.

These generalized latitudinal zones are useful concepts, but are not rigid delineations. The *Tropic of Cancer* (23.5° north parallel) and the *Tropic of Capricorn* (23.5° south parallel), discussed further in Chapter 2, are the most extreme north and south parallels that experience perpendicular (directly overhead) rays of the Sun at local noon on the first day of summer. The Arctic Circle (66.5° north parallel) and the Antarctic Circle

(66.5° south parallel) are the parallels farthest from the poles that experience 24 uninterrupted hours of night during local winter, or day during local summer.

Longitude

Longitude *is an angular distance east or west of a point on Earth's surface,* measured from the center of Earth (Figure 1-12a). On a map or globe, the lines designating these angles of longitude run north and south at right angles (90°) to the equator and to all parallels. A line connecting all points along the same longitude is a **meridian**. Thus, *longitude* is the name of the angle, *meridian* names the line, and both indicate distance east or west of an arbitrary **prime**

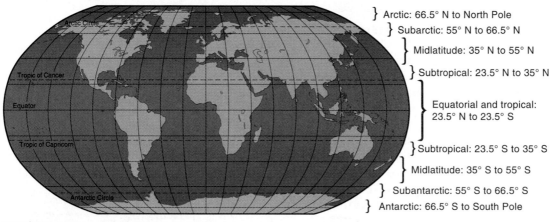

Arctic: 66.5° N to North Pole
Subarctic: 55° N to 66.5° N
Midlatitude: 35° N to 55° N
Subtropical: 23.5° N to 35° N
Equatorial and tropical: 23.5° N to 23.5° S
Subtropical: 23.5° S to 35° S
Midlatitude: 35° S to 55° S
Subantarctic: 55° S to 66.5° S
Antarctic: 66.5° S to South Pole

FIGURE 1-11
Latitudinal geographic zones. These generalized zones phase into one another over broad transitional areas.

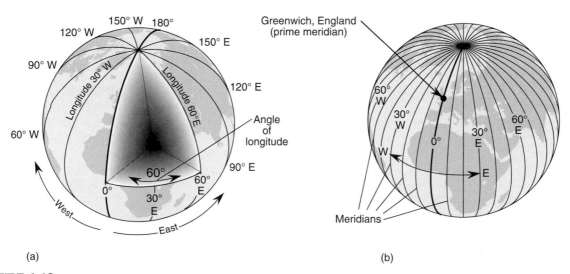

(a)   (b)

FIGURE 1-12
(a) Longitude is measured in degrees east or west of a 0° starting line. (b) Angles of longitude measured from this prime meridian, which is the line drawn from the North Pole through the observatory in Greenwich, England, to the South Pole, determine other meridians. (See FYI Report 2-1, "The Search for Longitude.")

F.Y.I. Report 1-2

The Search for Longitude

Unlike latitude, longitude cannot be readily determined from fixed celestial bodies. The problem is Earth's rotation, which constantly changes the apparent position of the Sun and stars. Determining longitude is particularly critical at sea, where no landmarks are visible. Accurately measuring time at sea was not possible until the late eighteenth century. Amerigo Vespucci, the explorer for whom America is named, made this note in his diary, as quoted by Daniel J. Boorstin in *The Discoverers:*

> 5 September 1499—As to longitude, I declare that I found so much difficulty in determining it that I was put to great pains to ascertain the east-west distance I had covered. (p. 247)

In his historical novel *Shogun*, author James Clavell expresses a similar perspective through his pilot Blackthorn:

> Find how to fix longitude and you're the richest man in the world. . . . the Queen, God bless her, 'll give you ten thousand pound and dukedom for answer to the riddle. . . . Out of sight of land you're always lost, lad. (p. 10)*

In the early 1600s Galileo explained that longitude could be measured by using two clocks. Any point on Earth takes 24 hours to travel around the full 360° of one rotation (one day). If you divide 360° by 24 hours, you find that any point on Earth travels through 15° of longitude every hour. Thus, if there were a way to measure time accurately at sea, a comparison of two clocks could give a value for longitude. One clock would indicate the time back at home port (Figure 1). The other clock would be reset at local noon each day, as determined by the highest Sun position in the sky (solar zenith). The time difference then would indicate the longitude difference traveled: one hour for each 15° of longitude.

---

*From *Shogun* by James Clavell, copyright © 1975 by James Clavell, reprinted by permission of Delacorte Press, a division of Bantam, Doubleday, Dell Publishing Group, Inc.

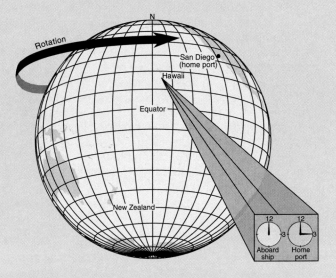

FIGURE 1

Using two clocks on a ship to determine longitude. The three-hour difference, at 15° per hour, indicates that the ship is 45° west of home port.

If the shipboard clock reads local noon and the clock set for home port reads 3:00 P.M., ship time is three hours earlier than home time. Therefore, calculating three hours at 15° per hour puts the ship at 45° west longitude from home port.

The principle was sound. All that was needed were accurate clocks. Unfortunately, the pendulum clock invented by Christian Huygens in 1656 did not work on the rolling deck of a ship at sea!

In 1707 the British lost four ships and 2000 men in a sea tragedy that was blamed specifically on the longitude problem. So, Parliament passed an act in 1714—"Publik Reward . . . to Discover the Longitude at Sea"—and authorized a prize of £20,000 sterling (equal to over $1 million today) to the first successful inventor of an accurate seafaring clock. The Board of Longitude was established to judge any devices submitted. John

Harrison, a self-taught country clockmaker, began work on the problem in 1728 and finally produced his marine chronometer, known as Number 4, in 1760. The clock was tested during a nine-week trip to Jamaica in November 1761. When taken ashore for testing against Jamaica's land-based longitude, Harrison's Number 4 was only five seconds slow, an error that translates to only 1.25', or 2.3 km (1.4 mi), well within Parliament's standard. Following many delays, Harrison finally received most of the prize money in his last years of life.

From that time it was possible to determine longitude accurately on land and sea, as long as everyone agreed upon a prime meridian to use as the reference for time comparisons. Figure 2 shows the Royal Observatory, Greenwich, England, established by international treaty in 1884 as Earth's prime meridian.

In this modern era of atomic clocks and satellites in mathematically precise orbits, we have far greater accuracy available for the determination of longitude on Earth's surface. Measurements of Earth's surface from satellites are accurate within millimeters, and surface laser surveys are precise within micrometers.

FIGURE 2
Courtyard of the old Royal Observatory, Greenwich, which is still used as Earth's prime meridian–0° longitude. [© the National Maritime Museum, Greenwich, England.]

meridian (Figure 1-12b). Earth's prime meridian (0°) passes through the old Royal Observatory at Greenwich, England.

Because meridians of longitude converge toward the poles, the actual distance on the ground spanned by a degree of longitude is greatest at the equator (where meridians separate to their widest distance apart) and diminishes to zero at the poles (where they converge). The distance covered by a degree of latitude, however, varies only slightly, owing to Earth's geoidal shape.

We have noted that latitude is easily determined by sighting the Sun, North Star, or Southern Cross, but a method of accurately determining longitude, especially at sea, remained a major difficulty until the mid-1700s. In addition, a specific determination of longitude was needed before world standard time

and time zones could be established. The interesting geographic quest for longitude is the topic of FYI Report 1-2.

Great Circles and Small Circles

Great circles and small circles are important concepts that help summarize latitude and longitude (Figure 1-13). A **great circle** is any circle of Earth's circumference whose center coincides with the center of Earth. An infinite number of great circles can be drawn on Earth. Every meridian is one-half of a great circle that passes through the poles. On flat maps, airline and shipping routes appear to arch their way across oceans and landmasses. These are *great circle routes,* the shortest distance between two points on Earth. Only one parallel is a great circle—the

"GPS: A Personal Locator"

The Global Positioning System (GPS) comprises 24 orbiting satellites that transmit navigational signals for Earth-bound use. Originally devised in the 1970s by the Department of Defense for military purposes, the present system is now commercially available worldwide. A hand-held receiver about the size of a pocket radio accesses three satellites at the same time and calculates and displays your latitude and longitude, accurate within 40 to 100 m. (Precision down to millimeters is possible for military applications.)

GPS is useful for such diverse applications as ocean navigation, land surveying, commercial fishing, trucking and the monitoring of highway taxes, mining and resource mapping, farming, environmental planning, and, as of 1994, airlines began using GPS to improve routes and increase fuel efficiency. GPS is also available for the back-packer and sportsperson. Sales of GPS systems are expected to reach $10 billion by the turn-of-the-century. The importance of GPS to geography is obvious because this precise technology reduces the need to maintain ground control points for location, mapping, and spatial analysis.

---

equatorial parallel. All other parallels diminish in length toward the poles and, along with any other non-great circle that one might draw, constitute **small circles**—circles whose centers do not coincide with Earth's center.

Prime Meridian and Standard Time

Today we take for granted standard time zones and an agreed-upon prime meridian. If you live in Oklahoma City and it is 3:00 P.M., it is 4:00 P.M. in Baltimore, 2:00 P.M. in Salt Lake City, and 1:00 P.M. in Seattle and Los Angeles. It is 9:00 P.M. in London and midnight in Riyadh, Saudi Arabia. (The designation A.M. is for *ante meridiem,* "before noon," whereas P.M. is for *post meridiem,* meaning "after noon.") Coordination of international trade, airline schedules, business activities, and daily living depends on this common time system. Such a standard time system is based on a prime meridian. Until 1884, the world was without a standard prime meridian.

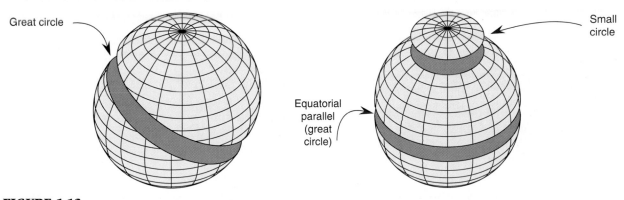

FIGURE 1-13
Great circles and small circles on Earth.

Most nations in the early 1800s used their own national prime meridians, creating confusion in global mapping and clock time. As a result, no standard time existed between or even within countries.

Because Earth revolves 360° every 24 hours, or 15° per hour (360° ÷ 24 = 15°), a time zone of one hour spans each 15° of longitude. Setting time was not so great a problem in small European countries, most of which are less than 15° wide. In North America, which covers more than 90° of longitude (the equivalent of six 15° time zones), the problem was serious. In 1870, travelers going from Maine to San Francisco made 22 adjustments to their watches to stay consistent with local railroad time. Sir Sanford Fleming led the fight in Canada for standard time and an international agreement upon a prime meridian. His struggle led the United States and Canada to adopt a standard time in 1883.

In 1884 the International Meridian Conference was held in Washington, DC, attended by 27 nations. After lengthy debate, most participating nations chose the Royal Observatory at Greenwich in England as the place for the prime meridian of 0° longitude.

The Observatory was highly respected, and more than 70% of the world's merchant ships already used the London prime meridian. (The Royal Observatory is pictured in FYI Report 1-2; note the lighted strip in the courtyard dividing the Eastern and Western hemispheres and the open roof used for sighting on the Sun.) Thus, a world standard was established–**Greenwich Mean Time (GMT)**–and the first international time system was set.

International Date Line

An important corollary of the prime meridian is the 180° meridian on the opposite side of the planet. This meridian is called the **International Date Line** and marks the place where each day officially begins (12:01 A.M.) and sweeps westward as Earth rotates. This *westward* movement of time is created by the planet's turning *eastward* on its axis. At the International Date Line, the west side of the line is always one day ahead of the east side. No matter what time of day it is when the line is crossed, the calendar changes a day (Figure 1-14).

Locating the date line in the sparsely populated Pacific Ocean minimizes most local confusion. However, the consternation of early explorers before the date-line concept was understood is interesting. For example, Magellan's crew returned from the first circumnavigation of Earth in 1522, confident from the ship's log that the day of their arrival was a Wednesday. They were shocked when informed by insistent local residents that it was actually a Thursday!

Coordinated Universal Time

Greenwich Mean Time was replaced by a universal time system in 1928, and this system was expanded in 1964 when **Coordinated Universal Time (UTC)** was instituted. Today, UTC is the reference for official time in all countries. Although the prime meridian still runs through Greenwich, UTC is based on average time calculations collected in Paris and broadcast worldwide.

Each time zone theoretically covers 7.5° on either side of a controlling meridian and represents one hour (Figure 1-15). However, as you can see in the figure, time zone boundaries are adjusted to keep certain areas together in the same time zone—see China, for

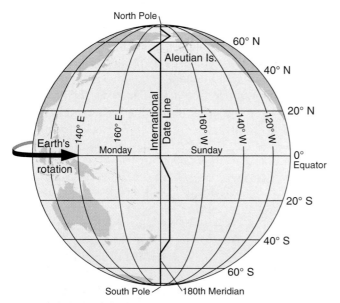

FIGURE 1-14
International Date Line location, the 180th meridian (see also Figure 1-15).

"A New Clock for UTC"

Coordinated Universal Time (UTC) and the official length of the second are measured today by the very regular vibrations of cesium atoms in five *primary standard clocks*. Three are operated in Ottawa, Canada, by the Time Standards group of the Institute for Measurement Standards, National Research Council. The other two are operated in Germany by Physikalisch-Technische Bundesanstalt (PTB). These five are now joined by the new *NIST-7* primary clock in Boulder, Colorado, built and tested by Time and Frequency Services of the National Institute for Standards and Technology. *NIST-7* began service as the sixth primary standard clock in 1994.

Added to these primary clocks, 200 *normal standard clocks* of commercial accuracy, operated in 70 nations, report to the Bureau International de l'Heure (BIH) in Paris, the international headquarters where computers bring together all measurements for keeping the world on time. This level of precision has permitted scientists to add a leap second each year since January 1, 1972, to maintain UTC accuracy in coordination with Earth's slightly irregular rate of rotation. In addition to these annual adjustments a leap second was added for the first time midyear during 1994.

example. From the map of global time zones, can you determine the present time in Moscow? London? Halifax? Chicago? Winnipeg? Denver? Los Angeles? Fairbanks? Honolulu? Tokyo? Singapore?

Daylight Saving Time

In many nations, time is set ahead one hour in the spring and set back one hour in the fall—a practice known as **daylight saving time**. The idea to extend daylight in the evening was first proposed by Benjamin Franklin, although not until World War I did Great Britain, Australia, Germany, Canada, and the United States adopt the practice.

Clocks in the United States were left advanced an hour from 1942 to 1945 during World War II, producing an added benefit of energy savings (one less hour of artificial lighting needed), and again during 1974–1975.

Daylight saving time was increased in length in the United States and Canada in 1986. Time is set ahead on the first Sunday in April and set back on the last Sunday in October—only Hawaii, Arizona, portions of Indiana and Saskatchewan exempt themselves. In Europe the last Sundays in March and September generally are used to begin and end daylight saving time.

Maps, Scales, and Projections

The earliest known graphic map presentations date to 2300 B.C., when the Babylonians used clay tablets to record information about the region of the Tigris and Euphrates rivers (the area of modern-day Iraq). Much later, when sailing vessels dominated the seas, a pilot's *rutter*—a descriptive diary of locations, places, coastlines, and collected maps—became a critical reference. *Portolan* charts, describing harbor locations and coastal features and showing compass directions, also were prepared from these many navigational experiences.

Today, the making of maps and charts is a specialized science as well as an art, blending aspects of geography, engineering, mathematics, graphics, computer science, and artistic specialties. It is similar in ways to architecture, in which aesthetics and utility are combined to produce an end product.

A *map* is a generalized view of an area, usually some portion of Earth's surface, as seen from above and greatly reduced in size. The part of geography that embodies mapmaking is called **cartography**. Maps are critical tools with which geographers depict spatial information and analyze spatial relationships. We all use maps at some time to visualize our

19

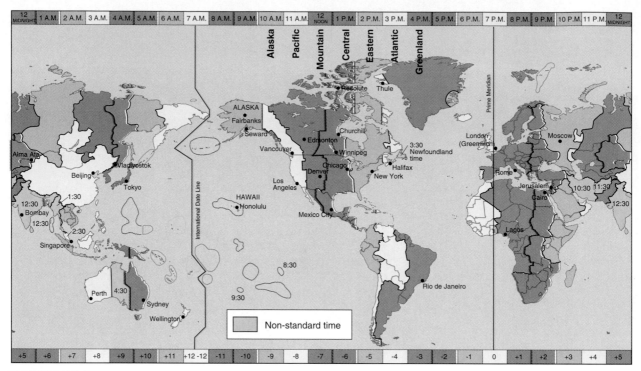

FIGURE 1-15
Modern international standard time zones. Numbers indicate how many hours each zone is earlier (minus sign) or later (plus sign) than the Coordinated Universal Time standard at the prime meridian. The United States has five time zones; Canada is divided into six zones. The new independent states of Eastern Europe recently changed their time zone by one hour to +2 hours.

location and our relation to other places, or maybe to plan a trip, or to coordinate commercial and economic activities. Have you found yourself looking at a map, planning real and imagined adventures to far-distant places? Maps are wonderful tools! Learning a few basics about maps is essential to our study of physical geography.

Map Scales

Architects and mapmakers have something in common: they both create *scale models*. They both reduce real things and places to the more convenient scale of a model, diagram, or map. An architect renders a blueprint of a building to guide the contractors, selecting a scale so that one centimeter (or inch) on the drawing represents so many meters (or feet) on the proposed building. Often, the drawing is 1/50 to 1/100 of real size.

The cartographer does the same thing in preparing a map. The ratio of the image on a map to the real world is called **scale**; it relates a unit on the map to a similar unit on the ground. A 1:1 scale means that a centimeter on the map represents a centimeter on the ground (although this certainly is an impractical map scale, for the map would be as large as the area mapped). A more appropriate scale for a local map is 1:24,000, in which 1 unit on the map represents 24,000 identical units on the ground.

Map scales are presented in several ways: as a written scale, a representative fraction, or a graphic scale (Figure 1-16). A *written scale* simply states the ratio—for example, "one centimeter to one kilometer" or "one inch to one mile." A *representative fraction* (RF, or fractional scale) can be expressed with either a : or a /, as in 1:125,000 or 1/125,000. No actual units of measurement are mentioned because any unit is applicable as long as both parts of the fraction are in the same unit: 1 cm to 125,000 cm, 1 in. to 125,000 in., or even 1 arm length to 125,000 arm lengths, and so on.

A *graphic scale,* or bar scale, is a bar graph with units to allow measurement of distances on the map. An important advantage of a graphic scale is that it changes along with the map if the map is enlarged or reduced. In contrast, written and fractional scales become incorrect with enlargement or reduction: you can shrink a map from 1:24,000 to 1:63,360, but the scale will still say "1 in. = 2000 ft," instead of the new correct scale of 1 in. = 1 mi.

Scales are called *small, medium,* and *large,* depending on the ratio described. Thus, in relative terms, a scale of 1:24,000 is a *large scale,* whereas a scale of 1:50,000,000 is a *small scale.* The greater the denominator in a fractional scale (or the number on the right in a ratio expression), the smaller the scale and the more abstract the map must be in relation to what is being mapped. Examples of selected representative fractions and written scales are listed in Table 1-1 for small-, medium-, and large-scale maps. In Chapter 9, Figure 9-12 presents a small-scale map of a portion of Pennsylvania. Enlarged from this in the figure is a *Landsat* image and topographic map at a medium scale. Note the increased detail at the larger scale.

If there are globes and maps available in your library or classroom, check to see the scale at which they are drawn. See if you can find examples of written, representative, and graphic scales on wall maps, highway maps, and in atlases.

TABLE 1-1

Sample Representative Fractions and Written Scales for Small-, Medium-, and Large-scale Maps

Units	Scale Size	Representative Fraction	Written Scale
English	Small	1:3,168,000	1 in. = 50 mi
		1:1,000,000	1 in. = 16 mi
		1:500,000	1 in. = 8 mi
		1:250,000	1 in. = 4 mi
	Medium	1:125,000	1 in. = 2 mi
		1:63,360 (62,500)	1 in. = 1 mi
		1:31,680	1 in. = 0.5 mi
	Large	1:24,000	1 in. = 2000 ft

Units	Representative Fraction	Written Scale
Metric	1:1,100,000	1 cm = 10.0 km
	1:50,000	1 cm = 0.50 km
	1:25,000	1 cm = 0.25 km

Map Projections

A globe is not always a helpful representation of Earth. Travelers, for example, need more detailed information than a globe can provide, and large globes don't fit well in cars and aircraft. Consequently, to provide localized detail, cartographers prepare large-scale flat maps, which are two-dimensional representations (scale models) of our three-dimensional Earth. This reduction of the spherical Earth to a flat surface is called a **map projection**.

Because a globe is the only true representation of distance, direction, area, shape, and proximity, the preparation of a flat version means that decisions must be made as to the type and amount of distortion that users will accept. To understand this problem, consider these important properties of a *globe.*

- Parallels always are parallel to each other, always are evenly spaced along meridians, and always decrease in length toward the poles.
- Meridians converge at both poles and are evenly spaced along any individual parallel.
- The distance between meridians decreases toward poles, with the spacing between meridians at the 60th parallel equal to one half the equatorial spacing.

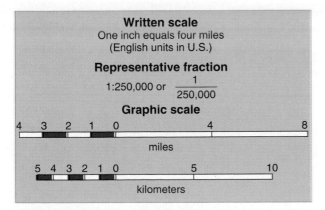

FIGURE 1-16
Three common expressions of map scale.

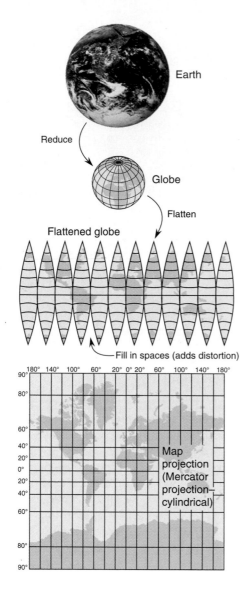

FIGURE 1-17
Conversion of the globe to a flat map projection requires decisions about which properties to preserve and the amount of distortion that is acceptable. [NASA photo.]

- Parallels and meridians always cross each other at right angles.

The problem is that all these qualities cannot be reproduced on a flat surface. Simply taking a globe apart and laying it flat on a table illustrates the difficulty faced by cartographers in constructing a flat map (Figure 1-17). You can see the empty spaces that open up

between the sections, or *gores*, of the globe. Thus, no flat map projection of Earth can ever have all the features of a globe. Flat maps always possess some degree of distortion—much less for large-scale maps representing a few kilometers; much more for small-scale maps covering individual countries, continents, or the entire world.

Properties of Projections. *The best projection is always determined by its intended use.* There are many possibilities, four of which are shown in Figure 1-18. The major decisions in selecting a map projection involve the properties of **equal area** (equivalence) and **true shape** (conformality). If a cartographer selects equal area as the desired trait, as for a map showing the distribution of world climates, then shape must be sacrificed by *stretching* and *shearing* (allowing parallels and meridians to cross at other than right angles). On an equal-area map, a coin covers the same amount of surface area, no matter where you place it on the map.

If, on the other hand, a cartographer selects the property of true shape, as for a map used for navigational purposes, then equal area must be sacrificed, and the scale will actually change from one region of the map to another. Two additional properties related to map projections are true direction (azimuth) and true distance (equidistance).

The Nature and Classes of Projections. Despite modern cartographic technology utilizing mathematical constructions and computer-assisted graphics, the word "projection" is still used. It comes from times past, when a globe literally was projected onto a surface. The globe was constructed of wire parallels and meridians. A light source then casts a pattern of latitude and longitude lines from the globe onto the desired geometric surfaces, such as a cylinder, plane, or, cone.

Figure 1-18 illustrates the derivation of the general classes of map projections and the perspectives from which they are generated. The classes shown include the *cylindrical, planar* (or *azimuthal*), and *conic*. Another class of projections that cannot be derived from this physical-perspective approach is the nonperspective *oval* shape. Still others are derived from purely mathematical calculations.

With all projections, the contact line or contact point between the wire globe and the projection sur-

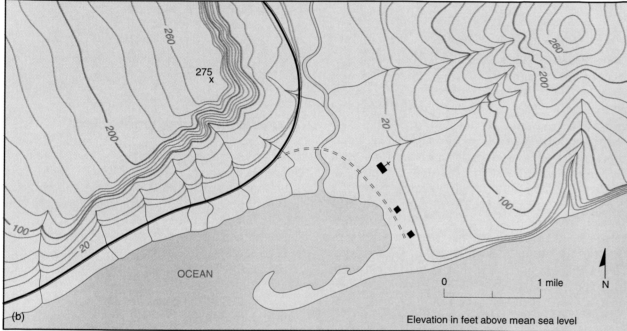

FIGURE 1-22
(a) Perspective view of a hypothetical landscape; (b) depiction of that landscape
on a topographic map (contours in feet). [(a) Reprinted with the permission of
Macmillan Publishing Company, from *The Earth: An Introduction to Physical
Geology,* 4th ed., by Edward J. Tarbuck and Frederick K. Lutgens. Copyright ©
1993 by Macmillan Publishing Company.]

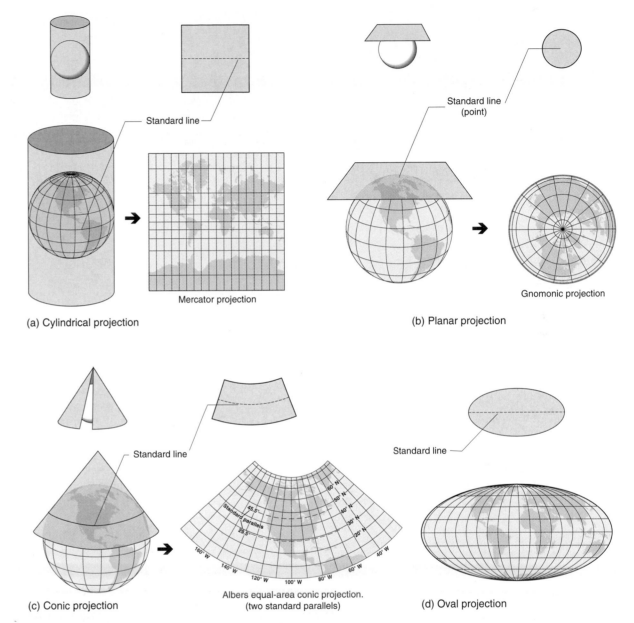

FIGURE 1-18
General classes and perspectives of map projections.

tures is added to portray the ups and downs of the
terrain. The most popular and widely used maps of
this type are the detailed **topographic maps** prepared by the USGS. Topographic maps portray *physical relief* (the vertical component of hills and valleys)
through the use of elevation **contour lines**. A contour line connects all points at the same elevation, as

you can see from the contour lines in Figure 1-22b.
A reference level—usually mean sea level—is called
the *vertical datum.* The *contour interval* is the vertical distance between two adjacent contour lines (20 ft
or 6.1 m in Figure 1-22).

The topographic map in Figure 1-22 shows a hypothetical landscape that demonstrates how contour

face—called a *standard line (or point)—is the only
place where all globe properties are preserved;* thus, a
standard parallel or *standard meridian* is a standard
line true to scale along its entire length without any
distortion. Areas away from this critical line or point
become increasingly distorted. Consequently, this
area of optimum spatial properties should be cen-

tered on the region of greatest interest so that greatest accuracy is preserved.

The commonly used **Mercator projection** (from
Gerardus Mercator, A.D. 1569) is a cylindrical projection (Figure 1-18a). We chose a Mercator projection
to show standard time zones in Figure 1-15 because
true shape and straight meridians are more important

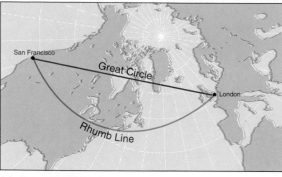

(a) Gnomonic Projection   (b) Mercator Projection

FIGURE 1-19

Comparison of rhumb lines and great-circle routes from San Francisco to London on gnomonic and Mercator projections. All great-circle routes are the shortest distance between any two points on a map.

than equal area in presenting this information. The Mercator is a true-shape projection, with meridians appearing as equally spaced straight lines and parallels appearing as straight lines that come closer together near the equator. The poles are infinitely stretched, with the 84th north parallel and 84th south parallel fixed at the same length as that of the equator. Locally, the shape is accurate and recognizable for navigation; however, the scale varies with latitude.

Unfortunately, Mercator classroom maps present false notions of the size (area) of midlatitude and poleward landmasses. A dramatic example on the cylindrical projection in Figure 1-18a is Greenland, which looks bigger than all of South America. In reality, Greenland is only one-eighth the size of South America and is actually 20% smaller than Argentina alone!

The advantage of the Mercator projection is that lines of constant direction, called **rhumb lines**, are straight

and thus facilitate plotting directions between two points (see Figure 1-19). Thus, the Mercator projection is useful in navigation and has been the standard for nautical charts prepared by the National Ocean Service (U.S. Coast and Geodetic Survey) since 1910.

The gnomonic or planar projection in Figure 1-18b is generated with a light source at the center of a globe projecting onto a plane touching the globe's surface. The resulting severe distortion prevents showing a full hemisphere on one projection. However, a valuable feature is derived: all great-circle routes, which are the shortest distance between two points on Earth's surface, are projected as straight lines. The great-circle routes plotted on a gnomonic projection then can be transferred to a true-direction projection, such as the Mercator, for determination of precise compass headings (Figure 1-19).

FIGURE 1-20

Goode's homolosine equal-area projection. [Copyright by the University of Chicago. Used by permission of the University of Chicago Press.]

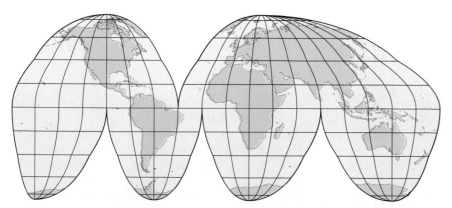

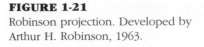

FIGURE 1-21

Robinson projection. Developed by Arthur H. Robinson, 1963.

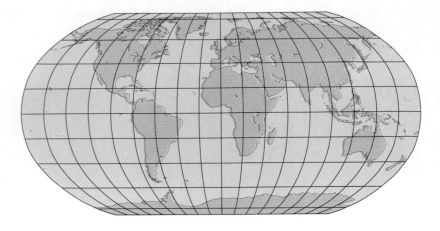

Maps Used in This Text. Several different projections are used in this text: the Goode's homolosine, Robinson, Mercator, and polyconic. Each was chosen to best present specific types of data. The two we use most are the Goode and the Robinson. **Goode's homolosine projection** is an interrupted world map designed in 1925 by Dr. J. Paul Goode of the University of Chicago. Goode's homolosine equal-area projection (Figure 1-20) is a combination of two oval projections (*homolographic* and *sinusoidal* projections). To improve the rendering of landmass shapes, two equal-area projections are cut and pasted together. Areal size relationships are preserved, so the map is excellent for mapping spatial distributions when interruptions do not pose a problem. We use the Goode's homolosine projection in this book for the world climate map (fold-out that accompanies your text), the soils map (Chapter 15), and the terrestrial ecosystems map (Chapter 16).

The other projection we often use is the **Robinson projection**, designed by Arthur Robinson in 1963 (Figure 1-21). This projection is neither equal area nor true shape, but is a compromise between both considerations. The North and South Poles appear as lines slightly more than half the length of the equator, thus higher latitudes are exaggerated less than on other oval and cylindrical projections. Examples in this text are the latitudinal geographic zones map (Figure 1-11), the world temperature maps in Chapter 3 and the maps of lithospheric plates of crust and volcanoes and earthquakes in Chapter 8. The National Geographic Society

adopted this projection for their primary world map in 1988.

Mapping and Topographic Maps

The westward expansion across the vast North American continent demanded a land survey. Maps were needed to subdivide the land and to guide travel, exploration, settlement, and transportation. The Public Lands Survey System (1785) delineated locations in the United States. Preparation and recording of this information was done by the U.S. Geological Survey (USGS), a branch of the Department of the Interior.

Today, the USGS depicts the information on quadrangle maps, which are rectangular maps bounded by parallels and meridians rather than by political boundaries. From the conic class of map projections, the Albers equal-area projection is used as a base for these quadrangle maps (Figure 1-18c). To improve the accuracy in conformality and scale for the 48 conterminous states, not one but two standard parallels are used (the line along which the cone intersects the globe's surface). These parallels are 29.5° N and 45.5° N latitudes. The standard parallels are shifted for conic projections of Alaska (55° N and 65° N) and Hawaii (8° N and 18° N).

In mapping, a basic **planimetric map** is first prepared, showing the horizontal position of boundaries, land-use aspects, and political, economic, and cultural features. A highway map is a common example of a planimetric map. Then, a vertical scale of physical fea-

lines and intervals depict slope and relief, the three-dimensional aspect of terrain. Slope is indicated by the pattern of lines and the space between them. The steeper a slope or cliff, the closer together the contour lines appear—in Figure 1-22b, note the narrowly spaced contours representing the cliffs. A more gradual slope is portrayed by a wider spacing of these contour lines, as you can see from the widely spaced lines on the beach. Actual USGS topographic maps appear in several chapters of this text because they are so useful in depicting features of the physical landscape.

A quadrangle system based on latitude and longitude coordinates is used for map classification and layout. Each map quadrangle is referred to by its angular dimensions (Figure 1-23). Thus, a map that is one-half a degree (30') on each side is a 30-minute quadrangle, and a map one-fourth of a degree (15') on each side is a 15-minute quadrangle (the USGS standard from 1910 to 1950). A map that is one-eighth of a degree (7.5') on each side is a 7.5-minute quadrangle, the most widely produced of all USGS topographic maps, and the standard since 1950.

The USGS National Mapping Program recently completed coverage of the entire country (except Alaska) on 7.5-minute maps. The standard 7.5-minute quadrangle is scaled at 1:24,000 (1 in. = 2000 ft, a large scale). It takes 53,838 separate 7.5-minute quadrangles to cover the lower 48 states, Hawaii, and U.S. territories. Alaska is covered with a series of more general 15-minute topographic maps.

The principal symbols used on USGS topographic maps are presented inside the front cover of this text. Colors are standard on all USGS topographic maps: black for human constructions, blue for water features, brown for relief features and contours, pink for urbanized areas, and green for woodlands, orchards, brush, and the like. The margins of a topographic map contain a wealth of information about the concept and content of the map: quadrangle name, names of adjoining quads, quad series and type, relative positioning on the latitude-longitude and other grid coordinate systems, title, legend, magnetic declination (alignment of magnetic north), datum plane, symbols used for roads and trails, the dates and history of the survey of that particular quad, and more. Many outdoor-product stores and state geological surveys sell topographic maps to assist people in planning their activities. These maps also may be purchased directly from the USGS.

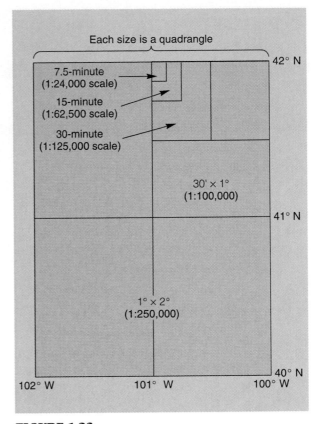

FIGURE 1-23
Quadrangle system of maps used by the U.S. Geological Survey.

Remote Sensing and GIS

Remote Sensing

In this era of observations from orbit outside the atmosphere and from aircraft within it, scientists are obtaining a wide array of remotely sensed data (Figure 1-24). Remote sensing is an important tool for the spatial analysis of Earth's environment. When we observe the environment with our eyes, we are sensing the shape, size, and color of objects from a distance, utilizing the visible-wavelength portion of the electromagnetic spectrum. Similarly, the film in a camera senses the wavelengths for which it was designed (visible light or infrared) and is exposed by the energy that is reflected and emitted from a scene. Aerial photographs have been used for years to improve the accuracy of surface maps at less expense

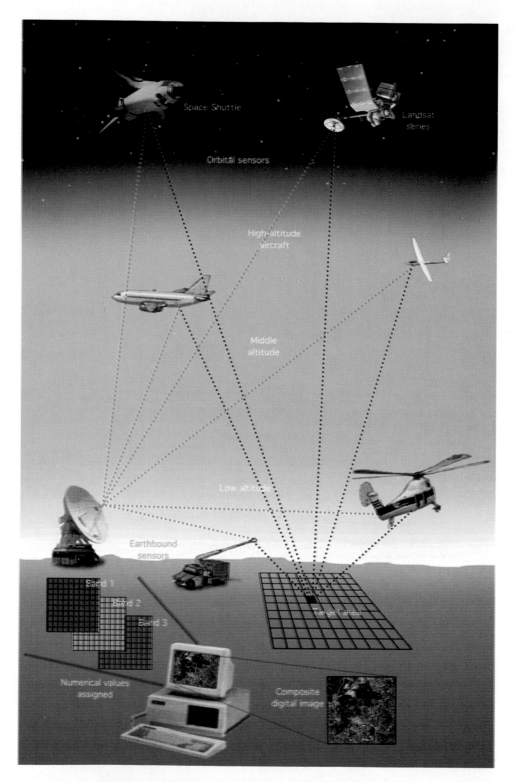

FIGURE 1-24
Remote sensing from orbiting spacecraft, aircraft in the atmosphere, and ground-based sensors measure and monitor Earth's systems. Computers process data to produce digital images for analysis. Many of the physical systems discussed in this text are studied using this technology. The Space Shuttle is shown in its inverted orbital flight mode (not to scale).

and with greater ease than is permitted by actual on-site surveys. Our eyes and cameras both represent familiar means of **remote sensing** information about a distant subject, without physical contact.

Remote sensors on satellites and other craft sense a broader range of wavelengths than our eyes and photographic films. This is because various surface materials absorb and reflect the Sun's radiation in different and characteristic ways. These differences can be remotely sensed by their reflected wavelengths: ultraviolet, visible light, reflected infrared (near- and shortwave), thermal infrared, and microwave radar (see Figure 2-7 for an illustration of these different wavelengths).

Many of the remotely sensed images used in this text were initially recorded in digital form for later processing, enhancement, and generation. Satellites do not take conventional film photographs. Rather, they record electronic *images* that are transmitted to Earth-based receivers in a manner similar to television broadcasts. A scene is scanned and broken into *pixels* (*picture elements*) identified by coordinates known as *lines* (horizontal rows) and *samples* (vertical columns). A grid of 6000 lines and 7000 samples produces a resolution of features on the ground of 30 to 120 meters depending on lenses used and the sensor's altitude above the ground. You can see that the pixel count for such an image runs well into the millions.

Digital data are processed in many ways to enhance their utility: false and simulated natural color, enhanced contrast, signal filtering, and different levels of sampling and resolution. Various wavelengths are recorded and categorized by spectral band and numerical values are assigned (see Figure 1-24). Two types of remote-sensing systems are used: active and passive.

Active Systems. Active systems direct a beam of energy at a surface and analyze the reflected energy. Radar (*Ra*dio *De*tection *a*nd *R*anging) is an example of an active remote sensor. It transmits radiation of relatively long wavelengths (1 to 10 m) in short bursts to the subject terrain, penetrating clouds and darkness. Reflected radiation, known as *backscatter,* is received by the sensing instrument and analyzed. NASA sent imaging radar systems into orbit on the *Seasat* satellite (1978) and on three Space Shuttles in 1981, 1984, and 1994. Oceanog-

raphy, landforms and geology, and biogeography were all subjects of study. The computer image of wind and sea-surface patterns over the Pacific used in Chapter 4 (Figure 4-4) was developed from 150,000 radar-derived measurements made on a single day. Side-Looking Airborne Radar (SLAR) also is such an active system. Its radar energy produces high-resolution images of surfaces it scans. Chapter 5 presents an analysis of Hurricane Gilbert using this active radar system.

Passive Systems. *Passive* remote sensing systems record energy radiated and reflected from a surface, as in the visible-light photograph of Earth on the back cover of this text. The U.S. *Landsat* series of five satellites, launched between 1972 and 1984, provided a variety of visible and infrared data, as shown in images of the Appalachians in Chapter 9, river deltas in Chapter 11, and the Hubbard glacier in Alaska featured in Chapter 14. The two *Landsats* (4 and 5) that remain operational are equipped with the Thematic Mapper (TM) as a principal sensor. The TM provides high resolution from seven spectral bands of visible and infrared wavelengths.

The National Oceanic and Atmospheric Administration (NOAA) series of polar-orbiting satellites carry the Advanced Very High Resolution Radiometer (AVHRR) sensors (*NOAA 10* and *11* are currently operating). AVHRR is sensitive in visible, near-infrared, and thermal-infrared wavelengths in five spectral bands. The incredible images of Hurricane Andrew in Chapter 5 are examples, among others in the text, produced by an AVHRR system. AVHRR is primarily used for sensing day or night clouds, snow and ice; monitoring forest fires, clouds, and surface temperature; and for determining vegetation patterns. In Chapter 16, an AVHRR image portrays clear-cutting of forests in the Pacific Northwest. These examples of resource analysis were impossible to do at such a scale just a few years ago.

The highest resolution commercial system, the French satellite called *SPOT* (Systeme Probatoire d' Observation de la Terre), can resolve from 10 to 20 m, depending on which of its sensors is used. The dramatic image in Chapter 12 of sand dunes in the Rub' al Khālī Erg, Saudi Arabia, and in Chapter 14 of floating ice in the Weddell Sea, Antarctica, are ex-

"Careers in GIS"

The entire GIS methodology involves great career opportunities in many fields. Careers are available in industry, government, business, marketing, teaching, sales, military and many other fields.

Geographers trained in GIS are analyzing ozone depletion, deforestation, soil erosion, and acid deposition. They are mapping ecosystems and monitoring plant and animal species and their declining diversity. They are designing and surveying urban developments and land planning, following the trends of global warming, studying the impact of human population, and analyzing air and water pollution, among many applications.

GIS-degree programs are now available at many colleges and universities. National Centers for Geographic Information and Analysis (NCGIA) are established at the University of California Santa Barbara (Department of Geography, UC Santa Barbara, CA 93106), University of Maine (Department of Surveying and Engineering, U of Maine, Orono, ME 04669), and State University of New York Buffalo (SUNY Buffalo, Buffalo, NY 14260). A brochure is available from these NCGIA schools entitled *Geographic Information and Your Future—Careers for a Fragile Planet.*

---

amples of *SPOT* images. Diverse applications for such remote sensing include agricultural and crop monitoring, terrain analysis, documentation of change in an ecosystem over time, river studies to reduce flood hazards, analysis of logging practices, identification of development and construction impacts, and surveillance.

The Geostationary Operational Environmental Satellite, known as *GOES-7* (Central), provides the daily infrared and visible images of weather in the Western Hemisphere that you see on television. Geostationary satellites stay in semipermanent positions because they keep pace with Earth's rotational speed at their altitude of 35,400 km (21,995 mi). A *GOES* image is used with the weather map for April 2, 1988, that appears in Chapter 5. A new generation of *GOES* weather satellites began with *GOES-8* that became operational in 1994. Images are considerably improved over previous satellites. Other specific weather satellites include the *GMS* weather satellite run by the Japan Weather Association which covers the Far East, and *METEOSAT* for Europe and Africa, operated by the European Space Agency.

Geographic Information System (GIS)

Remote sensing is an important tool for acquiring large volumes of spatial data. The next step is storage, processing, and retrieval of these data in useful ways. The value of remote sensing rests on an ability to devise powerful information handling systems. Computers have allowed integration of geographic information from direct surveys (on-the-ground mapping) and remote sensing in complex ways never before possible. A **geographic information system (GIS)** is a computer-based data-processing tool for gathering, manipulating, and analyzing geographic information. An example is shown in Figure 1-25.

GIS allows applications of almost unlimited variety. A common terrestrial reference system, such as the latitude and longitude coordinates provided on a map, is an essential beginning component for any GIS. Such maps can be converted into digital data of areas, points, and lines, depending on the task. Remotely sensed imagery and data are then transferred to the coordinate system on the reference map.

A GIS is capable of analyzing patterns and relationships within one data plane, such as the floodplain or soil layer in Figure 1-25. A GIS also can produce an *overlay analysis* where two or more data planes interact. Various assumptions, comparisons, and policies can be tested. For example, when the layers are combined the resulting synthesis, a *composite overlay*, follows specific points or areas through the complex of overlay planes. (As an example see the small composite digital image shown in Figure 1-24.)

Prior to the advent of computers, an environmental-impact analysis required someone to gather various data and painstakingly hand-produce overlays of information to determine positive and negative impacts of a project or event. Today, this layered information can be handled by a computer-driven GIS, which assesses the complex interconnections among different components. In this way, subtle changes in one element of a landscape may be identified as having a powerful impact elsewhere.

GIS applications are useful for the analysis of environmental events, both natural and human-caused. For example, the USGS completed a GIS to help analyze the spatial impact of the 1989 *Exxon Valdez* oil spill in Alaska. Scientists at NASA's Goddard Space Flight Center recently completed a three-year comprehensive GIS of Brazil in an effort to better understand land-use patterns—specifically, loss of the rain forest.

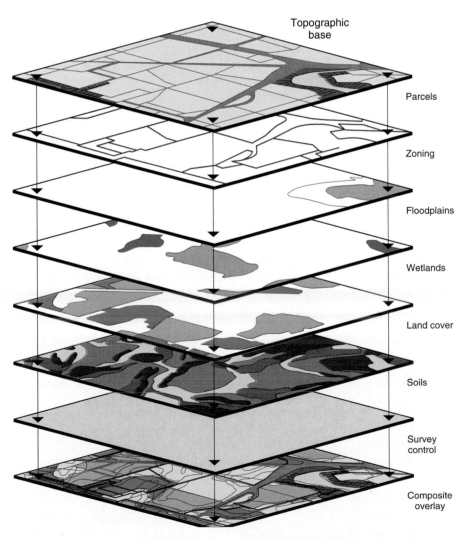

Topographic base

Parcels

Zoning

Floodplains

Wetlands

Land cover

Soils

Survey control

Composite overlay

FIGURE 1-25
A computer processes layered spatial data in a geographic information system (GIS) to produce a composite overlay for analysis. [After USGS.]

SUMMARY—Foundations of Geography

Geography is a science of method, a special way of analyzing physical and human phenomena. Its method is **spatial analysis**, used to study the interdependence among geographic areas, natural systems, society, and cultural activities over space. Geography integrates a wide range of subject matter. It brings together disciplines from the physical/life sciences with the cultural/human sciences to attain a holistic view of Earth. Geographic education recognizes five major themes: **location**, **place**, **human-Earth relationships** (including environmental concerns), **movement**, and **regions**.

Physical geography applies spatial analysis to all the physical elements and processes that make up the environment: energy, air, water, weather, climate, landforms, soils, animals, plants, and Earth itself. Understanding the complex relationships of these elements is important to human survival because Earth's physical systems and human society are so intertwined.

Systems analysis is an important tool used by geographers. A system is any ordered, related set of things and their attributes, as distinct from their surrounding environment. Earth is an **open system** for energy, receiving energy from the Sun, but essentially a **closed system** for matter and physical resources. Four immense open systems powerfully interact at Earth's surface: three nonliving abiotic systems (**atmosphere**, **hydrosphere**, and **lithosphere**) and a living biotic system (**biosphere**). Geographers often construct **models** of systems to better understand them.

Earth bulges slightly through the equator and is flattened at the poles, a misshapen spheroid that geographers call a **geoid**. Absolute location on Earth is described with a specific reference grid of parallels of **latitude** and meridians of **longitude**. A historic breakthrough in navigation and time keeping occurred with the establishment of an international **prime meridian** (0° through Greenwich, England) and the invention of precise chronometers that enabled accurate measurement of longitude. Today, **Coordinated Universal Time (UTC)** is the worldwide standard and the basis for international time zones.

Maps are used by geographers for the spatial portrayal of Earth's physical systems. Cartographers create **map projections** for specific purposes, selecting the best compromise of projection and scale for each application. Compromise is always necessary because Earth's round surface cannot be exactly duplicated on a flat map. **Equal area**, **true shape**, true direction, and true distance are all considerations in selecting a projection. The U.S. Geological Survey publishes maps using various projections and scales. Best known are **topographic maps**, which use contour lines to portray elevation and which show detailed physical and cultural features.

The operation of Earth's systems is observed through orbital and aerial **remote sensing**. Satellites do not take photographs but record images that are transmitted to Earth-based receivers for later processing, enhancement, and generation. This flood of data has lead to the development of **geographic information systems (GIS)**. Computers process geographic information as an important step in better understanding Earth's systems. GIS represents a vital career opportunity for geographers.

The science of physical geography is in a unique position to synthesize the spatial environmental and human aspects of our increasingly complex relationship with our home planet—Earth.

KEY TERMS

abiotic

atmosphere

biosphere

biotic

cartography

closed system

contour lines

Coordinated Universal Time (UTC)

daylight saving time

ecosphere

equal area

equilibrium

feedback loop

geodesy

geographic information system (GIS)

geography

geoid	parallel (of latitude)
Goode's homolosine projection	physical geography
great circle	place
Greenwich Mean Time (GMT)	planimetric map
human-Earth relationships	positive feedback
hydrosphere	prime meridian
International Date Line	process
latitude	region
lithosphere	remote sensing
location	rhumb lines
longitude	Robinson projection
map projection	scale
Mercator projection	small circle
meridian (of longitude)	spatial analysis
model	steady-state equilibrium
movement	system
negative feedback	topographic map
open system	true shape

REVIEW QUESTIONS

1. What is geography? Based on information in this chapter, define physical geography and give a specific instance of the geographic approach.
2. Assess your geographic literacy by examining available atlases and maps. What types of maps have you used—political? physical? topographic? Do you know what projections they employed? Do you know the names and locations of the four oceans, seven continents, and individual countries?
3. Suggest a representative example for each of the five geographic themes and use that theme in a sentence.
4. Define systems theory as an organizational strategy. What are open systems and closed systems, negative feedback, and a system in a steady-state equilibrium condition? What type of system (open or closed) is a human body? A lake? A wheat plant?
5. What are the three abiotic spheres that comprise Earth's environment? Relate these to the biosphere.
6. Describe Earth's shape and size with a diagram.
7. What are the latitude and longitude coordinates (in degrees, minutes, and seconds) of your present location? Where can you find this information?
8. Define latitude and parallel. Define longitude and meridian.
9. What do clocks have to do with longitude? Explain this relationship. How is standard time determined on Earth?
10. What and where is the prime meridian? How was it selected? Describe the meridian opposite the prime meridian.
11. What is map scale? How is it expressed on a map?
12. Describe the differences between the characteristics of a globe and a flat map.
13. What is remote sensing? What is it you are seeing when you observe a weather satellite image on the nightly television news? Explain.
14. If you were in charge of planning for a large tract of land, how would GIS assist you?

PART 1

THE ENERGY-ATMOSPHERE SYSTEM

CHAPTER 2
Solar Energy, Seasons, and the Atmosphere

CHAPTER 3
Atmospheric Energy and Global Temperatures

CHAPTER 4
Atmospheric and Oceanic Circulations

Sunrise at Mono Lake. [Photo by author.]

Our planet and our lives are powered by radiant energy from the star that is closest to Earth—the Sun. For more than 4.6 billion years, solar energy has traveled across interplanetary space to Earth, where a small portion of the solar output is intercepted. Because of Earth's curvature, the energy at the top of the atmosphere is unevenly distributed, creating imbalances from the equator to each pole—the equatorial region experiences energy surpluses; the polar regions experience energy deficits. This unevenness of energy receipt empowers circulations in the atmosphere and on the surface below. The pulse of seasonal change varies the distribution of energy during the year.

Earth's atmosphere acts as an efficient filter, absorbing most harmful radiation, charged particles, and space debris so that they do not reach Earth's surface. Surface energy balances are established, giving rise to global patterns of temperature, winds, and ocean currents. Each of us depends on many systems that are set into motion by energy from the Sun. These systems are the subjects of Part One.

Sun-baked sandstone fins rise to the atmosphere and storm clouds in Arches National Park, Utah. [Photo by author.]

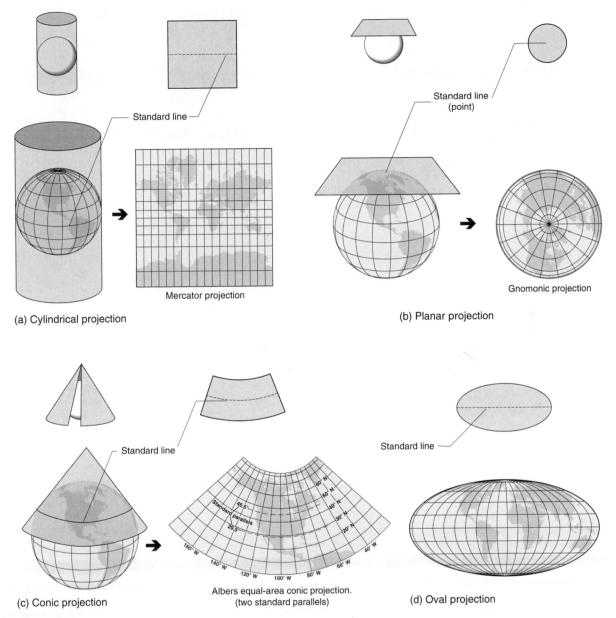

FIGURE 1-18
General classes and perspectives of map projections.

face—called a *standard line (or point)—is the only place where all globe properties are preserved;* thus, a *standard parallel* or *standard meridian* is a standard line true to scale along its entire length without any distortion. Areas away from this critical line or point become increasingly distorted. Consequently, this area of optimum spatial properties should be cen-

tered on the region of greatest interest so that greatest accuracy is preserved.

The commonly used **Mercator projection** (from Gerardus Mercator, A.D. 1569) is a cylindrical projection (Figure 1-18a). We chose a Mercator projection to show standard time zones in Figure 1-15 because true shape and straight meridians are more important

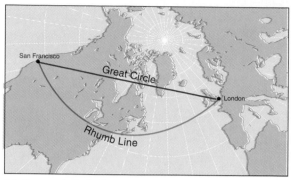

(a) Gnomonic Projection

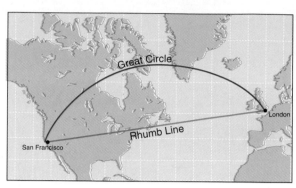

(b) Mercator Projection

FIGURE 1-19

Comparison of rhumb lines and great-circle routes from San Francisco to London on gnomonic and Mercator projections. All great-circle routes are the shortest distance between any two points on a map.

than equal area in presenting this information. The Mercator is a true-shape projection, with meridians appearing as equally spaced straight lines and parallels appearing as straight lines that come closer together near the equator. The poles are infinitely stretched, with the 84th north parallel and 84th south parallel fixed at the same length as that of the equator. Locally, the shape is accurate and recognizable for navigation; however, the scale varies with latitude.

Unfortunately, Mercator classroom maps present false notions of the size (area) of midlatitude and poleward landmasses. A dramatic example on the cylindrical projection in Figure 1-18a is Greenland, which looks bigger than all of South America. In reality, Greenland is only one-eighth the size of South America and is actually 20% smaller than Argentina alone!

The advantage of the Mercator projection is that lines of constant direction, called **rhumb lines**, are straight and thus facilitate plotting directions between two points (see Figure 1-19). Thus, the Mercator projection is useful in navigation and has been the standard for nautical charts prepared by the National Ocean Service (U.S. Coast and Geodetic Survey) since 1910.

The gnomonic or planar projection in Figure 1-18b is generated with a light source at the center of a globe projecting onto a plane touching the globe's surface. The resulting severe distortion prevents showing a full hemisphere on one projection. However, a valuable feature is derived: all great-circle routes, which are the shortest distance between two points on Earth's surface, are projected as straight lines. The great-circle routes plotted on a gnomonic projection then can be transferred to a true-direction projection, such as the Mercator, for determination of precise compass headings (Figure 1-19).

FIGURE 1-20

Goode's homolosine equal-area projection. [Copyright by the University of Chicago. Used by permission of the University of Chicago Press.]

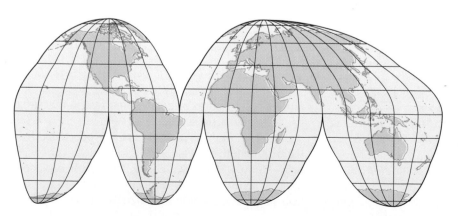

FIGURE 1-21
Robinson projection. Developed by
Arthur H. Robinson, 1963.

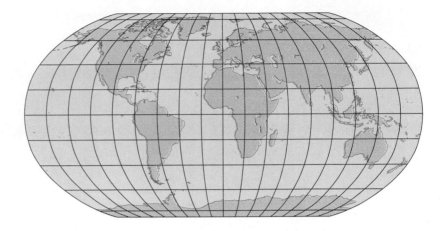

Maps Used in This Text. Several different projections are used in this text: the Goode's homolosine, Robinson, Mercator, and polyconic. Each was chosen to best present specific types of data. The two we use most are the Goode and the Robinson. **Goode's homolosine projection** is an interrupted world map designed in 1925 by Dr. J. Paul Goode of the University of Chicago. Goode's homolosine equal-area projection (Figure 1-20) is a combination of two oval projections (*homolographic* and *sin*usoidal projections). To improve the rendering of landmass shapes, two equal-area projections are cut and pasted together. Areal size relationships are preserved, so the map is excellent for mapping spatial distributions when interruptions do not pose a problem. We use the Goode's homolosine projection in this book for the world climate map (fold-out that accompanies your text), the soils map (Chapter 15), and the terrestrial ecosystems map (Chapter 16).

The other projection we often use is the **Robinson projection**, designed by Arthur Robinson in 1963 (Figure 1-21). This projection is neither equal area nor true shape, but is a compromise between both considerations. The North and South Poles appear as lines slightly more than half the length of the equator, thus higher latitudes are exaggerated less than on other oval and cylindrical projections. Examples in this text are the latitudinal geographic zones map (Figure 1-11), the world temperature maps in Chapter 3 and the maps of lithospheric plates of crust and volcanoes and earthquakes in Chapter 8. The National Geographic Society

adopted this projection for their primary world map in 1988.

Mapping and Topographic Maps

The westward expansion across the vast North American continent demanded a land survey. Maps were needed to subdivide the land and to guide travel, exploration, settlement, and transportation. The Public Lands Survey System (1785) delineated locations in the United States. Preparation and recording of this information was done by the U.S. Geological Survey (USGS), a branch of the Department of the Interior.

Today, the USGS depicts the information on quadrangle maps, which are rectangular maps bounded by parallels and meridians rather than by political boundaries. From the conic class of map projections, the Albers equal-area projection is used as a base for these quadrangle maps (Figure 1-18c). To improve the accuracy in conformality and scale for the 48 conterminous states, not one but two standard parallels are used (the line along which the cone intersects the globe's surface). These parallels are 29.5° N and 45.5° N latitudes. The standard parallels are shifted for conic projections of Alaska (55° N and 65° N) and Hawaii (8° N and 18° N).

In mapping, a basic **planimetric map** is first prepared, showing the horizontal position of boundaries, land-use aspects, and political, economic, and cultural features. A highway map is a common example of a planimetric map. Then, a vertical scale of physical fea-

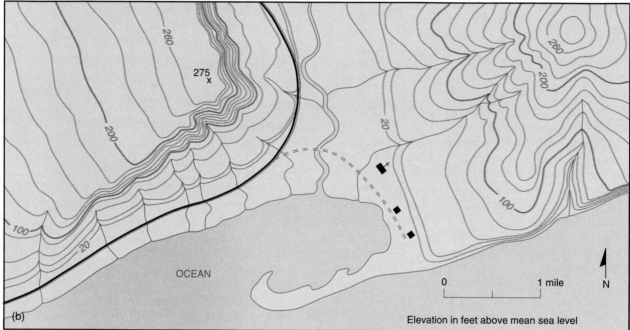

FIGURE 1-22
(a) Perspective view of a hypothetical landscape; (b) depiction of that landscape on a topographic map (contours in feet). [(a) Reprinted with the permission of Macmillan Publishing Company, from *The Earth: An Introduction to Physical Geology,* 4th ed., by Edward J. Tarbuck and Frederick K. Lutgens. Copyright © 1993 by Macmillan Publishing Company.]

tures is added to portray the ups and downs of the terrain. The most popular and widely used maps of this type are the detailed **topographic maps** prepared by the USGS. Topographic maps portray *physical relief* (the vertical component of hills and valleys) through the use of elevation **contour lines**. A contour line connects all points at the same elevation, as

you can see from the contour lines in Figure 1-22b. A reference level—usually mean sea level—is called the *vertical datum.* The *contour interval* is the vertical distance between two adjacent contour lines (20 ft or 6.1 m in Figure 1-22).

The topographic map in Figure 1-22 shows a hypothetical landscape that demonstrates how contour

lines and intervals depict slope and relief, the three-dimensional aspect of terrain. Slope is indicated by the pattern of lines and the space between them. The steeper a slope or cliff, the closer together the contour lines appear—in Figure 1-22b, note the narrowly spaced contours representing the cliffs. A more gradual slope is portrayed by a wider spacing of these contour lines, as you can see from the widely spaced lines on the beach. Actual USGS topographic maps appear in several chapters of this text because they are so useful in depicting features of the physical landscape.

A quadrangle system based on latitude and longitude coordinates is used for map classification and layout. Each map quadrangle is referred to by its angular dimensions (Figure 1-23). Thus, a map that is one-half a degree (30') on each side is a 30-minute quadrangle, and a map one-fourth of a degree (15') on each side is a 15-minute quadrangle (the USGS standard from 1910 to 1950). A map that is one-eighth of a degree (7.5') on each side is a 7.5-minute quadrangle, the most widely produced of all USGS topographic maps, and the standard since 1950.

The USGS National Mapping Program recently completed coverage of the entire country (except Alaska) on 7.5-minute maps. The standard 7.5-minute quadrangle is scaled at 1:24,000 (1 in. = 2000 ft, a large scale). It takes 53,838 separate 7.5-minute quadrangles to cover the lower 48 states, Hawaii, and U.S. territories. Alaska is covered with a series of more general 15-minute topographic maps.

The principal symbols used on USGS topographic maps are presented inside the front cover of this text. Colors are standard on all USGS topographic maps: black for human constructions, blue for water features, brown for relief features and contours, pink for urbanized areas, and green for woodlands, orchards, brush, and the like. The margins of a topographic map contain a wealth of information about the concept and content of the map: quadrangle name, names of adjoining quads, quad series and type, relative positioning on the latitude-longitude and other grid coordinate systems, title, legend, magnetic declination (alignment of magnetic north), datum plane, symbols used for roads and trails, the dates and history of the survey of that particular quad, and more. Many outdoor-product stores and state geological surveys sell topographic maps to assist people in planning their activities. These maps also may be purchased directly from the USGS.

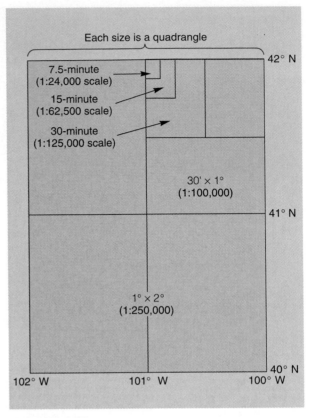

FIGURE 1-23
Quadrangle system of maps used by the U.S. Geological Survey.

Remote Sensing and GIS

Remote Sensing

In this era of observations from orbit outside the atmosphere and from aircraft within it, scientists are obtaining a wide array of remotely sensed data (Figure 1-24). Remote sensing is an important tool for the spatial analysis of Earth's environment. When we observe the environment with our eyes, we are sensing the shape, size, and color of objects from a distance, utilizing the visible-wavelength portion of the electromagnetic spectrum. Similarly, the film in a camera senses the wavelengths for which it was designed (visible light or infrared) and is exposed by the energy that is reflected and emitted from a scene. Aerial photographs have been used for years to improve the accuracy of surface maps at less expense

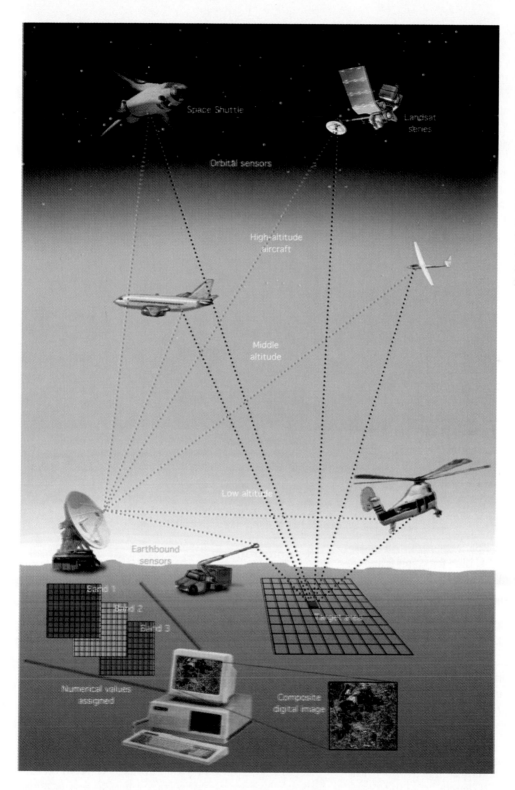

FIGURE 1-24
Remote sensing from orbiting spacecraft, aircraft in the atmosphere, and ground-based sensors measure and monitor Earth's systems. Computers process data to produce digital images for analysis. Many of the physical systems discussed in this text are studied using this technology. The Space Shuttle is shown in its inverted orbital flight mode (not to scale).

and with greater ease than is permitted by actual on-site surveys. Our eyes and cameras both represent familiar means of **remote sensing** information about a distant subject, without physical contact.

Remote sensors on satellites and other craft sense a broader range of wavelengths than our eyes and photographic films. This is because various surface materials absorb and reflect the Sun's radiation in different and characteristic ways. These differences can be remotely sensed by their reflected wavelengths: ultraviolet, visible light, reflected infrared (near- and shortwave), thermal infrared, and microwave radar (see Figure 2-7 for an illustration of these different wavelengths).

Many of the remotely sensed images used in this text were initially recorded in digital form for later processing, enhancement, and generation. Satellites do not take conventional film photographs. Rather, they record electronic *images* that are transmitted to Earth-based receivers in a manner similar to television broadcasts. A scene is scanned and broken into *pixels* (*picture elements*) identified by coordinates known as *lines* (horizontal rows) and *samples* (vertical columns). A grid of 6000 lines and 7000 samples produces a resolution of features on the ground of 30 to 120 meters depending on lenses used and the sensor's altitude above the ground. You can see that the pixel count for such an image runs well into the millions.

Digital data are processed in many ways to enhance their utility: false and simulated natural color, enhanced contrast, signal filtering, and different levels of sampling and resolution. Various wavelengths are recorded and categorized by spectral band and numerical values are assigned (see Figure 1-24). Two types of remote-sensing systems are used: active and passive.

Active Systems. Active systems direct a beam of energy at a surface and analyze the reflected energy. Radar (*Ra*dio *D*etection *a*nd *R*anging) is an example of an active remote sensor. It transmits radiation of relatively long wavelengths (1 to 10 m) in short bursts to the subject terrain, penetrating clouds and darkness. Reflected radiation, known as *backscatter,* is received by the sensing instrument and analyzed. NASA sent imaging radar systems into orbit on the *Seasat* satellite (1978) and on three Space Shuttles in 1981, 1984, and 1994. Oceanog-

raphy, landforms and geology, and biogeography were all subjects of study. The computer image of wind and sea-surface patterns over the Pacific used in Chapter 4 (Figure 4-4) was developed from 150,000 radar-derived measurements made on a single day. Side-Looking Airborne Radar (SLAR) also is such an active system. Its radar energy produces high-resolution images of surfaces it scans. Chapter 5 presents an analysis of Hurricane Gilbert using this active radar system.

Passive Systems. *Passive* remote sensing systems record energy radiated and reflected from a surface, as in the visible-light photograph of Earth on the back cover of this text. The U.S. *Landsat* series of five satellites, launched between 1972 and 1984, provided a variety of visible and infrared data, as shown in images of the Appalachians in Chapter 9, river deltas in Chapter 11, and the Hubbard glacier in Alaska featured in Chapter 14. The two *Landsats* (4 and 5) that remain operational are equipped with the Thematic Mapper (TM) as a principal sensor. The TM provides high resolution from seven spectral bands of visible and infrared wavelengths.

The National Oceanic and Atmospheric Administration (NOAA) series of polar-orbiting satellites carry the Advanced Very High Resolution Radiometer (AVHRR) sensors (*NOAA 10* and *11* are currently operating). AVHRR is sensitive in visible, near-infrared, and thermal-infrared wavelengths in five spectral bands. The incredible images of Hurricane Andrew in Chapter 5 are examples, among others in the text, produced by an AVHRR system. AVHRR is primarily used for sensing day or night clouds, snow and ice; monitoring forest fires, clouds, and surface temperature; and for determining vegetation patterns. In Chapter 16, an AVHRR image portrays clear-cutting of forests in the Pacific Northwest. These examples of resource analysis were impossible to do at such a scale just a few years ago.

The highest resolution commercial system, the French satellite called *SPOT* (Systeme Probatoire d' Observation de la Terre), can resolve from 10 to 20 m, depending on which of its sensors is used. The dramatic image in Chapter 12 of sand dunes in the Rub' al Khālī Erg, Saudi Arabia, and in Chapter 14 of floating ice in the Weddell Sea, Antarctica, are ex-

"Careers in GIS"

The entire GIS methodology involves great career opportunities in many fields. Careers are available in industry, government, business, marketing, teaching, sales, military and many other fields.

Geographers trained in GIS are analyzing ozone depletion, deforestation, soil erosion, and acid deposition. They are mapping ecosystems and monitoring plant and animal species and their declining diversity. They are designing and surveying urban developments and land planning, following the trends of global warming, studying the impact of human population, and analyzing air and water pollution, among many applications.

GIS-degree programs are now available at many colleges and universities. National Centers for Geographic Information and Analysis (NCGIA) are established at the University of California Santa Barbara (Department of Geography, UC Santa Barbara, CA 93106), University of Maine (Department of Surveying and Engineering, U of Maine, Orono, ME 04669), and State University of New York Buffalo (SUNY Buffalo, Buffalo, NY 14260). A brochure is available from these NCGIA schools entitled *Geographic Information and Your Future—Careers for a Fragile Planet.*

---

amples of *SPOT* images. Diverse applications for such remote sensing include agricultural and crop monitoring, terrain analysis, documentation of change in an ecosystem over time, river studies to reduce flood hazards, analysis of logging practices, identification of development and construction impacts, and surveillance.

The Geostationary Operational Environmental Satellite, known as *GOES-7* (Central), provides the daily infrared and visible images of weather in the Western Hemisphere that you see on television. Geostationary satellites stay in semipermanent positions because they keep pace with Earth's rotational speed at their altitude of 35,400 km (21,995 mi). A *GOES* image is used with the weather map for April 2, 1988, that appears in Chapter 5. A new generation of *GOES* weather satellites began with *GOES-8* that became operational in 1994. Images are considerably improved over previous satellites. Other specific weather satellites include the *GMS* weather satellite run by the Japan Weather Association which covers the Far East, and *METEOSAT* for Europe and Africa, operated by the European Space Agency.

Geographic Information System (GIS)

Remote sensing is an important tool for acquiring large volumes of spatial data. The next step is storage, processing, and retrieval of these data in useful ways. The value of remote sensing rests on an ability to devise powerful information handling systems. Computers have allowed integration of geographic information from direct surveys (on-the-ground mapping) and remote sensing in complex ways never before possible. A **geographic information system (GIS)** is a computer-based data-processing tool for gathering, manipulating, and analyzing geographic information. An example is shown in Figure 1-25.

GIS allows applications of almost unlimited variety. A common terrestrial reference system, such as the latitude and longitude coordinates provided on a map, is an essential beginning component for any GIS. Such maps can be converted into digital data of areas, points, and lines, depending on the task. Remotely sensed imagery and data are then transferred to the coordinate system on the reference map.

A GIS is capable of analyzing patterns and relationships within one data plane, such as the flood-

plain or soil layer in Figure 1-25. A GIS also can produce an *overlay analysis* where two or more data planes interact. Various assumptions, comparisons, and policies can be tested. For example, when the layers are combined the resulting synthesis, a *composite overlay*, follows specific points or areas through the complex of overlay planes. (As an example see the small composite digital image shown in Figure 1-24.)

Prior to the advent of computers, an environmental-impact analysis required someone to gather various data and painstakingly hand-produce overlays of information to determine positive and negative impacts of a project or event. Today, this layered information can be handled by a com-puter-driven GIS, which assesses the complex interconnections among different components. In this way, subtle changes in one element of a landscape may be identified as having a powerful impact elsewhere.

GIS applications are useful for the analysis of environmental events, both natural and human-caused. For example, the USGS completed a GIS to help analyze the spatial impact of the 1989 *Exxon Valdez* oil spill in Alaska. Scientists at NASA's Goddard Space Flight Center recently completed a three-year comprehensive GIS of Brazil in an effort to better understand land-use patterns—specifically, loss of the rain forest.

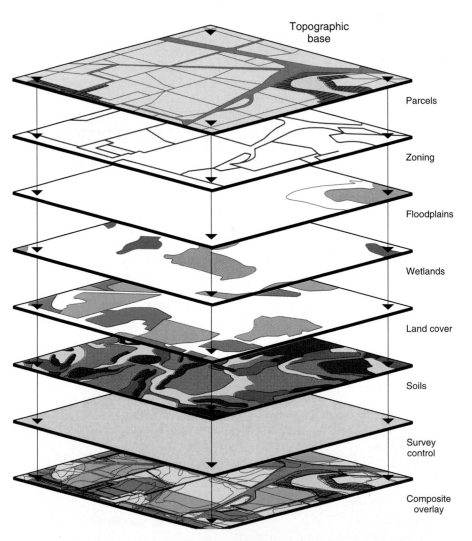

Topographic base

Parcels

Zoning

Floodplains

Wetlands

Land cover

Soils

Survey control

Composite overlay

FIGURE 1-25

A computer processes layered spatial data in a geographic information system (GIS) to produce a composite overlay for analysis. [After USGS.]

SUMMARY—Foundations of Geography

Geography is a science of method, a special way of analyzing physical and human phenomena. Its method is **spatial analysis**, used to study the interdependence among geographic areas, natural systems, society, and cultural activities over space. Geography integrates a wide range of subject matter. It brings together disciplines from the physical/life sciences with the cultural/human sciences to attain a holistic view of Earth. Geographic education recognizes five major themes: **location**, **place**, **human-Earth relationships** (including environmental concerns), **movement**, and **regions**.

Physical geography applies spatial analysis to all the physical elements and processes that make up the environment: energy, air, water, weather, climate, landforms, soils, animals, plants, and Earth itself. Understanding the complex relationships of these elements is important to human survival because Earth's physical systems and human society are so intertwined.

Systems analysis is an important tool used by geographers. A system is any ordered, related set of things and their attributes, as distinct from their surrounding environment. Earth is an **open system** for energy, receiving energy from the Sun, but essentially a **closed system** for matter and physical resources. Four immense open systems powerfully interact at Earth's surface: three nonliving abiotic systems (**atmosphere**, **hydrosphere**, and **lithosphere**) and a living biotic system (**biosphere**). Geographers often construct **models** of systems to better understand them.

Earth bulges slightly through the equator and is flattened at the poles, a misshapen spheroid that geographers call a **geoid**. Absolute location on Earth is described with a specific reference grid of parallels of **latitude** and meridians of **longitude**. A historic breakthrough in navigation and time keeping occurred with the establishment of an international **prime meridian** (0° through Greenwich, England) and the invention of precise chronometers that enabled accurate measurement of longitude. Today, **Coordinated Universal Time (UTC)** is the worldwide standard and the basis for international time zones.

Maps are used by geographers for the spatial portrayal of Earth's physical systems. Cartographers create **map projections** for specific purposes, selecting the best compromise of projection and scale for each application. Compromise is always necessary because Earth's round surface cannot be exactly duplicated on a flat map. **Equal area**, **true shape**, true direction, and true distance are all considerations in selecting a projection. The U.S. Geological Survey publishes maps using various projections and scales. Best known are **topographic maps**, which use contour lines to portray elevation and which show detailed physical and cultural features.

The operation of Earth's systems is observed through orbital and aerial **remote sensing**. Satellites do not take photographs but record images that are transmitted to Earth-based receivers for later processing, enhancement, and generation. This flood of data has lead to the development of **geographic information systems (GIS)**. Computers process geographic information as an important step in better understanding Earth's systems. GIS represents a vital career opportunity for geographers.

The science of physical geography is in a unique position to synthesize the spatial environmental and human aspects of our increasingly complex relationship with our home planet—Earth.

KEY TERMS

abiotic
atmosphere
biosphere
biotic
cartography
closed system
contour lines
Coordinated Universal Time (UTC)

daylight saving time
ecosphere
equal area
equilibrium
feedback loop
geodesy
geographic information system (GIS)
geography

geoid	parallel (of latitude)
Goode's homolosine projection	physical geography
great circle	place
Greenwich Mean Time (GMT)	planimetric map
human-Earth relationships	positive feedback
hydrosphere	prime meridian
International Date Line	process
latitude	region
lithosphere	remote sensing
location	rhumb lines
longitude	Robinson projection
map projection	scale
Mercator projection	small circle
meridian (of longitude)	spatial analysis
model	steady-state equilibrium
movement	system
negative feedback	topographic map
open system	true shape

REVIEW QUESTIONS

1. What is geography? Based on information in this chapter, define physical geography and give a specific instance of the geographic approach.
2. Assess your geographic literacy by examining available atlases and maps. What types of maps have you used—political? physical? topographic? Do you know what projections they employed? Do you know the names and locations of the four oceans, seven continents, and individual countries?
3. Suggest a representative example for each of the five geographic themes and use that theme in a sentence.
4. Define systems theory as an organizational strategy. What are open systems and closed systems, negative feedback, and a system in a steady-state equilibrium condition? What type of system (open or closed) is a human body? A lake? A wheat plant?
5. What are the three abiotic spheres that comprise Earth's environment? Relate these to the biosphere.
6. Describe Earth's shape and size with a diagram.
7. What are the latitude and longitude coordinates (in degrees, minutes, and seconds) of your present location? Where can you find this information?
8. Define latitude and parallel. Define longitude and meridian.
9. What do clocks have to do with longitude? Explain this relationship. How is standard time determined on Earth?
10. What and where is the prime meridian? How was it selected? Describe the meridian opposite the prime meridian.
11. What is map scale? How is it expressed on a map?
12. Describe the differences between the characteristics of a globe and a flat map.
13. What is remote sensing? What is it you are seeing when you observe a weather satellite image on the nightly television news? Explain.
14. If you were in charge of planning for a large tract of land, how would GIS assist you?

PART 1

THE ENERGY-
ATMOSPHERE SYSTEM

CHAPTER 2
Solar Energy, Seasons, and the Atmosphere

CHAPTER 3
Atmospheric Energy and Global Temperatures

CHAPTER 4
Atmospheric and Oceanic Circulations

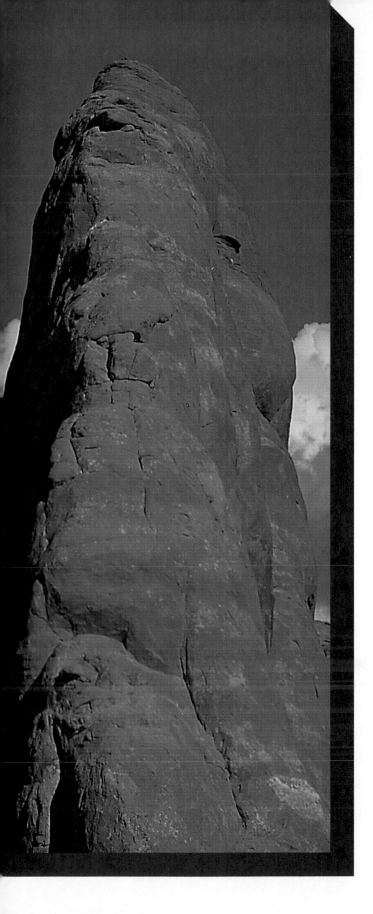

2

SOLAR ENERGY, SEASONS, AND THE ATMOSPHERE

SOLAR SYSTEM, SUN, AND EARTH
  Dimensions and Distances   Earth's Orbit
SOLAR ENERGY: FROM SUN TO EARTH
  Solar Wind
  Electromagnetic Spectrum of Radiant Energy
ENERGY AT THE TOP OF THE
  ATMOSPHERE
  Intercepted Energy
THE SEASONS
  Seasonality   Reasons for Seasons
  Annual March of the Seasons
EARTH'S ATMOSPHERE
  Heterosphere   Homosphere
VARIABLE ATMOSPHERIC
  COMPONENTS
  Natural Sources
  Natural Factors that Affect Air Pollution
  Anthropogenic Pollution

SUMMARY

FYI REPORT 2-1 THE SCIENTIFIC METHOD
FYI REPORT 2-2 STRATOSPHERIC OZONE LOSSES:
              A WORLDWIDE HAZARD

The Universe is populated with millions of galaxies. One of these is our own Milky Way Galaxy, consisting of billions of stars. Among these stars is an average yellow star we call the Sun. Our Sun radiates energy in all directions and upon its family of orbiting planets. Of special interest to us is the solar energy that falls on the third planet and on our immediate home North America, a major continent on planet Earth.

Solar energy at the top of the atmosphere sets into motion the winds, weather systems, and ocean currents that greatly influence our lives. This solar energy input to the atmosphere plus Earth's tilt and rotation produce daily, seasonal, and annual patterns of changing daylength and Sun angles.

Earth's atmosphere is a unique reservoir of gases, the product of nearly 5 billion years of development. We all participate in the atmosphere with each breath we take. In this chapter we examine the atmosphere's structure, function, and composition. Our consideration also must include the spatial aspects of human-induced problems that affect the atmosphere, such as air pollution and the stratospheric ozone predicament. We consider these critical topics carefully, for we are participating in producing the atmosphere of the future.

The Solar System, Sun, and Earth

Our Solar System is located on a remote, trailing edge of the **Milky Way Galaxy**, a flattened, disk-shaped mass estimated to contain up to 400 billion stars. From our Earth-bound perspective in the Milky Way, the galaxy appears to stretch across the night sky like a narrow band of hazy light. On a clear night the unaided eye can see only a few thousand of these billions of stars.

According to prevailing theory, our Solar System condensed from a large, slowly rotating, collapsing cloud of dust and gas called a **nebula**. *Gravity*, the natural force exerted by the mass of an object upon all other objects, was the key organizing force in this condensing solar nebula. The beginnings of the Sun and its Solar System are estimated to have occurred more than 4.6 billion years ago.

This same process—suns condensing from nebular clouds with planetesimals forming in orbits around their central masses—is called the **planetesimal hypothesis**, or dust-cloud hypothesis. Astronomers are observing this formation process under way in other parts of the galaxy. The planets appear to accrete from dust, gases, and icy comets that are drawn by gravity into collision and coalescence.

The development of such hypotheses and theories is an exercise of the **scientific method**, a methodology important to physical geography research. FYI Report 2-1 explains this essential process of science.

Dimensions and Distances

The **speed of light** is 299,792 km per second (186,282 mi/sec), which is about 9.5 trillion km or nearly 6.0 trillion miles per year. This tremendous distance that light travels in a year is known as a *light-year* and is used as a unit of measurement for the vast universe. For spatial comparison, our Moon is an average distance of 384,400 km (238,866 mi) from Earth, or about 1.28 seconds in terms of light speed. Our entire Solar System is approximately 11 hours in diameter, measured by light speed. In contrast, the Milky Way is about 100,000 light-years from side to side, and the known universe that is observable from Earth stretches approximately 12 billion light-years in all directions. Figure 2-1 compares Earth's size with that of the other planets in the Solar System.

Earth's Orbit

Earth's orbit around the Sun is presently elliptical—a closed, oval-shaped path. Earth's average distance from the Sun is approximately 150 million km (93 million mi), which means that light reaches Earth from the Sun in an average of 8 minutes and 20 seconds. Earth is at **perihelion** (its closest position to the Sun) on January 3 and at **aphelion** on July 4 (see Figure 2-2). A plane including all points of Earth's orbit is termed the *plane of the ecliptic*.

The Scientific Method

The term *scientific method* may have an aura of complexity that it should not. The scientific method is simply the application of common sense in an organized and objective manner. A scientist conducts observations, makes generalizations, formulates hypotheses, performs tests and experiments, and develops theories. Sir Isaac Newton (1642–1727) developed this method of discovering the patterns of nature, although the term scientific method was applied later.

The scientific method begins with a specific question to be answered by scientific inquiry. Scientists who study the physical environment turn to nature for clues that they can observe and measure. Then, observations and data are analyzed to identify coherent patterns that may be present. If patterns are discovered, the researcher may formulate a *hypothesis*—a general principle—deduced from a specific group of observations (e.g., planetesimal hypothesis, nuclear-winter hypothesis, moisture-benefits-from-hurricanes hypothesis). Further observations are related back to the general principles established by the hypothesis. The fact that a hypothesis can be made about things not yet observed produces a useful flexibility within the scientific method. Finally, if data gathered are found to support the hypothesis, and if predictions made according to it prove accurate, the hypothesis may be elevated to the status of a theory.

A *theory* is constructed on the basis of several hypotheses that have been extensively tested. It represents a truly broad general principle (e.g., theory of relativity, theory of evolution, atomic theory, Big Bang theory, or the theory of plate tectonics discussed in Chapter 8). A theory is a powerful device with which to understand both the orderliness and chaos in nature. Predictions based on a theory can be made about things not yet known, the effects of which can be tested and verified or disproved through tangible evidence. The ultimate value is the continued observation, testing, understanding, and pursuit of knowledge that the theory stimulates.

Important to consider is that pure science does not make value judgments. Instead, pure science provides people and their institutions with information on which to base their own value judgments. These social and political judgments about the applications of science are increasingly critical as Earth's natural systems respond to the impact of human civilization.

The structure of Earth's orbit is not a constant but instead exhibits changes over long periods. Earth's distance from the Sun varies more than 17.7 million km (11 million mi) during a 100,000-year cycle, placing it closer or farther at different periods in the cycle. This variation is thought to be one of several factors that create Earth's cyclical pattern of glaciations and interglacial periods, which are colder and warmer times respectively. Such cycles are the subject of FYI Report 14-1, "Deciphering Past Climates".

Solar Energy: From Sun to Earth

Our Sun is unique to us, and yet commonplace in our galaxy. It is only average in temperature, size, and color when compared to other stars, yet it is the ultimate energy source for most life processes in our biosphere.

The Sun captured about 99.9% of the matter (dust and gas) from the original nebula. The remaining 0.1% gave rise to all the planets, their satellites, asteroids, comets, and debris. Clearly, the dominant

FIGURE 2-1

A size comparison of planets in our Solar System. The size of each planet and the
Sun is approximately to scale.

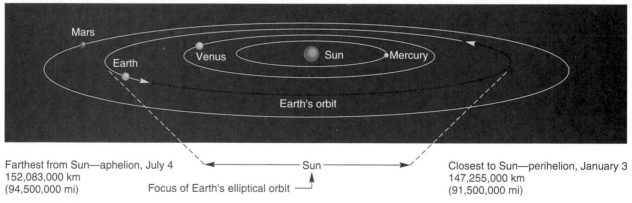

Farthest from Sun—aphelion, July 4
152,083,000 km
(94,500,000 mi)

Focus of Earth's elliptical orbit

Closest to Sun—perihelion, January 3
147,255,000 km
(91,500,000 mi)

FIGURE 2-2

Structure of Earth's elliptical orbit. The plane of Earth's orbit is called the *plane of
the ecliptic.*

object in our region of space is the Sun. In the entire Solar System, it is the only object having the
enormous mass needed to create the internal conditions that produce significant energy.

The solar mass produces tremendous pressure
and high temperatures deep in its dense interior.
Under these conditions, the Sun's abundant atoms
of hydrogen are forced together, and pairs of hydrogen nuclei are fused. This process is called **fusion**, and it liberates enormous quantities of energy.
A sunny day can seem so peaceful, certainly bely-
ing the violence proceeding on the Sun. The Sun's
principal outputs consist of the solar wind and radiant energy in portions of the electromagnetic spectrum. Let us trace each of these emissions across
space to Earth.

Sunrise at Mono Lake. [Photo by author.]

Our planet and our lives are powered by radiant energy from the star that is closest to Earth—the Sun. For more than 4.6 billion years, solar energy has traveled across interplanetary space to Earth, where a small portion of the solar output is intercepted. Because of Earth's curvature, the energy at the top of the atmosphere is unevenly distributed, creating imbalances from the equator to each pole—the equatorial region experiences energy surpluses; the polar regions experience energy deficits. This unevenness of energy receipt empowers circulations in the atmosphere and on the surface below. The pulse of seasonal change varies the distribution of energy during the year.

Earth's atmosphere acts as an efficient filter, absorbing most harmful radiation, charged particles, and space debris so that they do not reach Earth's surface. Surface energy balances are established, giving rise to global patterns of temperature, winds, and ocean currents. Each of us depends on many systems that are set into motion by energy from the Sun. These systems are the subjects of Part One.

Sun-baked sandstone fins rise to the atmosphere and storm clouds in Arches National Park, Utah. [Photo by author.]

"Scientists Discover the Atmosphere's Origin"

The evolution of Earth's atmosphere and surface, the formation of free oxygen gas, and the development of the biosphere all represent complex interactions that were in operation from the beginning of Earth's environment. A principal component of Earth's history is the evolution of its modern atmosphere.

This evolution occurred in four broad stages that blended, each into the next. Earth's *primordial atmosphere* was derived from the original solar nebula. This atmosphere and the second stage—*evolutionary atmosphere*—are thought to have persisted for relatively short periods. The third and fourth stages of the atmosphere's development have existed over much greater time spans.

Approximately 3.3 billion years ago, the first organism to perform *photosynthesis* evolved in Earth's shallow waters. Photosynthesis releases oxygen into the air, so this marked the beginning of the third stage, the *living atmosphere.* Cyanobacteria (blue-green algae) began this production of oxygen, but oxygen did not reach levels comparable to today until about 500 million years ago, marking commencement of the fourth stage, the *modern atmosphere.* Human civilization has such great impact that aspects of Earth's next atmosphere most likely will be *anthropogenic,* or human-influenced in character—a fifth distinct atmosphere.

Solar Wind

The Sun constantly emits ionized (electrically charged) particles (principally electrons and protons) that surge outward in all directions from the Sun's surface. This stream of energetic material travels at much less than the speed of light, taking approximately three days to reach Earth. The term **solar wind** was first applied to this phenomenon in 1958. The solar wind grows stronger during periods of increased sunspot activity. **Sunspots** are large magnetic storms which reveal unusual solar activity (Figure 2-3).

A regular cycle exists for sunspot occurrences, averaging 11 years from maximum to maximum; however, the cycle may vary from 7 to 17 years. In recent cycles, a solar minimum occurred in 1976 and a solar maximum took place during 1979, with over 100 sunspots visible. Another minimum was reached in 1986, and an extremely active solar maximum followed in 1990 with over 200 sunspots visible at some time during the year. In fact, the 1990–1991 maximum was the most intense ever observed.

Earth's outer defense against the charged particles of the solar wind is the **magnetosphere**, which is a magnetic field surrounding Earth, generated by dynamolike motions within our planet. The magnetosphere deflects the solar wind toward both poles so that only a small portion of it enters the atmosphere.

The extreme northern and southern polar regions of the upper atmosphere are the points of entry for the solar wind stream, far above Earth's protected surface. Because the solar wind does not reach the surface, research on this phenomenon must be conducted in space. The *Apollo XI* astronauts deployed a solar wind experiment on the lunar surface on July 20, 1969 (Figure 2-4). A piece of foil was exposed to the solar wind while the astronauts worked on the Moon. When examined back on Earth, the exposed foil exhibited particle impacts that confirmed the presence and character of the solar wind.

The solar wind creates **auroras** in the upper atmosphere when absorbed energy is reradiated as light energy of varying colors. These lighting effects are the: *aurora borealis* (northern lights) and *aurora*

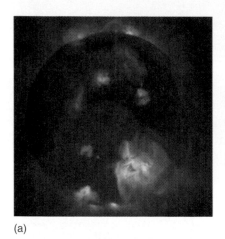
(a)

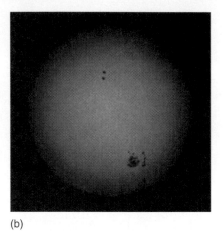
(b)

FIGURE 2-3

(a) The Sun in X-ray and (b) visible light wavelengths as imaged on the same day in 1992 by the *Yohkoh* satellite. Sunspots appear as visible dark patches across the Sun and as areas of intense X-ray activity. Individual sunspots range from 10,000 to 50,000 km in diameter (6,200 to 31,000 mi), and some have exceeded 12 times Earth's diameter. [Images courtesy of Dr. Keith T. Strong of Lockheed and Dr. Yutaka Uchida of the *Yohkoh* Science Committee, National Astronomical Observatory of Japan, and the University of Tokyo.

FIGURE 2-4

Without a protective atmosphere, the lunar surface receives direct solar wind and electromagnetic radiation. Here a solar wind experiment is being deployed by an *Apollo XI* astronaut. [NASA photo.]

australis (southern lights). The auroras generally are visible poleward of 65° latitude when the solar wind is active. They appear as folded sheets of green, yellow, blue, and red light that undulate across the skies of high latitudes as shown in Figure 2-5. In addition, the solar wind may disrupt certain radio broadcasts and some satellite transmissions, cause overloads on Earth-based electrical systems, and create possible effects on weather patterns.

Electromagnetic Spectrum of Radiant Energy

The key solar input to life is electromagnetic energy. Solar radiation occupies a portion of the **electromagnetic spectrum** of radiant energy. This radiant energy travels at the speed of light to Earth. The total spectrum of this radiant energy is made up of different wavelengths. A **wavelength** is the distance between corresponding points on any two successive waves (Figure 2-6). The number of waves passing a fixed point in one second is the *frequency*. The Sun emits radiant energy composed of 8% ultraviolet, X-ray, and gamma ray wavelengths; 47% visible light wavelengths; and 45% infrared wavelengths. A portion of the electromagnetic spectrum is illustrated in Figure 2-7; note the wavelengths at which various phenomena occur.

FIGURE 2-5
Aurora borealis in the night sky over Alaska. [Photo by Johnny Johnson. Allstock]

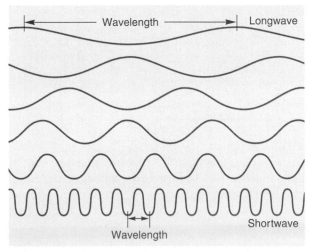

FIGURE 2-6
Wavelength and frequency are used to describe the electromagnetic spectrum. These are two ways of describing the same phenomenon—electromagnetic wave motion. Shorter wavelengths are higher in frequency, whereas longer wavelengths are lower in frequency.

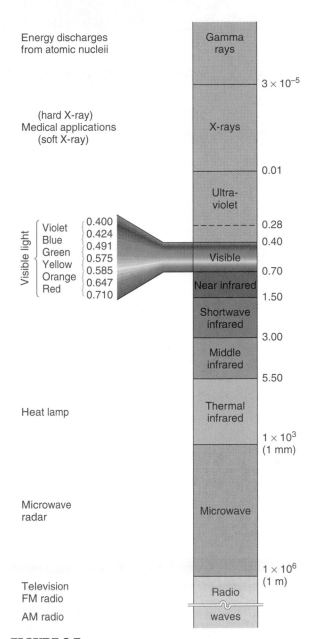

FIGURE 2-7
A portion of the electromagnetic spectrum of radiant energy.

An important physical law states that all objects radiate energy in wavelengths related to their individual surface temperatures: the hotter the object, the shorter the wavelengths emitted. Figure 2-8 shows that the hot Sun radiates energy primarily in shorter wavelengths. The Sun's surface temperature is about 6000°C (11,000°F). Earth, on the other hand, demonstrates that the cooler the radiating body, the longer are the wavelengths emitted. Figure 2-8 shows that the radiation emitted by Earth occurs in longer wavelengths.

To summarize, the solar spectrum is *shortwave radiation* that peaks in the visible wavelengths, and Earth's spectrum is *longwave radiation* because its outgoing energy is concentrated in infrared wave-

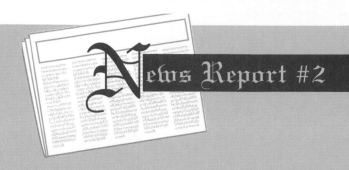

"Does Solar Wind Affect Earth's Weather Patterns?"

Scientists are studying possible effects of the solar wind on weather patterns. Why do wetter periods in some midlatitude areas tend to coincide with every other solar maximum? Why do droughts often occur near the time of every other solar minimum? For example, sunspot cycles during the 250 years from 1740 to 1990 coincide with periods of wetness and drought, as esti-

mated by an analysis of tree-growth rings for that period throughout the western United States. These variations in weather tend to occur within two or three years after the solar maximum or minimum.

Although the correlation is interesting—and controversial, and still undergoing research—the specific causative link in the atmosphere remains to be discovered. Speculation

as to the weather connection is centered on heating and wind changes in the upper atmosphere. Other explanations also have been proposed.

A remarkable failure in current planning worldwide is the lack of attention given these cyclical patterns of drought and wetness, regardless of their cause. If such patterns were prepared for, it would be possible to reduce property loss and

casualties. For instance, cyclical drought could be offset through implementation of widespread water conservation. Wet spells might require strengthening of levees along river channels and more careful reservoir management to reduce flooding. As knowledge of the solar wind/weather interaction increases, an ability to forecast the affect on Earth systems may become feasible.

---

lengths. In Chapter 3, we will see that Earth, clouds, sky, ground, and all things that are terrestrial are cool-body radiators.

Energy at the Top of the Atmosphere

The region at the top of the atmosphere, approximately 480 km (300 mi) above Earth's surface, is termed the **thermopause**. It is the outer boundary of Earth's energy system and provides a useful point at which to assess the arriving solar radiation before it is diminished by passage through the atmosphere.

Intercepted Energy

Earth's distance from the Sun results in its interception of only one two-billionth of the Sun's total energy output. Nevertheless, this tiny fraction of the Sun's overall output represents an enormous amount

of energy input into Earth's systems. Intercepted solar radiation is called **insolation**. Insolation at the top of the atmosphere is expressed as the *solar constant*.

Solar Constant. Knowing the amount of insolation intercepted by Earth is important to climatologists and other scientists. The **solar constant** is the average value of insolation received at the thermopause when Earth is at its average distance from the Sun. That value of the solar constant is 1372 watts per square meter (W/m²).* As we follow insolation through the atmosphere to Earth's surface in

---

*A *watt* is equal to one joule (a unit of energy) per second and is the standard unit of power in the SI-metric system. (See the inside back cover of this text for more information on measurement conversions.) In nonmetric calorie heat units, the solar constant is expressed as approximately 2 calories per cm² per minute, or 2 *langleys* per minute (a langley being 1 cal per cm²). A *calorie* is the amount of energy required to raise the temperature of one gram of water (at 15°C) one degree Celsius and is equal to 4.184 joules.

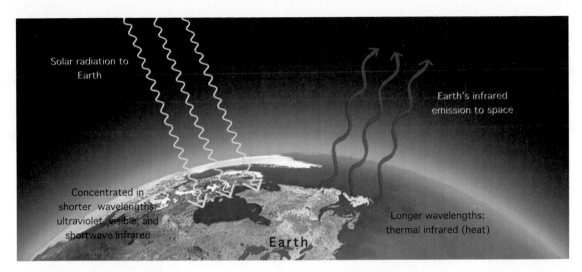

FIGURE 2-8
Solar radiation is concentrated in shorter wavelengths. Earth emits longer
wavelengths of infrared to the atmosphere and eventually to space.

Chapter 3, we see the value of the solar constant reduced by half or more through reflection and absorption of shortwave radiation.

The constancy of this value over time is important, for small variations of up to 1.0% could prove
dramatic for Earth's energy system. Paleontologists,
who deal with the life of past geologic periods, estimate that solar energy levels have varied slightly,
perhaps less than 10%, over the past several billion
years. The Solar Maximum Mission (1980–1989),
dubbed Solar Max, was a satellite launched to measure total solar output. Solar Max found average variations of ±0.04% in the solar constant, with the
largest changes ranging up to 0.3%. Since 1978, the
Earth-atmosphere energy budget has been monitored by the Earth Radiation Budget package on
board the *Nimbus-7* satellite.

Uneven Distribution of Insolation. Earth's
curved surface presents a continually varied angle
to the incoming parallel rays of insolation (Figure 2-
9). The latitudinal variation in the angle of solar rays
results in an uneven global distribution of insolation.
The only point receiving insolation perpendicular to
the surface (from directly overhead) is the **subsolar point**. This occurs at lower latitudes, where the
energy received is more concentrated. All other
places receive insolation at an angle less than 90°

and thus experience more diffuse energy; this effect
is pronounced at higher latitudes. During a year's
time, the thermopause above the equatorial region
receives 2.5 times more insolation than the thermopause above the poles. Of lesser importance is
the fact that the lower-angle solar rays toward the
poles must pass through a greater thickness of atmosphere, resulting in greater losses of energy due
to scattering, absorption, and reflection.

Global Net Radiation. The Earth Radiation Budget (ERB) instrument aboard *Nimbus-7* measures
shortwave and longwave flows of energy at the top
of the atmosphere. ERB sensors collected the data
used to develop the map in Figure 2-10. This map
shows *net radiation*, or the balance between incoming shortwave and outgoing longwave radiation.
First note the latitudinal energy imbalance in net radiation—positive values in lower latitudes and negative values toward the poles.

In middle and high latitudes, approximately poleward of 36° north and south latitudes, net radiation
is negative. This happens because Earth's climate
system loses more heat to space than it gains from
the Sun as measured at the top of the atmosphere
for these higher latitudes. In the lower atmosphere,
these polar energy deficits are offset by flows of heat
from tropical energy surpluses, as we shall see in

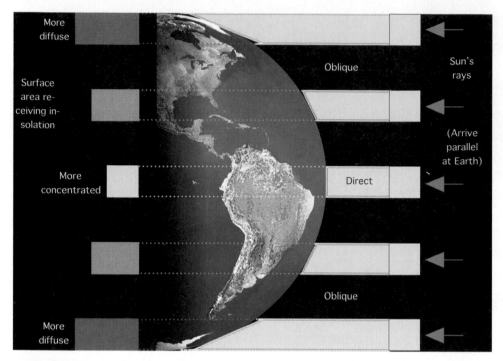

FIGURE 2-9
Solar insolation angles determine the concentration of energy received at each latitude.
Lower latitudes receive more concentrated energy from a more direct solar beam. Higher
latitudes receive slanting (oblique) rays and more diffuse energy.

Chapters 3 and 4. The largest net radiation values are above the tropical oceans along a narrow equatorial zone, averaging 80 W/m². Net radiation minimums are lowest over Antarctica.

Of interest is the –20 W/m² area over the Sahara desert region, where usually clear skies—which permit large longwave radiation losses from Earth's surface—and reflective (light-colored) surfaces work together to reduce net radiation values measured at the thermopause. Clouds in the lower atmosphere also affect net radiation patterns at the top of the atmosphere by reflecting greater amounts of shortwave energy to space.

Figure 2-11 summarizes this radiation balance for all shortwave and longwave energy by latitude. In the equatorial zone, energy surpluses dominate, for in those areas more energy is received than is lost. The solar angle is high, with consistent daylength. However, deficits exist in the polar regions, where more energy is lost than gained. At the poles, the Sun is extremely low in the sky, snow and ice surfaces are light and reflective, and for six months during the year no insolation is received.

This overall imbalance of insolation and heating from equator to the poles is the basis for major circulations within the lower atmosphere and in the ocean. The transfer agents are global wind circulation, ocean currents, weather systems, and related phenomena. Tropical cyclones (hurricanes and typhoons) represent dramatic, concentrated examples of such energy and mass transfers. As you go about your daily activities, use these dynamic natural systems as reminders of the constant flow of solar energy.

Having examined the flow of solar energy to Earth, let us now look at the nature of seasonal change as it affects energy receipts.

The Seasons

The periodic rhythm of warmth and cold, rain and drought, dawn and daylight, twilight and night have fascinated humans for centuries. In fact, many ancient societies demonstrated a greater awareness of seasonal change than modern peoples and formally com-

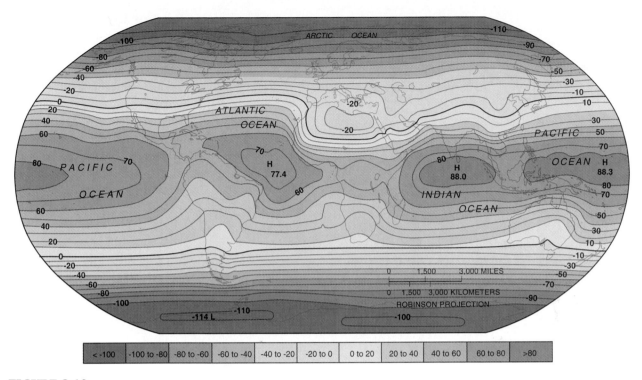

FIGURE 2-10
Average daily net radiation flows for a 9-year period (1979 to 1987) measured at the top of the atmosphere by the Earth Radiation Budget (ERB) instrument aboard the *Nimbus-7* satellite. Units are watts per square meter (W/m²). [Map courtesy of Dr. H. Lee Kyle, Goddard Space Flight Center, NASA.]

memorated natural energy rhythms with festivals and monuments. Figure 2-12 shows such a monument at Stonehenge in England. Here, rocks weighing 25 metric tons (28 tons) were hauled 480 km (300 mi) and placed in patterns that evidently mark seasonal changes—specifically, the sunrise on or about June 21 at the summer solstice. Other seasonal events are predicted by this 3500-year-old calendar monument. Such ancient seasonal monuments occur worldwide, including thousands of sites in North America.

Seasonality

Seasonality refers to both the seasonal variation of the Sun's position above the horizon and the changing daylengths during the year. Seasonal variations are a response to changes in the Sun's **altitude**, or the angle between the horizon and the Sun. At sunrise or sunset, the Sun is at the horizon, so its altitude is 0°. If during the day the Sun reaches halfway between the horizon and directly overhead, it is at 45°, and if

directly overhead it is at 90° altitude (called the *zenith*). The Sun is directly overhead only at the subsolar point. Here the insolation is maximum, and all other surface points receive a lower Sun angle and therefore receive more diffuse insolation.

The Sun's **declination** is the latitude of the subsolar point. Declination annually migrates through 47° of latitude between the *Tropic of Cancer* at 23.5° N and the *Tropic of Capricorn* at 23.5° S latitude. The subsolar point does not reach the U.S. or Canadian mainland. Other than Hawaii, all other states and provinces are too far north.

Seasonality also means a changing **daylength**, or duration of exposure. Daylength varies during the year depending on latitude. The equator always receives equal hours of day and night, whereas people living along 40° N or S latitude experience about 6 hours' difference in daylight hours between winter and summer. Those at 50° N or S latitude experience almost 8 hours of annual daylength variation. At the polar extremes, the range extends from a six-month

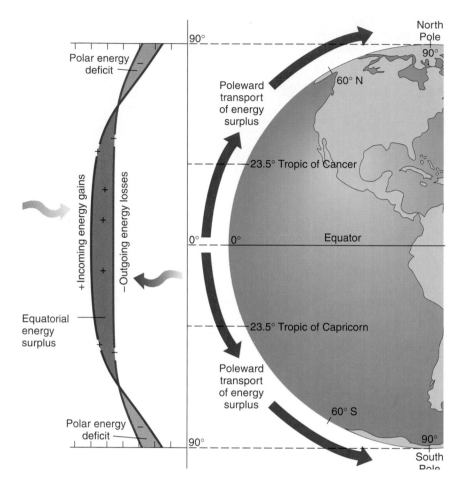

FIGURE 2-11
Earth's energy surpluses (equatorial regions) and deficits (polar regions). The overall imbalance of energy from equator to poles sets major atmospheric and oceanic circulations in motion.

period of no insolation and darkness to a six-month period of continuous 24-hour days.

Reasons for Seasons

Seasons result from variations in the Sun's altitude above the horizon, declination latitude, and daylength. These in turn are created by several physical factors that operate together: Earth's revolution in orbit around the Sun, its daily rotation on its axis, its tilted axis, the unchanging orientation of its axis, and its sphericity. We now briefly describe these factors as they appear in summary in Table 2-1. Of course, the essential ingredient is having a single source of radiant energy—the Sun.

Revolution. The structure of Earth's orbit and **revolution** about the Sun is shown in Figure 2-2. Note the distinction between orbital *revolution* and *rotation*—the

FIGURE 2-12
Sunrise on the first day of summer at Stonehenge, Salisbury Plain, England. [Photo by Robert Llewellyn.]

Table 2-1

Five Reasons for Seasons

Factor	Description
Revolution	Orbit around the Sun; requires 365.24 days to complete at 107,280 kmph (66,660 mph)
Rotation	Earth turning on its axis; takes approximately 24 hours to complete at 1675 kmph at equator (1041 mph)
Tilt	Axis is aligned at a 23.5° angle from a perpendicular to the plane of the ecliptic (the plane of Earth's orbit)
Axis	Remains in a fixed alignment of axial parallelism, with Polaris directly overhead at the North Pole throughout the year
Sphericity	Appears as an oblate spheroid to the Sun's parallel rays; the geoid

spinning of Earth on its axis. At an average distance from the Sun of 150 million km (93 million mi), Earth completes its annual orbit in 365.24 days at speeds averaging 107,280 kmph (66,660 mph) in a counterclockwise direction when viewed from above Earth's North Pole (Figure 2-13). Earth's revolution determines the *length of the year* and therefore the *duration of the seasons*.

Rotation. Earth's **rotation**, or turning, is a complex motion that averages 24 hours in duration. A true day varies slightly from 24 hours, but by international agreement a day is considered to be exactly 24 hours (86,400 seconds) in length. Rotation determines daylength, causes the apparent deflection of winds and ocean currents, and produces the twice-daily rise and fall of the ocean tides in relation to the gravitational pull of the Sun and the Moon. Earth rotates about its **axis**, an imaginary line extending through the planet from the geographic North Pole to the South Pole. When viewed from above the North Pole, Earth rotates counterclockwise around this axis; viewed from above the equator it moves west to east, or eastward. This west-to-east rotation creates the Sun's *apparent* daily journey from east to west, even though the Sun remains in a fixed position in the center of the Solar System. Similarly, the Moon revolves around Earth and rotates counterclockwise on its axis (Figure 2-13).

Earth's rotation produces the diurnal pattern of day and night. The dividing line between day and night is called the **circle of illumination**, as illustrated in Figure 2-15. Because this day-night dividing circle of illumination intersects the equator, *daylength at the equator is always evenly divided*—12 hours of day and 12 hours of night. All other latitudes experience uneven daylength through the seasons, except for two days a year, on the equinoxes.

Tilt of Earth's Axis. To understand Earth's tilt, imagine Earth's elliptical orbit about the Sun as a plane, with half of the Sun and Earth above the plane and half below. (It may help to envision two spheres floating in water, where the water's surface forms a plane.) This flat surface is termed the **plane of the ecliptic**. Now, imagine a perpendicular line passing through the plane. From this perpendicular, Earth's axis is tilted 23.5°. It forms a 66.5° angle from the plane itself (Figure 2-14). The axis through Earth's two poles points just slightly off Polaris, which is appropriately called the North Star.

Hypothetically, if Earth was tilted on its side with its axis parallel to the plane of the ecliptic, we would experience a maximum variation in seasons worldwide. On the other hand, if Earth's axis was perpendicular to the plane of its orbit—that is, with no tilt—we would experience no seasonal changes, just a perpetual spring/fall season, and all latitudes would have 12-hour days and nights.

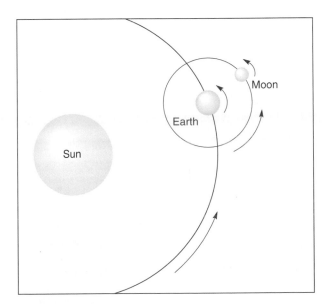

FIGURE 2-13

Earth's revolves about the Sun and rotates on its axis. The Moon revolves about Earth and rotates on its axis. This view is from above Earth's North Pole. Earth and the Moon revolve and rotate counterclockwise from this perspective.

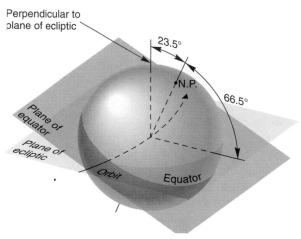

FIGURE 2-14
The plane of Earth's equator is 23.5° offset from the plane of the ecliptic. This is because Earth is tilted 23.5° from the plane of the ecliptic.

Axial Parallelism. Throughout our annual journey around the Sun, Earth's axis *maintains the same alignment* relative to the plane of the ecliptic and to Polaris and the other stars. You can see this in Figure 2-15: note that in each position of Earth shown revolving about the Sun the axis is oriented identically, or parallel to itself. This a condition known as **axial parallelism**.

Annual March of the Seasons

The combined effect of all these physical factors is the annual march of the seasons on Earth. Daylength is the most evident way of sensing changes in season at latitudes away from the equator. The extremes of daylength occur in December and June. The times around December 21 and June 21 are termed the *solstices*. They mark the times of the year when the

Sun's declination places it directly over one of the two *tropics,* parallels of latitude that represent the Sun's farthest northerly or southerly position. ("Tropic" is from the Latin *tropicus,* meaning a turn or change.) Table 2-2 presents the key seasonal anniversary dates, their names, and the subsolar point location (declination).

Figure 2-15 demonstrates the annual march of the seasons and illustrates Earth's relationship to the Sun during each month of the year. Let us begin with December (the globe to the far right in the figure). On December 21 or 22, at the **winter solstice** (literally, "winter Sun stance") or **December solstice**, the circle of illumination excludes the North Pole region from sunlight and includes the South Pole region. The subsolar point is at 23.5° S latitude, a parallel known as the **Tropic of Capricorn**, at the moment of the solstice. The Northern Hemisphere is tilted away from these more direct rays of sunlight, thereby creating a lower angle for the incoming solar rays and a more diffuse pattern of insolation, thus causing our northern winter. Examine the photo on the back cover of this book. During what month do you think it was taken?

From 66.5° N latitude to 90° N (the North Pole), the Sun remains below the horizon the entire day. This latitude (66.5° N) marks the *Arctic Circle,* the southernmost parallel (in the Northern Hemisphere) that experiences a 24-hour period of darkness. During the following three months, daylength and solar angles gradually increase in the Northern Hemisphere as Earth completes one-fourth of its orbit.

The **vernal equinox**, or **March equinox**, occurs on March 20 or 21. At that time, the circle of illumination passes through both poles so that all locations on Earth experience a 12-hour day and a 12-hour night. Those living around 40° N latitude (New York, Denver) have gained three hours of day-

Table 2-2
Annual March of the Seasons

Approximate Date	Northern Hemisphere Name	Location of the Subsolar Point
December 21–22	Winter solstice (December solstice)	23.5° S latitude (Tropic of Capricorn)
March 20–21	Vernal equinox (March equinox)	0° (equator)
June 20–21	Summer solstice (June solstice)	23.5° N latitude (Tropic of Cancer)
September 22–23	Autumnal equinox (September equinox)	0° (equator)

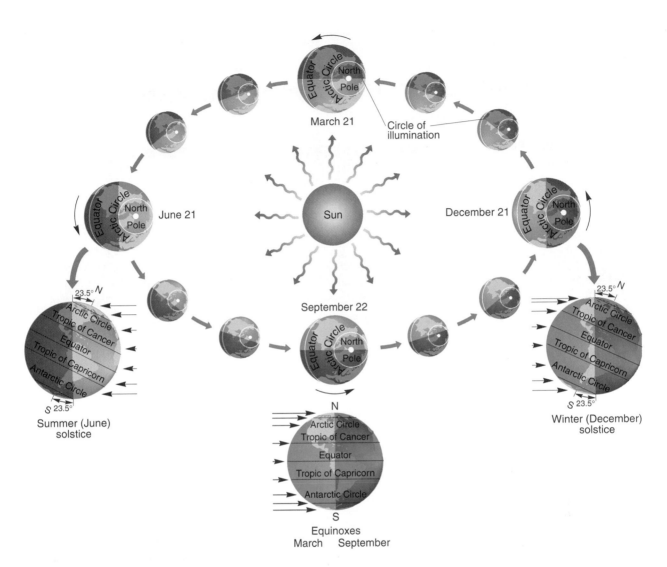

FIGURE 2-15
Annual march of the seasons as Earth revolves about the Sun. Shading indicates the changing position of the circle of illumination in the Northern Hemisphere (polar views) and in both Northern and Southern hemispheres (three profile views).

light since the December solstice. At the North Pole, the Sun peeks above the horizon for the first time since the previous September; at the South Pole the Sun is setting.

From March, the seasons move on to June 20 or 21 and the **summer solstice**, or **June solstice**. The subsolar point has now shifted from the equator to 23.5° north latitude, the **Tropic of Cancer**. Because the circle of illumination now includes the

North Pole region, everything north of the Arctic Circle receives 24 hours of daylight—the "Midnight Sun" (Figure 2-16). In contrast, the region from the *Antarctic Circle* to the South Pole (66.5°–90° S latitude) is in darkness the entire day.

Note that the orientation of Earth's axis has remained fixed relative to the heavens and parallel to its previous position 6 months earlier. The Northern Hemisphere in June is tilted toward the Sun, receives

FIGURE 2-16
Multiple exposure photo taken north of the Arctic Circle during the northern summer. The lowest image of the Sun in the series is the "Midnight Sun" demonstrating that the Sun never sets during this time of year. [Photos © by Braasch-Wiancko/Allstock.]

higher Sun angles, and experiences longer days—and therefore more insolation—than the Southern Hemisphere. Those living at 40° N latitude now receive more than 15 hours of sunlight a day, which is 6 hours more than in December.

September 22 or 23 is the time of the **autumnal equinox**, or **September equinox**, when Earth's orientation is such that the circle of illumination again passes through both poles so that all parts of the globe receive a 12-hour day and a 12-hour night. The subsolar point has returned to the equator, with days growing shorter to the north and longer to the south. Researchers stationed at the South Pole see the disk of the Sun just rising, ending their six months of night. In the Northern Hemisphere, fall arrives, a time of many colorful changes in the landscape (Figure 2-17).

Seasonal Observations. For most of you reading this text, the point of sunrise migrates along the horizon from the southeast in December to the northeast in June. The point of sunset migrates from the southwest to the northwest during those same times. The Sun's altitude at local noon at 40° N latitude migrates from a 26° angle above the horizon at the winter (De-

FIGURE 2-17
Fall foliage and seasonal change in a mixed forest create a scene of colorful beauty in Olympic National Park, Washington. [Photo by Steve Lambros/Lambros Photography.]

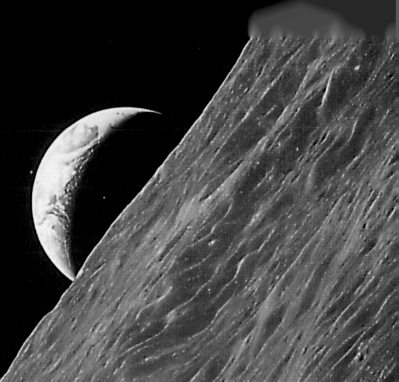

Viewed from the distance of the Moon, the astonishing thing about Earth, catching the breath, is that it is alive. The photographs show the dry, pounded surface of the Moon in the foreground, dead as an old bone. Aloft, floating free beneath the moist, gleaming membrane of a bright blue sky, is the rising Earth, the only exuberant thing in this part of the cosmos. . . . It has the organized, self-contained look of a live creature, full of information, marvelously skilled in handling the sun. . . . When the Earth came alive it began constructing its own membrane, for the general purpose of editing the sun. . . . Taken all in all, the sky is a miraculous achievement. It works, and for what it is designed to accomplish it is as infallible as anything in nature.

FIGURE 2-18
Earthrise over the lunar surface. [NASA photo. Quotation from "The World's Biggest Membrane" from *The Lives of a Cell* by Lewis Thomas. Copyright © 1973 by the Massachusetts Medical Society. Originally published in the *New England Journal of Medicine*.]

cember) solstice to a 73° angle above the horizon at the summer (June) solstice—a range of 47°.

Dawn and Twilight Concepts. The diffused light that occurs before sunrise and after sunset represents useful work time for humans. Light is scattered by the molecules of atmospheric gases and reflected by dust and moisture so that the atmosphere is illuminated. Such effects may be enhanced by the presence of pollution and other suspended particles, such as those from volcanic eruptions or forest fires. The duration of dawn and twilight is a function of latitude, because the angle of the Sun's path above the horizon determines the thickness of the atmosphere through which the Sun's rays must pass. Lower Sun angles produce longer dawn and twilight periods.

At the equator, where the Sun's rays are nearly perpendicular to the horizon throughout the year, dawn and twilight are limited to 30–45 minutes each. These times increase to 1–2 hours at 40° latitude, and at 60° latitude they range upward from 2.5 hours, with little true night in summer. The poles ex-

perience about 7 weeks of dawn and 7 weeks of twilight, leaving only 2.5 months of near darkness during the six months when the Sun is completely below the horizon.

Earth's Atmosphere

Insolation cascades through the atmosphere toward Earth's surface, powering the physical systems in the atmosphere, the oceans, and on land. Along the way the atmosphere works as an efficient filter, removing harmful radiation from sunlight. Let's now examine this unique atmosphere—its structure, composition, and function.

As explained in NEWS Report #1, the modern atmosphere probably is the fourth general atmosphere in Earth's history. This modern atmosphere is a gaseous mixture of ancient origin, the sum of all the exhalations and inhalations of life on Earth throughout time. The principal substance of this atmosphere is air, the medium of life as well as a major industrial and chemical raw material. *Air* is a simple ad-

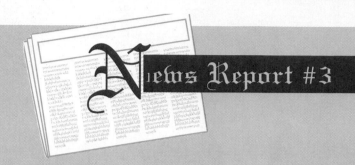

"Earth's Atmosphere Unique Among Planets"

Earth's atmosphere is formed by gases from the crust and interior and the exhalations of all life over time. Using a variety of satellites and remote sensing techniques, scientists are studying the atmospheres of other planets. In our Solar System, Earth's life-supporting atmosphere is unique, for the atmospheres of the other three inner planets bear no resemblance to Earth's.

Mars has a cold, thin atmosphere of carbon dioxide, equivalent in its low pressure to Earth's atmosphere 32 km (20 mi) above the surface. Venus also has an atmosphere dominated by carbon dioxide and is shrouded in clouds of sulfuric acid, with surface pressures about 90 times those on Earth, and surface temperatures averaging 500°C (932°F). Mercury has no appreciable atmosphere and appears similar to the Moon, pockmarked by innumerable craters.

---

ditive mixture of gases that is naturally odorless, colorless, tasteless, and formless, blended so thoroughly that it behaves as if it were a single gas.

In his book *The Lives of a Cell*, physician and self-styled "biology-watcher" Lewis Thomas compared the atmosphere of Earth to an enormous cell membrane. The membrane around a cell regulates the interactions between the cell's delicate inner workings and the potentially disruptive outer environment. Each cell membrane is very selective as to what it will and will not allow to pass through. The modern atmosphere acts as Earth's protective membrane, as Thomas describes so vividly (Figure 2-18).

Earth's modern atmosphere is arranged in a series of imperfectly shaped concentric "shells" or "spheres" that grade into one another, all bound to the planet by gravity. As critical as the atmosphere is to us, it represents only the thinnest envelope, amounting to less than one-millionth of Earth's total mass. Let's begin our study of the atmosphere by analyzing its composition, structure, and function.

Important to the following discussion is Figure 2-19, a vertical cross-section, or sideview, of Earth's atmosphere. We simplify our discussion of the atmosphere by looking at it in three ways: its composition, temperature, and function. These categories are shown along the left side of Figure 2-19. On the basis of chemical *composition*, the atmosphere is divided into two broad regions, the heterosphere and the homosphere. Based on *temperature*, it is divided into four distinct zones, the thermosphere, mesosphere, stratosphere, and troposphere. Finally, two functional zones are identified for their role in removing most of the harmful wavelengths of solar radiation: the ionosphere and the ozonosphere (ozone layer).

We divide the following discussion by composition, and examine the heterosphere and homosphere. Note as you read that our description follows the same path that incoming solar energy travels down through the atmosphere to Earth's surface.

Heterosphere

The name **heterosphere** indicates that the gases in this part of the atmosphere are *not evenly mixed*, a condition quite different from the blended gases that we breath closer to Earth in the homosphere. Gases in the heterosphere are distributed in distinct layers, sorted by gravity, with the lightest elements (hydrogen and helium) at the margins of outer space, and heavier elements (oxygen, nitrogen) toward Earth.

FIGURE 2-19

(a)An integrated chart of our modern atmosphere. Columns show division of the atmosphere by composition, temperature, and function. The chart spans from Earth's surface to the thermopause at 480 km. In (b), Space Shuttle astronauts captured a dramatic sunset through various atmospheric layers across the edge of our planet–called Earth's limb. A silhouetted cumulonimbus thunderhead cloud is seen rising to the tropopause. [Space Shuttle photo from NASA.]

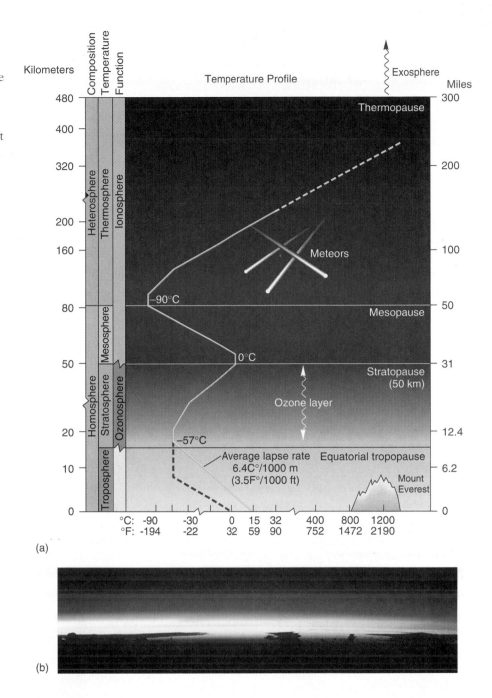

(a)

(b)

The heterosphere begins at around 80 km (50 mi) altitude and extends outward some 10,000 km (6,000 mi). However, as a practical matter, we consider the top of the atmosphere at around 480 km (300 mi), the same altitude we use for measuring the solar constant. Above that point, the atmosphere is rarefied (nearly a vacuum) and is called the **exos-phere,** which means "outer sphere." It contains individual atoms of the light gases hydrogen and helium, weakly bound by gravity.

In addition to its distinctive layered-gas composition, the heterosphere has a thermal region (called the thermosphere) and within it functional layers (called the ionosphere).

Thermosphere. Based on temperature, the atmosphere is divided into four temperature regions. Dominating the heterosphere portion of the atmosphere is the **thermosphere** (the "heat sphere"). The upper thermosphere is a transition zone into space, termed the *thermopause* (the suffix *-pause* means "to change"). High temperatures are generated in the thermosphere because the gas molecules absorb shortwave solar radiation.

The temperature curve in Figure 2-19 shows that temperatures rise sharply in the thermosphere, up to 1200°C (2200°F) and higher. Despite such high temperatures, the thermosphere is not "hot" in the way you might expect. Temperature and heat are two different things. The intense solar radiation in this portion of the atmosphere excites individual molecules and atoms (principally nitrogen and oxygen) to high levels of vibration. This **kinetic energy**, the energy of motion, is the vibrational energy that we measure, stated as *temperature*. However, the density of the molecules is so low that little actual *heat* is produced (heat is the quantity of thermal energy). Heating in the lower atmosphere near Earth's surface is different because the greater number of molecules in the denser atmosphere transmit their kinetic energy as **sensible heat**, meaning that we can sense it.

Ionosphere. The atmosphere has two functional layers, so-called because both function to filter harmful wavelengths of solar radiation, protecting Earth's surface from bombardment in any significant quantity. The upper functional layer, the **ionosphere**, coincides with the thermosphere. It is composed of atoms that acquired electrical charges when they absorbed cosmic rays, gamma rays, X-rays, and shorter wavelengths of ultraviolet radiation. These charged atoms are called *ions*, giving the ionosphere its name. Radiation bombards the ionosphere constantly, producing a constant flux (flow) of electrons and charged atoms.

Homosphere

Below the heterosphere is the second compositional region of the atmosphere, the **homosphere**. It extends from Earth's surface up to an altitude of 80 km (50 mi). Even though the atmosphere rapidly decreases in density with increasing altitude, the blend of gases is nearly uniform throughout the homosphere. The only exceptions are the concentration of ozone (O_3) in the "ozone layer" from 19 to 50 km (12–31 mi), and the variations in water vapor and pollutants in the lowest portion of the atmosphere near Earth's surface.

The stable mixture of gases throughout the homosphere has evolved slowly. The present proportion, which includes oxygen, was attained approximately 600 million years ago.

Figure 2-20 reveals the homosphere to be a vast reservoir of relatively inert nitrogen, principally originating from volcanic sources. Nitrogen is a key element of life, yet we exhale all the nitrogen that we inhale. This apparent contradiction is explained because the input of nitrogen to our body systems comes instead through compounds in food. In the soil, nitrogen is bound into these compounds by nitrogen-fixing bacteria and is returned to the atmosphere by denitrifying bacteria that remove nitrogen from organic materials.

Oxygen, a by-product of photosynthesis, is essential for life processes. Although it forms about one-fifth of the atmosphere, oxygen forms compounds that compose about half of Earth's crust. Both nitrogen and oxygen reserves in the atmosphere are so extensive that, at present, they far exceed human capabilities to disrupt or deplete them.

FIGURE 2-20

Stable components of the modern atmosphere (percentage concentration by volume).

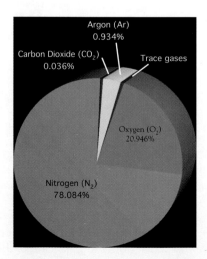

Argon (Ar) 0.934%

Carbon Dioxide (CO_2) 0.036%

Trace gases

Oxygen (O_2) 20.946%

Nitrogen (N_2) 78.084%

The gas argon, constituting about 1% of the atmosphere, is completely inert and therefore is unusable in life processes. Argon is a residue from the radioactive decay of an isotope (form) of potassium. Slow accumulation over millions of years accounts for all the argon present in the modern atmosphere. Because industry has found uses for inert argon (in light bulbs, welding, and some lasers), it is extracted or "mined" from the atmosphere, in addition to nitrogen and oxygen, for commercial and industrial uses (Figure 2-21).

Carbon dioxide is a natural product of life processes and is essentially a stable atmospheric component. Although its percentage in the atmosphere is small (0.036 %), it is an important player in maintaining global temperatures. Its percentage has been increasing gradually over the past 200 years as a result of human activities. The implications of these CO_2 increases to global warming and our future is explained in Chapter 7.

Shifting our view of the homosphere to temperature, we can subdivide the homosphere into three layers: the mesosphere, the stratosphere, and the one closest to Earth's surface, the troposphere.

Mesosphere. The **mesosphere** is the area from 50 up to 80 km (30 to 50 mi). Its upper boundary, the mesopause, is the coldest portion of the atmosphere, averaging –90°C (–130°F), although that temperature may vary by ±25° to 30C° (±45° to 54F°); see the temperature curve in Figure 2-19.

The *Upper Atmosphere Research Satellite (UARS)* is detecting large, continent-sized windstorms in the mesosphere. The very rarefied air is moving in vast waves at speeds in excess of 320 kmph (200 mph). The importance of this new discovery to surface weather patterns and to the Earth-atmosphere energy budget is being considered.

Stratosphere and Ozonosphere. The **stratosphere** extends from 20 to 50 km (12.4–31 mi) above Earth's surface. Temperatures increase throughout the stratosphere (see curve in Figure 2-19). The heat source in the stratosphere is the other functional layer, called the **ozonosphere**, or **ozone layer**. Ozone is a highly reactive oxygen molecule made up of three oxygen atoms (O_3) instead of the usual two (O_2) that make up most oxygen gas. Ozone ab-

FIGURE 2-21
Air is a major industrial and chemical raw material that is extracted from the atmosphere by "air mining" companies. [Photo by author.]

"New UV Index Announced to Help Save Your Skin"

The television weather reporter says: "the temperature is presently 25°C (78°F), relative humidity is 55%, light winds are from the south, and the *UV Index* is 9." This last item is what you are hearing in many cities where the National Weather Service and the Environmental Protection Agency are providing a new ultraviolet radiation index. An ultraviolet index value of 0 to 2 is "minimal," 5 to 6 "moderate," 7 to 10 "high," and over 10 "very high" in terms of UV exposure levels at Earth's surface as they relate to sunburn. As an example, a moderate 6 on the index means a burn in 10 minutes for the most susceptible skin to a burn in 50 minutes for the least susceptible skin. A similar UV Index has been used in Canada since 1992.

As stratospheric ozone levels continue to thin—midlatitude ozone losses of 21% during 1992-93 set a record—surface exposure to cancer-causing radiation continues to climb. The public now is alerted when to take extra precautions in the form of sunscreens, hats, and sunglasses. Remember, damage is cumulative and it may be decades before you experience the ill effects triggered by last summer's sunburn. (For more information call the American Cancer Society, 800-227-2345; or write the American Academy of Dermatology, P.O. Box 681069, Shaumburg, IL 60168-1069.)

sorbs wavelengths of ultraviolet light and subsequently reradiates that energy at longer wavelengths, as infrared (heat) energy. Through this process, most harmful ultraviolet radiation is effectively "filtered" from the incoming solar radiation, safeguarding life at Earth's surface and heating the stratosphere.

The ozone layer, presumed to be relatively stable over the past several hundred million years (allowing, of course, for daily and seasonal fluctuations), is now in a state of continuous change. FYI Report 2-2 analyzes the developing crisis in this critical section of our atmosphere. The spatial implications for humanity of losses in stratospheric ozone make this an important topic for applied studies in physical geography and atmospheric sciences.

Troposphere. The fourth temperature region of the atmosphere nearest Earth's surface is the **troposphere**. It is the home of the biosphere, the atmospheric layer that supports life on Earth. Approximately 90% of the total mass of the atmosphere and the bulk of all water vapor, clouds, weather, and air pollution are contained within the troposphere. The *tropopause*, its upper limit, is defined by an average temperature of −57°C (−70°F), but its exact elevation varies with the season, latitude, and surface temperatures and pressures. Near the equator, because of intense heating from below, the tropopause occurs at 18 km (11 mi); in the middle latitudes, it occurs at 13 km (8 mi); and at the North and South Poles it is only 8.0 km (5.0 mi) or less above Earth's surface.

Figure 2-22 illustrates the normal temperature gradient within the troposphere during the daytime. As the graph shows, temperatures decrease with increasing altitude at an average of 6.4C° per km (3.5F° per 1000 ft), a rate known as the **normal lapse rate**. The *actual* lapse rate at any particular time and place under local weather conditions, called the **environmental lapse rate**, may vary greatly from the normal lapse rate. This variance in temperature gradient in the lower troposphere is important to our discussion of weather processes (Chapter 5).

Stratospheric Ozone Losses: A Worldwide Health Hazard

Consider these news reports:

- Antarctic scientists experience the equivalent of "standing under an ultraviolet lamp 24 hours a day." Exposed skin quickly burns and swells, and certain protective clothing deteriorates rapidly.
- During the months of Antarctic spring, ozone concentrations approach total depletion in the stratosphere over an area twice the size of the Antarctic continent. The protective ozone layer is being depleted over southern South America, southern Africa, Australia, and New Zealand.
- Antarctic researchers have recorded steady losses in the ozone layer since 1970, noting a huge temporary decrease every October. Each year, the "ozone hole" widens and deepens. In addition to ground-based measurements NOAA satellites also record this alarming situation (Figure 1).
- At Earth's opposite pole, a similar ozone depletion over the Arctic exceeded 35% in 1989. Losses have increased in each subsequent year. Since 1992, the Canadian government regularly reports an "ultraviolet index."
- Environment Canada measured an 8% loss in normal ozone levels in 1993, with greatest thinning between January and April when it dropped to 14% below normal. Extreme lows of 22% below normal were achieved in March 1993 over Toronto and Edmonton; this produced increases in ultraviolet radiation at the surface.
- In 1994, Environment Canada opened a new scientific observatory to monitor the ozone losses over Canada. This high-Arctic facility is at a remote weather station on Ellesmere Island, N.W.T., about 1000 km from the North Pole.
- Overall losses in the midlatitudes are continuing at 6 to 8% per decade. In North America, related skin cancers are increasing at an alarming 10% per year, now totaling more than 700,000 cases annually with over 10,000 malignant melanoma deaths. Most affected are light-skinned persons who live at higher elevations and those who work principally outdoors.
- Other potential problems related to increased ultraviolet radiation include eye tissue damage, reduced immunity, crop damage, and injury to some aquatic life-forms. Losses of 6 to 12% now are measured in the primary production of small light-using organisms (phytoplankton) in Antarctic water.

Given these news items, clearly, ultraviolet radiation (UV) from the Sun is breaking through Earth's protective ozone layer. Why is this happening, and how are people and their governments responding?

Monitoring Earth's Fragile Safety Screen

A sample from the ozone layer's densest part (at 29 km, 18 mi) contains only 1 part ozone per 4 million parts of air. Yet this rarefied zone has been in steady-state equilibrium for at least 500 million years, absorbing intense UV radiation and permitting life to proceed safely on Earth.

The ozone layer has been monitored since the 1920s, measured with instrumented balloons, aircraft, orbiting satellites, and a 30-station ozone monitoring network (mostly in North America) established in 1957. The Total Ozone Mapping Spectrometer (TOMS) has been operating since 1978, and the *Upper Atmosphere Research Satellite* (*UARS*) is now monitoring ozone losses and the invasion of chlorine compounds in detail.

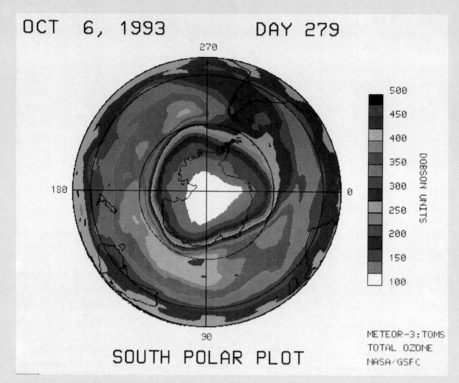

OCT 6, 1993   DAY 279

SOUTH POLAR PLOT

METEOR-3:TOMS
TOTAL OZONE
NASA/GSFC

FIGURE 1

Nimbus-7 TOMS (Total Ozone Mapping Spectrometer) measurements for October 1993. The color scale represents ozone concentrations in Dobson units with blues and purples for amounts of less than 200. (One Dobson unit equals 2.69 x 10[16] molecules of O_3 per cm[2].) Measurements dropped as low as 105 Dobson units in 1992. The "ozone hole" has grown larger since 1979 covering a record 25 million km[2] (9 million mi[2]) in 1993. This is an area larger than Canada, the United States, and Mexico combined. [Images courtesy of Goddard Space Flight Center, NASA.]

Ozone Losses Explained

In 1974, two researchers, Drs. Rowland and Molina, hypothesized that some synthetic chemicals are releasing chlorine atoms that decompose ozone. These **chlorofluorocarbon compounds,** or *CFCs*, are complex synthetic molecules of chlorine, fluorine, and carbon. They are very stable (inert) under conditions at Earth's surface and they possess remarkable heat properties, making them valuable as refrigerants and as propellants in aerosol sprays. In addition some 45% of CFCs are used as solvents in the electronics industry and as foaming agents.

Being stable, CFC molecules do not dissolve in water and do not break down in biological processes. (In contrast, chlorine derived from volcanic eruptions and the ocean are soluble and rarely reach the stratosphere.) The researchers hypothesized that stable CFC molecules slowly migrate into the stratosphere where intense ultraviolet radiation splits them, freeing chlorine (Cl) atoms. These combine with oxygen to form chlorine monoxide (ClO). This stimulates a complex reaction that breaks up ozone molecules (O_3). The effect is severe, for a single chlorine atom destroys tens of thousands of ozone molecules. The long residence time of chlorine atoms in the ozone layer (40 to 100 years) means potential long-term consequences through the next century from the chlorine already in place. By 1991, 18 million metric tons (19.8 million tons) of CFCs had been sold worldwide and subsequently released into the atmosphere.

If correct, the hypothesis meant an increase in ClO molecules in the ozonosphere, and a corresponding decline of O_3 molecules. Subsequent evidence confirms this, as shown in Figure 2. This model now is accepted by research scientists. Chemical manufacturers once claimed that no hard evidence existed to prove the ozone-depletion model and successfully delayed remedial action for 15 years. Today, following extensive scientific evidence and verification of losses, even the industry admits that the problem is serious.

With such political delays in response to actual measurements, Dr. Rowland expressed his frustration:

What's the use of having developed a science well enough to make predictions, if in the end all we are willing to do is stand around and wait for them to come true. . . . Unfortunately, this means that if there is a disaster in the making in the stratosphere, we are probably not going to avoid it.*

---

*Roger B. Barry, "The Annals of Chemistry," *The New Yorker* (9 June 1986): 83.

How do Northern Hemisphere CFCs become concentrated over the South Pole? Evidently, chlorine freed in the Northern Hemisphere midlatitudes is redistributed by atmospheric winds, concentrating over Antarctica. Persistent cold temperatures over the South Pole, and the presence of thin, icy clouds in the stratosphere, promote development of the hole. Over the North Pole, conditions are more changeable, so the hole is smaller although growing in size each year.

An International Response and the Future

Between 1976 and 1979, Canada, Sweden, Norway, and the state of Oregon banned CFC propellants in aerosols, and a federal ban followed in 1978. However, more than half of U.S. production was exempted, including CFCs used as air conditioning refrigerants and to make polyurethane foam. CFC sales initially dropped in the late 1970s, but rose again under a 1981 presidential order that permitted the export of banned products. Sales increased and hit a new peak in 1987 at 1.2 million metric tons (1.32 million tons). Sales now are declining as many nations reduce demand and as industry phases in alternative chemicals. By 1991, CFC production had declined to 682,000 metric tons (750,000 tons)— about one-half the record-sales year. The reason for this decline is public and government pressure for action on an international scale.

In 1985, a short-term CFC production freeze and a consensus for long-term phase out was agreed to by 31 nations. This led to the Montreal Protocol, which by 1992 had been endorsed by the 93 nations that produce 90% of CFCs. These nations agreed to freeze production at 1986 levels. The Montreal Protocol currently calls for an accelerated complete ban of CFCs by the year 1996. The Protocol also bans other chemicals damaging to the atmosphere using different target dates.

The holes in the polar ozone layer and general depletion overall provide an early warning to civilization about this long-term worldwide health problem. Hopefully, by taking action now, the hazard to life in the next century will be reduced.

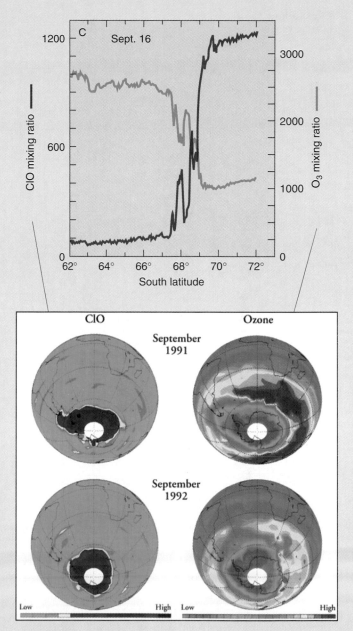

FIGURE 2

(a) The anticorrelation between ClO and O_3 over the Antarctic continent poleward of 68° S latitude. The data were collected from flights during September 1987 at stratospheric altitudes. (b) Chlorine monoxide (ClO) and ozone (O_3) concentrations above 20 km (12.5 mi) measured by the *Upper Atmosphere Research Satellite* (*UARS*) from September 1991 and 1992—the beginning of the Antarctic spring. [(a) Data from NASA. (b) Microwave Limb Sounder, developed and operated by Jet Propulsion Laboratory aboard Goddard Space Flight Center's *UARS*.]

61

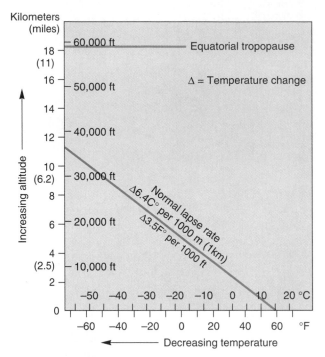

FIGURE 2-22
The temperature profile of the troposphere. During daytime, temperature decreases with increasing altitude at a rate known as the *normal lapse rate*.

The marked warming in the stratosphere with increasing altitude (Figure 2-19) causes the tropopause to act like a lid, essentially preventing whatever is in the cooler air below from mixing into the stratosphere. However, the tropopause is disrupted above the midlatitudes wherever jet streams occur (Chapter 4), and the resulting vertical turbulence does allow some interchange between the troposphere and the stratosphere. Also, hurricanes occasionally inject moisture above the tropopause, and powerful volcanic eruptions may push ash and sulfuric acid mists into the stratosphere, as did the Mount Pinatubo eruptions in 1991.

Variable Atmospheric Components

Within the troposphere, both natural and human-generated variable gases, particles, and other chemicals are part of the atmosphere. The spatial aspects of these variable pollution components are a significant applied topic in physical geography.

Air pollution is not a new problem. Romans complained over 2000 years ago about the foul air of their cities. The stench of open sewers, fires, kilns, and smelters filled the air. In human experience, cities are always the place where the environment's natural ability to process and recycle waste is most taxed. English diarist John Evelyn recorded in 1684 that the air in London was so filled with smoke that ". . . one could hardly see across the streets, and this filling the lungs with its gross particles exceedingly obstructed the breast so as one could hardly breathe."* Air pollution is closely related to our production and consumption of energy, and as human occupation of a region increases in duration and complexity, the environmental impact on the troposphere also increases.

Natural Sources

Natural sources produce a greater quantity of nitrogen oxides, carbon monoxide, and carbon dioxide than do human sources. Table 2-3 lists some of these natural sources and the substances they contribute to the air. However, any attempt to diminish the impact of human-made air pollution through a comparison with natural sources is irrelevant, for we have coevolved with and adapted to the *natural* ingredients in the air. We have not evolved in relation to the comparatively recent concentrations of anthropogenic (human-caused) contaminants in our metropolitan areas.

---

*W. Wise, *Killer Smog: The World's Worst Air Pollution Disaster*, Chicago: Rand McNally, 1968.

Sources	Contribution
Volcanoes	Sulfur oxides, particulates
Forest fires	Carbon monoxide and dioxide, nitrogen oxides, particulates
Plants	Hydrocarbons, pollens
Decaying Plants	Methane, hydrogen sulfides
Soil	Dust and viruses
Ocean	Salt spray and particulates

Table 2-3
Sources of Natural Variable Gases and Materials

A dramatic natural source of pollution was produced by the 1991 eruption of Mount Pinatubo in the Philippines that injected between 15 and 20 million tons of sulfur dioxide (SO_2) into the stratosphere. The spread of these emissions is shown in a sequence of satellite images presented in Chapter 4. This event gives scientists an opportunity to study naturally produced sulfuric acid mists in the atmosphere.

Natural Factors That Affect Air Pollution

Augmenting the problems resulting from both natural and human-made atmospheric contaminants are several important natural factors. Among these are wind, local and regional landscape characteristics, and temperature inversions in the troposphere.

Winds gather and move pollutants from one area to another, sometimes reducing the concentration of pollution in one location while increasing it in another. Such air movements make the atmosphere's condition an international issue. Indeed, the movement of air pollution from the United States to Canada is the subject of much complaint and negotiation between these two governments. In Europe, the cross-boundary drift of pollution is a major issue because of the close proximity of nations and has led in part to Europe's unification.

Wind can produce dramatic episodes of dust movement (Figure 2-23). (Dust is defined as particles less than 62 microns, or 0.0025 in.). In 1977, a *GOES* weather satellite followed a 1 million km² (400,000 mi²) dust cloud from Colorado and New Mexico eastward. Former soil was lofted 5000 m (3 mi) in altitude. Elsewhere, dust from Africa contributes to the soils of South America and Texas dust ends up in Europe. Chemical analysis is frequently employed by scientists to track such dust to its source area. An example of a point-source dust storm is shown in Figure 2-23b. Silt and alkaline dust are picked up by high-altitude winds in the Andes of South America and eventually carried out over the Atlantic.

Local and regional landscapes are another important factor in air pollution. Surrounding mountains and hills can form barriers to air movement or can direct pollutants from one area to another. Some of the worst incidents have resulted when local landscapes have trapped and caused a concentration of air pollution.

Temperature Inversion. Vertical temperature distribution in the troposphere also can promote pollution. A **temperature inversion** occurs when the normal decrease of temperature with increasing altitude (normal lapse rate) reverses at any point from

FIGURE 2-23
Natural variable pollution. (a) A dust storm in central Nevada. (b) Wind-blown alkali dust rising from high interior-drainage basins in the Andes Mountains of Chile and Argentina and blowing far over the Atlantic Ocean. [(a) Photo by author. (b) Space Shuttle photo from NASA.]

(a)

(b)

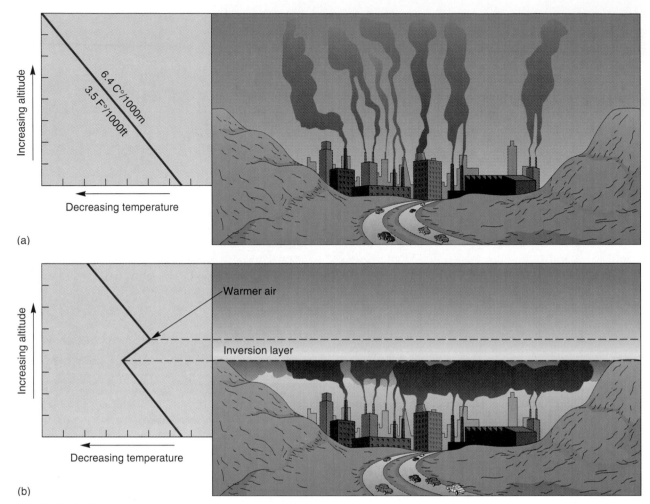

FIGURE 2-24
(a)A comparison of a normal temperature profile with (b) a temperature inversion in the lower atmosphere. Air pollution episodes are worsened by temperature inversions conditions.

ground level up to several thousand meters. Figure 2-24 compares a normal temperature profile with that of a temperature inversion. The normal profile illustrated in part (a) permits lifting air to ventilate surface pollution, but the warm air inversion shown in part (b) prevents the vertical mixing of pollutants with other atmospheric gases. Thus, instead of being carried away, pollutants are trapped under the inversion layer.

Inversions most often result from certain weather conditions, such as when the air near the ground is radiatively cooled on clear nights. Also, inversions may form where topographic situations produce cold air drainage into valleys.

Anthropogenic Pollution

Anthropogenic, or human-caused, air pollution remains most prevalent in urbanized regions. Table 2-4 lists the names, symbols, and principal sources of variable anthropogenic components in the air. The first seven components in Table 2-4 result from com-

Table 2-4
Anthropogenic Gases and
Materials in the Lower
Atmosphere

Name	Symbol	Source
Carbon Monoxide	CO	Incomplete combustion of fuels
Nitrogen oxides	NO_x (NO, NO_2)	High temperature/pressure combustion
Hydrocarbons	HC	Incomplete combustion of fuels
Ozone	O_3	Photochemical reactions
Peroxyacetyl nitrates	PAN	Photochemical reactions
Sulfur oxides	SO_x (SO_2, SO_3)	Combustion of sulfur-containing fuels
Particulates	—	Dust, dirt, soot, salt, metals, organics
Exotics	—	Fission products, other toxics
Carbon dioxide	CO_2	Complete combustion
Water vapor	H_2O vapor	Combustion processes, steam
Methane	CH_4	Organic processes

bustion of fossil fuels in transportation (specifically automobiles) and at stationary sources such as power plants and factories. Overall, automobiles contribute about 60% of American and 50% of Canadian human-caused air pollution. Specifically, transportation produces 70% of the carbon monoxide, 60% of hydrocarbons, and 50% of nitrogen oxides in the United States. Stationary sources and industrial processes using fossil fuels contribute the most sulfur oxides and particulates.

The last three gases shown in Table 2-4 are discussed elsewhere in this text: water vapor is examined with weather and the hydrologic cycle (Chapters 5 and 6), carbon dioxide with temperature and climate (Chapter 7), and methane with greenhouse gases (Chapter 3).

Carbon Monoxide. Carbon monoxide (CO) is an odorless, colorless, and tasteless combination of one atom of carbon and one of oxygen. CO is produced by incomplete combustion of fuels or other carbon-containing substances. A log decaying in the woods produces CO, as does a forest fire or other organic decomposition. Natural sources produce up to 90% of existing carbon monoxide, whereas anthropogenic sources, principally transportation, produce the other 10%. In the tropics of south-central Africa, an interesting source of CO is widespread burning of biomass (trees and brush) during the dry season from August to October. This CO spreads throughout the Southern Hemisphere—Africa, Australia, Antarctica, and the south Atlantic. Generally,

though, the carbon monoxide from human sources is concentrated in urban areas, where it directly impacts human health.

The toxicity of CO is well known; one sometimes hears of a death from carbon monoxide poisoning in an enclosed space with a running automobile engine or in a tent with an unvented heater. The toxicity of CO is due to its affinity for blood hemoglobin, which is the oxygen-carrying pigment in red blood cells. In the presence of CO, the oxygen is displaced and the blood becomes deoxygenated. Elevated CO levels (50 to 100 ppm) in cities are dangerous because of the resultant reduction of oxygen in the blood and the health symptoms experienced, including headaches, some vision and judgment loss, and impaired performance of simple tasks.

CO pollution did not become a problem until many fossil fuel–burning cars, trucks, and buses became concentrated in urban areas. Normal CO levels range from 10 to 30 parts per million (ppm) in metropolitan areas, but freeways and parking garages can reach 50 ppm CO, with an additional 30 ppm possible under a temperature inversion.

Photochemical Smog Reactions. Photochemical smog is another type of pollution that was not generally experienced in the past but developed with the advent of the automobile. Today it is the major component of anthropogenic air pollution. **Photochemical smog** results from the interaction of sunlight and the products of automobile exhaust.

(a)

(b)

FIGURE 2-25

A rare day of clear air (a) in contrast to a more typical day of photochemical smog pollution (b) in the skies over Mexico City. [Photos by Larry Reider/PIX/Sipa Press.]

Although the term "smog"—a combination of the words "smoke" and "fog"—is a misnomer, it is generally used to describe this phenomenon. Smog is responsible for the hazy sky and reduced sunlight in many of our cities.

Mexico City is notorious for poor air quality as its 20 million inhabitants work, commute, and live in the world's second largest metropolitan region. Conditions are worsened by frequent subtropical high-pressure systems that act as effective air traps over the Valley of Mexico. New laws were enacted in 1990 to reduce these unhealthy photochemical smog conditions: more public rapid transportation, controls on automobiles, limitations on factory operations, and more pollution-absorbing park space and trees. The contrast between a rare, clear day and frequent polluted days is dramatic (Figure 2-25).

The connection between automobile exhaust and smog was not determined until 1953, well after society had established its dependence upon individualized transportation. Despite this discovery, widespread mass transit has declined, the railroads have dwindled, and the polluting individual automobile remains America's preferred transportation mode.

Please examine Figure 2-26 as you read this paragraph. The increased temperatures in modern automobile engines cause reactions that produce **nitrogen dioxide** (NO_2). This nitrogen dioxide derived from automobiles, and to a lesser extent from power plants, is highly reactive with ultraviolet light. The reaction liberates atomic oxygen (O) and a nitric oxide (NO) molecule from the NO_2 (Figure 2-26). The free oxygen atom combines with an oxygen molecule (O_2) to form the oxidant ozone (O_3); this same gas that is so beneficial in the stratosphere is an air-pollution hazard at Earth's surface. In addition, the nitric oxide (NO) molecule reacts with hydrocarbons (HC) to produce a whole family of chemicals generally called **peroxyacetyl nitrates (PAN)**. PAN produces no known health effect in humans but is particularly damaging to plants, including both agriculture and forests. Damage in California is estimated to exceed 1 billion dollars a year in the farming sector.

Nitrogen dioxide (NO_2) is a reddish-brown gas that damages and inflames human respiratory systems, destroys lung tissue, and damages plants. Concentrations of 3 ppm are dangerous enough to require alerting parents to keep children indoors; a level of 5 ppm is extremely serious. Worldwide, the

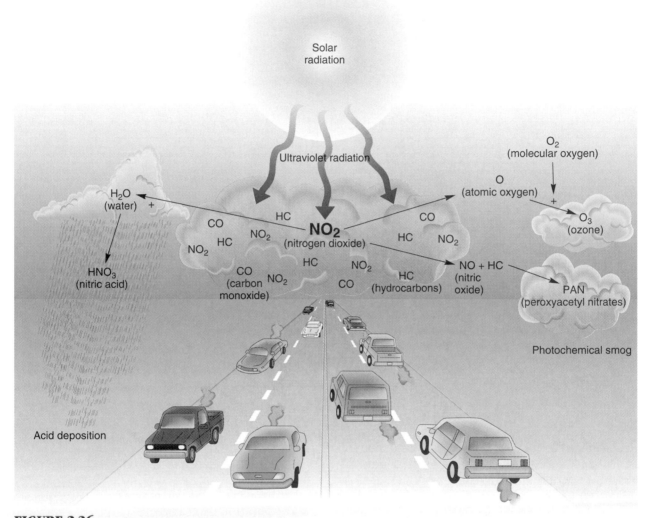

FIGURE 2-26
Photochemical reactions are produced through the interaction of automobile exhaust (NO₂, HC, CO) and ultraviolet radiation in sunlight. The resulting photochemical pollution includes ozone and PAN.

problem with nitrogen dioxide production is its concentration in metropolitan regions. North American urban areas may have from 10 to 100 times higher nitrogen dioxide concentrations than nonurban areas. Nitrogen dioxide interacts with water vapor to form nitric acid (HNO_3), a contributor to acid deposition by precipitation, the subject of the FYI Report in Chapter 6. Emission levels of nitrogen oxides and their deposition by rain and snow during 1991 are shown in Figure 2-27.

Ozone Pollution. Highly reactive ozone causes paint to oxidize, painted surfaces to peel, elastic and rubber to dry and crack, paper to dry and yellow, and plants to be damaged at levels of only 0.01–0.09 ppm. Exposure to 0.1 ppm for a few

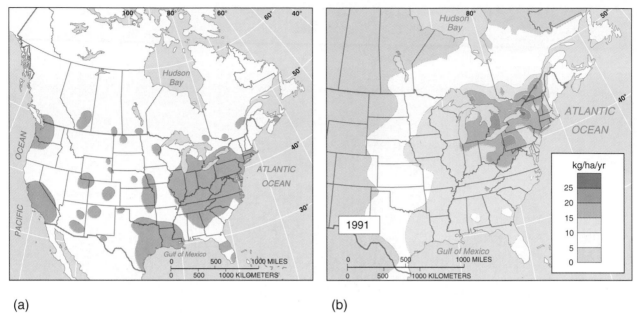

(a)   (b)

FIGURE 2-27

Spatial portrayal of (a) representative emissions of nitrogen oxides in the 1980s, and (b) annual wet (rain and snow) deposition of nitrogen oxides (principally NO_3) on the landscape for the year 1991. [Data courtesy of National Atmospheric Chemistry Data Base (NAtChem), Air Quality Measurements and Analysis Research Division, Environment Canada and the U.S. Environmental Protection Agency, National Atmospheric Deposition Program/National Trends Network.]

hours a day for several days to several weeks reduces yields as much as 50% for a wide variety of agricultural crops. At 0.3 ppm eye, nose, and throat irritation is noticeable. As concentration increases to 1.0–3.0 ppm, extreme fatigue and poor coordination can appear in humans.

Industrial Smog and Sulfur Oxides. Over the past 300 years, coal has slowly replaced wood as the basic fuel used by society. The industrial revolution required high-grade energy to run machines and to facilitate the conversion from animate energy (energy from living sources, such as animals powering farm equipment) to inanimate energy (energy from nonliving sources, such as coal, steam, and water). The air pollution associated with coal-burning industries is known as **industrial smog** (Figure 2-28). The term smog was coined by a London physician at the turn of this century to describe the

FIGURE 2-28

Industrial smog. Pollution generated by industry differs from that produced by transportation. Industrial pollution has high concentrations of sulfur oxides, particulates, and carbon dioxide. [Photo by author.]

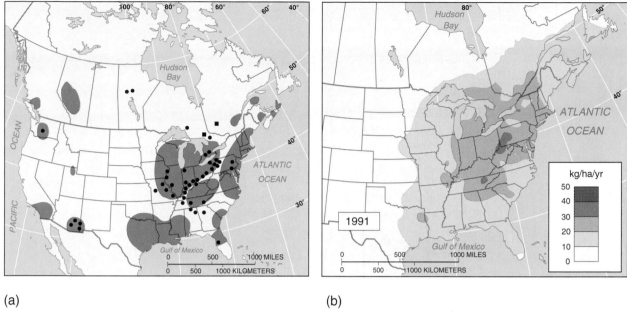

(a)   (b)

FIGURE 2-29

Spatial portrayal of (a) representative emissions of sulfur dioxide in the 1980s, and (b) wet (rain and snow) deposition of sulfur oxides (principally SO_4) on the landscape for 1991. In (a), the dots locate sources producing 100-500 metric kilotons (110,000 and 550,000 tons) per year; the squares locate sources producing 500 metric kilotons (500,000 tons) or more per year. [Data courtesy of National Atmospheric Chemistry Data Base (NAtChem), Air Quality Measurements and Analysis Research Division, Environment Canada and the U.S. Environmental Protection Agency, National Atmospheric Deposition Program/National Trends Network.]

combination of fog and smoke containing sulfur, an impurity in fossil fuels.

Sulfur dioxide (SO_2), which forms during combustion, is colorless but can be detected by its pungent odor. At low concentrations of from 0.1 to 1 ppm, sulfur dioxide-polluted air is dangerous to breathe; at 0.3 ppm the air takes on a definite taste. Human respiratory systems can become irritated, especially in individuals with asthma, chronic bronchitis, and emphysema.

Once in the atmosphere, SO_2 reacts with oxygen to form sulfur trioxide (SO_3), which is highly reactive and, in the presence of water or water vapor, forms sulfuric acid (H_2SO_4). Sulfuric acid can occur in even moderately polluted air, and its synthesis does not require high temperatures. In this manner, a sulfur dioxide–laden atmosphere is dangerous to health,

corrodes metals, and deteriorates stone building materials at accelerated rates. Sulfuric acid deposition has increased in severity since it was first described in the 1970s and is a growing blight in the biosphere. FYI Report 6-1 discusses this vital atmospheric issue.

In the United States, coal-burning electric utilities and steel manufacturing are the main sources of sulfur dioxide. And, because of the prevailing movement of air masses, they are the main sources of sulfur dioxide in Canada, too. This conclusion comes from detailed studies that show as much as 70% of Canadian sulfur dioxide is initiated within the United States.

Figure 2-29 portrays representative production levels of sulfur dioxide in the 1980s and rain and snow deposition of sulfur oxides during 1991. In 1988 alone, 20,000 industrial facilities poured 1.1 billion

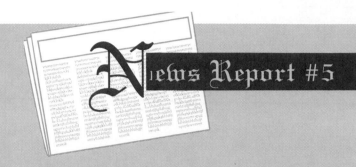

"Air Pollution Abatement–Costs vs. Benefits"

Are pollution controls too expensive? To determine the relevancy of such costs they must be weighed against the financial benefits derived from reducing pollution damage.

The cost of ozone damage to plants was studied by the U.S. National Crop Loss Assessment Network (NCLAN). Other studies have examined photochemical smog damage to human health and property, effect on mortality rate, hospital usage, and indirect health consequences. An 11-year UCLA study found significant lung deterioration after chronic exposure to oxidants and air pollution.

The benefits of air-pollution abatement far outweigh their cost. The South Coast Air Quality Management District, which oversees the Los Angeles Basin, estimated in 1991 that full implementation of their clean air plan would cost about $6.1 billion, but would generate at least $9.4 billion in benefits from ozone reduction alone. Thus, taking action is cheaper than allowing the situation to further degrade. Similar money savings occur in most cost-benefit studies of air, water, and toxic pollution abatement.

---

kilograms (2.4 billion pounds) of chemicals into the atmosphere in the United States.

Certainly, society cannot simply halt two centuries of industrialization; the resulting economic chaos would be devastating. However, neither can it permit pollution production to continue unabated, for catastrophic environmental feedback inevitably will result. People have been on Earth for such a brief fragment of Earth's history, and yet our activities can be a major influence on the future of the atmosphere. We are now contributing significantly to the creation of the *anthropogenic atmosphere*, a tentative label for Earth's next (fifth) atmosphere. The urban air we breathe today may be just a preview.

SUMMARY—Solar Energy, Seasons, and the Atmosphere

The Solar System, planets, and Earth began to condense from clouds of dust, gas, debris, and icy comets approximately 4.6 billion years ago. Over the span of these years Earth's **atmosphere** evolved in four stages—primordial, evolutionary, living, and modern. **Fusion** processes in the Sun's interior generate incredible quantities of energy, which travel outward in all directions. A tiny portion of this energy reaches Earth as solar energy in the form of charged particles of **solar wind** and the **electromagnetic spectrum** of radiant energy.

Solar wind is deflected by Earth's **magnetosphere**, producing various effects in the upper atmosphere, including spectacular **auroras** that surge across the skies at higher latitudes. The intensity of the solar wind is related to **sunspot** activity on the Sun's surface. The solar wind–atmosphere interaction disrupts certain radio transmissions and appears to influence Earth's weather. Electromagnetic radiation from the Sun passes through Earth's magnetosphere to the top of the atmosphere—the **thermopause**. The average incoming solar radiation at the thermopause, called **insolation**, is 1372 W/m^2—a value called the **solar constant**.

The dynamic flows in the atmosphere and ocean, the patterns of temperature and weather, and the vibrant biosphere result from the uneven distribution of insolation. The Sun's rays are more direct and more concentrated at lower latitudes; the angle is more oblique and the rays more diffuse at higher latitudes, which receive less energy. Insolation received at any specific place varies from day to day.

Seasonality is the change in daylength and varying altitude of the Sun above the horizon. Earth's distinct seasons are produced by interactions of its **revolution, rotation, sphericity,** axial tilt, and **axial parallelism** throughout the year. The annual march of the seasons regulates many aspects of our lives as the calendar marks the year between the two **solstice** and two **equinox** dates.

Each breath we take is a reminder of Earth's physical and natural history, for the atmosphere is a product of the planet and the inhalations and exhalations of all life that has existed. Our modern atmosphere is a gaseous mixture of ancient origins so evenly mixed it behaves as if it was a single gas, naturally odorless, colorless, tasteless, and formless. The principal substance of this atmosphere is air—the medium of life.

Our atmosphere is unique among the planets. Earth is the home of the only biosphere in the Solar System. The atmosphere is defined by various criteria. Using *composition*, we can divide the atmosphere into the **heterosphere** and the **homosphere.** With *temperature*, we can divide it into the **thermosphere, mesosphere, stratosphere,** and **troposphere.** Within the troposphere occur the biosphere and almost all weather, water vapor, and the bulk of pollution. Using *function* as a criterion, we can distinguish two regions: the **ionosphere** and the **ozonosphere,** both of which absorb life-threatening radiation and convert it into heat energy. The overall reduction of stratospheric ozone during the past several decades is actively being measured and analyzed by scientists. This ozone loss represents a hazard to society and many natural systems.

We coevolved with natural "pollution" and thus are adapted to it. However, we are not adapted to cope with our own anthropogenic pollution that constitutes a major health threat, particularly where people concentrate in cities. Photochemical and industrial smog does not respect political boundaries and is carried by winds far from its point of origin. The distribution of human-produced **carbon monoxide, nitrogen dioxide, ozone, PAN,** and **sulfur dioxide** over North America, Europe, and Asia is related to transportation and energy production. Therefore, energy conservation and efficiency, and cleaning of emissions are essential strategies for abating air pollution. Earth's next atmosphere most likely will be anthropogenic, or human-influenced in character. The air we breathe in our cities just might be a preview of the future.

KEY TERMS

altitude (Sun's)	thermosphere
aphelion	troposphere
atmosphere	aurora
exosphere	axial parallelism
heterosphere	axis
homosphere	carbon dioxide
ionosphere	carbon monoxide
mesosphere	chlorofluorocarbon compounds
ozone layer	(CFCs)
ozonosphere	circle of illumination
stratosphere	daylength

declination

electromagnetic spectrum

environmental lapse rate

equinoxes

    autumnal (September) equinox

    vernal (March) equinox

fusion

industrial smog

insolation

kinetic energy

magnetosphere

Milky Way Galaxy

nebula

nitrogen dioxide

normal lapse rate

perihelion

peroxyacetyl nitrates (PAN)

photochemical smog

plane of the ecliptic

planetesimal hypothesis

revolution

rotation

scientific method

sensible heat

solar constant

solar wind

solstices

    summer (June) solstice

    winter (December) solstice

speed of light

subsolar point

sulfur dioxide

sunspots

temperature inversion

thermopause

Tropic of Cancer

Tropic of Capricorn

wavelength

REVIEW QUESTIONS

1. Describe the Sun's status among stars in the Milky Way Galaxy. Describe the Sun's location, size, and relationship to its planets.
2. How far is Earth from the Sun in terms of light speed? In terms of kilometers and miles? Relate this distance to the shape of Earth's orbit during the year.
3. Define the scientific method and give an example of its application.
4. What is the sunspot cycle? At what stage in the cycle were we in 1990?
5. Summarize the presently known and hypothesized effects of the solar wind relative to Earth's environment.
6. Describe the segments of the electromagnetic spectrum, from shortest to longest wavelength.
7. What is the solar constant? Why is it important to know?
8. The concept of seasonality refers to what specific factors? How do the variables of seasonality change for each of these latitudes: 0°? 40°? 90°?
9. Differentiate between the Sun's altitude and its declination.
10. For the latitude at which you live, how does daylength vary during the year? How does the Sun's altitude vary? Does your local newspaper publish a weather calendar containing such information?
11. List the five physical factors that operate together to produce seasons and define each.
12. Describe revolution and rotation, and differentiate between them.
13. Describe seasonal conditions at each of the four key seasonal anniversary dates during the year. What are the solstices and equinoxes?
14. In view of the analogy by Lewis Thomas, characterize the various functions the atmosphere performs.

15. What three distinct criteria are employed in dividing the atmosphere for study?
16. Describe the overall temperature profile of the atmosphere, from surface to the exosphere.
17. Name the four most prevalent stable gases in the homosphere. Where did each originate? Is the prevalence of any of these changing at this time?
18. Why is stratospheric ozone so important? Summarize the ozone predicament and the present findings and international efforts to deal with it.
19. What is the difference between industrial smog and photochemical smog?
20. Describe the relationship between automobiles and the production of ozone and PAN in city air. What are the principal negative impacts of these gases?
21. In what ways does a temperature inversion worsen an air pollution episode?
22. In your opinion, what are the solutions to the problems developing in a potential anthropogenic atmosphere?

Spring break-up on the Turnagain Arm inlet, Alaska. [*Photo by Galen Rowell.*]

3

ATMOSPHERIC ENERGY AND GLOBAL TEMPERATURES

ENERGY BALANCE IN THE
   TROPOSPHERE
   Energy in the Atmosphere: Some Basics
   Earth-Atmosphere Radiation Balance
ENERGY AT EARTH'S SURFACE
   Daily Radiation Curves
   Simplified Surface Energy Balance
PRINCIPAL TEMPERATURE CONTROLS
   Latitude   Cloud Cover
   Altitude   Land-Water Heating Differences
EARTH'S TEMPERATURE PATTERNS
   January Temperature Map
   July Temperature Map
   Annual Range of Temperatures
   Air Temperature and the Human Body
   The Urban Environment
SUMMARY
FYI REPORT 3-1 SOLAR ENERGY COLLECTION AND
   CONCENTRATION
FYI REPORT 3-2 TEMPERATURE CONCEPTS, TERMS,
   AND MEASUREMENTS

Earth's biosphere pulses with flows of energy. Think for a moment of your own life and activities—all reflect atmosphere and surface energy patterns, such as the seasons, climate, and daily weather. This chapter examines the cascade of energy through the atmosphere to Earth's surface, as insolation is absorbed and redirected along various pathways in the troposphere. An important output produced by this energy system is the present pattern of global temperatures.

Air temperature has a remarkable influence upon our lives. What temperature is it now (both indoors and outdoors) as you read these words? How is it measured and what does the value mean? How is today's air temperature controlling your plans for the day? Our bodies sense temperature and subjectively judge comfort, reacting to changing temperatures with predictable responses. The cities in which we live alter surface energy characteristics, so the temperatures and climates of our urban areas differ from those of surrounding rural areas.

A variety of temperature regimes worldwide affect cultures, decision making, and resources consumed. Global temperature patterns presently are changing in a warming trend that is affecting us all and is the subject of much scientific, geographic, and political interest. A detailed discussion of global changes in climate is in Chapter 7.

Energy Balance in the Troposphere

Earth's atmosphere and surface are heated by solar energy, which is unevenly distributed by latitude and which fluctuates seasonally. Our budget of atmospheric energy comprises shortwave radiation *inputs* (ultraviolet light, visible light, and near-infrared wavelengths) and longwave radiation *outputs* (thermal infrared, or heat). *Transmission* refers to the passage of shortwave and longwave energy through either the atmosphere or water. The atmosphere and surface eventually reradiate heat energy back to space, and this energy, together with reflected energy, equals the initial solar input. Thus a *balance* of energy input and output exists in our atmosphere.

Energy in the Atmosphere: Some Basics

When viewing a photograph of Earth taken from space, the pattern of surface response to incoming insolation is clearly visible (see the back cover of this text). Solar energy is intercepted by land and water surfaces, clouds, and atmospheric gases and dust. The flows of energy are evident in weather patterns, oceanic currents, and the distribution of vegetation. Specific energy patterns differ for deserts, oceans, mountain tops, rain forests, and ice-covered landscapes. In addition, clear or cloudy weather might mean a 75% difference in the amount of energy reaching the surface.

Figure 3-6 is a detailed flow diagram of the energy balance between Earth and the atmosphere discussed in the next few pages. You will find it helpful to refer ahead to this figure as you read the following descriptions of what happens to energy in the atmosphere.

Insolation Input. Insolation is the single energy input driving the Earth-atmosphere system. The world map in Figure 3-1 shows the distribution of average annual solar energy at Earth's surface. It includes all radiation arriving at Earth's surface, both direct and diffuse (or downward scattered).

Several patterns are notable on the map. Insolation decreases poleward from about 25° latitude in both the Northern and Southern Hemispheres. Consistent daylength and high Sun altitude produce average annual values of 180–200 W/m² throughout the equatorial and tropical latitudes. In general, greater insolation of 240–280 W/m² occurs in low-latitude deserts worldwide because of frequently cloudless skies. Note the energy pattern in the cloudless subtropical deserts in both hemispheres (for example, the Sonoran, Saharan, Arabian, Gobi, Atacama, Namib, Kalahari, and Australian deserts).

Albedo and Reflection. A portion of arriving energy bounces directly back into space without being converted into heat or performing any work. This returned energy is called **reflection**, a term that applies to both visible and ultraviolet light. The reflective quality of a surface is its **albedo**. Albedo is the percentage of insolation that is reflected. Albedo is an extremely important control over the amount of insolation that is available to heat a surface.

In the visible wavelengths, darker colors have lower albedos, and lighter colors have higher albedos. On water surfaces, the angle of the solar rays also affects albedo values; lower angles produce a greater reflection than do higher angles (Figure 3-2).

In addition, smooth surfaces increase albedo, whereas rougher surfaces reduce it.

Figure 3-3 illustrates albedo values for various surfaces. Specific locations experience highly variable albedo during the year in response to changes in cloud and ground cover. Earth Radiation Budget (ERB) sensors aboard the *Nimbus-7* satellite measure average albedos of 19–38% between the tropics (23.5° N to 23.5° S) to as high as 80% in the polar regions.

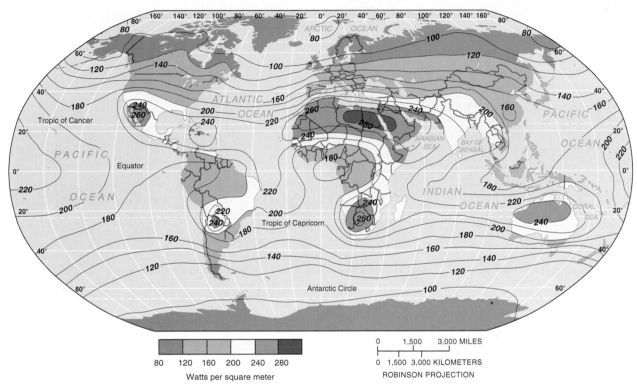

FIGURE 3-1
Average annual solar radiation receipt on a horizontal surface at ground level in watts per square meter (100 W/m² = 75 kcal/cm²/yr). [After M. I. Budyko, 1958, and *Atlas Teplovogo Balansa*, Moscow, 1963.]

FIGURE 3-2
An astronaut's view of reflected sunlight off the Mozambique Channel between the east coast of Africa and the island nation of Madagascar. Albedo values increase with lower sun angles and calmer seas. [Space shuttle photo from NASA.]

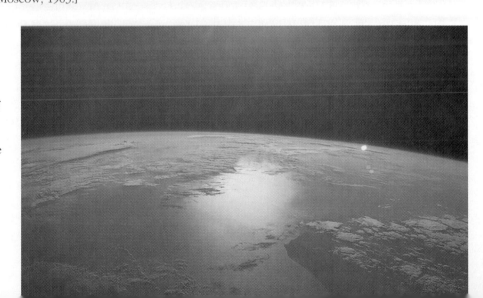

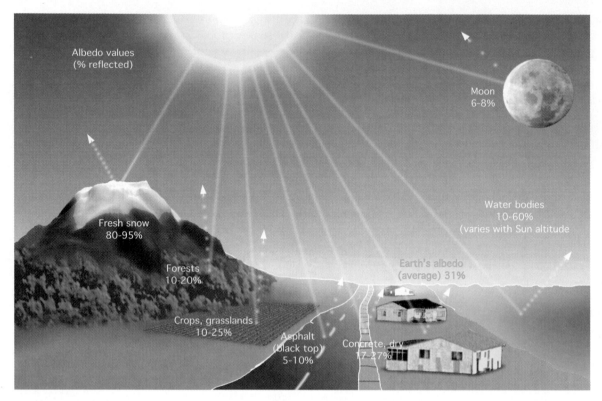

FIGURE 3-3
Selected albedos for different surfaces. Generally, light surfaces are more reflective and thus have higher albedo values. Darker surfaces are less reflective. [Data from M.I. Budyko, 1958, *The Heat Balance of the Earth's Surface*, Washington, DC, U.S. Department of Commerce, p. 36.]

Earth and its atmosphere reflect 31% of all insolation when averaged over a year. Figure 3-6 shows that Earth's average albedo is a combination of light reflected by clouds, reflected by the ground (combined land and oceanic surfaces), and reflected and scattered by the atmosphere. By comparison, a full Moon, which is bright enough to read by under clear, night skies, has only a 6–8% albedo value. Thus, with earthlight being four times brighter than moonlight (four times the albedo), and with Earth being four times larger than the Moon, it is no surprise that astronauts report how startling our planet looks from space.

A major uncertainty in the tropospheric energy budget, and therefore in refining climatic models, is the role of clouds. The high albedo of clouds reflects *insolation,* thus cooling Earth's surface. Yet, clouds act as *insulation,* thus trapping longwave radiation and raising minimum temperatures. Important is not merely the percentage of cloud cover but the cloud type,

height, and thickness (water content and density). High-altitude, ice-crystal clouds have albedos of about 50%, whereas thick, lower cloud cover reflects about 90%. Understanding the nature of global cloud cover is crucial in refining computer models of predicted global warming. A section on clouds is in Chapter 5.

Other mechanisms affect atmospheric albedo and therefore temperature patterns. The eruption of Mount Pinatubo in the Philippines, beginning explosively during June 1991, illustrates how Earth's internal processes can affect the atmosphere. Approximately 15–20 megatons of SO_2 and ash was injected into the stratosphere; winds rapidly spread this aerosol worldwide. As a result, atmospheric albedo increased and produced a worldwide average temperature decrease of 0.5C° (0.9F°).

Scattering (Diffuse Reflection). Insolation encounters an increasing density of atmospheric gases as

FIGURE 3-4
A rainbow is produced when raindrops refract and reflect light. [Photo by author.]

it travels down to the surface. These molecules redirect the insolation, changing the direction of light's movement without altering its wavelengths. This phenomenon is known as **scattering** and is a part of Earth's albedo (Figure 3-6). Dust particles, ice, cloud droplets, and water vapor produce further scattering.

Why is Earth's sky blue? And why are sunsets and sunrises often red? These simple questions have an interesting explanation, based upon a principle known as Rayleigh scattering (named for English physicist Lord Rayleigh, who stated the principle in 1881). This principle relates wavelength to the size of molecules or particles that cause the scattering. The general rule is: *the shorter the wavelength, the greater the scattering,* and *the longer the wavelength, the less the scattering.* Shorter wavelengths of light are scattered by small gas molecules in the air. Thus, the shorter wavelengths of visible light—the blues and violets—are scattered the most and dominate the lower atmosphere. And because there are more blue than violet wavelengths in sunlight, a blue sky prevails. A sky filled with smog and haze appears almost white because the larger particles associated with air pollution act to scatter all wavelengths of the visible light.

The angle of the Sun's rays determines the thickness of atmosphere they must pass through to reach the surface. Direct rays experience less scattering and absorption than do low, oblique-angle rays that must travel farther through the atmosphere. As for red sunsets and sunrises, insolation from the low-altitude Sun must pass through an increasing mass of atmosphere. The effective scattering of shorter wavelengths leaves the residual oranges and reds to reach the observer (see Part 1 opening photograph).

Figure 3-6 shows that some incoming insolation is diffused by clouds and atmosphere and is transmitted to Earth as **diffuse radiation**, the downward component of scattered light. This light is multidirectional and thus casts shadowless light on the ground.

Refraction. When insolation enters the atmosphere, it passes from one medium to another (from virtually empty space to atmospheric gas) and is subject to a change of speed, which also shifts its direction, a bending action called **refraction**. In the same way, a crystal or prism refracts light passing through it, bending different wavelengths to different degrees, separating the light into its component colors to display the spectrum. A rainbow is created when visible light passes through myriad raindrops and is refracted and reflected toward the observer at a precise angle (Figure 3-4). Another example of refraction is a *mirage,* an

FIGURE 3-5
The distorted appearance of the Sun, nearing sunset over the ocean, is produced by refraction of the Sun's image in the atmosphere. [Photo by author.]

image that appears near the horizon where light waves are refracted by layers of air of differing temperatures (and resulting differing densities) on a hot day.

An interesting function of refraction is that it adds approximately eight minutes of daylight for us. The Sun's image is refracted in its passage from space through the atmosphere, and so, at sunrise, we see the Sun about four minutes before it actually peeks over the horizon. Similarly, the Sun actually sets at sunset, but its image is refracted over the horizon for about four minutes afterward. To this day, modern science cannot predict the exact time of sunrise or sunset within these four minutes because the degree of refraction continually varies with temperature, moisture, and pollutants. The distortion of the setting Sun in Figure 3-5 is a product of refraction.

Absorption. **Absorption** is the *assimilation* of radiation and its *conversion* from one form to another. Insolation (both direct and diffuse) that is not part of the 31% reflected from Earth's surfaces is absorbed and converted into infrared radiation (heat) or is used by plants in photosynthesis. The temperature of the absorbing surface is raised in the process, causing that surface to radiate more total energy at shorter wavelengths. In addition to absorption by land and water surfaces, absorption also occurs in atmospheric gases, dust, clouds, and stratospheric ozone. Figure 3-6 summarizes the pathways of insolation and the flow of heat in the atmosphere and at the surface.

Earth Reradiation and the Greenhouse Effect. Previously we characterized Earth as a cool-body radiator, emitting energy in infrared (heat) wavelengths from its surface and atmosphere back toward space. However, some of this infrared radiation is absorbed by carbon dioxide, water vapor, methane, CFCs, and other gases in the lower atmosphere and is then reradiated to Earth, thus delaying heat loss to space. This counterradiation process is an important factor in warming the lower atmosphere. The approximate similarity between this process and the way a greenhouse operates gives the process its name—the **greenhouse effect**.

In a greenhouse, the glass is transparent to short-wave insolation, allowing light to pass through to the soil, plants, and materials inside. The absorbed energy is then radiated as infrared energy back toward the glass, but the glass effectively traps both the infrared wavelengths and the warmed air inside the greenhouse. Thus, the glass acts as a one-way filter, allowing the light in but not allowing the heat out. The same process also can be observed in a car parked in direct sunlight.

Opening a roof vent in a greenhouse or a window in a car allows the air inside to mix with the outside environment, thereby removing heat by moving air *physically* from one place to another. Such turbulent movement is called **convection** if it involves a strong vertical motion and **advection** if a lateral (horizontal) motion is dominant. Heat in a greenhouse also can be conducted by exposed surfaces. **Conduction** is the molecule-to-molecule transfer of heat as it diffuses through a substance. As molecules warm, their vibration increases, causing collisions that produce motion in neighboring molecules and thus transferring heat energy from warmer to cooler materials.

In the atmosphere, the greenhouse analogy is not fully applicable because infrared radiation is not trapped as in a greenhouse. Rather, its passage to space is delayed as the heat is radiated and reradiated back and forth between Earth's surface and certain gases and particulates in the atmosphere. Today's increasing carbon dioxide concentration is causing more infrared radiation absorption in the lower atmosphere, thus forcing a warming trend and disruption of the Earth-atmosphere energy system, according to many scientists.

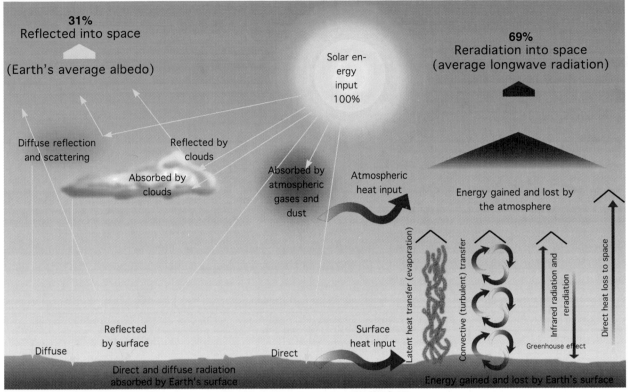

Earth-atmosphere energy budget

FIGURE 3-6
The Earth-atmosphere energy balance. Solar energy cascades through the lower
atmosphere, where it is absorbed, reflected, and scattered. Eventually, Earth reradiates
absorbed energy in infrared wavelengths that, when added to Earth's average albedo
(reflected energy), yields a total energy output equal to the input of 100 units.

Earth-Atmosphere Radiation Balance

The Earth-atmosphere energy system naturally balances in a steady-state equilibrium. The natural energy balance occurs through energy transfers that
are *radiative* and *nonradiative*. Radiative transfer
is by infrared radiation between the surface and the
atmosphere. Nonradiative transfers include convection, conduction, and the latent heat of evaporation (heat absorbed by water vapor when water
evaporates).

Figure 3-6 summarizes the Earth-atmosphere radiation balance. It brings together all the elements discussed to this point in the chapter by following 100%
of energy through the system. To summarize: of
100% of the solar energy arriving, Earth's average

albedo reflects 31% into space. Absorption by atmospheric clouds, dust, gases, and stratospheric
ozone comprises the atmospheric heat input. This
leaves less than half of the incoming insolation to
actually reach Earth's surface as direct and diffuse
radiation. Earth eventually reradiates the remaining
69% back into space as heat.

Energy at the Surface

Solar energy is the principal heat source at Earth's
surface. The direct and diffuse radiation and infrared
radiation arriving daily at the ground surface are of
great interest to geographers. Examine the terms
along the ground surface in Figure 3-6.

Daily Radiation Curves

The fluctuating daily pattern of incoming shortwave energy and outgoing longwave energy at the surface, and resultant air temperature, are shown in Figure 3-7. This graph represents conditions for bare soil on a cloudless day in the middle latitudes. Incoming energy arrives during daylight, beginning at sunrise, peaking at noon, and ending at sunset. The shape and height of this insolation curve varies with season and latitude. The highest trend for such a curve occurs at the time of the summer solstice (around June 21 in the Northern Hemisphere and December 21 in the Southern Hemisphere).

The temperature plot also responds to seasons and variations in input. Within a 24-hour day, air temperature peaks at around 3:00–4:00 P.M. and dips to its lowest point right at or slightly after sunrise.

The relationship between the insolation curve and the temperature curve on the graph is interesting—they do not align. As long as the incoming energy exceeds the outgoing energy, temperature continues

to increase during the day, not peaking until the incoming energy begins to diminish in the afternoon as the Sun loses altitude.

The warmest time of day occurs not at the moment of maximum insolation but at that moment *when a maximum of insolation is absorbed*. Thus, this temperature lag places the warmest time of day three to four hours after solar noon as absorbed heat is supplied to the atmosphere from the ground. Then, as the insolation input decreases toward sunset, the amount of heat lost exceeds the input, and temperatures begin to drop until the surface has radiated away the maximum amount of energy, just at dawn.

The annual pattern of insolation and temperature exhibits a similar lag. For the Northern Hemisphere, January is usually the coldest month, occurring after the winter solstice, the shortest day in December. Similarly, the warmest months of July and August in the Northern Hemisphere occur after the summer solstice, the longest day in June.

Simplified Surface Energy Balance

Earth's surface is supplied with energy that daily and seasonally varies. The climate at or near Earth's surface is generally termed the *boundary layer climate*. **Microclimatology** is the study of this portion of the atmosphere. The following discussion will be more meaningful if you keep in mind an actual surface— perhaps a park, a front yard, or a place on campus.

The surface receives light and heat, and it reflects light and radiates heat according to this basic scheme (SW = shortwave, LW = longwave):

$$+SW\downarrow \ - \ SW\uparrow \ + \ LW\downarrow \ - \ LW\uparrow \ = \ NET\ R$$

(Insolation, (Albedo   (Heat)   (Heat)   (Net
   light)   value of   ⏟   Radiation)
            surface) (Infrared radiation)

Along the surface, as illustrated in Figure 3-8, are the components of a surface energy balance. The column of soil continues to a depth at which energy exchange with surrounding materials becomes negligible, usually less than a meter.

Energy moving toward the surface is regarded as positive (a gain), and energy moving away from the surface is considered negative (a loss). The amount of insolation to the surface ($+SW\downarrow$) varies with sea-

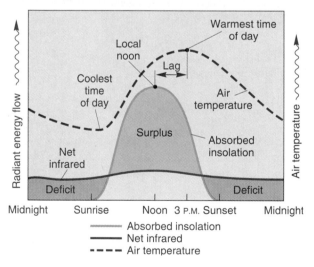

FIGURE 3-7
Sample radiation curves for a typical day show the changes in insolation (orange), reradiated infrared (red), and air temperature (dashed). Comparing the curves demonstrates a lag between local noon (the insolation peak for the day) and the warmest time of day. The gradual heating of the air by convective transfer from the ground produces a lag of several hours following maximum insolation input.

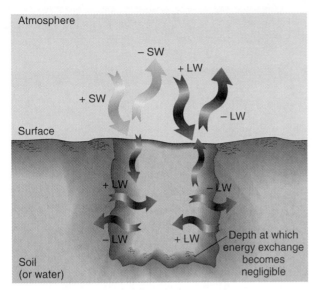

FIGURE 3-8
Input and output components of a surface energy balance for a typical soil column.

son, cloudiness, and latitude. The albedo dictates the amount of insolation reflected and therefore not absorbed at the surface (–SW ↑). Darker and rougher surfaces reflect less energy, absorb more energy, and therefore produce more net radiation. Lighter and smoother surfaces reflect more short-wave energy, leading to less net radiation. Snow-covered surfaces reflect most of the light reaching them, whereas urban areas tend to be darker in color and therefore convert more insolation to infrared radiation.

Adding and subtracting the heat flow at the surface completes the calculation of **net radiation (NET R)**, or the balance of all radiation at Earth's surface—shortwave and longwave.

Net Radiation. Net radiation (NET R), the net radiation (of all wavelengths) available at Earth's surface, is the final outcome of the entire radiation-balance process discussed in this chapter. Net radiation is expended from a nonvegetated surface through three pathways:

- *Turbulent sensible heat transfer*, the quantity of sensible heat energy physically transferred from air to surface, and surface to air, through convection and conduction.

- *Latent heat of evaporation*, the heat energy that is stored in water vapor as water evaporates. Large quantities of latent heat are absorbed into water vapor during its change of state from liquid to gas.
- *Ground heating and cooling*, energy that flows into and out of the ground surface by conduction.

Understanding the net radiation balance is essential to solar energy technologies that concentrate shortwave and longwave energy for use. Solar energy offers great potential worldwide and will see expanded use in the near future. FYI Report 3-1 briefly reviews solar energy strategies.

Variation in the expenditure of NET R among sensible heat (heat we can feel), latent heat, and ground heating and cooling, produces the variety of environments we experience in nature. Figure 3-9a illustrates a desert environment where the absence of clouds, few plants, and little water result in high sensible heat production and hot ground. Figure 3-9b shows the area near Pitt Meadows, British Columbia, featuring a moist, vegetated environment of irrigated mixed orchard and rye grass. Here, high amounts of energy are expended for latent heat of evaporation with more moderate amounts spent as sensible heat.

Now that we have examined the flow of solar energy across space, through the atmosphere, to Earth's surface, we turn our attention to the some of the work that this energy accomplishes: the pattern of global temperatures, and, in the next chapter, the flow of atmospheric and oceanic circulation.

Principal Temperature Controls

Temperature, a measure of sensible heat energy present in matter, indicates the average kinetic energy of individual molecules. FYI Report 3-2 introduces important temperature concepts that will help you understand the following material. Principal controls and influences upon temperature patterns include latitude, altitude, cloud cover, land-water heating differences, ocean currents and sea-surface temperatures, and general surface conditions. Let us begin by examining each of the principal temperature controls.

Solar Energy Collection and Concentration

Insolation not only warms Earth's surface; it provides the closest thing to an inexhaustible supply of energy for humanity. Yet, it is essentially untapped. Sunlight is direct, pervasive, and renewable and has been collected for centuries through various technologies. Turn-of-the-century photographs of residences in Southern California and Florida show solar (flat-plate) water heaters on many rooftops. Early twentieth-century ads in newspapers and merchandise catalogs featured the Climax solar water heater (1905) and the Day and Night water heater (1909). However, low-priced natural gas and oil displaced many of these early applications.

By the early 1970s, the depletion of fossil fuels, their rising prices, and our growing dependence upon foreign sources spurred a reintroduction of tried-and-true solar technologies and energy-efficiency strategies. Cities such as Davis, California, passed energy-conservation and solar-utilization ordinances to guide construction and to encourage solar-energy installations. The country seemed to be rediscovering old, proven techniques. Unfortunately, during the early 1980s, much of this progress was halted by political decisions and severe budget cuts made by President Reagan's administration. A new energy plan in 1992 promises to renew the push for solar applications.

Collecting and Concentrating Solar Energy

Any surface that receives light from the Sun is a solar collector. In fact, an average building in the United States receives 6 to 10 times more energy from the Sun than is required to heat the inside! But the diffuse nature of solar energy received at the surface requires that it be collected, transformed, and stored to be useful. Space heating is the simplest application. Windows that are carefully designed and placed allow sunlight to shine into the building, where it is absorbed and converted into sensible heat. This is an everyday application of the greenhouse effect.

A *passive solar system* captures heat energy and stores it in a "thermal mass," such as water-filled tanks, adobe, tile, or concrete. Three key features of passive solar design are (1) large areas of glass facing south (or facing north in the Southern Hemisphere) (2) a thermal storage medium, and (3) shades or screens to prevent light entry in the warm summer months. At night, shades or other devices can cover glass areas to prevent heat loss from the structure. An *active solar system* involves heating water or air in a collector and then pumping it through a plumbing system to a tank where it can provide hot water for direct use or for space heating.

Solar energy systems are capable of generating low-grade heat of an appropriate scale for approximately half the present domestic applications in the United States (space heating and water heating). In marginal climates, solar assisted water and space heating is feasible as a backup; even in New England and the Northern Plains states, solar-efficient collection systems have proven effective.

Reflective surfaces, such as focusing mirrors or parabolic troughs, can be used to attain very high temperatures to heat water or other fluids. Kramer Junction, California, about 225 km (140 mi) northeast of Los Angeles, is the location of the world's largest operating solar electric-generating facility, with 194 megawatts capacity. Long troughs of computer-guided curved mirrors concentrate sunlight to create temperatures of 390°C (735°F) in vacuum-sealed tubes filled with synthetic oil. This is then used to heat water, which produces steam that rotates turbines to generate cost-effective electricity. Thirty years of contracts to buy this electrical power already are signed, and utility companies are seeking the construction of more installed capacity (Figure 1). However, even the continued success of this private venture appears tied to the political arena.

Electricity is also being produced by photovoltaic cells, a technology that has been used in spacecraft since 1958. When sunlight shines upon a semiconductor material in these cells, it stimulates a flow of electrons (an electrical current). The efficiency of these cells has gradually improved and they now are cost-competitive. A drawback of both solar heating and solar electric sys-

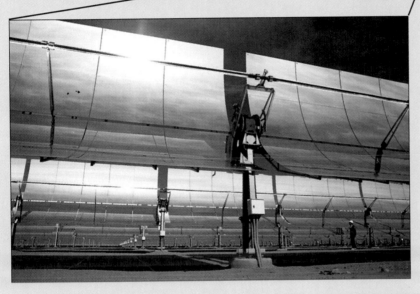

FIGURE 1
Kramer International solar-thermal
energy installation in southern
California. [Photos courtesy of
Kramer International Limited, Los
Angeles.]

tems is that periods of cloudiness and night inhibit operations. Research is going forward to enhance energy storage techniques, such as hydrogen fuel production (using energy to extract hydrogen from water for later use in producing more energy) and improved battery technology.

The Promise of Solar Energy

Solar energy is a wise choice for the future because it is directly available to the consumer, it is based upon renewable sources of an appropriate scale for end-use needs, and most solar strategies are labor intensive (rather than capital intensive). Solar seems preferable to further development of our decreasing fossil fuel reserves, oil imports, and troubled nuclear power. Continued dependence on large centralized power production is less desirable in that it is indirect, nonrenewable, polluting, high-grade (high temperature and cost) energy that is capital intensive.

Whether the alternative path of solar energy is taken depends principally upon political decisions, for many of the technological innovations are presently ready for installation, requiring only leadership from the government.

85

(a)

(b)

FIGURE 3-9

(a) In the California desert east of Los Angeles (at about 35° N), the daily expenditures of net radiation produce a desert landscape of high temperature and sparse vegetation. (b) Irrigated blueberry orchards are characteristic of the agricultural activity near Pitt Meadows, British Columbia (at about 49° N), where a moist environment and moderate temperatures occur. [Photos by author.]

Latitude

Insolation is the single most important influence on temperature variations. Figures 2-9 and 2-10 showed how insolation intensity decreases as we move away from the subsolar point, a point that migrates between the Tropic of Cancer and Tropic of Capricorn—that is, between 23.5° N and 23.5° S—during the year. In addition, daylength and Sun angle change throughout the year, increasing in seasonal effect with increasing latitude. From equator to poles, Earth ranges from continually warm, to seasonally variable, to continually cold.

Altitude

Within the troposphere, temperatures decrease with increasing altitude above Earth's surface (recall that the normal lapse rate of temperature change with altitude is 6.4C°/1000 m or 3.5F°/1000 ft—see Figure 2-22). Thus, worldwide, mountainous areas experience lower temperatures than do regions nearer sea level, even at similar latitudes. The density of the atmosphere also diminishes with increasing altitude. In fact, the density of the atmosphere at an elevation of 5500 m (18,000 ft) is about half of that at sea level. As the atmosphere thins, its ability to absorb and radiate heat is reduced.

The consequences are that average air temperatures at higher elevations are lower, nighttime cooling increases, and the temperature range between day and night and between areas of sunlight and shadow also is greater. Temperatures may decrease noticeably in the shadows and shortly after sunset. Surfaces both gain heat rapidly and lose heat rapidly to the thinner atmosphere. Also, at higher elevations, the insolation received is more intense because of the reduced mass of atmospheric gases. As a result of this intensity, the ultraviolet energy component makes sunburn a distinct hazard.

The snowline seen in mountain areas indicates where winter snowfall exceeds the amount of snow lost through summer melting and evaporation. The snowline's location is a function both of elevation and latitude, which means that glaciers can exist even at equatorial latitudes if the elevation is great enough. In equatorial mountains, the snowline occurs at approximately 5000 m (16,400 ft) and permanent ice fields and glaciers exist on equatorial mountain summits in the Andes and East Africa. With increasing latitude toward the poles, snowlines gradually descend in elevation from 2700 m (8850 ft) in the midlatitudes to lower than 900 m (2950 ft) in southern Greenland.

Temperature Concepts, Terms, and Measurements

The amount of heat energy present in any substance can be measured and expressed as its temperature. Temperature actually is a reference to the speed at which atoms and molecules that make up a substance are moving. Changes in temperature are caused by the absorption or emission (gain or loss) of energy. The temperature at which all motion in a substance stops is called 0° absolute temperature. Its equivalent in different temperature-measuring schemes is –273° Celsius (C), –459.4° Fahrenheit (F), or 0 Kelvin (K).

The Celsius scale (formerly called centigrade) is named after Swedish astronomer Anders Celsius, 1701–1744. It divides the difference between the freezing and boiling temperatures of water at sea level into 100 degrees on a decimal scale, and labels them 0° and 100°, respectively.

The Fahrenheit scale places the freezing point of water at 32°F and boiling point of water at 212°F, a span of 180° degrees. The United States remains the *only* major country still using the Fahrenheit scale, named after the German physicist who invented the alcohol and mercury thermometers and who developed this thermometric scale in the early 1700s.

The Kelvin scale, named after British physicist Lord Kelvin (born William Thomson, 1824–1907) who proposed the scale in 1848, is used in science because temperature readings are proportional to the actual kinetic energy in a material. The freezing point and boiling point of water are 273K and 373K, respectively.

Most countries use the Celsius scale to express temperature. This text presents Celsius (with Fahrenheit equivalents in parentheses) throughout in an effort to bridge this transitional era in the United States. The continuing pressure of the scientific community and corporate interests, not to mention the rest of the world, makes

FIGURE 1
Scales for expressing temperature in degrees Celsius and Fahrenheit.

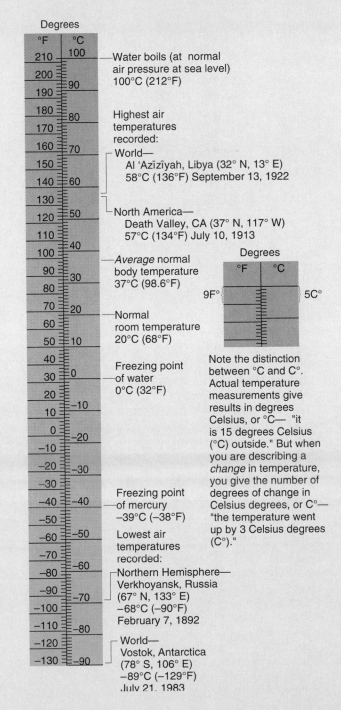

Degrees

°F °C

Water boils (at normal air pressure at sea level) 100°C (212°F)

Highest air temperatures recorded:
World—
Al 'Azīzīyah, Libya (32° N, 13° E) 58°C (136°F) September 13, 1922

North America—
Death Valley, CA (37° N, 117° W) 57°C (134°F) July 10, 1913

Average normal body temperature 37°C (98.6°F)

Normal room temperature 20°C (68°F)

Freezing point of water 0°C (32°F)

Freezing point of mercury –39°C (–38°F)

Lowest air temperatures recorded:

Northern Hemisphere—
Verkhoyansk, Russia (67° N, 133° E) –68°C (–90°F) February 7, 1892

World—
Vostok, Antarctica (78° S, 106° E) –89°C (–129°F) July 21, 1983

Degrees

°F °C

9F° 5C°

Note the distinction between °C and C°. Actual temperature measurements give results in degrees Celsius, or °C— "it is 15 degrees Celsius (°C) outside." But when you are describing a *change* in temperature, you give the number of degrees of change in Celsius degrees, or C°— "the temperature went up by 3 Celsius degrees (C°)."

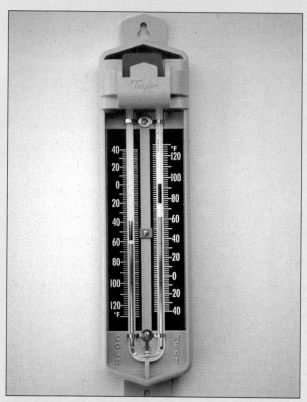

FIGURE 2
Minimum-maximum thermometer. [Photo by Bobbé Christopherson.]

FIGURE 3
Standard weather instrument shelter. [Belfort Instruments, Inc.]

adoption of the Celsius and the SI (International System of Units) system inevitable for the United States. Figure 1 compares these two scales. Formulas for converting between Celsius, metric, SI, and English units are on the inside back cover of this text.

Outdoor temperature usually is measured with a mercury thermometer or alcohol thermometer in a sealed glass tube. Alcohol is preferred in cold climates because it freezes at –112°C (–170°F), whereas mercury freezes at –39°C (–38.2°F). The principle of these thermometers is simple: when these fluids are heated, they expand; upon cooling, they contract. A thermometer stores fluid in a small reservoir at one end and is marked with calibrations to measure the expansion or contraction of the fluid, which reflects the temperature of the thermometer's environment.

Figure 2 shows a mercury minimum-maximum thermometer, designed to record the day's highest and lowest temperatures and preserve the readings until it is reset. Resetting is accomplished by moving the markers with a magnet. Another type, the recording thermometer, creates an inked record on a turning drum; it is usually set for one full rotation every 24 hours or every 7 days.

Thermometers for standardized official readings are deployed inside white (for high albedo) louvered (for ventilation) weather stations (Figure 3), designed to avoid excessive heating of the instruments inside. The instrument shelter is placed at least 1.2–1.8 m (4–6 ft) above a surface, usually on turf; in the United States 1.2 m (4 ft) is standard. Official temperatures are measured in the shade to prevent the effect of direct insolation. Temperature readings are taken daily, sometimes hourly, at more than 10,000 weather stations worldwide. Some stations with recording equipment also report the duration of temperatures, rates of rise or fall, and variation over time throughout the day and night.

Daily minimum-maximum readings are averaged to get the daily mean temperature. The monthly mean temperature is the total of daily mean temperatures for the month, divided by the number of days in the month. An annual temperature range expresses the difference between the lowest and highest monthly mean temperatures for a given year. If you install a thermometer for outdoor temperature reading, be sure to avoid direct sunlight on the instrument and place it in an area of good ventilation.

Two cities in Bolivia illustrate the interaction of the two temperature controls, latitude and altitude. Figure 3-10 displays temperature data for the cities of Concepción and La Paz, which are near the same latitude (about 16° S). Note the elevation, average annual temperature, and precipitation for each location. The hot, humid climate of Concepción at its much lower elevation stands in marked contrast to the cool, dry climate of highland La Paz. People living around La Paz actually grow wheat, barley, and potatoes—crops characteristically grown in the cooler midlatitudes—despite the fact that La Paz is 4103 m (13,461 ft) above sea level. (For comparison, the summit of Pikes Peak in Colorado is at 4301 m or 14,111 ft.) The combination of elevation and equatorial location guarantee La Paz nearly constant daylength and moderate temper-

atures, averaging about 9°C (48°F) for every month. Such moderate temperature and moisture conditions lead to the formation of more fertile soils than those found in the warmer, wetter climate of Concepción.

Cloud Cover

The type, height, and density of cloud cover are important to temperature patterns. Orbital surveys reveal that approximately 50% of Earth is cloud-covered at any one time. Clouds are the most variable factor influencing Earth's radiation budget, and they are the subject of much investigation and effort to improve computer models of atmospheric behavior. The International Satellite Cloud Climatology Project, part of the World Climate Research Program, is presently in the midst of

FIGURE 3-10
Comparison of temperature patterns in La Paz and Concepción, Bolivia.3

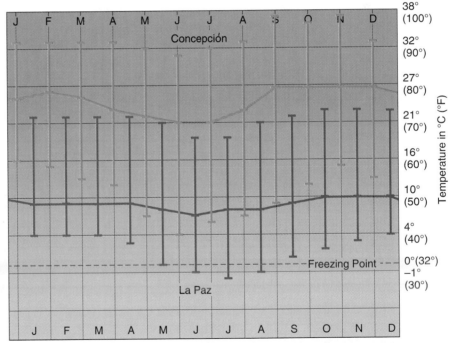

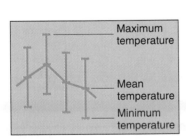

	Station	
	Concepción, Bolivia	**La Paz, Bolivia**
Latitude/longitude	16°15′S, 62°03′W	16°30′S, 68°10′W
Elevation	490 m (1607.6 ft)	4103 m (13,461 ft)
Avg. annual temperature	24°C (75.2°F)	9°C (48.2°F)
Ann. temperature range	5C° (9F°)	3C° (5.4F°)
Annual precipitation	121.2 cm (47.7 in.)	55.5 cm (21.9 in.)
Population	768,000	993,000

such research. As part of the Earth Radiation Budget Experiment (ERBE), cloud effects on longwave, shortwave, and net radiation patterns are assessed.

Clouds are moderating influences on temperature, producing lower daily maximums and higher nighttime minimums. Acting as insulation, clouds hold heat energy below them at night, preventing more rapid radiative losses, whereas during the day, clouds reflect insolation as a result of their high albedo values. The moisture in clouds both absorbs and liberates large amounts of heat energy, yet another factor in moderating temperatures at the surface. Overall, the Earth-atmosphere energy system responds with slightly lower temperatures as a result of cloud cover.

Land-Water Heating Differences

The irregular arrangement of landmasses and water bodies on Earth contributes to the overall pattern of temperature. The physical nature of the substances themselves—rock and soil vs. water—is the reason for these **land-water heating differences.** More moderate temperature patterns are associated with water bodies, compared to more extreme temperatures inland. Figure 3-11 visually summarizes the following discussion.

Evaporation. More of the energy arriving at the ocean's surface is expended for evaporation than is expended over a comparable area of land. An estimated 84% of all evaporation on Earth is from the oceans. When water evaporates and thus changes to water vapor, *heat energy is absorbed in the process and is stored in the water vapor.* This stored heat energy is called *latent heat* and is discussed fully in Chapter 5. You can experience this evaporative heat loss (cooling) by wetting the back of your hand and then blowing on the moist skin. Sensible heat energy is drawn from your skin to supply some of the energy for the water's evaporation, and you feel the cooling. Similarly, as surface water evaporates, substantial energy is absorbed, resulting in a lowering of nearby air temperatures. The land, containing far less water, experiences far less evaporation, and therefore is moderated less by evaporative cooling.

Transparency. The transmission of light obviously differs between soil and water; solid ground is opaque, water is transparent. Consequently, light striking a soil surface does not penetrate but is absorbed, heating the ground surface. That heat is accumulated during times of exposure and is rapidly lost at night or in shadows. Maximum and minimum temperatures generally are experienced right at ground level. Below the surface, even at shallow depths, temperatures remain about the same throughout the day. This situation often exists at a beach, where surface sand may be painfully hot to your feet but the sand a few centimeters below the surface feels cooler and offers relief.

In contrast, when light reaches a body of water, it penetrates the surface because of water's **transparency,** transmitting light to an average depth of 60 m (200 ft) in the ocean. This illuminated zone is known as the *photic layer* and has been recorded in

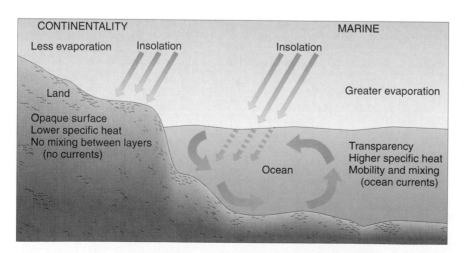

FIGURE 3-11
The differential heating of land and water produce contrasting marine and continental temperature regimes.

some ocean waters to depths of 300 m (1000 ft). *This characteristic of water results in the distribution of available heat energy over a much greater depth and volume,* forming a larger heat reservoir than that made up of the surface layers of the land.

Specific Heat. When equal volumes of water and land are compared, water requires far more heat to raise its temperature than does land. In other words, *water can hold more heat than can soil or rock,* and therefore water is said to have a higher **specific heat**. A given volume of water represents a more substantial heat reservoir than an equal volume of land, so that changing the temperature of the oceanic heat reservoir is a slower process than changing the temperature of land.

Movement. Land is a rigid, solid material, whereas water is a fluid and is capable of movement. Differing temperatures and currents result in a mixing of cooler and warmer waters, and that *mixing spreads the available heat over an even greater volume* than if the water were still. Surface water and deeper waters mix, redistributing heat energy. Both ocean and land surfaces radiate heat at night, but land loses its heat energy more rapidly than does the moving mass of the oceanic heat reservoir.

Ocean Currents and Sea-Surface Temperatures. Although ocean currents are discussed in greater detail in the next chapter, the influence of currents on temperature and weather requires mention here. In an air mass, water vapor content is affected by ocean temperatures, for warm water tends to energize overlying air through high evaporation rates and transfers of latent heat.

As a specific example, the **Gulf Stream** (described in Chapter 4) moves northward off the east coast of North America, carrying warm water far into the North Atlantic (Figure 3-12). As a result, the southern third of Iceland experiences much milder temperatures than would be expected for a latitude of 65° N, just below the Arctic Circle (66.5°). In Reykjavik, on the southwestern coast of Iceland, monthly temperatures average above freezing during all months of the year. The Gulf Stream affects Scandinavia and northwestern Europe in the same manner.

In the western Pacific Ocean, the *Kuroshio* or Japan Current, similar to the Gulf Stream, functions much

FIGURE 3-12

Satellite image of the warm Gulf Stream as it flows northward along the North American east coast—the streamlike flow in red, orange, and yellow colors. Instruments sensitive to infrared wavelengths produced this remote-sensing image. Temperature differences are distinguished by computer-enhanced coloration: reds/oranges = 25° to 29°C (76° to 84°F); yellows/greens = 17° to 24°C (63° to 75°F); blues = 10° to 16°C (50° to 61°F); and purples = 2° to 9°C (36° to 48°F). [Imagery by NASA and NOAA.]

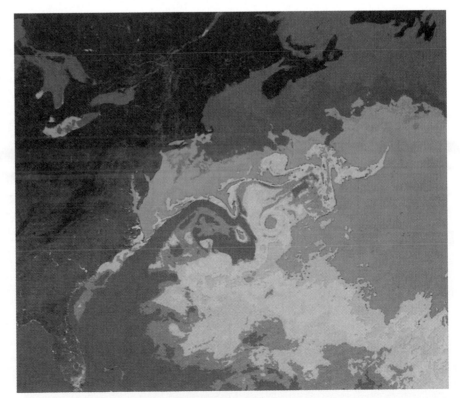

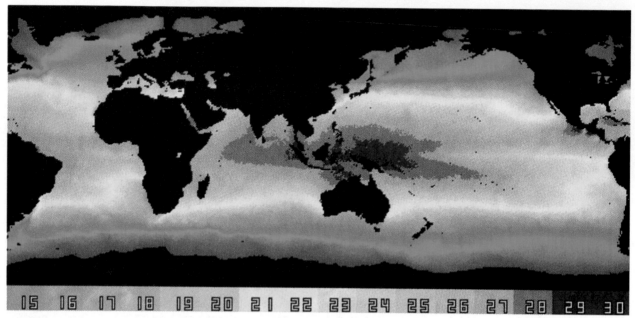

| 15 | 16 | 17 | 18 | 19 | 20 | 21 | 22 | 23 | 24 | 25 | 26 | 27 | 28 | 29 | 30 |

FIGURE 3-13

Average annual sea-surface temperatures for the period 1982–1991 as measured by
TIROS-N and *NOAA* satellites. The Western Pacific Warm Pool, warmest area of all
oceans, is well defined. These remotely sensed data correlate closely with actual
measurements of the ocean's surface temperature. [Satellite image processed by
Dr. Xiao-Hai Yan, Center for Remote Sensing, University of Delaware. Reprinted by
permission.]

the same in its warming effect on Japan and the Aleutians. In contrast, along midlatitude west coasts, cool
ocean currents moderate air temperatures.

Remote sensing from satellites is providing sea-
surface temperature (SST) data that correlate well
with actual sea-surface measurements. This correlation permits a thorough global assessment of SSTs
in programs such as Tropical Ocean Global Atmosphere (TOGA) and Coupled Ocean-Atmosphere Response Experiment (COARE).

Figure 3-13 displays ten years of SST data measured
by several satellites. The red and orange area in the
southwestern Pacific Ocean with temperatures above
28°C (82.4°F), occupying a region larger than the
United States, is called the *Western Pacific Warm Pool*.
Sea-surface temperatures are affected by recurring
weather events and by volcanic eruptions such as El
Chichón in 1982 and Mount Pinatubo in 1991. Understanding the linkage between ocean and atmosphere
systems is important in establishing accurate general
circulation models for forecasting temperatures.

**Summary of Marine vs. Continental Condi-
tions.** Figure 3-11 summarizes the operation of all
the land-water temperature controls: evaporation,
transparency, specific heat, movement, and ocean currents and sea-surface temperatures. The term **marine**,
or maritime, is used to describe locations that exhibit
the moderating influences of the ocean, usually along
coastlines or on islands. **Continentality** refers to the
condition of areas that are less affected by the sea and
therefore have a greater range between maximum
and minimum temperatures diurnally and seasonally.

The cities of San Francisco and Wichita, Kansas,
provide us with a comparison of marine and continental conditions (Figure 3-14). Both cities are at approximately 37° 40′ N latitude and are at 5 m (16.4
ft) and 403 m (1321 ft) in elevation, respectively.
Summer fog plays a role in delaying until September the warmest summer month in San Francisco. In
90 years of weather records there, summer maximum temperatures have risen above 32.2°C (90°F)
an average of only once per year and have dropped

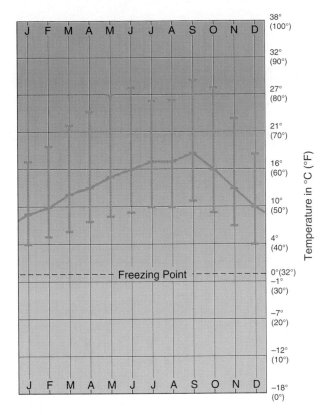

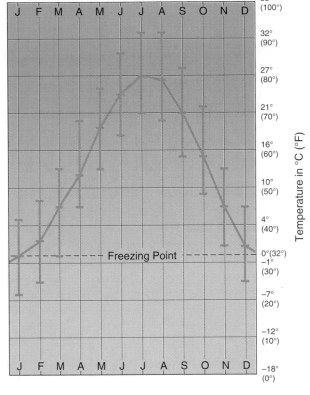

Station: San Francisco, California
Lat/long: 37°37'N, 122°23'W
Avg. Ann. Temp.: 14°C (57.2°F)
Total Ann. Precip.:
   47.5 cm (18.7 in.)

Elevation: 5 m (16.4 ft)
Population: 750,000
Ann. Temp. Range:
   9C° (16.2F°)

Station: Wichita, Kansas
Lat/long: 37°39'N, 97°25'W
Avg. Ann. Temp.: 13.7°C (56.6°F)
Total Ann. Precip.:
   72.2 cm (28.4 in.)

Elevation: 402.6 m (1321ft)
Population: 280,000
Ann. Temp. Range:
   27C° (48.6F°)

FIGURE 3-14
Comparison of temperatures in coastal San
Francisco and continental Wichita, Kansas.

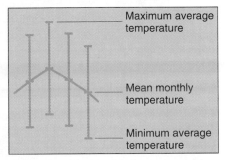

below freezing an average of once per year. In contrast, Wichita is susceptible to freezes from late October to mid-April, with diurnal variations slightly
enhanced by its elevation. Wichita has temperatures
of 32.2°C (90°F) or higher at least 65 days each year.
West of Wichita, winters increase in severity with increasing distance from the moderating influences of
invading air masses from the Gulf of Mexico.

Earth's Temperature Patterns

Both the pattern and variety of temperatures on Earth
result from the global energy system. We covered
the factors that control temperatures. Now let's look
at their combined effect, which is portrayed in
the temperature maps in this section. These show
worldwide mean monthly air temperatures for January

(Figure 3-15) and July (Figure 3-17). These maps, along with Figure 3-18, which shows January–July temperature ranges, also are useful in identifying areas that experience the greatest annual extremes. We use maps for January and July instead of the solstice months (December and June) because, as explained earlier, a lag occurs between insolation received and maximum or minimum temperatures experienced.

The lines on temperature maps are known as isotherms. An **isotherm** connects points of equal temperature and portrays the temperature pattern, just as a contour line on a topographic map portrays points of equal elevation. Geographers are concerned with spatial analysis of temperatures, and isotherms help with this analysis.

With each map, begin by finding your own city or town and noting the temperatures indicated by the isotherms (the small scale of these maps will permit only a general determination). Record the informa-

tion from these maps in your notebook. As you work through the different maps throughout this text, note atmospheric pressure and winds, precipitation, climate, landforms, soil orders, vegetation, and terrestrial biomes. By the end of the course you will have recorded a complete profile of your specific locale.

January Temperature Map

Figure 3-15 is a map of mean January temperatures. In the Southern Hemisphere, the higher Sun altitude means longer days and summer weather conditions, whereas in the Northern Hemisphere, January marks winter's shortened daylength and lower Sun angles. Isotherms generally trend east-west (zonal), parallel to the equator, marking the general decrease in insolation and net radiation with distance from the equator. The **thermal equator**, a line connecting all points of highest mean temperature (the red line on

FIGURE 3-15

Worldwide mean sea level temperatures for January. Temperatures are in Celsius (converted to Fahrenheit by means of the scale) and are equated to sea level to compensate for the effects of landforms.

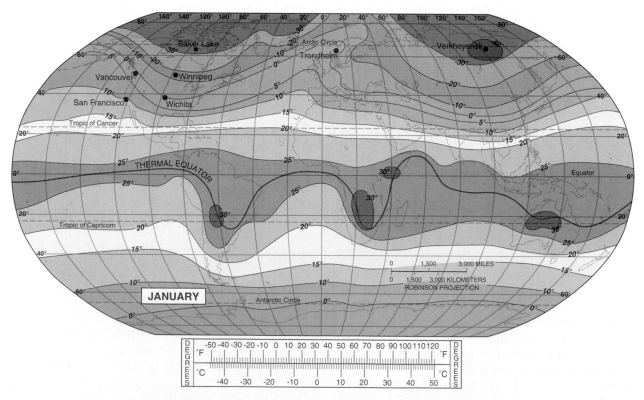

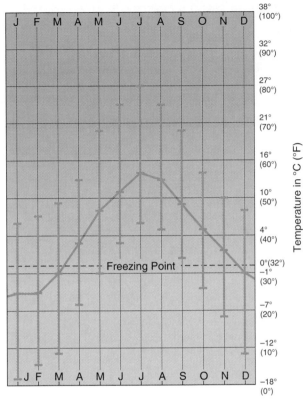

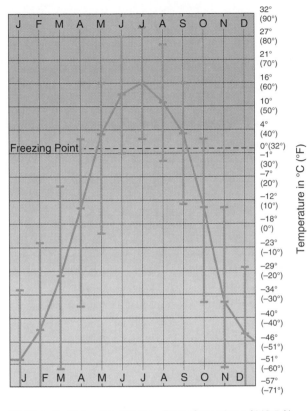

Station: Trondheim, Norway   **Elevation:** 115 m (377.3 ft)
Lat/long: 63°25'N, 10°27'E   **Population:** 135,000
Avg. Ann. Temp.: 5°C (41°F)   **Ann. Temp. Range:**
Total Ann. Precip.:   17C° (30.6F°)
    85.7 cm (33.7 in.)

Station: Verkhoyansk, Russia   **Elevation:** 137 m (449.5 ft)
Lat/long: 67°33'N, 135°23'E   **Population:** 1400
Avg. Ann. Temp.: −15°C (5°F)   **Ann. Temp. Range:**
Total Ann. Precip.:   63C° (113.4F°)
    15.5 cm (6.1 in.)

FIGURE 3-16
Comparison of temperatures in coastal Trondheim, Norway, and continental Verkhoyansk, Russia.

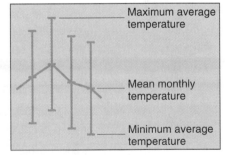

the map), trends southward into the interior of South America and Africa, indicating higher temperatures over landmasses. In the Northern Hemisphere, isotherms trend equatorward as cold air chills the continental interiors. The oceans, on the other hand, appear more moderate, with warmer conditions extending farther north than over land at comparable latitudes.

The coldest area on the map is in northeastern Siberia in Russia. The cold experienced there relates to consistent clear, dry, calm air, small insolation input, and an inland location far from any moderating maritime effects. Verkhoyansk and Oymyakon, Russia, each have experienced a minimum temperature of −68°C (−90°F) and a daily average of −50.5°C (−58.9°F) for January (Figure 3-16). Verkhoyansk experiences at least seven months of temperatures below freezing, including at least four months below −34°C (−30°F)! People do live and work in Verkhoyansk, which has a population of 1400; it has been continuously occupied since 1638 and is today a minor mining district. In contrast, these same towns experience maximum temperatures of +37°C (+98°F) in July.

Trondheim, Norway, is near the latitude of Verkhoyansk and at a similar elevation. However, Trondheim's coastal location moderates its annual temperature regime (Figure 3-16): January minimum and maximum temperatures range between −17° and +8°C (+1.4° and +46°F), and the minimum/maximum range for July is from +5° to +27°C (+41° to +81°F). The most extreme minimum and maximum temperatures ever recorded in Trondheim are −30° and +35°C (−22° and +95°F)—quite a difference from the extremes at Verkhoyansk.

July Temperature Map

Average July temperatures are presented in Figure 3-17. The longer days of summer and higher Sun altitude now are in the Northern Hemisphere. Winter dominates the Southern Hemisphere, although it is milder there because large continental landmasses are absent and oceans and seas dominate. The thermal equator shifts northward with the high summer Sun and reaches the Persian Gulf–Pakistan–Iran area. The Persian Gulf is the site of the highest recorded sea-surface temperature of 36°C (96°F), difficult to imagine for a sea or ocean body.

July is a time of 24-hour-long nights in Antarctica. The lowest natural temperature reported on Earth occurred on July 21, 1983, at the Russian research base at Vostok, Antarctica (78° 27′ S, elevation 3420 m or 11,220 ft): a frigid −89.2°C (−128.6°F). For comparison, such a temperature is 11C° (19.8F°) colder than dry

FIGURE 3-17

Worldwide mean sea level temperatures for July. Temperatures are in Celsius (converted to Fahrenheit by means of the scale) and are equated to sea level to compensate for the effects of landforms. (Compare to Figure 3-15.)

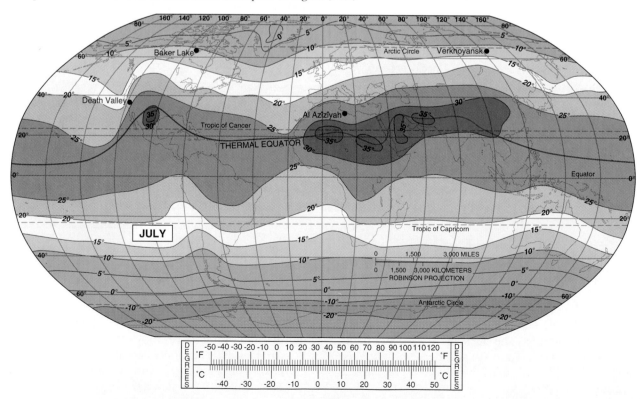

ice (solid carbon dioxide)! During July in the Northern Hemisphere, isotherms trend poleward over land, as higher temperatures dominate continental interiors. July temperatures in Verkhoyansk average more than 13°C (56°F), which represents a 63C° (113F°) seasonal variation between winter and summer averages. The Verkhoyansk region of Siberia is probably the greatest example of continentality on Earth.

The hottest places on Earth occur in Northern Hemisphere deserts during July. These deserts are areas of clear skies and strong surface heating, with virtually no surface water and few plants. Locations such as the Sonoran Desert of North America and the Sahara of Africa are prime examples. Africa has recorded shade temperatures in excess of 58°C (136°F), a record set on September 13, 1922 at Al 'Azīzīyah, Libya (32° 32′ N; 112 m or 367 ft elevation). The highest maximum and annual average temperatures in North America occurred in Death Valley, California, where the Greenland Ranch Station (37° N; –54.3 m or –178 ft below sea level) reached 57°C

(134°F) in 1913. During June 1994 temperatures hit a high of 53.3°C (128°F) in Death Valley. Such arid and hot lands are discussed further in Chapter 12.

Annual Range of Temperatures

The pattern of marine vs. continental influence emerges dramatically when the range in averages is mapped for a full year (Figure 3-18). As you might expect, the largest temperature ranges occur in subpolar locations in North America and Asia, where average ranges of 64C° (115F°) are recorded. The Southern Hemisphere, on the other hand, produces little seasonal variation in mean temperatures, although the landmasses present there do produce some increase in temperature range.

For example, in January (see Figure 3-15) Australia is dominated by isotherms of 20° to 30°C (68° to 86°F), whereas in July (see Figure 3-17) Australia experiences isotherms of 10° to 20°C (50° to 68°F). Generally speaking, Southern Hemisphere patterns

FIGURE 3-18
Annual range of global temperatures in Celsius (Fahrenheit) degrees.

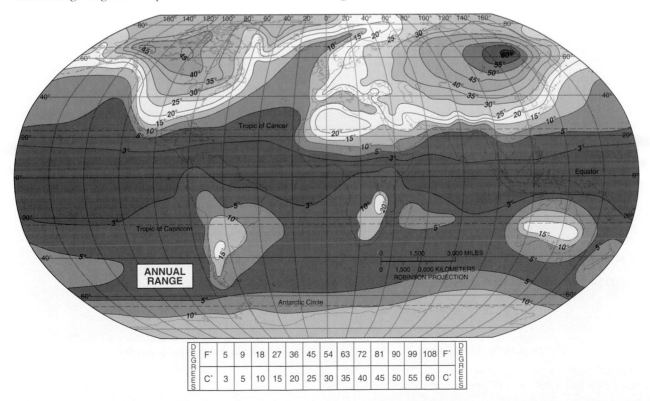

DEGREES	F°	5	9	18	27	36	45	54	63	72	81	90	99	108	F°	DEGREES
	C°	3	5	10	15	20	25	30	35	40	45	50	55	60	C°	

are more marine, whereas Northern Hemisphere patterns are more continental. The Northern Hemisphere, with greater land area overall, registers a slightly higher average surface temperature than does the Southern Hemisphere: 15°C (59°F) as opposed to 12.7°C (55°F).

A logical question to ask about these patterns of world temperatures is what effect they have on the human body. What would you experience if you went to some of these locations?

Air Temperature and the Human Body

We humans are capable of sensing small changes in temperature in the environment around us. Our perception of temperature is described by the terms **apparent temperature**, or sensible temperature. This perception of temperature varies among individuals and cultures. Through complex mechanisms, our bodies maintain an average internal temperature ranging

within a degree of 37°C (98.6°F), slightly lower in the morning or in cold weather and slightly higher at emotional times or during exercise and work.

The water vapor content of air, wind speed, and air temperature taken together affect each individual's sense of comfort. High temperatures, high humidity, and low winds produce the most heat discomfort, whereas low humidity and strong winds enhance cooling rates. Although modern heating and cooling systems can reduce the impact of uncomfortable temperatures indoors, the danger to human life from excessive heat or cold persists outdoors. When changes occur in the surrounding air, the human body reacts in various ways to maintain its core temperature and to protect the brain at all cost.

The wind chill index is important to those who experience winters with freezing temperatures (Figure 3-19). The **wind chill factor** indicates the enhanced rate at which body heat is lost to the air. As wind speeds increase, heat loss from the skin in-

Actual Air Temperatures in °C (°F)

Wind speed, kmph (mph)	10° (50°)	4° (40°)	−1° (30°)	−7° (20°)	−12° (10°)	−18° (0°)	−23° (−10°)	−29° (−20°)	−34° (−30°)	−40° (−40°)	−46° (−50°)	−51° (−60°)
					Temperature Effects on Exposed Flesh							
Calm	10° (50°)	4° (40°)	−1° (30°)	−7° (20°)	−12° (10°)	−18° (0°)	−23° (−10°)	−29° (−20°)	−34° (−30°)	−40° (−40°)	−46° (−50°)	−51° (−60°)
8 (5)	9° (48°)	3° (37°)	−2° (28°)	−9° (16°)	−14° (6°)	−21° (−5°)	−26° (−15°)	−32° (−26°)	−38° (−36°)	−44° (−47°)	−49° (−57°)	−56° (−68°)
16 (10)	4° (40°)	−2° (28°)	−9° (16°)	−16° (4°)	−23° (−9°)	−29° (−21°)	−36° (−33°)	−43° (−46°)	−50° (−58°)	−57° (−70°)	−64° (−83°)	−71° (−95°)
24 (15)	2° (36°)	−6° (22°)	−13° (9°)	−21° (−5°)	−28° (−18°)	−38° (−36°)	−43° (−45°)	−50° (−58°)	−58° (−72°)	−65° (−85°)	−73° (−99°)	−74° (−102°)
32 (20)	0° (32°)	−8° (18°)	−16° (4°)	−23° (−10°)	−32° (−25°)	−39° (−39°)	−47° (−53°)	−55° (−67°)	−63° (−82°)	−71° (−96°)	−79° (−110°)	−87° (−124°)
40 (25)	−1° (30°)	−9° (16°)	−18° (0°)	−26° (−15°)	−34° (−29°)	−42° (−44°)	−51° (−59°)	−59° (−74°)	−64° (−83°)	−76° (−104°)	−81° (−113°)	−92° (−133°)
48 (30)	−2° (28°)	−11° (13°)	−19° (−2°)	−28° (−18°)	−36° (−33°)	−44° (−48°)	−53° (−63°)	−62° (−79°)	−70° (−94°)	−78° (−109°)	−87° (−125°)	−96° (−140°)
56 (35)	−3° (27°)	−12° (11°)	−20° (−4°)	−29° (−20°)	−37° (−35°)	−45° (−49°)	−53° (−64°)	−63° (−82°)	−72° (−98°)	−84° (−119°)	−89° (−129°)	−98° (−145°)
64 (40)	−3° (26°)	−12° (10°)	−21° (−6°)	−29° (−21°)	−38° (−37°)	−47° (−53°)	−56° (−69°)	−65° (−85°)	−74° (−102°)	−82° (−116°)	−91° (−132°)	−100° (−148°)

Low danger with proper clothing   Exposed flesh in danger of freezing

FIGURE 3-19
Wind chill factor for various temperatures and wind velocities.

creases. For example, if the air temperature is –1°C (30°F) and the wind is blowing at 32 kmph (20 mph), skin temperatures will be –16°C (4°F). The lower wind chill values present a serious freezing hazard to exposed flesh.

The *heat index (HI)* indicates the human body's reaction to air temperature and water vapor. The water vapor in air is expressed as relative humidity, a concept presented in Chapter 5. For now, we simply can assume that the amount of water vapor in the air affects the evaporation rate of perspiration on the skin because the more water vapor in the air (the higher the humidity), the less water from perspiration the air can absorb through evap-

oration. The heat index indicates how the air *feels to an average person*—in other words, its apparent temperature.

Figure 3-20 is an abbreviated version of the heat index used by the National Weather Service (NWS) and now included in its daily weather summaries during appropriate months. The table beneath the graph describes the effects of heat-index categories on higher-risk groups. A combination of high temperature and humidity can severely reduce the body's natural ability to regulate internal temperature. There is a distinct possibility that future humans may experience greater temperature-related challenges due to complex changes now underway in the lower atmosphere.

FIGURE 3-20

Heat index graph for various temperatures and relative humidity levels. [Courtesy of National Weather Service.]

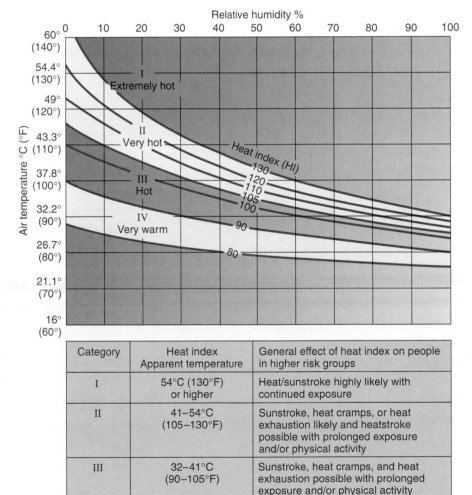

Category	Heat index Apparent temperature	General effect of heat index on people in higher risk groups
I	54°C (130°F) or higher	Heat/sunstroke highly likely with continued exposure
II	41–54°C (105–130°F)	Sunstroke, heat cramps, or heat exhaustion likely and heatstroke possible with prolonged exposure and/or physical activity
III	32–41°C (90–105°F)	Sunstroke, heat cramps, and heat exhaustion possible with prolonged exposure and/or physical activity
IV	27–32°C (80–90°F)	Fatigue possible with prolonged exposure and/or physical activity

The Urban Environment

For most of you reading this book, an urban landscape effects the temperatures you feel each day. Let's conclude this chapter with a brief look at these urban conditions. It is well established that urban microclimates generally differ from those of nearby nonurban areas. In fact, the surface energy characteristics of urban areas possess unique properties similar to desert locations. Because more than 60% of the world's population will be living in cities by the year 2000, the study of **urban heat islands** and other specific environmental effects related to cities is an important topic for physical geographers.

At least six factors contribute to urban microclimates:

1. *Urban surfaces typically are metal, glass, asphalt, concrete, or stone.* These city surfaces conduct up to three times more heat than wet sandy soil. The heat storage capacity of these materials also exceeds that of most natural surfaces making cities *heat islands*. During the day and evening, higher temperatures exist above these surfaces. At night such surfaces rapidly radiate this stored heat to the atmosphere, producing minimum temperatures some 5°–8C° (9°–14F°) warmer in urban areas as compared to rural areas, especially on calm, clear nights. As a result, both maximum and minimum temperatures are higher than in nonurban areas, although the effect of heat-island characteristics is more profound during nighttime cooling.

2. *Urban surfaces behave differently from natural surfaces in energy balance.* Albedo values are lower, leading to a higher net radiation value. However, urban surfaces expend more of that energy as sensible heat than nonurban areas; more than 70% of the net radiation in an urban setting is spent in this way.

3. *The irregular geometric shapes of a modern city affect radiation patterns and winds.* Incoming insolation is caught in mazelike reflection and radiation "canyons," which tend to trap energy for conduction into surface materials, thus increasing temperatures (Figure 3-21). In natural settings, insolation is more readily reflected, stored in plants, or converted into latent heat through evaporation. Buildings tend to interrupt wind flows, thereby diminishing heat loss through advective (horizontal) movement. Winds average 25% lower velocity in cities than in rural areas, although buildings create local

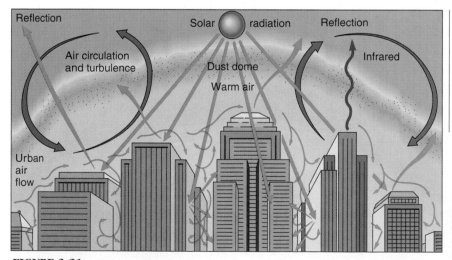

FIGURE 3-21
Insolation, wind movements, and dust dome in city environments. The inset photo shows an all-to-often sight of an urban afternoon. [Photo inset by author.]

turbulence and funneling effects. Wind flows can significantly reduce heat-island effects by increasing turbulent heat loss. Thus, maximum heat-island effects in the most built-up areas of a city occur on calm, clear days and nights.

4. *Human activity alters the heat characteristics of cities.* In New York City, for example, during the summer months, production of electricity and use of fossil fuels release an amount of energy equivalent to 25–50% of the arriving insolation. In the winter, urban-generated heat is on the average 250% greater than arriving insolation! In fact, for many northern cities in North America, winter heating requirements are actually reduced by this enhanced community heat-island effect.

5. *Urban surfaces generally are sealed (built on and paved) so that water cannot reach the soil.* Central business districts are on the average 50% sealed, whereas suburbs are about 20% sealed. Because urban precipitation cannot penetrate the soil, water runoff increases. Urban areas respond much as a desert landscape: a storm may cause a flash flood over the hard, sparsely vegetated surfaces, only to be followed by a return to dry conditions a few hours later. Little of the net radiation is expended for evaporation on such a surface.

6. *The amount of air pollution, including gases and aerosols, is greater in urban areas than in comparable natural settings.* Pollution actually increases reflection (albedo) in the atmosphere above the city, thus reducing insolation reaching the ground. The pollution blanket also absorbs infrared radiation, radiating the heat downward. Every major city produces its own **dust dome** of airborne pollution, which can be blown from the city in elongated plumes, depending on wind direction and speed (Figure 3-21). The increased particulates present in pollution act as condensation nuclei for water vapor. Convection created by the heat island lifts the air and particulates, producing increased cloud formation and the potential for increased precipitation. Although these precipitation effects of the urban heat island are difficult to isolate and prove, research suggests that urban-stimulated precipitation occurs downwind from cities (Figure 3-22).

FIGURE 3-22
Urbanized areas, such as Chicago, Illinois, possess unique physical attributes different from surrounding natural areas. Studies in the 1970s established the effects of the Chicago metropolitan region on weather downwind from the city. [Photo by author.]

Table 3-1

Average Differences in Climatic Elements Between
Urban and Rural Environments

Element	Urban Compared to Rural Environs
Contaminants	
Condensation nuclei	10 times more
Particulates	10 times more
Radiation	
Total on horizontal surface	0–20% less
Sunshine duration	5–15% less
Cloudiness	
Clouds	5–10% more
Precipitation	
Amounts	5–15% more
Snowfall, downwind (lee) of city	10% more
Thunderstorms	10–15% more
Temperature	
Annual mean	0.5–3.0C° (0.9–5.4F°) warmer
Winter minima (average)	1.0–2C° (1.8–3.6F°) warmer
Summer maxima	1.0–3C° (1.8–3.0F°) warmer

SOURCE: Helmut E. Landsberg, 1981, *The Urban Climate,* Vol. 28. International Geophysics Series, p. 258. Reprinted by permission from Academic Press.

Table 3-1 compares climatic factors of rural and urban environments.

SUMMARY—Atmospheric Energy and Global Temperatures

Earth's biosphere is powered by radiant energy from the Sun that cascades through complex circuits to the surface. This energy links Earth's hydrosphere, atmosphere, and lithosphere together in an intricate system that supports the biosphere—all life. Satellite measurements are a valuable contribution to understanding our planet's energy system.

Physical geographers use spatial models that effectively budget energy inputs, pathways, and outputs in the troposphere, on the ground, and within the ocean. Approximately 31% of all insolation is **reflection** to space without performing any work. This value represents Earth's average **albedo**. Insolation is partially absorbed by atmospheric clouds, gases, and dust, with the remaining energy eventually reaching land and water surfaces. This **absorption** of shortwave energy is transformed by Earth and emitted as longwave infrared radiation back to space. Carbon dioxide, water vapor, and other gases, absorb heat and delay its loss to space producing a **greenhouse effect**. The input of insolation and outputs of reflected light and emitted heat from our atmosphere and surface environments determine the energy available to perform work.

Surface energy balances are used to summarize the energy patterns for various locations. Surface energy measurements are used as analytical tools of **microclimatology**, an important subdiscipline of physical geography. A natural Earth-atmosphere energy balance occurs through a combination of nonradiative energy transfers from the surface as latent heat of evaporation, convection, and conduction, and radiative transfers of infrared radiation. Surface energy balances produce a **net radiation (NET R)** value. NET R is expended as sensible heat, latent heat, ground heating, and conversion of heat energy to biochemical energy through photosynthesis.

Probably the first thing you notice when you step outside is the air temperature. The principal factors that operate to produce the pattern of world temperatures include latitude, altitude, cloud cover, and the numerous physical properties that create **land-water heating differences** in heating and cooling rates.

The pattern of present Earth temperatures is best portrayed on **isotherm** maps, which permit geographic analysis. The annual range of mean temperatures between January and July demonstrates the interaction of the principal temperature controls. The large continental regions experience a greater annual range in temperature, or **continentality**, as compared with areas nearer a coastline, where conditions are more **marine**.

As humans, we need to be aware of our sensitivity to temperature differences and the physiological responses produced in our bodies by exposure to **apparent temperature** extremes. The *heat-index* and **wind-chill** charts are two useful indicators of apparent temperature.

A growing percentage of Earth's human inhabitants live in cities. These cities possess a unique set of altered microclimatic effects, distinct from natural landscapes: increased conduction, lower albedos, higher NET R values, increased moisture runoff, complex radiation and reflection patterns, anthropogenic heating, and the gases, dusts, and aerosols that form a **dust dome**. A cityscape produces an **urban heat island**, where temperatures are warmer compared to surrounding, cooler, rural areas.

Sunlight is direct, pervasive, and renewable and is a viable energy resource for society through known technologies. With development, solar energy applications can provide an inexhaustible, nonpolluting source of heat. Yet we find sunlight an important resource waiting for development, for reasons of poor leadership and politics.

KEY TERMS

absorption	marine
advection	microclimatology
albedo	net radiation (NET R)
apparent temperature	reflection
conduction	refraction
continentality	scattering
convection	specific heat
diffuse radiation	temperature
dust dome	thermal equator
greenhouse effect	transparency
Gulf Stream	urban heat island
isotherm	wind chill factor
land-water heating differences	

REVIEW QUESTIONS

1. Diagram a simple energy balance for the troposphere. Label each shortwave and longwave component and the directional aspects of related flows.
2. List several types of surfaces and their albedos. Explain the differences. What determines the reflectivity of a surface?
3. Why is the lower atmosphere blue? What would you expect the sky color to be at 50 km (30 mi) altitude? Why?
4. Define conduction, convection, absorption, transmission, and diffuse radiation.
5. What are the similarities and differences between an actual greenhouse and the gaseous atmospheric greenhouse? Why is Earth's greenhouse changing?
6. Why is there a temperature lag between the highest Sun altitude and the warmest time of day? Relate your answer to the insolation and temperature curves.
7. In terms of surface energy balance, explain the term *net radiation*.

8. What are the expenditure pathways for surface net radiation? What kind of work is accomplished?

9. What does air temperature indicate about energy in the atmosphere?

10. Explain the effect of altitude upon air temperature.

11. Describe the effect of cloud cover with regard to Earth's temperature patterns.

12. List the physical aspects of land and water that produce their different responses to heating. Differentiate between marine and continental temperatures. Give geographical examples of each from the text.

13. Explain the extreme temperature range experienced in north-central Siberia between January and July.

14. From the maps in Figures 3-15, 3-17, and 3-18, determine the average temperature values and annual range of temperatures for your location.

15. Describe the interaction between air temperature and wind speed and their affect on skin chilling. Select several temperatures and wind speeds from Figure 3-19 and determine the wind chill factor.

16. What is the basis for the urban heat-island concept? Describe the climatic effects attributable to urbanization as compared to nonurban environments.

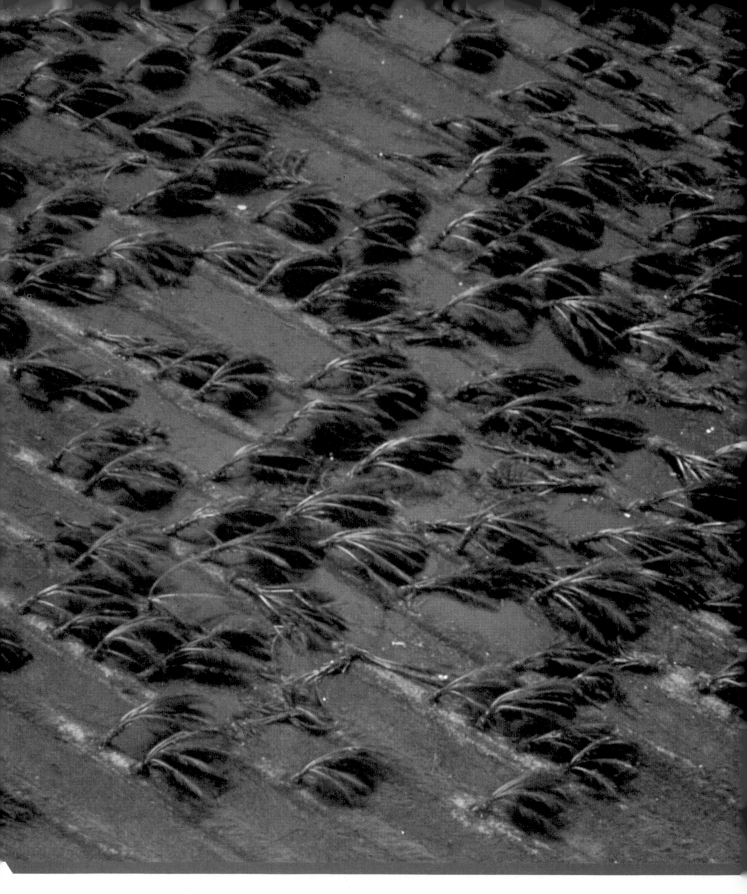

The powerful winds of Hurricane Andrew blew down thousands of trees in southern Florida. [Photo by John Lopinot/Palm Beach Post/Sygma.]

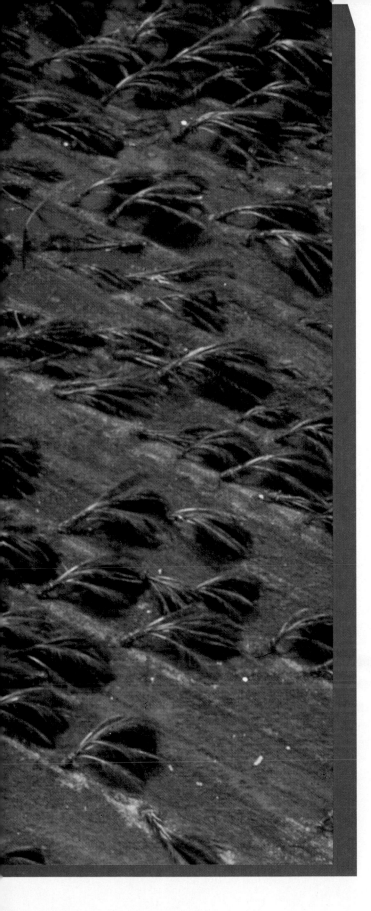

4

ATMOSPHERIC AND OCEANIC CIRCULATION

WIND ESSENTIALS
  Wind: Description and Measurement
  Global Winds
  Air Pressure

DRIVING FORCES WITHIN THE ATMOSPHERE
  Pressure Gradient Force
  Coriolis Force
  Friction Force

ATMOSPHERIC PATTERNS OF MOTION
  Primary High-Pressure and Low-Pressure Areas
  Upper Atmospheric Circulation
  Local Winds

OCEANIC CURRENTS
  Surface Currents
  Deep Currents

SUMMARY

FYI REPORT 4-1 WIND POWER: AN ENERGY RESOURCE

Early in April 1815, on an island named Sumbawa in present-day Indonesia, the volcano Tambora erupted violently, releasing an estimated 150 km³ (36 mi³) of material, 25 times the volume produced by the 1980 Mount Saint Helens eruption in Washington State. Some materials from Tambora—the ash and acid mists—were carried worldwide by global atmospheric circulation, creating a stratospheric dust veil. The result was both a higher atmospheric albedo and absorption of energy by the particulate materials injected into the stratosphere. Remarkable optical and meteorological phenomena resulting from the spreading dust were noted worldwide for months and years after the eruption.

Scientists in 1815 lacked the remote-sensing capability of satellite technology, and had no way of knowing the global impact of Tambora's eruption. Today, technology permits a depth of analysis unknown in the past—satellites now track the atmospheric effects from dust storms, forest fires, industrial haze, warfare aftereffects, and the dispersal of volcanic explosions, among many things.

After 635 years of dormancy, Mount Pinatubo in the Philippines erupted on June 15, 1991, an event of tremendous atmospheric impact. This volcanic explosion lofted 15–20 million tons of ash, dust, and SO_2 into the atmosphere. (The volume of material blasted from Tambora was about eight times greater than that from Pinatubo.) As the SO_2 rose into the stratosphere, it quickly formed sulfuric acid (H_2SO_4), which became concentrated at 16—25 km (10–15.5 mi) altitude. The increase in atmospheric albedo (about 1.3%) produced by these aerosols is related to the *aerosol optical thickness* (AOT) of the atmosphere. Thus, albedo measurements gave scientists an estimate of the overall aerosol load generated by Mount Pinatubo and a unique insight into the dynamics of atmospheric circulation.

The AVHRR (Advanced Very High Resolution Radiometer) instrument aboard *NOAA-11* monitored the reflected solar radiation from these aerosols as they were swept around Earth by global winds. Figure 4-1 shows combined images made at two-week intervals that clearly track this handiwork of our atmospheric circulation. The earliest image (a), near the time of the eruption, shows winds blowing dust westward from Africa, smoke from the Kuwaiti oil well fires of the Persian Gulf War, and some haze off the east coast of the United States. Also visible is the Pinatubo aerosol layer beginning to emerge north of Indonesia.

Atmospheric winds spread the debris worldwide in just 21 days—(a) to (c)—dusting skies in a band from 15° S to 25° N in width. By the last image in the sequence (d), some 60 days after the eruption, the aerosol cloud spanned from 20° S to 30° N and covered about 42% of the globe. Most of the world's population experienced spectacular sunrises and sunsets and a lowering of average temperatures during the two years that followed.

In this chapter we examine the dynamic circulation of Earth's atmosphere that carried Tambora's debris and Mount Pinatubo's acid aerosols worldwide, and that carries the common ingredients oxygen, carbon dioxide, and water vapor around the globe. We also consider Earth's wind-driven oceanic currents. The energy driving all this movement comes from one source: the Sun.

Wind Essentials

Earth's atmospheric circulation is an important transfer mechanism for both energy and mass. In the process, the imbalance between equatorial energy surpluses and polar energy deficits is partly resolved, Earth's weather patterns are generated, and ocean currents are produced. Human-caused pollution also is spread by this circulation, far from its points of origin. The fluid movement of the atmosphere socializes humanity more than any other natural or cultural factor. Our atmosphere makes all the world a spatially linked society—one person's or nation's exhalation is another's breath intake.

Atmospheric circulation is generally categorized at three levels: *primary* (general) *circulation, secondary circulation* of migratory high-pressure and

FIGURE 4-1

These false-color images show aerosols from Mount Pinatubo, smoke from fires, and dust storms, all swept about the globe by the general atmospheric circulation. The dramatic increase in aerosols from mid-June 1991 (a) to mid-August (d) is due to the Mount Pinatubo eruption. See the text for details. (The false-color denotes the atmosphere's *aerosol optical thickness*, or AOT. Lowest AOT values are dull yellow, medium values are bright yellow, and highest values are white.) Image (e) is a satellite image of the volcano as it erupted. [(a-d) AVHRR satellite images of aerosols courtesy of Dr. Larry L. Stowe, National Environmental Satellite, Data, and Information Service, National Oceanic and Atmospheric Administration. Used by permission. (e) AVHRR satellite eruption image courtesy of EROS Data Center.]

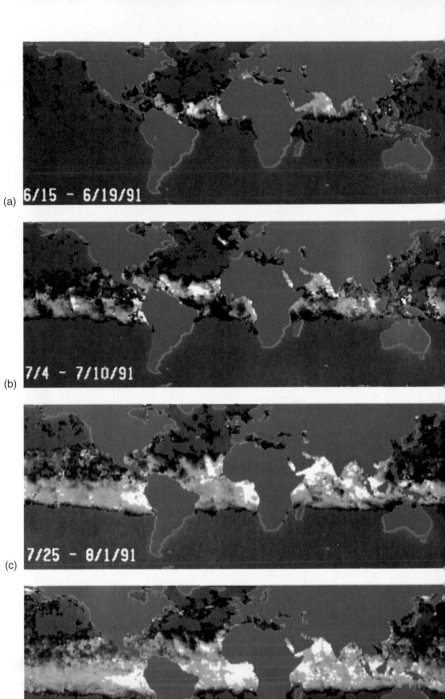

low-pressure systems, and *tertiary circulation* that includes local winds and temporal weather patterns. Winds that move principally north or south along meridians are known as *meridional flows*. Winds moving east or west along parallels of latitude are called *zonal flows*.

Wind: Description and Measurement

Simply stated, **wind** is the horizontal motion of air across Earth's surface. *It is produced by differences in air pressure from one location to another.* Wind's two principal properties are speed and direction, and instruments are used to measure each. Wind speed is measured with an **anemometer** and is expressed in kilometers per hour (kmph), miles per hour (mph), meters per second (m/sec), or knots. (A *knot* is a nautical mile per hour, equivalent to 1.85 kmph or 1.15 mph.) Wind direction is determined with a **wind vane**; the standard measurement is taken 10 m (33 ft) above the ground to avoid, as much as possible, local effects of topography upon wind direction (Figure 4-2).

Winds are named for the direction *from which they originate*. For example, a wind from the west is a *westerly* wind (it blows eastward); a wind out of the south is a *southerly* wind (it blows northward). Figure 4-3 illustrates a simple wind compass, naming the 16 principal wind directions.

Global Winds

The primary circulation of winds across Earth has fascinated travelers, sailors, and scientists for centuries, although only in the modern era is a true picture emerging of the pattern and cause of global winds. With the assistance of scientific breakthroughs in space-based observations and earthbound computer technology, models of total atmospheric circulation are becoming more refined.

A NASA *Seasat* satellite image was painstakingly assembled by scientists at the Jet Propulsion Laboratory and the University of California–Los Angeles from 150,000 measurements made during a single day (Figure 4-4). Wind drives waves on the ocean surface that combine to form patterns of currents. *Seasat* used radar to measure the motion and direction of waves from their backscatter to the satellite, producing this

FIGURE 4-2
Instruments used to measure wind direction (wind vane) and wind speed (anemometer) at a weather station installation. [Photo by Belfort Instruments.]

FIGURE 4-3
A wind compass.

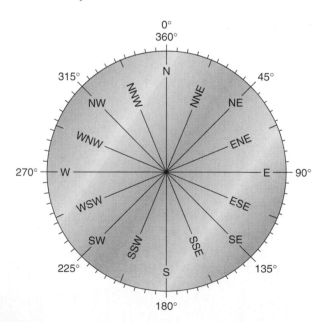

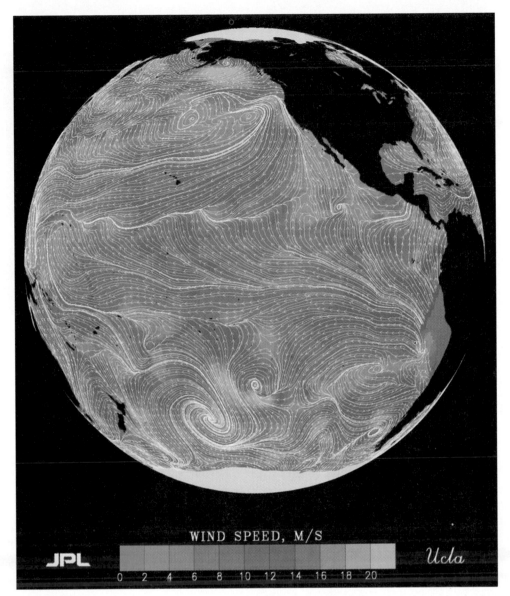

FIGURE 4-4

Surface wind measured by radar scatterometer aboard the *Seasat* satellite on September 14, 1978. Scientists analyzed 150,000 measurements to produce this image. Colors correlate to wind speeds, and the white arrows note wind direction. [Image courtesy of Dr. Peter Woiceshyn, Jet Propulsion Laboratory, Pasadena.]

portrait of surface wind fields over the Pacific Ocean. The patterns you see are the response to specific forces at work in the atmosphere, and these forces are our next topic. As we progress through this chapter, you may want to refer to this *Seasat* image to identify the winds, eddies, and vortexes it portrays.

Air Pressure

Important to an understanding of wind is the concept of air pressure, its measurement and expression. The gases that comprise air create pressure through their motion, size, and number. This pres-

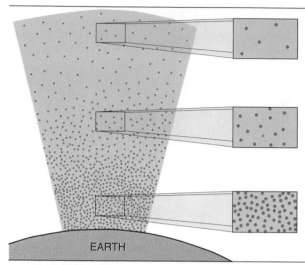

FIGURE 4-5

The atmosphere, denser near Earth's surface, rapidly decreases in density with altitude. This difference in density is easily measured because air exerts its weight as pressure.

sure is exerted on all surfaces in contact with the air. The weight of the atmosphere, or **air pressure**, crushes in on all of us. Fortunately, that same pressure also exists inside us, pushing outward; otherwise we would be crushed by the mass of air around us.

The atmosphere exerts an average force of approximately 1 kg/cm² (14.7 lb/in.²) at sea level. Under the influence of gravity, air is compressed and therefore denser near Earth's surface (Figure 4-5). It rapidly thins with increased altitude, a decrease that is measurable because air exerts its weight as pressure. Consequently, over half the total mass of the atmosphere is compressed below 5500 m (18,000 ft), 75% is compressed below 10,700 m (35,100 ft), and 90% is below the tropopause (16,000 m, 52,500 ft). All but 0.1% is accounted for at an altitude of 50 km (31 mi), as shown in the pressure profile (orange line) in Figure 4-6.

In A.D. 1643, a pupil of Galileo's named Evangelista Torricelli was working on a mine drainage problem. This led him to discover a method for measuring air pressure (Figure 4-7a). He knew that pumps in the mine were able to draw water upward only about 10 m (33 ft), but did not know why. Careful observation led him to discover that this lim-

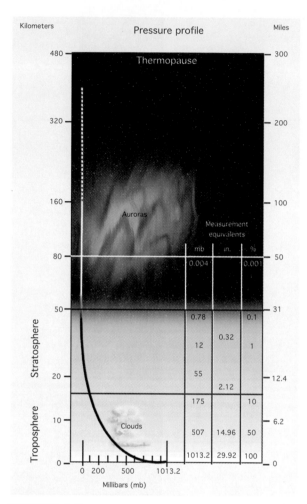

FIGURE 4-6

Pressure profile of the atmosphere. Note that about 90% of the atmospheric mass resides in the troposphere. Clouds in the troposphere and the auroras in the thermosphere are illustrated to help you relate this profile to the discussion of the atmosphere in Chapter 2.

itation was not caused by the pumps but by the atmosphere itself. Torricelli noted that the water level in the vertical pipe fluctuated from day to day. He surmised correctly that air pressure varied over time and with changing weather conditions. A pump of the type shown works by creating a vacuum above the water, which allows the pressure of the atmosphere that is pushing down on water in the mine to force water up the pipe.

To simulate the problem at the mine, Torricelli devised an instrument at the suggestion of Galileo.

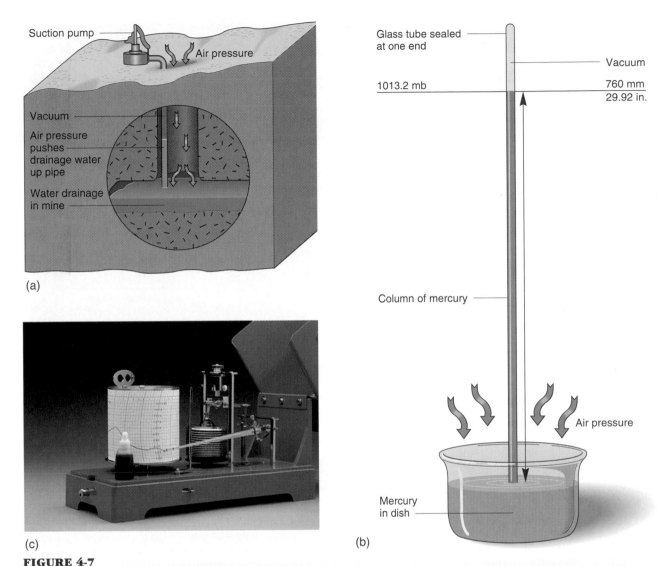

FIGURE 4-7

Evangelista Torricelli developed the barometer to measure air pressure as a by-product of trying to solve a mine drainage problem (a). Instruments used to measure atmospheric pressure include: (b) a mercury barometer and (c) an aneroid barometer. [(c) Courtesy of Qualimetrics, Inc., Sacramento, CA.]

Instead of water, he used a much denser fluid, mercury (Hg), so that he could use a glass laboratory tube only 1 m high. He sealed the glass tube at one end, filled it with mercury, and then inverted it into a dish of mercury (Figure 4-7b). Torricelli established that the average height of the column of mercury in the tube was 760 mm (29.92 in.). He concluded that the column of mercury was counterbalanced by the mass of surrounding air exerting an equivalent pressure on the mercury in the

dish. Thus, the 10 m (33 ft) limit to which the mine pumps could draw water was controlled by the mass of the atmosphere. Using similar instruments today, normal sea level pressure is expressed as 1013.2 millibars of mercury (a way of expressing force per square meter of surface area), or 29.92 in. of mercury, or 101.32 kPa (kilopascal; 1 kilopascal = 10 millibars).

Any instrument that measures air pressure is called a barometer. Torricelli developed a **mercury**

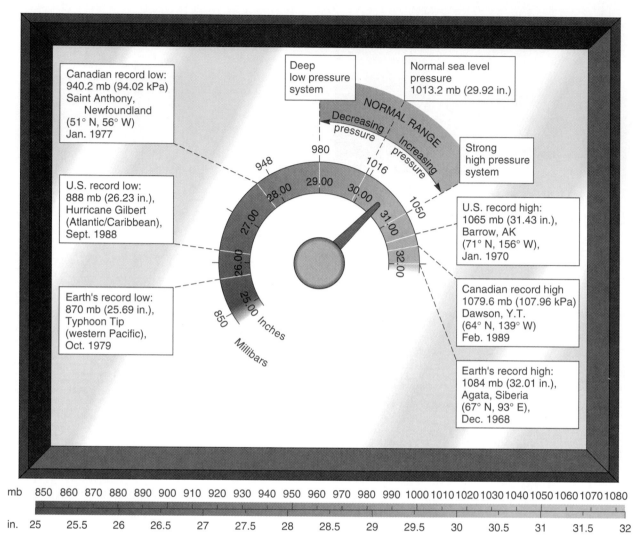

FIGURE 4-8
Scales for expressing barometric air pressure in millibars and inches, with average
air pressure values and recorded pressure extremes. Canadian values include
kilopascal equivalents (10 mb = 1 kPa).

barometer. A more compact design that works
without a meter-long tube of mercury is called an
aneroid barometer (Figure 4-7c). Inside this type
of barometer, a small chamber is partially emptied
of air, sealed, and connected to a mechanism that is
sensitive to changes in the chamber. As air pressure
varies, the mechanism responds to the difference
and moves the needle on the dial. Altimeters used
in aircraft are a type of aneroid barometer that is able
to accurately measure altitude because air pressure

diminishes with elevation above sea level.

Figure 4-8 illustrates comparative scales in millibars
and inches used to express air pressure and its relative
force. Strong high pressure to deep low pressure can
range from 1050 to 980 mb. The figure also indicates
the extreme highest and lowest pressures ever recorded in the United States, Canada, and on Earth. The concept of air pressure, its measurement, and its application
to atmospheric circulation and weather are further explained in the following section and in Chapter 5.

Driving Forces Within the Atmosphere

Several forces determine both speed and direction of winds:

- Earth's *gravitational force* on the atmosphere is practically uniform, equally compressing the atmosphere near the ground worldwide. Density decreases as altitude increases.
- The **pressure gradient force** drives air from areas of higher barometric pressure to areas of lower barometric pressure, causing winds.
- The **Coriolis force**, a deflective force, appears to deflect wind from a straight path in relation to Earth's rotating surface—to the right in the Northern Hemisphere and to the left in the Southern Hemisphere.
- The **friction force** drags on the wind as it moves across surfaces; it decreases with height above the surface.

All four of these forces operate on moving air and water and affect the circulation patterns of global winds. The following sections describe the actions of the pressure gradient, Coriolis, and friction forces.

Pressure Gradient Force

High- and low-pressure areas that exist in the atmosphere principally result from unequal heating at Earth's surface and from certain dynamic forces in the atmosphere. A *pressure gradient* is the difference in atmospheric pressure between areas of higher pressure and lower pressure. Lines plotted on a weather map connecting points of equal pressure are called **isobars**; the distance between isobars indicates the degree of pressure difference. Isobars on a weather map are used to make a spatial analysis of pressure patterns (Figure 4-9).

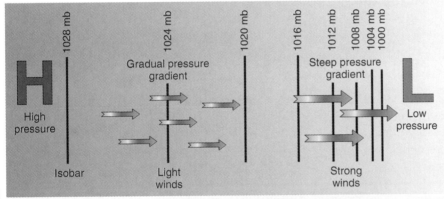

(a)

(b)

FIGURE 4-9

Pressure gradient (a). On a weather map (b) the closer spacing of isobars represents a steeper pressure gradient that produces strong winds; wider spacing of isobars denotes a gradual pressure gradient that leads to light winds.

Just as closer contour lines on a topographic map mark a steeper slope, so do closer isobars denote steepness in the pressure gradient. In Figure 4-9a, note the spacing of the isobars (purple lines). A steep gradient causes faster air movement from a high-pressure area to a low-pressure area. Isobars spaced at greater distances from one another mark a more gradual pressure gradient, one that creates a slower air flow. Along a horizontal surface, the pressure gradient force acts at right angles to the isobars. Note the location of steep and gradual pressure gradients and their relationship to wind intensity on the map in Figure 4-9b.

Figure 4-10 illustrates the combined effect of the forces that direct the wind. The pressure gradient force acting alone is shown in Figure 4-10a from two perspectives. As air descends in the high-pressure area, a field of subsiding, or sinking, air develops. Air moves out of the high-pressure area in a flow described as diverging. High-pressure areas feature *descending, diverging* air flows. On the other hand, air moving into a low-pressure area does so with a converging flow. Thus, low-pressure areas feature *converging, ascending* air flows.

Coriolis Force

You might expect surface winds to move in a straight line from areas of higher pressure to areas of lower pressure, but they do not. The reason for this is the Coriolis force. The *Coriolis force* deflects from a straight path any object that flies or flows across Earth's surface, be it the wind, a plane, or ocean currents. The deflection produced by the Coriolis force is caused by Earth's rotation.

Earth's rotational speed varies with latitude, increasing from 0 kmph at the poles to 1675 kmph (1041 mph) at the equator (Figure 4-11b). Because Earth rotates eastward, objects that move in an absolute straight line over a distance (such as winds and ocean currents) appear to curve to the right in the Northern Hemisphere and to the left in the Southern Hemisphere. The effect of the Coriolis force increases as the speed of the moving object increases; thus, the higher the wind speed, the greater its apparent deflection.

Viewed from an object passing over Earth's surface, the surface rotates beneath the object. But viewed from the surface, the object appears to curve off-course. The object does not actually deviate from a straight path, but it *appears* to because of the rotational movement of Earth's surface beneath it (note the curved path in Figure 4-11b). The deflection occurs regardless of the direction in which the object is moving. The Coriolis force is zero along the equator, increasing to half the maximum deflection at 30° N and S latitudes, and to maximum deflection flowing away from the poles (Figure 4-11a).

As an example of the effect of this force, a pilot leaves the North Pole and flies due south toward Quito, Ecuador, traveling a direct route (Figure 4-11b). If Earth were not rotating, the aircraft would simply travel along a meridian of longitude and arrive at Quito. However, Earth is rotating beneath the aircraft's flight path. If the pilot does not allow for this rotation, the plane will reach the equator over the ocean far to the west of the intended destination. Pilots must correct for this Coriolis deflection in their navigation calculations.

How does the Coriolis force affect wind? As air rises from the surface through the lowest levels of the atmosphere, leaving the drag of surface friction behind, its speed increases, thus increasing the Coriolis force, spiraling the winds to the right in the Northern Hemisphere or to the left in the Southern Hemisphere. In the upper troposphere the Coriolis force just balances the pressure gradient force. Consequently, the winds between high-pressure and low-pressure areas aloft flow parallel to the isobars.

Figure 4-10b illustrates the combined effect of the pressure gradient force and the Coriolis force on air currents aloft, producing **geostrophic winds**. Geostrophic winds are characteristic of upper tropospheric circulation. The air does not flow directly from high to low, but *around* the pressure areas instead, remaining parallel to the isobars and producing the characteristic pattern shown on the upper-air weather chart in Figure 4-12. Note the inset illustration showing the effects of the pressure gradient and Coriolis forces that produce a geostrophic flow of air.

Friction Force

The effect of surface friction extends to a height of about 500 m (1650 ft) and varies with surface texture, wind speed, time of day and year, and atmospheric conditions. Generally, rougher surfaces

FIGURE 4-10

Three physical forces that integrate to produce wind patterns at the surface and aloft: (a) the pressure gradient force, (b) the Coriolis force added to the effects of the pressure gradient force produces a geostrophic wind flow in the upper atmosphere, and (c) the friction force considered together with the other forces produces characteristic surface winds. Note the reverse circulation pattern in the Southern Hemisphere. (The gravitational force is assumed.)

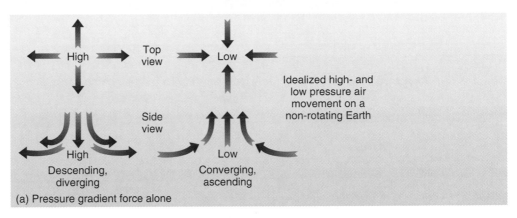

Idealized high- and low pressure air movement on a non-rotating Earth

(a) Pressure gradient force alone

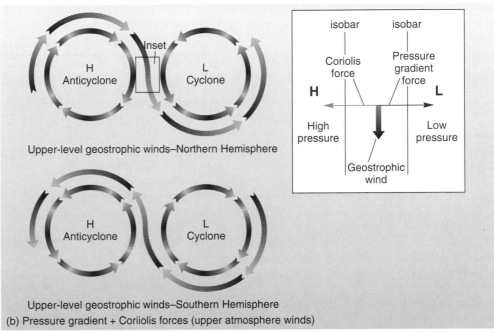

(b) Pressure gradient + Coriiolis forces (upper atmosphere winds)

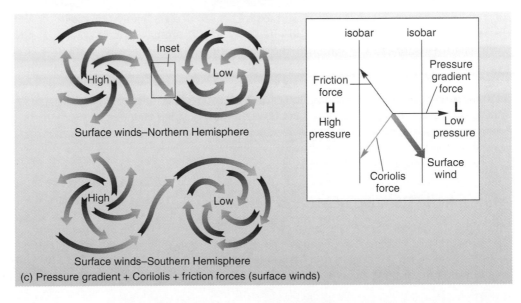

(c) Pressure gradient + Coriiolis + friction forces (surface winds)

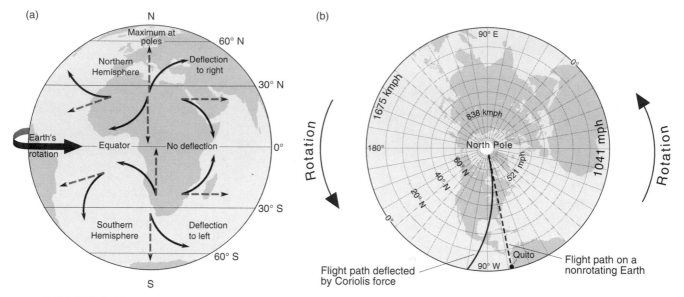

FIGURE 4-11

Distribution of the Coriolis force on Earth: (a) apparent deflection to the right of a straight line in the Northern Hemisphere; apparent deflection to the left in the Southern Hemisphere. (b) Coriolis deflection of a flight path between the North Pole and Quito, Ecuador. Note the latitudinal variations in speed of rotation.

produce more friction. Figure 4-10c adds the effect of friction to the Coriolis and pressure gradient forces on wind movements.

Near the surface, friction disrupts the equilibrium established in geostrophic wind flows between the pressure gradient and Coriolis forces—note the inset illustration in Figure 4-10c. Because surface friction decreases wind speed, it reduces the effect of the Coriolis force and causes winds in the Northern Hemisphere to move across isobars at an angle, spiraling out from a high-pressure area clockwise to form an **anticyclone** and spiraling into a low-pressure area counterclockwise to form a **cyclone**. In the Southern Hemisphere these circulation patterns are reversed, flowing out of high-pressure cells counterclockwise and into low-pressure cells clockwise.

Atmospheric Patterns of Motion

With these forces and motions in mind, we are ready to understand the *Seasat* image of winds (see Figure 4-4) and build a general model of total atmospheric circulation. If Earth did not rotate, then the warmer, less

dense air of the equatorial regions would rise, creating low pressure at the surface, and the colder and denser air of the poles would sink, creating high pressure at the surface. The net effect would be a meridional flow of winds established by this simple pressure gradient. But in reality, because Earth does rotate, a much more complex system exists for the transfer of energy and air masses from equatorial surpluses to polar deficits— one with waves, streams, and eddies on a planetary scale. And instead of this hypothetical meridional flow, the flow is predominantly zonal (latitudinal) at the surface and aloft: westerly (eastward) in the middle and high latitudes and easterly (westward) in the low latitudes of both hemispheres.

Primary High-Pressure and Low-Pressure Areas

The following discussion of Earth's pressure and wind patterns refers often to Figure 4-13, which shows January and July isobaric maps of average surface barometric pressure. These maps indicate

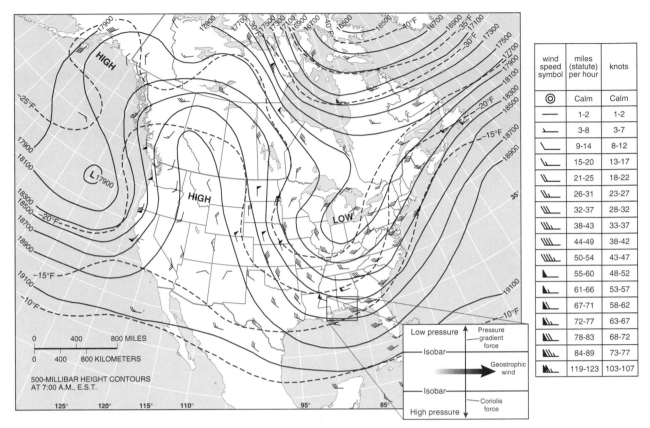

The following table appears within the figure:

wind speed symbol	miles (statute) per hour	knots
◎	Calm	Calm
—	1-2	1-2
⌐	3-8	3-7
⌐	9-14	8-12
⌐	15-20	13-17
⌐	21-25	18-22
⌐	26-31	23-27
⌐	32-37	28-32
⌐	38-43	33-37
⌐	44-49	38-42
⌐	50-54	43-47
⌐	55-60	48-52
⌐	61-66	53-57
⌐	67-71	58-62
⌐	72-77	63-67
⌐	78-83	68-72
⌐	84-89	73-77
⌐	119-123	103-107

FIGURE 4-12

Isobaric chart for April 25, 1977, 7:00 A.M. EST. Contours show height at which 500 mb pressure occurs (in feet). The pattern of contours reveals geostrophic wind patterns in the troposphere at approximately 5500 m (18,000 ft) altitude. Note the "ridge" of high pressure over the Intermountain West and the "trough" of low

prevailing surface winds, suggested by the isobars. The primary high- and low-pressure areas on Earth appear on these isobaric maps as interrupted cells or uneven belts of similar pressure that stretch across the face of the planet. Between these areas flow the primary winds, which have been noted in adventure and myth throughout human experience.

Secondary highs and lows, from a few hundred to a few thousand kilometers in diameter and hundreds to thousands of meters high, are formed within the primary pressure areas. These pressure systems seasonally migrate to produce changing weather and climate patterns in the regions over which they pass.

Four identifiable pressure areas cover the North-ern Hemisphere; a similar set is found in the Southern Hemisphere. In each hemisphere, two of the pressure areas are specifically stimulated by thermal (temperature) factors: the **equatorial low-pressure trough** and the weak **polar high-pressure cells**, both north and south. The other two areas are formed by dynamic factors: the **subtropical high-pressure cells** and **subpolar low-pressure cells**. Table 4-1 summarizes the characteristics of these pressure areas, and we now examine each.

Equatorial Low-Pressure Trough. The equatorial low-pressure trough is an elongated, narrow band of low pressure that nearly girdles Earth, following an undulating linear axis. Constant high

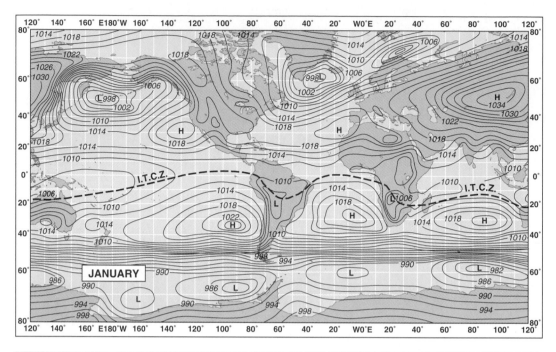

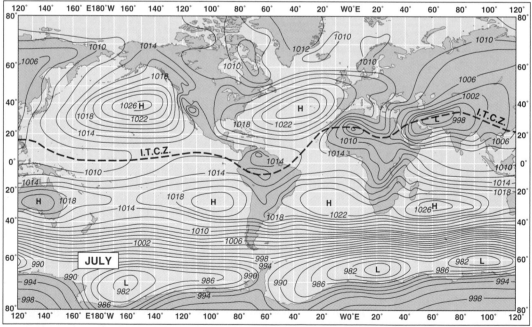

FIGURE 4–13

Average surface barometric pressure (millibars) for January and July. Dashed line marks the intertropical convergence zone (ITCZ). [Adapted from National Climatic Data Center, *Monthly Climatic Data* for the World, 46, no. 1, January and July 1993. Prepared in cooperation with the World Meteorological Organization. Washington, DC: National Oceanic and Atmospheric Administration.]

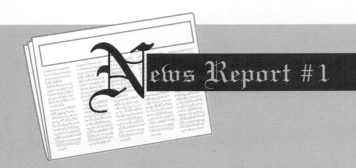

News Report #1

"Coriolis, a Forceful Effect on Drains?"

A common misconception about the Coriolis force is that it affects water draining out of a sink, tub, or toilet in a noticeable way. Does the Coriolis force cause this twist? When a ship crosses the equator, does the direction of a draining spiral of water in a sink suddenly reverse?

Moving water or air must cover some distance across space and time before they are noticeably deflected by the Coriolis force. Long-range artillery shells and guided missiles do exhibit small amounts of deflection that must be corrected for accuracy. So, such small water movements down a drain are not noticeably affected by this force.

You note that the text calls this a force. The label *force* is appropriate because, as the physicist Sir Isaac Newton (1643–1727) stated, when something is accelerating over a space, *a force is in operation*. This *apparent force* (in classical mechanics, an inertial force) *acts as an effect* on moving objects. It is named for Gaspard Coriolis, a French mathematician and researcher of applied mechanics, who first described the phenomenon in 1831.

Sun altitude and consistent daylength make large amounts of energy available in this region throughout the year. The warming creates lighter, less-dense, ascending air, with winds converging all along the extent of the trough. This converging air is extremely moist and full of latent heat energy. Vertical cloud columns frequently reach the tropopause, and precipitation is heavy throughout this zone. The combination of heating and convergence forces air aloft and forms the **intertropical convergence zone (ITCZ)**. Figure 4-14 is a satellite image of Earth showing the equatorial low-pressure trough. The ITCZ is identified by bands of clouds associated with the convergence of winds along the equator.

During summer, a marked wet season accompanies the shifting ITCZ over various regions. The maps in Figure 4-13 show the location of the ITCZ (dashed line) in January and July. In January, the zone crosses northern Australia and dips southward in eastern Africa and South America, whereas by July the zone shifts northward in the Americas and as far north as Pakistan and southern Asia.

Table 4-1
Four Hemispheric Pressure Areas

Name	Cause	Location	Air Temperature/ Moisture
Polar high-pressure cells	Thermal	90° N, 90° S	Cold/dry
Subpolar low-pressure cells	Dynamic	60° N, 60° S	Cool/wet
Subtropical high-pressure cells	Dynamic	20° to 35° N and S	Hot/dry
Equatorial low-pressure trough	Thermal	10° N to 10° S	Warm/wet

FIGURE 4–14
An interrupted band of clouds denotes the intertropical convergence zone (ITCZ), flanked on either side by several regions dominated by subtropical high-pressure systems. This natural color image was taken by the Solid State Imaging instrument (violet, green, and red filters) aboard the *Galileo* spacecraft during its December 8, 1990, flyby of Earth on its voyage to Jupiter. [Image courtesy of Dr. W. Reid Thompson, Laboratory of Planetary Studies, Cornell University. Used by permission.]

Figure 4-15 shows two views of the **Hadley cell**, named for the 18th-century English scientist who described the trade winds. Such a cell is generated by this equatorial low-pressure system of converging, ascending air. These winds converging on the equatorial low-pressure trough are known as the *northeast trade winds* (in the Northern Hemisphere) and *southeast trade winds* (in the Southern Hemisphere), or generally as **trade winds**. The trade winds pick up large quantities of moisture as they return through the Hadley circulation cell for another cycle of uplift and condensation. Within the ITCZ, winds are calm or mildly variable because of the even pressure gradient and the strong vertical ascent of air.

The rising air from the equatorial low-pressure area ~ls upward into a geostrophic flow to the north and ~. ~hese upper-air winds turn eastward, flowing ~to east, beginning at about 20° N and S,

forming descending anticyclonic flows above the subtropics. In each hemisphere, this circulation pattern appears most vertically symmetrical near the equinoxes.

Subtropical High-Pressure Cells. Between 20° and 35° latitude in both hemispheres, a broad high-pressure zone of hot, dry air is evident across the globe (see Figures 4-13 and 4-15). It is represented by the clear, frequently cloudless skies of the satellite imagery in Figure 4-14. The dynamic cause of these subtropical anticyclones is too complex to detail here, but generally they form as air in the Hadley cell descends in these latitudes, constantly supported by a dynamic mechanism in the upper-air circulation.

Air above the subtropics is mechanically pushed downward and is heated as it is compressed on its descent to the surface, as illustrated in Figure 4-15. Warmer air has a greater capacity to hold water vapor than does cooler air, making this descending

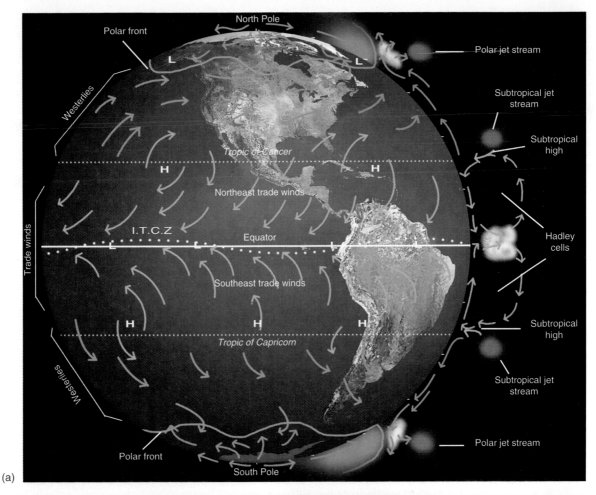

(a)

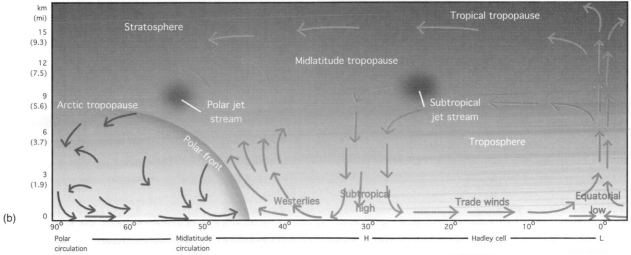

(b)

FIGURE 4-15

Two views of the general atmospheric circulation: (a) general circulation
schematic; (b) equator-to-pole cross-section for the Northern Hemisphere. Both
views show Hadley cells, subtropical highs, polar front, the subpolar low-pressure
cells, and approximate locations of the subtropical and polar jet streams.

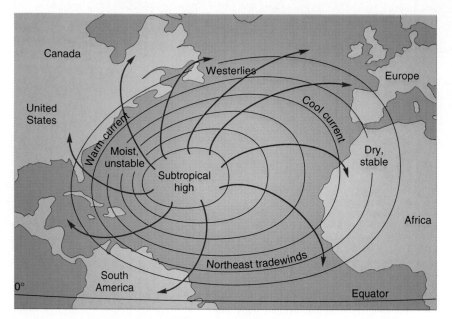

FIGURE 4-16
Characteristic circulation and climate conditions related to Atlantic subtropical high-pressure anticyclone in the Northern Hemisphere.

warm air relatively dry. The air is also dry because heavy precipitation along the equatorial circulation removed moisture.

Surface air diverging from the subtropical high-pressure cells generates Earth's principal surface winds: the westerlies and the trade winds. The **westerlies** are the dominant surface winds from the subtropics to high latitudes. They diminish somewhat in summer and are stronger in winter.

If you examine Figure 4-13, you will find several high-pressure areas. In the Northern Hemisphere, the Atlantic subtropical high-pressure cell is called the *Bermuda high* (in the western Atlantic) or the *Azores high* (when it migrates to the eastern Atlantic). The area under this subtropical high features clear, warm waters and large quantities of *Sargassum* (a seaweed), which gives the area its name—the Sargasso Sea. The *Pacific high*, or Hawaiian high, dominates the Pacific in July, retreating southward in January. In the Southern Hemisphere, three large high-pressure centers dominate the Pacific, Atlantic, and Indian oceans and tend to move along parallels in shifting zonal positions.

The entire high-pressure system migrates with the summer high Sun, fluctuating about 5–10° in latitude. eastern sides of these anticyclonic systems are more stable, and feature cooler ocean cur- the western sides. Thus, subtropical tude west coasts are influenced by the

eastern side of these systems and therefore experience dry-summer conditions (Figure 4-16). In fact, Earth's major deserts generally occur within the subtropical belt and extend to the west coast of each continent except Antarctica.

The western sides of subtropical high-pressure cells tend to be moist and unstable, overlying warm ocean currents, as characterized in Figure 4-16. This causes warm, moist conditions in Hawaii, Japan, southeastern China, and the southeastern United States.

Subpolar Low-Pressure Cells. The January map in Figure 4-13 shows two low-pressure cyclonic cells over the oceans around 60° N latitude. The North Pacific *Aleutian low* is near the Aleutian Islands; the North Atlantic *Icelandic low* is near Iceland. Both cells are dominant in winter and weaken significantly or disappear altogether in the summer with the strengthening of high pressure systems in the subtropics. The area of contrast between cold and warm air masses forms a contact zone known as the **polar front**, which encircles Earth at this latitude and is focused in these low-pressure areas.

Figures 4-15a and b illustrate this confrontation between warmer, moist air from the westerlies and colder, drier air from the polar and Arctic regions. The warmer air is displaced upward by the heavier cold air, forcing cooling and condensation in the

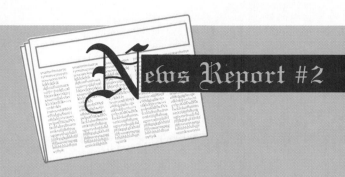

"Winds Can be 'Foolish' and 'Uncertain' "

The mildly changing winds and variable calms of the intertropical convergence zone (ITCZ) caused many problems to ships under sail. These equatorial calms were named the *doldrums* by 16th- and 17th-century sailors. The origin of the name appears to come from an older English word meaning foolish, because of the difficulty encountered by crews caught in this zone.

Other terms from the age of sail persist to this day. Because the core of the subtropical highs is near 25° N and S latitudes, these areas sometimes are known as the *calms of Cancer* and the *calms of Capricorn*. The name *horse latitudes* also is used for these zones of windless, hot, dry air, so deadly to sailing ships. Becalmed and thirsty crews stalled in the horse latitudes appear in classic sea-faring stories. This label is popularly attributed to the destruction of horses on board, because the struggling crew did not want to share food or water with the livestock. The term's true origin may never be known; the *Oxford English Dictionary* calls its origin "uncertain."

---

lifted air. Low-pressure cyclonic storms migrate out of the Aleutian and Icelandic frontal areas and may produce precipitation in North America and Europe, respectively.

In the Southern Hemisphere, a noncontinuous belt of subpolar cyclonic pressure systems surrounds Antarctica. The spiraling cloud patterns produced by these cyclonic systems are visible on the satellite image in Figure 4-17. Severe cyclonic storms can cross Antarctica, producing strong winds and new snowfall.

Polar High-Pressure Cells. The weakness of the polar high-pressure cells seems to argue against their inclusion among pressure areas on Earth. The polar atmospheric mass is small, receiving little energy to put it into motion. The variable cold, dry winds moving away from the polar region are anticyclonic, descending and diverging clockwise in the Northern Hemisphere (counterclockwise in the Southern Hemisphere) and forming weak, variable winds called **polar easterlies**. Of the two polar regions, the **Antarctic high** is significant in terms of both strength and persistence. In the Arctic, a polar high-pressure cell is

less pronounced, and when it does form, it tends to locate over the colder northern continental areas in winter (Canadian and Siberian highs) rather than directly over the relatively warmer Arctic Ocean.

Upper Atmospheric Circulation

Middle and upper tropospheric circulation is an important component of the atmosphere's general circulation. As described previously, these upper atmosphere winds tend to blow west to east from the subtropics to the poles.

Within the westerly flow of geostrophic winds are great waving undulations called **Rossby waves**, named for meteorologist C. G. Rossby, who first described them mathematically. The polar front is the contact between colder air to the north and warmer air to the south (Figure 4-18). The Rossby waves bring tongues of cold air southward, with warmer tropical air moving northward. The development of Rossby waves in the upper-air circulation is shown in the figure. As these disturbances mature, distinct cyclonic circulations form, with warmer air and colder air mixing along distinct fronts. The development of cyclonic

FIGURE 4-17
Centered on Antarctica, this satellite image shows a series of subpolar low-pressure cyclones in the Southern Hemisphere. Antarctica is fully illuminated by a mid-summer Sun. This natural color image was taken by the Solid State Imaging instrument (violet, green, and red filters) aboard the *Galileo* spacecraft during its December 8, 1990, flyby of Earth on its voyage to Jupiter. [Image from of Dr. W. Reid Thompson, Laboratory of Planetary Studies, Cornell University. Used by permission.]

storm systems at the surface is supported by these wave-and-eddy formations and other upper-air flows. These Rossby waves develop along the flow axis of a jet stream.

Jet Streams. The most prominent movement in these upper-level westerly wind flows is the **jet stream**, an irregular, concentrated band of wind occurring at several different locations. Rather flattened in vertical cross section, the jet streams normally are 160–480 km (100–300 mi) wide by 900–2150 m (3000–7000 ft) thick, with core speeds that can exceed 300 kmph (190 mph).

The *polar jet stream* is located at the tropopause along the polar front, at altitudes between 7600 and 10,700 m (24,900–35,100 ft), meandering between 30° and 70° N latitude. The polar jet stream can migrate as far south as Texas, steering colder air masses North America and influencing surface storms moving eastward. In the summer, the polar jet exerts less influence on storms by staying far poleward. The polar jet stream is commonly shown on television weather broadcasts. Shifting of daily and seasonal locations supports surface weather patterns. Figure 4-19 shows a stylized view of a polar jet stream.

In subtropical latitudes, near the boundary between tropical and midlatitude air, another jet stream flows near the tropopause. This *subtropical jet stream* ranges from 9100 to 13,700 m (29,850–45,000 ft) in altitude, and although it is generally weaker, it can reach greater speeds than those of the polar jet stream. The subtropical jet stream meanders from 20° to 50° latitude and may occur over North America simultaneously with the polar jet stream. The cross section of the troposphere in Figure 4-15b depicts the general location of these two jet streams. For review, these atmospheric patterns of highs, lows, fronts, and jet streams are summarized in Figure 4-15.

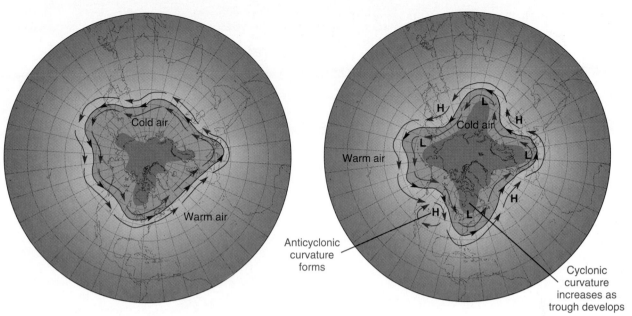

(a) Upper air circulation and jet stream begin to gently undulate.

(b) Long-wave patterns begin to form Rossby waves.

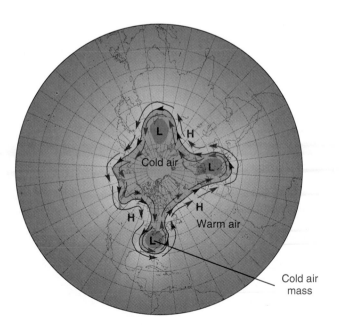

(c) Strong development of waves produce cells of cold and warm air—high-pressure ridges and low-pressure troughs.

Warm air   Cool air

Polar stereographic projections azimuthal and conformal

FIGURE 4-18

Development of longwaves in the upper-air circulation first described by C. G. Rossby in 1938 and detailed by J. Namias in 1952. [Adapted from J. Namias, National Oceanic and Atmospheric Administration.]

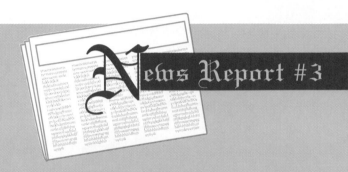
"Jet Streams Affect Flight Times"

As airplane travel increased during World War II, flight crews reported strong headwinds on routes to the west. In some cases planes leaving San Francisco for the Pacific were turned back by opposing wind in the upper troposphere. These were the jet streams.

Next time you plan a flight, note that airline schedules reflect the presence of these upper-level westerly winds, for they allot shorter flight times from west to east and longer flight times from east to west. Also important to both military and civilian aircraft is the effect the jet streams have on fuel consumption and the existence of air turbulence. Seasonal adjustments are necessary because the jet streams in both hemispheres tend to weaken during each hemisphere's summer and strengthen during winter as the streams shift closer to the equator.

Local Winds

Several winds that form in response to local terrain deserve mention in our discussion of atmospheric circulation. These local effects can, of course, be overwhelmed by weather systems passing through an area.

Land-sea breezes occur on most coastlines (Figure 4-20). The different heating characteristics of land and water surfaces create these winds (Chapter 3). During the day, land heats faster and becomes warmer than the water offshore. Because warm air is less dense, it rises and triggers an onshore flow of cooler marine air, usually stronger in

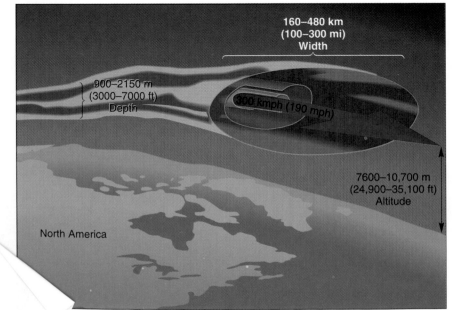

FIGURE 4-19

Stylized portrait of a polar jet stream.

the afternoon. At night, inland areas cool (radiate heat) faster than offshore waters. As a result, the cooler air over the land subsides and flows offshore over the warmer water, where the air is lifted. This night pattern reverses the process that developed during the day.

Well inland from the Pacific Ocean—160 km (100 mi)—Sacramento, California, demonstrates the sea-breeze effect. The city is at 39° N and 5 m (17 ft) elevation. The average July maximum and minimum temperatures are 34°C and 14°C (93°F and 57°F).

The evening cooling by the natural flow of marine air establishes a monthly mean of only 24°C (75°F), despite high daytime temperatures.

Mountain-valley breezes are created in a somewhat similar exchange. Mountain air cools rapidly at night, and valley air heats rapidly during the day (Figure 4-21). Thus, warm air rises upslope during the day, particularly in the afternoon; at night, cooler air subsides downslope into the valleys. Other downslope winds are discussed in Chapter 5.

Katabatic winds, or gravity drainage winds, are

FIGURE 4-20
Land-sea breezes characteristic of day and night.

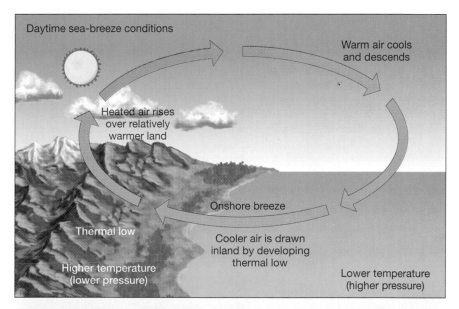

Daytime sea-breeze conditions

Warm air cools and descends

Heated air rises over relatively warmer land

Onshore breeze

Thermal low

Cooler air is drawn inland by developing thermal low

Higher temperature (lower pressure)

Lower temperature (higher pressure)

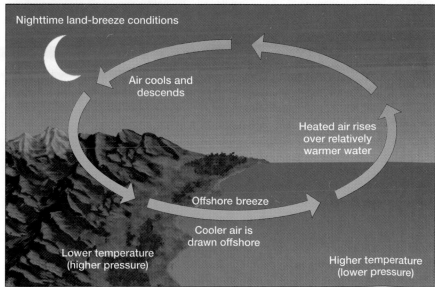

Nighttime land-breeze conditions

Air cools and descends

Heated air rises over relatively warmer water

Offshore breeze

Cooler air is drawn offshore

Lower temperature (higher pressure)

Higher temperature (lower pressure)

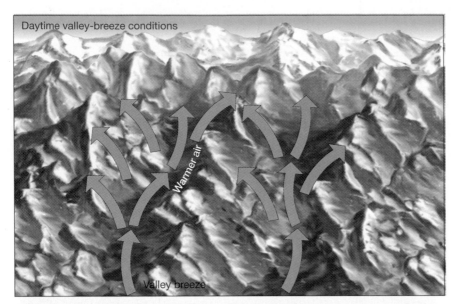

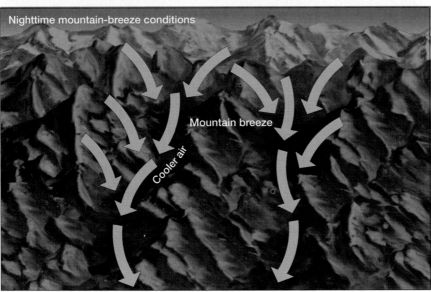

FIGURE 4-21
Pattern of mountain-valley breezes during day and night.

significant on a larger regional scale than mountain-valley breezes, under certain conditions. These winds usually are stronger than mountain-valley winds. An elevated plateau or highland is essential, where layers of air at the surface cool, become denser, and flow downslope. The ferocious winds that can blow off the ice sheets of Antarctica and Greenland are katabatic in nature. Worldwide, a variety of terrains produce such winds and bear many local names. The *mistral* of the Rhône Valley in southern France can cause frost damage to vineyards as the cold north winds move over the region to the Gulf of Lions and the Mediterranean Sea. The frequently stronger *bora,* driven by the cold air of winter high-pressure systems inland, flows across the Adriatic Coast to the west and south. In Alaska such winds are called the *taku.*

Santa Ana winds are another local wind type, generated by high pressure over the Great Basin of the western United States. A strong, dry wind is pro-

Wind Power: An Energy Resource

The principles of wind power may be old, but the technology is modern and the prospects favorable. In developed nations, energy sources are focused more upon nonrenewable fuels—coal, gas, and oil. However, more than half of the world's population in developing countries relies on renewable energy, derived from resources that are not depleted in the span of a human lifetime. This energy—for cooking, heating, and pumping—comes principally from small hydroelectric plants, wind-power systems, and wood resources. In developed countries, some electricity now is generated at wind farms, where groups of wind machines are massed, and at solar installations.

More than 50,000 wind turbines have been installed worldwide since 1974, with 13,000 of them in California alone (Figure 1). Denmark and California lead the world in this technology, with Denmark moving forward under its own official National Wind Strategy. Both California and Denmark have a goal to provide 10% of their electrical generating capacity with wind energy technology by A.D. 2000. Sweden is implementing plans to dismantle nuclear power plants and no doubt will increase use of its abundant wind resources.

Power generation from wind is site-specific, meaning that the localized conditions that produce adequate winds are confined to certain areas. In places where wind is reliable less than 25–30% of the time, only small-scale uses are economically feasible. Wind resources are greatest along coastlines influenced by trade winds and westerly winds; where mountain passes constrict the flow of air and interior valleys develop thermal low-pressure areas, thus drawing air across the landscape; or where localized winds such as katabatic and monsoonal flows occur. The mountains of North America, northwestern Europe, southwestern Australia, and the Arctic coastlands of Russia all offer favorable sites.

Politics of Delay and the Fossil-fuel Alternatives

Problems with further deployment of wind farms in the United States arose because of political decisions in the 1980s. All alternative energy programs, including those involving wind, were cut or canceled, and tax incentives eliminated, even though tax incentives and research for nonrenewable fossil fuels and nuclear power were increased.

One factor that may accelerate wind applications to replace oil-fired electrical generation at appropriate sites is an estimate of remaining domestic oil reserves made by the U.S. Geological Survey in 1989. With 51 billion barrels of recoverable oil left, including expected undiscovered reserves, and with the rate of domestic consumption at over 5.4 billion barrels per year, the United States has only a 16-year supply left. Imports at the 50% or greater level will stretch this reserve, but with unknown economic and military consequences. And the domestic natural gas supply estimate is not much better, with a 35-year supply left until depletion, at present consumption rates.

FIGURE 1
Wind farm in the Tehachapi Mountains of southern California. [Photo by Steve Mulligan.]

In this discussion of fossil fuels and the wind alternative, we must also look at coal. Coal is the main source of energy for about 55% of the steam-electric-power generation in the U.S. This consumes approximately 70% of the annual production of mixed coal. Estimates of how long known reserves of coal will last vary and are based on assumptions regarding the annual rate of increase in the consumption of coal. There are several hundred years of coal remaining in the United States and Canada if extraction and consumption rates stay about where they were in the 1980s (about 1 billion tons a year). However, if these reserves are subjected to an annual production growth of 3–5%, the lifetime of the reserves will be reduced to less than 100 years.

Coal is a valuable source of chemicals for many important uses, including medicines, so its depletion for the production of electricity would be a tragedy for future generations. In addition, an expanded use of coal will cause more air pollution, carbon dioxide emissions (a greenhouse gas), sulfur dioxide emissions (a contributor to acid deposition), and further strip mining and damage to the landscape.

The Benefits of the Wind Resource

These economic realities no doubt will override any further delaying actions, especially in regions where peak winds are in concert with peak electrical demand for air cooling, space heating, or agricultural water pumping. Long-term and marginal costs also favor wind-energy deployment, for it does not produce carbon dioxide, radioactive waste, sulfur dioxide, toxic wastes, or ash, nor does it require mining or the importation of foreign fuels—all of which represent significant costs with nonrenewable fuels. A Wind Energy Association study determined that the wind resource in 12 mid-continent states exceeds by 300% the total electrical consumption in the United States in 1987. The utility of wind power will be enhanced by improvements in energy storage. In addition, wind farm sites can have multiple uses—for example, for pasture and farming.

History is rich with the accomplishments of wind-powered sails. Experiments are going forward today to see whether the wind can still drive ships across the sea as in days past—an obvious answer awaits! In an experiment more than 10 years ago, the Japanese oil tanker *Shin Aitoku Maru* installed two computer-guided sails of 300 m² (3228 ft²) each. Fuel consumption was reduced by 50% in early tests. If we consider the world's cargo fleet of more than 25,000 vessels, consuming 4–5 million barrels of oil a day, a savings of just 10–15% would be significant.

Cash-poor, energy-poor developing countries already have a decentralized demand for energy, so for them large, centralized, capital-intensive electrical generation seems inappropriate. The key issue appears to be whether societies will develop energy resources that are nonrenewable-centralized (fossil-fuel power plants) or renewable-decentralized (solar, wind, and efficiency strategies). As for sails on ships: wind is a renewable resource that is always there, waiting to be realized as in the sailing days of old.

---

duced that flows out across the desert to southern California coastal areas. The air is heated by compression as it flows from higher to lower elevations and with increasing speed it moves through constricting valleys to the southwest. These winds irritate the population with their dust, dryness, and heat.

Locally, wind represents a source of renewable energy. FYI Report 4-1 briefly explores the potential for development of wind resources.

Monsoonal Winds. Regional wind systems that seasonally change direction are important in some areas. Examples occur in the tropics over Southeast Asia, Indonesia, India, northern Australia, and equatorial Af-rica. These winds involve an annual cycle of returning precipitation with the summer Sun and are named after the Arabic word for season, *mausim*, or **monsoon**.

The monsoons of southern and eastern Asia (Figure 4-22) are driven by the location and size of the Asian landmass and its proximity to the Indian Ocean. Also important to the generation of monsoonal flows are wind and pressure patterns in the upper-air circulation.

The extreme temperature range from summer to winter over the Asian landmass is due to its conti-

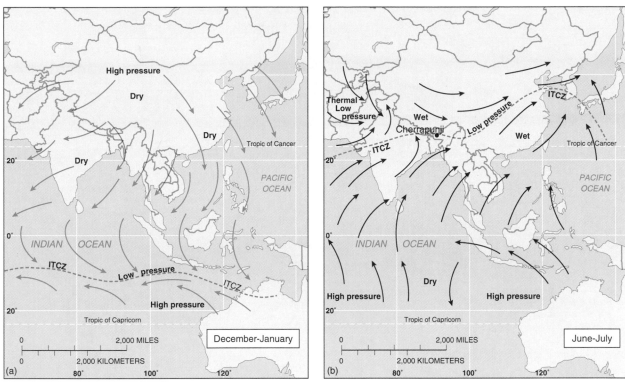

FIGURE 4-22
Asian monsoon patterns: (a) winter pressure and wind patterns; (b) summer
pressure and wind patterns. Note location of the ITCZ on each map. [Adapted
from Joseph E. Van Riper, *Man's Physical World*, p. 215. Copyright 1971 by
McGraw-Hill. Adapted by permission.]

nentality, reflecting its isolation from the modifying
effects of the ocean. This continental landmass is
dominated by an intense high-pressure anticyclone
in winter (Figures 4-13a), whereas the central area
of the Indian Ocean is dominated by the equatorial
low-pressure trough. Resultant cold, dry winds blow
from the Asian interior over the Himalayas, downslope, and across India, producing average temperatures between 15° and 20°C (60° to 68°F) at lower
elevations. These dry winds desiccate (dehydrate)
the landscape and then give way to hot weather
from March through May.

During the June-September wet period, the subsolar point shifts northward to the Tropic of Cancer,
near the mouths of the Indus and Ganges rivers. The
intertropical convergence zone is shifted northward
over southern Asia and the Asian continental inte-

rior develops a thermal low pressure, associated with
high average temperatures. Meanwhile, the Indian
Ocean, with surface temperature of 30°C (86°F), is
under the influence of subtropical high pressure. As
a result, hot, dry subtropical air sweeps over the
warm ocean, producing extremely high evaporation
rates (Figure 4-22b).

By the time this mass of air reaches India and the
convergence zone, it is laden with moisture in thun-
derous, dark clouds. The warmth of the land lends
additional lifting to the incoming air, as do the Himalayas, forcing the air mass to higher altitudes.
These conditions produce the wet monsoon of
India, where world-record rainfalls have occurred.
When the monsoonal rains arrive from June to September, they are welcome relief from the dust, heat,
and parched land of Asia's springtime.

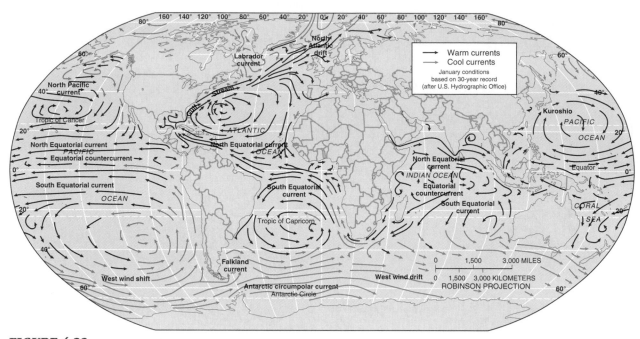

FIGURE 4-23
Major ocean currents.

Oceanic Currents

The annual monsoon is an integral part of Indian culture and life, as writer Khushwant Singh describes:

> To know India and her peoples, one has to know the monsoon. . . . It has to be a personal experience because nothing short of living through it can fully convey all it means. . . . What the four seasons of the year mean to the European, the one season of the monsoon means to the Indian. It is preceded by desolation; it brings with it the hopes of spring; it has the fullness of summer and the fulfillment of autumn all in one. . . . much of India's art, music, and literature is concerned with the monsoon. . . . First it falls in fat drops; the earth rises to meet them. She laps them up thirstily and is filled with fragrance. It brings the odor of the earth and of green vegetation to the nostrils.*

Oceanic Currents

Earth's atmospheric and oceanic circulations are closely interrelated. The driving force for ocean cur-

*Khushwant Singh, *I Hear the Nightingale*. New York: Grove Press, 1959, pp. 101-102.

rents is the frictional drag of the winds. Also important in shaping these currents is the interplay of the Coriolis force, density differences caused by temperature and salinity, the configuration of the continents and ocean floor, and astronomical forces (the tides). You may be surprised to learn that the ocean surface is not level, for it features small height differences (up to a few meters) in response to these currents, atmospheric pressure differences, slight variations in gravity, and the presence of waves.

Surface Currents

The general patterns of major ocean currents are shown in Figure 4-23. Because ocean currents flow over distance and through time, they are deflected by the Coriolis force. However, their pattern of deflection is not as tightly circular as that of the atmosphere. Compare this map with that of Earth's pressure and wind systems (Figure 4-13), and you can see that ocean currents are driven by the circulation around subtropical high-pressure cells in both hemispheres. (Remember: in the Northern Hemisphere, winds and ocean currents move *clockwise* about high pressure cells—note the currents in the

North Pacific and North Atlantic on the map. In the Southern Hemisphere the *counterclockwise* circulation about a high is evident on the map.)

These circulation systems are known as **gyres**. To understand these gyres and their western margins, which feature slightly stronger currents, we must understand a process known as **western intensification**.

Along the full extent of ocean areas adjoining the equator, the trade winds drive the oceans westward in a concentrated channel. These currents are kept near the equator by a Coriolis force influence that weakens near the equator. As the surface current approaches the western margins of the oceans, the water actually piles up an average of 15 cm (6 in.). From this western edge, ocean water then spills northward and southward in strong currents, flowing in tight channels along the western edges of the ocean basins (eastern shorelines of continents). Additionally, a strong countercurrent is generated that flows through the full extent of the Pacific, Atlantic, and Indian oceans. This occasionally strong, somewhat sporadic eastward **equatorial countercurrent** may be alongside or just beneath the surface current, at depths of 100 m (300 ft).

In the Northern Hemisphere, the Gulf Stream and the Kuroshio move forcefully northward as a result of western intensification, with their speed and depth increased by the constriction of the area they occupy. The warm, deep-blue water of the ribbonlike Gulf Stream usually is 50–80 km (30–50 mi) wide and 1.5–2.0 km (0.9–1.2 mi) deep, moving at 3–10 kmph (1.8–6.2 mph). In 24 hours, ocean water can move 70–240 km (40–150 mi) in the Gulf Stream, although a complete circuit around an entire gyre may take a year.

Deep Currents

Where surface water is swept away from a coast, an **upwelling current** occurs. This cool water generally is nutrient-rich and rises from great depths to replace the vacating water. Such cold upwelling currents occur off the Pacific coasts of North and South America and the subtropical and midlatitude west coast of Africa, and are some of Earth's prime commercial fishing areas. In other portions of the sea where there is an accumulation of water—as at the western end of an equatorial current, or the Labrador Sea, or along the margins of Antarctica—excess water gravitates downward in a **downwelling current**. Important mixing currents along the ocean floor are generated from such downwelling zones and travel the full extent of the ocean basins, carrying heat energy and salinity.

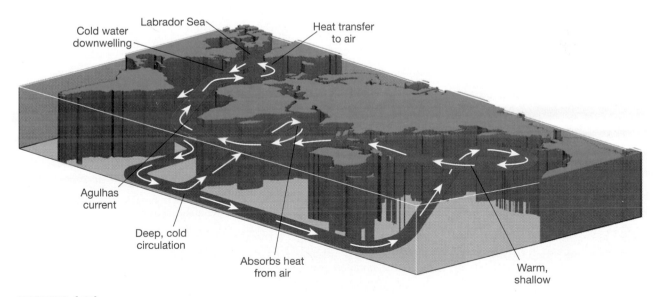

FIGURE 4-24
Centuries-long deep circulation in the oceans is being deciphered by scientists. This global circulation mimics a vast conveyor belt of water drawing heat from some regions and transporting it for release in others.

Imagine a continuous channel of water beginning with cold water downwelling in the North Atlantic, flowing deep and strong to upwellings in the Indian Ocean and North Pacific (Figure 4-24). Here it warms and then is carried in surface currents back to the North Atlantic. A complete circuit of the current may require 1000 years from downwelling in the Labrador Sea off Greenland to its reemergence in the southern Indian Ocean and return. Even deeper Antarctic bottom water flows northward in the Atlantic Basin beneath these currents.

SUMMARY—Atmospheric and Oceanic Circulations

Volcanic eruptions such as Tambora in 1815 and Mount Pinatubo in 1991 dramatically demonstrate the power of global winds to disperse aerosols worldwide in a matter of weeks. Atmospheric circulation facilitates important transfers of energy and air mass on Earth, thus maintaining Earth's natural energy balances. Earth's atmospheric and oceanic circulations are a vast heat engine powered by the Sun.

Wind is the horizontal movement of air across Earth's surface, directed and driven by the **pressure gradient force**, the **Coriolis force**, and **friction force**. The weight of the atmosphere produces **air pressure** that crushes in on all of us and averages about 1kg/cm²; we measure it with an **aneroid barometer** or **mercury barometer**.

The pattern of high and low pressures on Earth in generalized belts in each hemisphere produces the distribution of specific wind systems. These primary pressure regions are the: **equatorial low-pressure trough** and the weak **polar high-pressure cells**, both north and south, and the **subtropical high-pressure cells** and **subpolar low-pressure cells**. A generalized model of the overall atmospheric circulation system is helpful in visualizing wind circulation both at the surface and in the upper troposphere. The prominent movement in these upper level, westerly winds are formed by the polar and subtropical **jet streams**.

Different heating characteristics of land and water surfaces create *land-sea breezes. Mountain-valley* breezes are caused by changing temperature differences during the day and evening between valleys and mountain slopes and summits. Ocean currents are primarily caused by the frictional drag of wind and occur worldwide at varying intensities, temperatures, and speeds, both along the surface and at great depths in the oceanic basins.

Wind itself is an important energy source and is a renewable resource with potential for increasing development and application in the future. Earth's dynamic atmosphere and ocean bring together all the elements of Part 1 in this text. More than any other spatial factor, natural or cultural, the atmosphere brings together humanity—one person's or nation's exhalation is another's breath intake.

KEY TERM

air pressure	gyre
anemometer	Hadley cell
aneroid barometer	intertropical convergence zone (ITCZ)
Antarctic high	isobar
anticyclone	jet stream
Coriolis force	katabatic wind
cyclone	mercury barometer
downwelling current	monsoon
equatorial countercurrent	polar easterlies
equatorial low-pressure trough	polar front
friction force	polar high-pressure cell
geostrophic wind	pressure gradient force

Rossby waves

subpolar low-pressure cell

subtropical high-pressure cell

trade winds

upwelling current

westerlies

western intensification

wind

wind vane

REVIEW QUESTIONS

1. Explain the statement, "the atmosphere socializes humanity, making all the world a spatially linked society." Illustrate your answer with some examples.
2. Define wind. How is it measured? How is its direction determined?
3. Distinguish among primary, secondary, and tertiary classifications of global atmospheric circulation.
4. Describe the horizontal and vertical air motions in a high-pressure cell and in a low-pressure cell.
5. Describe the effect of the Coriolis force. How does it apparently deflect atmospheric and oceanic circulations? Explain.
6. What are geostrophic winds, and where are they encountered in the atmosphere?
7. Construct a simple diagram of Earth's general circulation, including the four principal pressure belts or zones and the three principal wind systems.
8. How does the intertropical convergence zone (ITCZ) relate to the equatorial low-pressure trough? How might it appear on a satellite image?
9. People living along coastlines generally experience variations in day-night winds. Explain the factors that produce these changing wind patterns.
10. Describe the seasonal pressure patterns that produce the Asian monsoonal wind and precipitation patterns.
11. What is the relationship between global atmospheric circulation and ocean currents? Relate oceanic gyres to patterns of subtropical high pressure.
12. What is the present status of wind-energy technologies? Describe deployment delays. How do these delays relate to the nature of the technology?

PART 2

WATER, WEATHER, AND CLIMATE

CHAPTER 5
Atmospheric Water and Weather

CHAPTER 6
Water Resources

CHAPTER 7
Earth's Climates

Cumulonimbus clouds and thunderstorm in eastern Washington. [Photo by author.]

Earth is the water planet. Its surface waters are unique in the solar system. Chapter 5 explains why water exists here in such quantity and describes the remarkable qualities and properties it possesses. The dynamics of daily weather phenomena follow: the powerful interaction of moisture and energy in the atmosphere, the interpretation of cloud forms, conditions of stability or instability, the interaction of air masses, and the occurrence of violent weather. In Chapter 6, the specifics of the hydrologic cycle and how water circulates over Earth are explained. We examine the water-balance concept, which is useful in understanding water-resource relationships, whether global, regional, or local. Important water resources include rivers, lakes, groundwater, and oceans. In Chapter 7, we see the spatial implications over time of the water-weather system that generates Earth's climate patterns.

Lightning over the Grand Canyon. [*Photo by Dick Dietrich.*]

5

ATMOSPHERIC WATER AND WEATHER

WATER ON EARTH
  Quantity Equilibrium
  Distribution of Earth's Water

UNIQUE PROPERTIES OF WATER
  Heat Properties

HUMIDITY
  Relative Humidity
  Expressions of Relative Humidity

ATMOSPHERIC STABILITY
  Adiabatic Processes
  Stable and Unstable Atmospheric Conditions

CLOUDS AND FOG
  Cloud Types and Identification   Fog

AIR MASSES

ATMOSPHERIC LIFTING MECHANISMS
  Convectional Lifting   Orographic Lifting
  Frontal Lifting

MIDLATITUDE CYCLONIC SYSTEMS
  Life Cycle of a Midlatitude Cyclone
  Daily Weather Map and the Midlatitude Cyclone

VIOLENT WEATHER
  Thunderstorms   Tornadoes   Tropical Cyclones

SUMMARY

FYI REPORT 5-1: 1992—THE YEAR OF HURRICANES

Walden is blue at one time and green at another, even from the same point of view. Lying between the earth and the heavens, it partakes of the color of both. . . . A lake is the landscape's most beautiful and expressive feature. It is earth's eye; looking into which the beholder measures the depth of his own nature. . . . Sky water. It needs no fence. Nations come and go without defiling it. It is a mirror which no stone can crack . . . Nature continually repairs . . . a mirror in which all impurity presented to it sinks, swept and dusted by the sun's hazy brush. A field of water . . . is continually receiving new life and motion from above. It is intermediate in its nature between land and sky.*

Thus did Thoreau speak of the water so dear to him—Walden Pond in Massachusetts, along whose shore he lived.

Water is critical to our daily lives and a principal compound in nature. It covers 71% of Earth (by area) and in the Solar System occurs in such significant quantities only on our planet. Pure water is naturally colorless, odorless, and tasteless and weighs 1 gram per cm³ or 1 kg per liter (62.3 lb/ft³ or 8.337 lb/gal).

Water constitutes nearly 70% of our bodies by weight and is the major ingredient in plants, animals, and our food. A human can survive 50 to 60 days without food, but only 2 or 3 days without water. The water we use must be adequate in quantity as well as quality for its many tasks—everything from personal hygiene to vast national water projects. Water indeed occupies that place between land and sky, mediating energy and shaping both the lithosphere and atmosphere, as Thoreau revealed.

Water has a leading role in the vast drama played out daily on Earth's stage. In this ongoing play, large air masses come into conflict, moving and shifting to dominate different regions in response to the controls of insolation, latitude, the arrangement of land, water, and mountains, and migrating cyclonic systems that cross the middle latitudes and the tropics. Water and its ability to absorb and release vast quantities of heat drives these systems.

Water in the air (relative humidity), temperature, air pressure, wind speed and direction, daylength,

and Sun angle are important measurable elements that contribute to the weather. **Weather** is the short-term condition of the atmosphere, as compared to *climate*, which reflects long-term atmospheric conditions and extremes. We tune to a local station for the day's weather report from the National Weather Service (in the United States) or Atmospheric Environment Service (in Canada) to see the current satellite images and to hear tomorrow's forecast.

Meteorology is the scientific study of the atmosphere (*meteor* means "heavenly" or "of the atmosphere"). Embodied within this science is a study of the atmosphere's physical characteristics and motions, related chemical, physical, and geological processes, the complex linkages of atmospheric systems, and weather forecasting. The spatial implications of atmospheric science and its relation to human activities strongly link meteorology to physical geography. We begin our study of weather by examining water and the dynamics of atmospheric moisture.

Water on Earth

Water on Earth was formed within the planet, reaching Earth's surface in an ongoing process called **outgassing**, by which water and water vapor emerge from deep within the crust—25 km (15.5 mi) or more below Earth's surface (Figure 5-1). In the early atmosphere, massive quantities of outgassed water vapor condensed and fell in torrents, only to vaporize again because of high temperatures at Earth's surface. For water to remain on Earth's surface, land temperatures had tox drop below water's boiling point of 100°C (212°F), something that occurred about 3.8 billion years ago.

The lowest places across the face of Earth then began to fill with water: first ponds, then lakes and seas, and eventually ocean-sized bodies of water. Massive flows of water washed over the landscape, carrying both dissolved and undissolved elements to these early seas and oceans.

Quantity Equilibrium

Today, water is the most common compound on the surface of Earth, having achieved the present volume of 1.36 billion km³ (326 million mi³) approximately 2 billion years ago. This quantity has remained relatively

---

*Reprinted by permission of Merrill, an imprint of Macmillan Publishing Company, from *Walden* by Henry David Thoreau, pp. 192, 202, 204. Copyright © 1969 by Merrill Publishing. Originally published 1854.

FIGURE 5-1

Outgassing of water from Earth's crust in the Chilean Andes. [Photo by Stephen Cunha.]

constant, even though water is continuously being lost from the system, escaping to space or breaking down and forming new compounds with other elements. Lost water is replaced by pristine water not previously at the surface, which emerges from within Earth. The net result of these inputs and outputs to water quantity is a steady-state equilibrium in Earth's hydrosphere.

Despite this overall net balance in quantity, worldwide changes in sea level do occur, called **eustasy**. Eustatic sea-level changes are specifically related to changes in volume of water in the oceans. Some of these changes are explained by the amount of water stored on Earth as ice. As more water is bound up in glaciers (in mountains worldwide) and in ice sheets (Greenland and Antarctica), sea level lowers, whereas a warmer era reduces the quantity of water stored as ice and thus raises sea level. Some 18,000 years ago,

during the most recent ice age, sea level was more than 100 m (330 ft) lower than today; 40,000 years ago it was 150 m (490 ft) lower. Over the past 100 years, mean sea level has steadily risen and is still rising worldwide at this time. Glacial effects on sea level are discussed further in Chapter 13. Apparent changes in sea level also are related to actual physical changes in landmasses, such as continental uplift or subsidence. (This is called *isostasy* and is discussed in Chapter 8).

Distribution of Earth's Water

If you briefly examine a globe, it becomes obvious that most of Earth's continental land is in the Northern Hemisphere, whereas the Southern Hemisphere is dominated by water. The present location of all of Earth's water is illustrated in Figure 5-2. The oceans

FIGURE 5-2

Ocean and freshwater distribution on Earth: all water, freshwater, and surface water.

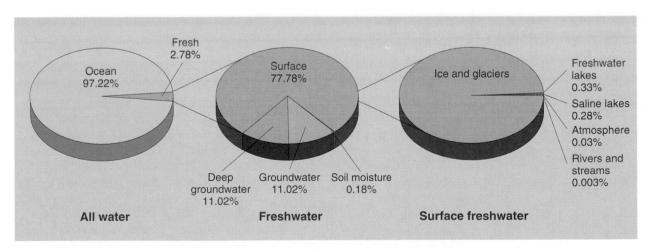

Ocean 97.22%   Fresh 2.78%

Surface 77.78%   Deep groundwater 11.02%   Groundwater 11.02%   Soil moisture 0.18%

Ice and glaciers   Freshwater lakes 0.33%   Saline lakes 0.28%   Atmosphere 0.03%   Rivers and streams 0.003%

All water   **Freshwater**   **Surface freshwater**

143

contain 97.22% of all water. The remaining 2.78% is classified as fresh, or nonoceanic water.

Of all the freshwater on Earth, about 22% exists as groundwater and soil moisture. However, the greatest single repository of surface freshwater is ice. Ice sheets and glaciers account for 77.14% of all freshwater on Earth. Add to this the subsurface groundwater, and we have accounted for 99.36% of all freshwater. The remaining freshwater, although very familiar to us, and present in seemingly huge amounts in lakes, rivers, and streams, actually represents but a small quantity, less than 1%.

Of all freshwater lakes in the world about 50% of the total volume is contained in just 7 lakes. The greatest volume, some 20% of all lake water, resides in 25-million-year-old Lake Baykal in Siberian Russia. This lake contains almost as much water as all the U.S. Great Lakes combined. Africa's Lake Tanganyika contains the next largest volume, followed by the five U.S. Great Lakes.

Saline lakes and inland seas are neither fresh nor connected to the ocean. Such lakes as Utah's Great Salt Lake, California's Mono Lake, and Southwest Asia's Caspian Sea exist as remnants of past wetter climatic eras and are usually in regions of interior river drainage.

Think of the weather occurring in the atmosphere worldwide at this moment, and the many flowing rivers and streams, which combined amount to only 0.033% of all freshwater! This small amount is very dynamic. A water molecule traveling along atmospheric and surface-water paths moves through the entire ocean-atmosphere-precipitation-runoff cycle in less than two weeks, whereas a water molecule located in deep-ocean circulation, groundwater, or a glacier moves slowly, taking thousands of years to migrate through the system.

Unique Properties of Water

Earth's distance from the Sun places it within a most remarkable temperate zone, compared to the other planets. This temperate location allows all states of water—ice, water, and water vapor—to occur naturally on Earth. Water is composed of two atoms of hydrogen and one of oxygen, which readily bond (see Figure 5-3). The resulting water molecule exhibits a

unique stability, is a versatile solvent, and possesses extraordinary heat characteristics. As the most common compound on the surface of Earth, water exhibits the most uncommon of properties.

The nature of the hydrogen-oxygen bond results in the hydrogen side having a positive charge and the oxygen side a negative charge. This polarity of the water molecule explains why water "acts wet" and dissolves so many other molecules and elements. Because of this solvency ability, pure water is rare in nature.

Water molecules are attracted to each other because of their polarity; the positive (hydrogen) side of one water molecule is attracted to the negative (oxygen) side of another. (Note this bonding pattern in the molecular illustrations in Figure 5-3b and c.) This bonding between water molecules is called *hydrogen bonding*. The effects of hydrogen bonding in water are observable in everyday life. Hydrogen bonding creates the *surface tension* that allows you to slightly overfill a glass with water, so that the water surface actually is above the rim of the glass, retained by millions of hydrogen bonds.

Hydrogen bonding also is the cause of capillarity, which you observe when you use a paper towel. The towel draws water upward through its fibers because each molecule is pulling on its neighbor. Capillary action is an important component of soil moisture processes, discussed in Chapters 6 and 15. It is important to note that, without hydrogen bonding, water would be a gas at normal surface temperatures.

Heat Properties

For water to change from one state to another (solid, liquid, gas), *heat energy must be added to it or released from it*. To cause a change of state, the amount of heat energy must be sufficient to affect the hydrogen bonds between the molecules. This relationship between water and heat energy is an important driving force in producing weather. In fact, the heat exchanged in the phase changes of water provides over 30% of the energy powering the general circulation of the atmosphere.

Figure 5-3 presents the three states of water and the terms used to describe each **phase change**. The term **sublimation** refers to the direct change of water vapor to ice or ice to water vapor; sometimes

FIGURE 5-3

The three physical states of water: (a) water vapor, (b) water, and (c) ice. Note the molecular arrangement in each state and the terms that describe the changes from one phase to another as energy is either absorbed or released. Also note how the polarity of water molecules bonds them to one another, loosely in the liquid state and firmly in the solid state.

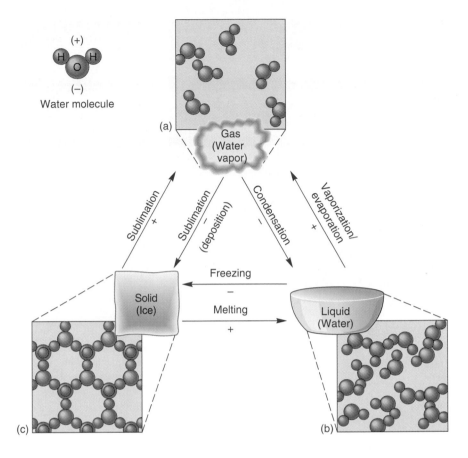

the term *deposition* is used if water vapor attaches itself directly to an ice crystal. The deposition of water vapor to ice may form *frost* on surfaces. *Melting* and *freezing* describe the phase change between solid and liquid. The terms *condensation* and *evaporation*, or *vaporization*, at boiling temperature, apply to the change between liquid and vapor.

Ice, the Solid Phase. As water cools, it behaves like most compounds and contracts in volume, reaching its greatest density at 4°C (39°F). But below that temperature, water behaves very differently from other compounds, and begins to expand as more hydrogen bonds form among the slower-moving molecules, creating the hexagonal structures shown in Figure 5-3c. This expansion continues to a temperature of −29°C (−20°F), with up to a 9% increase in volume possible. As shown in Figure 5-4a, the rigid internal structure of ice dictates the six-sided appearance of all ice crystals, which can loosely com-

bine to form snowflakes. This six-sided preference applies to ice crystals of all shapes: plates, columns, needles, and dendrites (branching or treelike forms)—a unique interaction of randomness and the determinism of physical principles.

The expansion in volume that accompanies the freezing process results in a decrease in density (the same number of molecules occupy greater space). Specifically, ice has 0.91 times the density of water, and so it floats. Without this change in density, much of Earth's freshwater would be bound in masses of ice on the ocean floor. Instead, we have floating icebergs, with approximately 1/11 (9%) of their mass exposed and 10/11 (91%) hidden beneath the ocean's surface (Figure 5-4b).

Water, the Liquid Phase. As a liquid, water assumes the shape of its container and is a noncompressible fluid, quite different from solid, rigid ice. For ice to melt, heat energy must increase the motion of

FIGURE 5-4

(a) Computer enhanced ice-crystal patterns are dictated by the internal structure between water molecules. This structure also explains the lower density of ice and why ice floats. (b) Icebergs (pinnacled and saddle-shaped forms) floating in Disko Bay, Greenland, calved from the Jakobshavn glacier. (c) Ice crystals forming on these branches exhibit a remarkable struggle between physical principles of randomness and order. [(a) Photo enhancement © Scott Camazine after W.A. Bentley; (b) photo by George Hunter; (c) photo by author.]

the water molecules to break some of the hydrogen bonds (Figure 5-3b). Despite the fact that there is no change in sensible temperature between ice at 0°C and water at 0°C, 80 calories of heat must be added for the phase change of 1 gram of ice to 1 gram of water (Figure 5-5, upper left). This heat is called **latent heat** because it is stored within the water and is liberated whenever phase reverses and a gram of water freezes. These 80 calories are the *latent heat of fusion,* or the *latent heat of melting* and *of freezing.*

To raise the temperature of 1 gram of water from freezing at 0°C (32°F) to boiling at 100°C (212°F), we must add 100 additional calories, gaining an increase of 1C° (1.8F°) for each calorie added (Figure 5-5, top center).

Water Vapor, the Gas Phase. Water vapor is an invisible and compressible gas in which each molecule moves independently (Figure 5-3a). To accomplish the phase change from liquid to vapor at boiling temperature, under normal sea-level pressure, 540 calories must be added to 1 gram of boiling water to achieve a phase change to water vapor (Figure 5-5, upper right). Those calories are the *latent heat of vaporization.* When water vapor condenses to a liquid, each gram gives up its hidden 540 calories as the **latent heat of condensation**.

To summarize, taking 1 gram of ice at 0°C and changing it to water vapor at 100°C—changing it from a solid, to a liquid, to a gas—*absorbs* 720 calo-

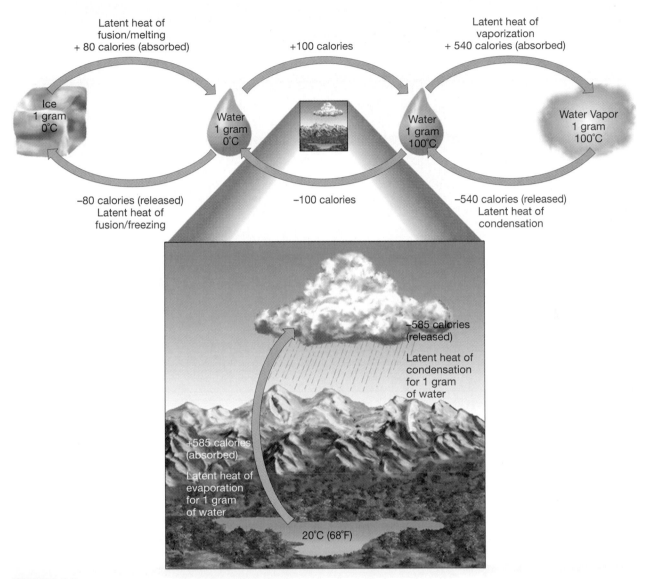

FIGURE 5-5
Significant latent heat energy is involved in the phase changes from ice to water
and from water to water vapor. The inset landscape illustrates average conditions
in the environment.

ries (80 cal + 100 cal + 540 cal). Or, reversing the
process, changing 1 gram of water vapor at 100°C
to ice at 0°C *liberates* 720 calories.

Heat Properties of Water in Nature. In a lake,
stream, or in soil water, at 20°C (68°F), every gram
of water that breaks away from the surface through
evaporation must absorb from the environment ap-

proximately 585 calories as the **latent heat of evaporation** (see the natural scene in Figure 5-5). This
is slightly more energy than would be required if the
water were boiling (540 cal). You can feel this absorption of latent heat as evaporative cooling on
your skin when it is wet. This latent heat exchange
is the dominant cooling process in Earth's energy
budget.

"Breaking Roads and Pipes and Sinking Ships"

Road crews are out in the summer in many parts of the country repairing streets and freeways. Winter damage to highways is often the result of the phase change from water to ice. Rainwater seeps into roadway cracks and then expands as it freezes, thus breaking up the pavement. Perhaps you have noticed that bridges are subjected to the greatest roadbed damage. This is a result of cold-air access beneath a bridge that produces more freeze-thaw cycles.

The expansion of freezing water exerts a tremendous force—enough to crack plumbing or an automobile engine block. People living in very cold climates use antifreeze and engine heaters to avoid such damage.

Wrapping water pipes with insulation is a common winter task in many places.

A major shipping hazard in higher latitudes is posed by floating ice. Since ice is 0.91 the density of water, an iceberg sits with approximately 10/11 of its mass below water level. The irregular edges of submarine ice can slash the side of a passing ship. This is what happened to the *HMS Titanic* in 1912 on its maiden voyage.

Historically, this physical property of water was put to useful work in quarrying rock for building materials. Holes were drilled and filled with water before winter, so that when cold weather arrived the water would freeze and expand, cracking the rock into manageable shapes.

The process reverses when air that contains water vapor is cooled. The vapor eventually condenses back into the liquid state, forming moisture droplets and thus liberating 585 calories as the *latent heat of condensation* for every gram of water.

The *latent heat of sublimation* absorbs 680 calories as a gram of ice transforms into vapor. A comparable amount of energy is released as vapor transforms back to ice. Because of these unique heat properties water is a major contributor of energy to the atmosphere. Let's now take a look at water vapor in the atmosphere.

Humidity

The water vapor content of air is termed **humidity**. The capacity of air to hold water vapor is primarily a function of the temperature of both the air and the water vapor, which are usually the same. Warmer air has a greater capacity for water vapor, whereas cooler air has a lesser capacity.

We are all aware of humidity in the air, for its relationship to air temperature determines our sense of comfort. North Americans spend billions of dollars a year to adjust humidity, either with air conditioning (extracting water vapor and cooling) or with air humidifying (adding water vapor). To determine the energy available to fuel weather activities, it is essential to know the water vapor content of air.

Relative Humidity

Next to air temperature and barometric pressure, the most common piece of information in local weather broadcasts is **relative humidity**. It is a ratio (expressed as a percentage) of the amount of water vapor that is actually in the air (*content*), compared to the maximum water vapor the air could hold at a given

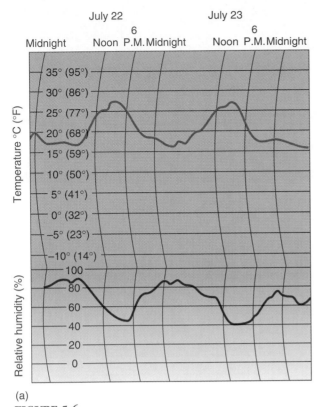

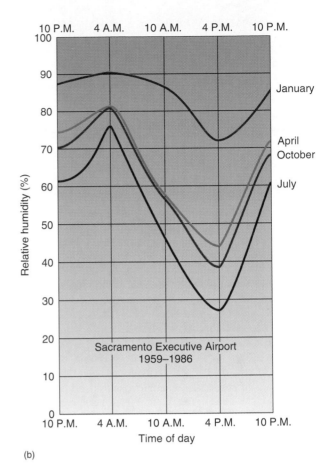

(a)

(b)

FIGURE 5-6

Sacramento, California: (a) typical daily variations in temperature and relative humidity, and (b) seasonal daily variations in relative humidity.

temperature (*capacity*). If the air is relatively dry in comparison to its capacity, the percentage is lower; if the air is relatively moist, the percentage is higher; and if the air is holding all the moisture it can for its temperature, the percentage is 100%. The formula is:

$$\text{relative humidity} = \frac{\text{actual water vapor content of the air}}{\text{max. water vapor capacity of the air}} \times 100$$

Relative humidity varies because of evaporation, condensation, or temperature changes, all of which affect both the moisture content (the numerator) and the capacity (the denominator) of the air to hold water vapor. Relative humidity is an expression of an ongoing process between air and moist surfaces, for condensation and evaporation operate continuously—water molecules move back and forth between air and water.

Air is said to be **saturated**, or full, if it is holding all the water vapor that it can hold at a given temperature (100% relative humidity). In saturated air, the net transfer of water molecules between a moist surface and air is in equilibrium. Saturation indicates

that any further addition of water vapor (increase in content) or any decrease in temperature (reduction in capacity) will result in active condensation (clouds, fog, or precipitation).

The temperature at which a given mass of air becomes saturated is termed the **dew-point temperature**. In other words, *air is saturated when the dew-point temperature and the air temperature are the same.* A cold drink in a glass provides a common example of these conditions: the water droplets that form on the outside of the glass condense from the air because the air layer next to the glass is chilled to below its dew-point temperature and thus is saturated.

During a typical day, air temperature and relative humidity relate inversely—as temperature rises, relative humidity falls (Figure 5-6a). Relative humidity is highest at dawn, when air temperature is lower and the capacity of the air to hold water vapor is less. Relative humidity is lowest in the late afternoon, when higher air temperatures increase the capacity of the air to hold water vapor. The actual water vapor content in the air may have remained the same throughout the day, but because the temperature varied, relative humidity percentages changed from morning until afternoon.

Weather records for Sacramento, California, show the seasonal variation in relative humidity by time of day, confirming the relationship of temperature and relative humidity (Figure 5-6b). Beyond the expected similarity in overall daily pattern for the four months shown, January readings are higher than July readings because air temperatures are lower overall in winter. Similar relative humidity records at most weather stations demonstrate the same relationship among season, temperature, and relative humidity.

Expressions of Relative Humidity

Several ways are used to express humidity and relative humidity, each with its own utility and application; two of these are vapor pressure and specific humidity.

Vapor Pressure. One way of describing humidity is related to air pressure. As free water molecules evaporate from a surface into the atmosphere, they become water vapor, one of the gases in air. That portion of total air pressure that is made up of water vapor molecules is termed **vapor pressure**. Like air pressure, it is expressed in millibars (mb). Water vapor molecules continue to evaporate from a moist surface, slowly diffusing into the air, until the increasing vapor pressure in the air causes some molecules to return to the surface. Saturation is reached when the movement of water molecules between surface and air is in equilibrium. The maximum capacity of the air at a given temperature is termed the *saturation vapor pressure* and indicates the maximum pressure that water vapor molecules can exert. Any increase or decrease in temperature causes the saturation vapor pressure to change. Figure 5-7 graphs the saturation vapor pressure at varying air temperatures. The graph illustrates that, for every temperature increase of 10C° (18F°), the vapor pressure capacity of air nearly doubles.

The inset in Figure 5-7 compares saturation vapor pressure over water and over ice surfaces at subfreezing temperatures. You can see that saturation vapor pressure is greater above a water surface than over an ice surface—that is, it takes more water vapor

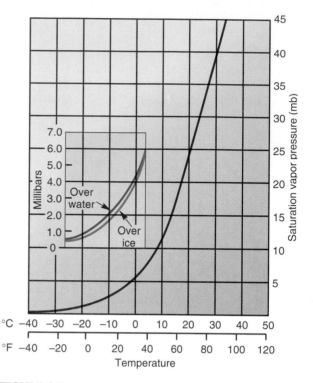

FIGURE 5-7
Saturation vapor pressure of air at various temperatures. Inset compares saturation vapor pressure over water surfaces and over ice surfaces at subfreezing temperatures.

molecules to saturate air above water than it does above ice. This fact is important to condensation processes and rain droplet formation, both of which are discussed in the section on clouds later in this chapter.

Specific Humidity. A useful humidity measure is one that remains constant as temperature and pressure change. **Specific humidity** refers to the mass of water vapor (in grams) per mass of air (in kilograms) at any specified temperature. Because it is measured in mass, specific humidity is not affected by changes in temperature or pressure such as occur when an air parcel rises to higher elevations, and is therefore more valuable in forecasting weather. Specific humidity stays constant despite volume changes.

The maximum mass of water vapor that a kilogram of air can hold at any specified temperature is termed the *maximum specific humidity* and is plotted in Figure 5-8. Specific humidity is useful in de-

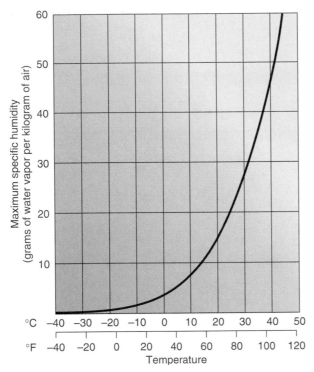

FIGURE 5-8
Maximum specific humidity for a mass of air at various temperatures.

scribing the moisture content of large air masses that are interacting in a weather system.

Instruments for Measurement. Relative humidity is measured with various instruments. The *hair hygrometer* uses the principle that human hair changes as much as 4% in length between 0 and 100% relative humidity. The instrument connects a standardized bundle of human hair through a mechanism to a gauge. As the hair absorbs or loses water in the air, it changes length indicating relative humidity.

Another instrument used to measure relative humidity is a *sling psychrometer*. This device has two thermometers mounted side-by-side on a metal holder. One is called the *dry-bulb thermometer;* it simply records the ambient (surrounding) air temperature. The other thermometer is called the *wet-bulb thermometer.* This bulb is covered by a cloth wick, which is moistened. The psychrometer is then spun by its handle. (Other versions use a motor to spin the thermometers or they are placed where a fan forces air over them.)

The rate at which water evaporates from the wick depends on the relative saturation of the surrounding air. If the air is dry, water evaporates quickly, absorbing the latent heat of evaporation from the wet-bulb thermometer, causing the temperature to drop (wet-bulb depression). In an area of high humidity, much less water evaporates from the wick. After a minute or two, the temperature of each bulb is compared on a relative humidity (psychrometric) chart, from which relative humidity can be determined. Now that we know something about atmospheric moisture, dew point, and relative humidity, let's examine the concept of stability in the atmosphere.

Atmospheric Stability

Temperature and humidity conditions in the atmosphere affect the ability of a parcel of air to lift off the ground. Meteorologists use the term *parcel* to describe a small body of air that is affected by these physical conditions. **Stability** refers to the tendency of a parcel of air with its water vapor cargo either to remain in place or to change its initial position by ascending (rising) or descending (falling). An air parcel is termed *stable* if it resists displacement upward or, when disturbed, it tends to return to its starting place. On the other hand, an air parcel is considered *unstable* when it continues to rise until it reaches an altitude where the surrounding air has a density similar to its own. This difference between an air parcel and the surrounding environment produces a buoyancy that contributes to further lifting.

Determining the degree of stability or instability involves measuring simple temperature relationships between the air parcel and the surrounding air. Such temperature measurements are made daily with balloon soundings (instrument packages called *radiosondes*) at thousands of weather stations.

Adiabatic Processes

The **normal lapse rate**, as introduced in Chapter 2, is the average decrease in temperature with increasing altitude, a value of 6.4C° per 1000 m (3.5F° per 1000 ft). This rate of temperature change is for still, calm air, but it can differ greatly under varying weather conditions, and so the actual lapse rate at a particular place and time is labeled the **environmental lapse rate**.

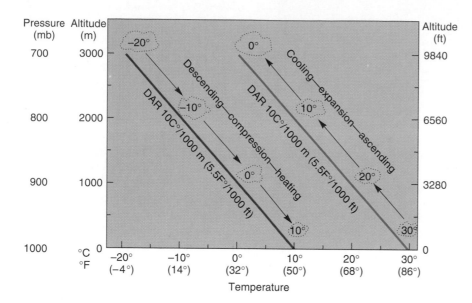

FIGURE 5-9

Vertically moving air parcels expand when they rise (because air pressure is less with increasing altitude) and are compressed when they descend. An *ascending parcel is cooled by expansion;* a *descending parcel is heated by compression.* In air parcels that are *not saturated,* temperature change in both expansion and compression occurs at the *dry adiabatic rate* (DAR)—10C° per 1000 m or 1C° per 100 m (5.5F° per 1000 ft).

In contrast, an ascending parcel of air tends to cool by expansion, responding to the reduced pressure at higher altitudes. Descending air tends to heat by compression (Figure 5-9).

Both ascending and descending temperature changes are assumed to occur *without any heat exchange between the surrounding environment and the vertically moving parcel of air.* The warming and cooling rates for a parcel of expanding or compressing air are termed **adiabatic.** (*Diabatic* means occurring *with* an exchange of heat; *adiabatic* means occurring *without* a loss or gain of heat.)

Adiabatic rates are measured with one of two specific rates, depending on moisture conditions in the parcel: *dry* adiabatic rate (DAR) and *moist* adaibatic rate (MAR).

Dry Adiabatic Rate (DAR). The **dry adiabatic rate** is the rate at which "dry" air cools by expansion (if ascending) or heats by compression (if descending). "Dry" air is less than saturated, with a relative humidity less than 100%. The DAR is 10C° per 1000 m (5.5F° per 1000 ft), as illustrated in Figure 5-9. For example, consider an unsaturated parcel of air at the surface, whose temperature measures 27°C (80°F). It is lifted, expands, and cools at the DAR as it rises from the ground to 2500 m (approximately 8000 ft). The temperature of the parcel at that altitude is 2°C (36°F), cooling by expansion as it is lifted:

10C° per 1000 m x 2500 m = 25C° of total cooling
(5.5F° per 1000 ft x 8000 ft = 44F°)

Subtracting the 25C° of cooling from the starting temperature of 27°C gives the temperature at 2500 m of 2°C. (Note that the parcel cooled adiabatically, without a loss of heat to the environment.)

Moist Adiabatic Rate (MAR). The **moist adiabatic rate** is the average rate at which ascending air that is moist (saturated) cools by expansion. The *average* MAR is 6C° per 1000 m (3.3F° per 1000 ft), or roughly 4C° (2F°) less than the DAR. However, the MAR varies with moisture content and temperature, and can range from 4C° to 10C° per 1000 m (2F° to 6F° per 1000 ft). The reason is that, in a saturated air parcel, latent heat of condensation is liberated as sensible heat, which reduces the adiabatic rate of cooling. The release of latent heat may vary, which affects the MAR. The MAR is much lower than the DAR in warm air, whereas the two rates are more similar in cold air.

Stable and Unstable Atmospheric Conditions

The relationship among the dry adiabatic rate (DAR), moist adiabatic rate (MAR), and the environmental (actual) lapse rate is complex and determines the stability of the atmosphere over an area. You can see the range of possible relationships in Figure 5-10.

Let's apply this concept to a specific situation. Figure 5-11 illustrates two of these temperature relationships in the atmosphere that lead to different conditions: unstable and stable. For the sake of illustration, both examples begin with a parcel of air

FIGURE 5-10

The relation between dry and moist adiabatic rates and environmental lapse rates produces three conditions: (a) unstable atmosphere—environmental lapse rate exceeds the DAR; (b) conditionally unstable atmosphere—environmental lapse rate is between the DAR and MAR; (c) stable atmosphere—environmental lapse rate is less than DAR and MAR.

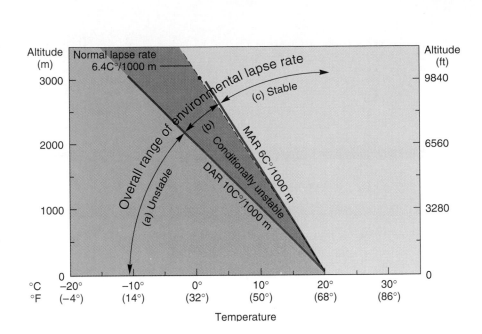

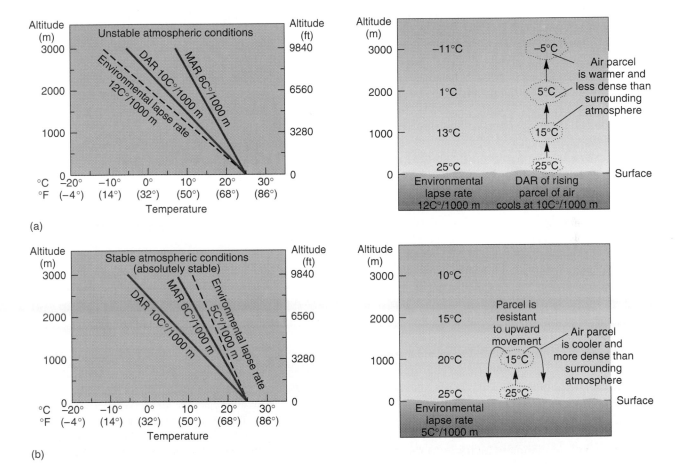

(a)

(b)

FIGURE 5-11

Specific examples of (a) unstable and (b) stable conditions in the lower atmosphere.

at the surface at 25°C (77°F). In each example, examine the temperature relationships between the parcel of air and the surrounding environment. Assume that a lifting mechanism is present (we will examine lifting mechanisms shortly).

Under the unstable conditions in Figure 5-11a, the air parcel continues to rise through the atmosphere because it is warmer (and therefore less dense) than the surrounding environment. The air surrounding the lifting parcel is cooler by an environmental lapse rate of 12C° per 1000 m (6.6F° per 1000 ft). By 1000 m (3280 ft), the parcel has adiabatically cooled by expansion to 15°C (59°F), but the surrounding air is at 13°C (55.4°F). The temperature in the parcel is 2C° (3.6F°) warmer than the surrounding air, and therefore less dense, so it continues to lift, cool, and approach saturation (100% relative humidity and cloud formation).

On the other hand, the stable conditions in Figure 5-11b are caused by an environmental lapse rate of 5C° per 1000 m (3F° per 1000 ft), which is less than both the DAR and the MAR, thus forcing the parcel of air to settle back to its original position because it is cooler (denser) than the surrounding environment.

With these stability relationships in mind, let's look at the most visible expressions of stability in the atmosphere: clouds and fog.

Clouds and Fog

Clouds are beautiful indicators of atmospheric stability, moisture content, and weather conditions. A **cloud** is an aggregation, or grouping, of moisture droplets and ice crystals that are suspended in air and are great enough in volume and density to be visible to the human eye. **Fog** is simply a cloud in contact with the ground. Although cloud types are too numerous to fully describe in this text, their classification scheme is simple and is summarized here.

Clouds are not initially composed of raindrops. Instead they are made up of a multitude of **moisture droplets**, each individually invisible to the human eye without magnification. An average raindrop, at 2000 μm diameter (0.2 cm or 0.078 in.), is made up of a million or more of these moisture droplets. After

more than a century of debate, scientists have determined how these droplets form during condensation and what processes cause them to coalesce into raindrops.

Under unstable conditions, a parcel of air may rise to an altitude where it becomes saturated—i.e., the air cools to the dew-point temperature and 100% relative humidity. Further cooling of the air parcel due to lifting produces active condensation of water vapor to water. This condensation requires microscopic particles called **condensation nuclei**, which always are present in the atmosphere.

In continental air masses, which average 10 billion nuclei per cubic meter, condensation nuclei are typically derived from ordinary dust, volcanic and forest-fire soot and ash, and particles from fuel combustion. Given the air composition over cities, great concentrations of such nuclei are available. In maritime air masses, which average 1 billion nuclei per cubic meter, the nuclei are formed from a high concentration of sea salts derived from ocean sprays. Salts are particularly attracted to moisture and thus are called *hygroscopic nuclei*. The lower atmosphere never lacks such nuclei.

Given the preconditions of saturated air, availability of condensation nuclei, and the presence of cooling (lifting) mechanisms in the atmosphere, active condensation occurs. Two principal processes account for the majority of the world's raindrops and snowflakes. These are summarized in Figure 5-12.

Cloud Types and Identification

In 1803 an English biologist named Luke Howard established a classification system for clouds and coined Latin names for them that we still use today.

Clouds usually are classified by *altitude* and by *shape*. They come in three basic forms—flat, puffy, and wispy—which occur in four primary altitude classes and ten basic types. Clouds that are developed horizontally—flat and layered—are called *stratiform* clouds. Those that are developed vertically—puffy and globular—are termed *cumuliform*. Wispy clouds usually are quite high in altitude, composed of ice crystals, and are labeled *cirroform*. These three basic forms occur in four altitudinal classes: low, middle, high, and those that are vertically de-

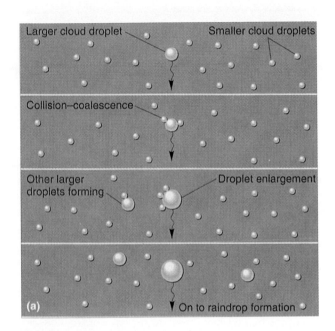

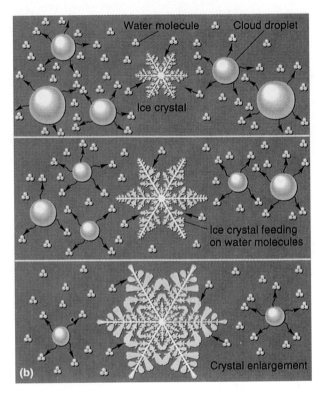

FIGURE 5-12

Principal processes for raindrop and snowflake formation: (a) the collision-coalescence process; (b) the ice-crystal process. (a) The collision-coalescence process predominates in clouds that form at above-freezing temperatures, principally in the warm clouds of the tropics. Initially, simple condensation takes place on small nuclei, some of which are larger and produce larger water droplets. As those larger droplets respond to gravity and fall through a cloud, they combine with smaller droplets, gradually coalescing into a raindrop. (b) Supercooled water droplets (minute droplets of water that are below freezing and still in liquid form) will evaporate rapidly near ice crystals, which then absorb the vapor. The ice crystals feed on the supercooled cloud droplets, grow in size, and eventually fall as snow or rain. Precipitation in middle and high latitudes begins as ice and snow high in the clouds, then melts and gathers moisture as it falls through the warmer portions of the cloud. [From Frederick K. Lutgens and Edward J. Tarbuck, *The Atmosphere: An Introduction to Meteorology*, 3d. ed., copyright © 1986, p. 127. Reprinted by permission of Prentice Hall, Inc., Englewood Cliffs, NJ.]

veloped through these altitude classes. Figure 5-13 illustrates the general appearance of the basic classes and types of clouds, and Figure 5-14 includes representative photographs.

Low clouds, ranging from the surface up to 2000 m (6500 ft) in the middle latitudes, are simply called **stratus** or **cumulus** (Latin for "layer" and "heap," respectively). Stratus clouds appear dull, gray, and featureless. When they yield precipitation, they are called **nimbostratus** (*nimbo-* denotes precipitation or rain bearing), and their showers typically fall as drizzling rain (Figures 5-13 and 5-14b).

FIGURE 5-13

Principal cloud types, classified by form (cirroform, stratiform, and cumuliform) and altitude (low, middle, high, and vertically developed across altitude).

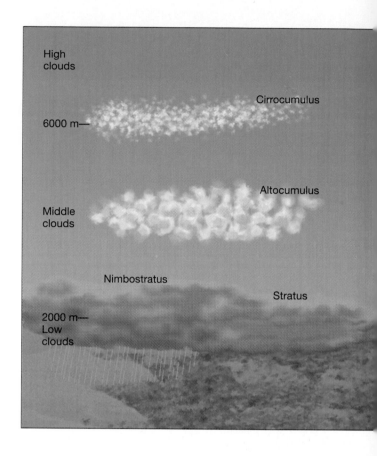

FIGURE 5-14

Principal cloud types: (a) stratus, (b) nimbostratus, (c) altostratus, (d) altocumulus, (e) cirrus, (f) cirrostratus, (g) cumulus, (h) cumulonimbus. [Photos by author.]

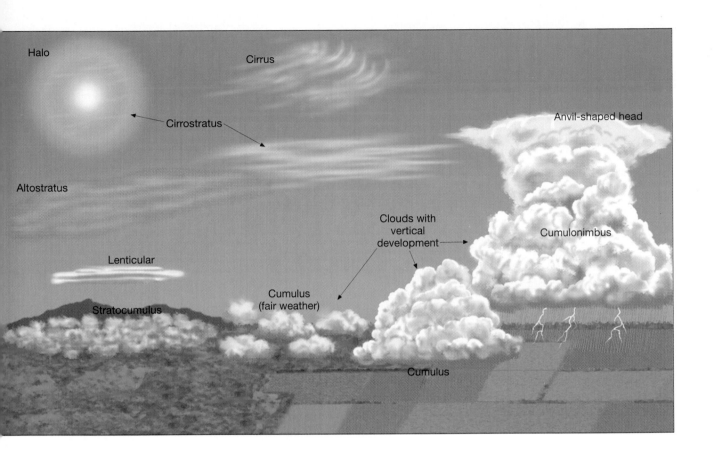

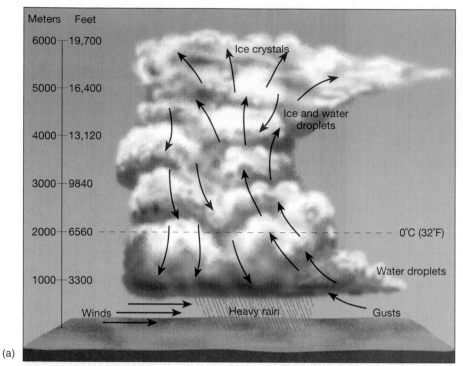

(a)

(b)

FIGURE 5-15
(a) Structure and form of a cumulonimbus cloud. Violent updrafts and downdrafts mark the circulation within the cloud. Blustery wind gusts occur along the ground.
(b) Shuttle astronauts capture a dramatic cumulonimbus thunderhead as it moves over Galveston Bay, Texas. [Space Shuttle photo from NASA.]

Cumulus clouds appear bright and puffy, like cotton balls (Figure 5-14g). When they do not cover the sky, they float by in infinitely varied shapes. Vertically developed cumulus clouds extend beyond low altitudes into middle and high altitudes and are illustrated to the far right in Figure 5-13.

Sometimes near the end of the day, lumpy, grayish, low-level clouds called **stratocumulus** may fill the sky in patches. Near sunset, these spreading puffy stratiform remnants may catch and filter the Sun's rays, sometimes indicating clearing weather.

Stratus and cumulus middle-level clouds are denoted by the prefix *alto-*. They are made of water droplets and, when cold enough, can be mixed with ice crystals. *Altocumulus* clouds, in particular, represent a broad category that occurs in many different styles.

Clouds occurring above 6000 m (20,000 ft) are composed principally of ice crystals in thin concentrations. These wispy filaments, usually white except when colored by sunrise or sunset, are termed **cirrus** clouds (Latin for "curl of hair"), sometimes dubbed mare's-tails. Cirrus clouds look as if an artist has taken a brush and placed delicate feathery strokes high in the sky (Figure 5-14e). Often, cirrus clouds are associated with an oncoming storm, especially if they thicken and lower in elevation. The prefix *cirro-* (cirrostratus, cirrocumulus) is used for other high clouds.

A cumulus cloud can develop into a towering giant called **cumulonimbus** (again, *-nimbus* denotes precipitation; Figure 5-14h). Such clouds are called thunderheads because of their shape and their associated lightning and thunder (Figure 5-15). Note the surface wind gusts, updrafts and downdrafts, heavy rain, and the presence of ice crystals at the top of the rising cloud column. High-altitude winds may shear the top of the cloud into the characteristic anvil shape of the mature thunderhead.

FIGURE 5-16
San Francisco's Golden Gate Bridge shrouded by an advection fog characteristic of summer conditions along a western coast. [Photo by author.]

Fog

By international agreement, fog is officially described as a cloud layer on the ground, with visibility restricted to less than 1 km (3300 ft). The presence of fog tells us that the air temperature and the dew-point temperature at ground level are nearly identical, producing saturated conditions. Generally, fog is capped by an inversion layer, with as much as 30C° (50F°) difference in air temperature between the ground under the fog and the clear, sunny skies above.

Two principal forms of fog related to cooling are advection fog and radiation fog. As the name implies, **advection fog** forms when air in one place migrates to another place where saturated conditions exist. When warm, moist air overlays cooler ocean currents, lake surfaces, or snow masses, the layer of air directly above the surface is chilled to the dew point. Off all subtropical west coasts in the world, summer fog forms in this manner (Figure 5-16).

A type of advection fog, involving the movement of air, forms when moist air is forced to higher elevations along a hill or mountain. This upslope lifting leads to cooling by expansion as the air rises. The resulting **upslope fog** forms a stratus cloud at the level of saturation. Along the Appalachians and the eastern slopes of the Rockies, such fog is common in winter and spring. Another fog associated with topography is called **valley fog**. Because cool air is denser, it settles in low-lying areas, producing a fog in the chilled, saturated layer near the ground.

Another type of fog forms when cold air flows over the warm water of a lake, ocean surface, or even a swimming pool. An **evaporation fog**, or steam fog, may form as the water molecules evaporate from the water surface into the cold overlying air, effectively humidifying the air. When visible at sea, the term *sea smoke* is applied to this shipping hazard.

Radiation fog forms when radiative cooling of a surface chills the air layer directly above that surface to the dew-point temperature, creating saturated conditions and fog. This fog occurs especially on clear nights over moist ground; it does not occur over water, because water does not cool appreciably

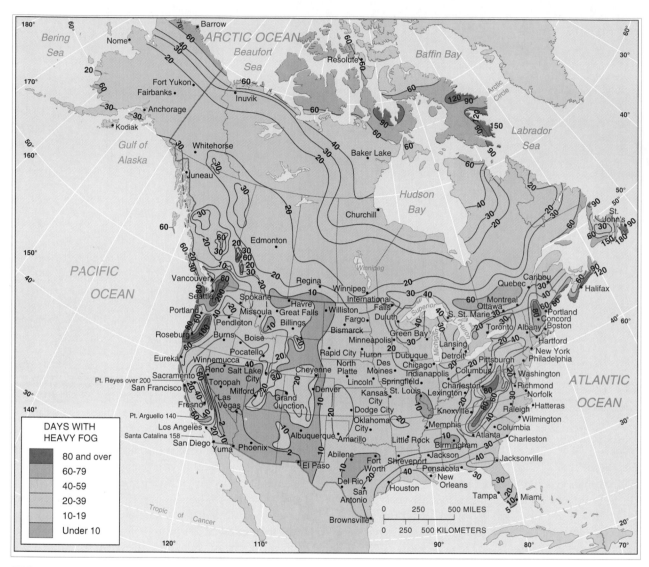

FIGURE 5-17

Mean annual number of days with heavy fog in the United States and Canada. Officially fog is declared if visibility is less than 1 km (3300 ft). The foggiest spot in the United States is the mouth of the Columbia River where it enters the Pacific Ocean at Cape Disappointment, Washington. One of the foggiest places in the world is Newfoundland's Avalon Peninsula, specifically Argentia and Belle Isle, which regularly exceed 200 days. [Data courtesy of National Weather Service; Map Series 3, *Climate Atlas of Canada*, Atmospheric Environment Service; and *The Climates of Canada,* compiled by David Philips, Senior Climatologist, Environment Canada, 1990.]

"Harvesting Fog"

Desert organisms have adapted to the presence of coastal fog along western coastlines in subtropical latitudes. For example, sand beetles live in the Namib Desert in extreme southwestern Africa. In an otherwise dry environment, they harvest water from the fog by holding up their wing surfaces so condensation collects and runs down to their mouths. As the air warms during the day and the fog evaporates, the tiny beetles retreat beneath the sand until early in the morning and the next fog—water harvesting opportunity.

In the Atacama Desert of Chile and Peru, residents stretch large nets to intercept the fog; moisture condenses on the netting and drips into barrels. Chungungo, Chile, receives water from fog harvesting in a pilot program developed by Canadian and Chilean interests.

Large sheets of plastic mesh along a ridge of the El Tofo mountains harvest water from advection fog. The captured water flows through pipes to the fishing village below. At least 22 countries distributed on six continents, experience conditions suitable for this water resource technology.

overnight. Slight movements of air deliver even more moisture to the cooled area for more fog formation of greater depth.

The prevalence of fog throughout the United States and Canada is shown in Figure 5-17. The spatial implications of fog occurrence on a local scale, related to the hazards of reduced visibility, should represent a key component in any location analysis for a proposed airport or harbor facility—a consideration that is often overlooked.

Air Masses

Specific conditions of humidity, stability, and cloud coverage occur in regional, homogenous masses of air. These air masses interact to produce weather patterns.

Each area of Earth's surface imparts its varying characteristics to the air it touches. Such a distinctive body of air is called an **air mass**, and it initially reflects the characteristics of its *source region*. The longer an air mass remains stationary over a region, the more definite its physical attributes become. Within each air mass there is a homogeneity of temperature and humidity that sometimes extends through the lower half of the troposphere.

Air masses generally are classified according to the moisture and temperature characteristics of their source regions:

1. Moisture—designated **m** for maritime (wetter) and **c** for continental (drier).

2. Temperature (latitude)—designated **A** (arctic), **P** (polar), **T** (tropical), **E** (equatorial), and **AA** (antarctic).

The principal air masses are detailed in Table 5-1, and those that affect North America in winter and summer are mapped in Figure 5-18.

Continental polar (**cP**) air masses form only in the Northern Hemisphere and are most developed in winter when they dominate cold weather conditions. The Southern Hemisphere lacks the necessary continental masses (continentality) at high latitudes to produce continental polar characteristics.

Marine polar (**mP**) air masses in the Northern Hemisphere are northwest and northeast of the

161

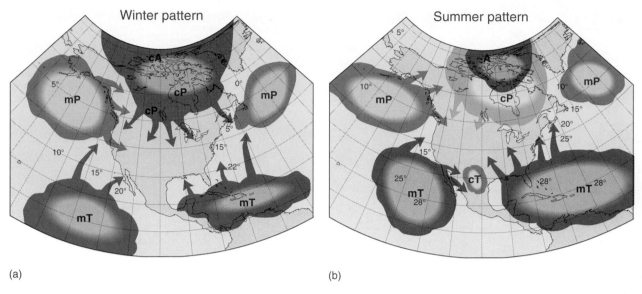

(a)

(b)

FIGURE 5–18
Air masses that influence North America and their characterisitics in (a) winter and
(b) summer. Temperatures shown are sea-surface temperatures (SSTs) in ˚C.

Table 5-1
Principal Air Masses

| Latitudinal Source Region | Moisture Source Region | |
	c: Continental	m: Marine
A: Arctic, 60–90° N	**cA:** Continental Arctic *Winter:* very cold, very dry, stable (avg. SH = 0.1 g/kg) *Summer:* cold, dry	(not applicable)
P: Polar, 40–60° N or S	**cP:** Continental Polar *Winter:* (N.H. only) cold, dry, stable, with high pressure (avg. SH = 1.4 g/kg) *Summer:* warm to cool, dry, moderately stable	**mP:** Maritime polar Cool, moist, unstable all year (avg. SH = 4.4 g/kg)
T: Tropical 15–35° N or S	**cT:** Continental tropical Hot, very dry, stable aloft, unstable at surface, turbulent in summer (avg. SH = 10 g/kg)	**mT:** Maritime tropical **mT** Gulf/Atlantic *Winter:* warm, wet, unstable *Summer:* warm, wet, very unstable (avg. SH = 17 g/kg) **mT** Pacific Weak, conditionally stable
E: Equatorial, 15° N to 15° S	(not applicable)	**mE:** Maritime equatorial Unstable, very warm, very wet, winterless (avg. SH = 19 g/kg)
AA: Antarctic, 60–90° S	**cAA:** Continental Antarctic Very cold, very dry, very stable	(not applicable)

SH = specific humidity, g/kg   N.H. = Northern Hemisphere

North American continent, and within them cool, moist, unstable conditions prevail throughout the year. The Aleutian and Icelandic subpolar low-pressure cells reside within these **mP** air masses, especially in their well-developed winter state.

North America also is influenced by two maritime tropical (**mT**) air masses—the **mT** *Gulf/Atlantic* and the **mT** *Pacific*. The humidity experienced in the East and Midwest is created by the **mT** *Gulf/Atlantic* air mass, which is particularly unstable and active from late spring to early fall. In contrast, the **mT** *Pacific* is stable to conditionally unstable and generally much lower in moisture content and available energy. As a result, the western United States, influenced by this weaker Pacific air mass, receives lower average precipitation than the rest of the country.

Please review Figure 4-16 and the discussion of subtropical high-pressure cells off the coast of North America—the moist, unstable conditions on the western edge (east coast), and the drier, stable conditions on the eastern edge (west coast). These conditions, coupled respectively with warmer and cooler ocean currents, produce the characteristics of each air mass source region.

As air masses migrate from their source regions, their temperature and moisture characteristics are modified. For example, an **mT** *Gulf/Atlantic* air mass may carry humidity to Chicago and on to Winnipeg, but gradually will lose its initial characteristics with each day's passage northward.

Atmospheric Lifting Mechanisms

If air masses are to cool adiabatically (by expansion), and if they are to reach the dew-point temperature and saturate, condense, form clouds, and perhaps precipitate, then they must be lifted. Three principal lifting mechanisms operate in the atmosphere: *convectional lifting* (stimulated by local surface heating), *orographic lifting* (when air is forced over a barrier such as a mountain range), and *frontal lifting* (along the leading edges of contrasting air masses). Descriptions of all three mechanisms follow.

Convectional Lifting

When an air mass passes from a maritime source region to a warmer continental region, heating from the warmer land causes lifting in the air mass. If conditions are *unstable*, initial lifting is sustained and clouds develop. Figure 5-19 illustrates convectional action stimulated by local heating, with unstable conditions present in the atmosphere. The rising parcel of air continues its ascent because it is warmer (less dense) than the surrounding environment.

In the figure, the MAR is used above the *lifting condensation level,* which is visible at the elevation of the cloud base where the rising air mass becomes saturated. Buoyancy is added at this level through the release of latent heat by condensation. Continued lifting above this altitude produces cooling of

FIGURE 5-19

Local heating and convection under unstable atmospheric conditions. Given specific humidity and temperature conditions within the lifting parcel of air and the unstable conditions in the environment, the dew point is reached at 1400 m (4600 ft). Note that the DAR is used when the air parcel is less than saturated, changing to the MAR above the condensation level.

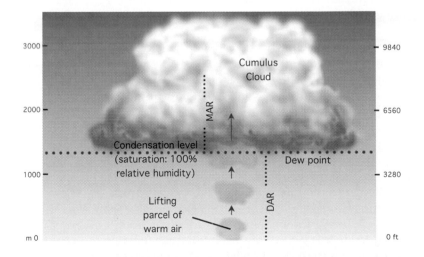

FIGURE 5-20
Convectional activity over the Florida peninsula. Cumulus clouds cover the land, with several cells developing into cumulonimbus thunderheads. Warm, moist air from the Gulf of Mexico is lifted by local heating as it passes over the land. [Project Gemini photo from NASA.]

the air temperature and the dew-point temperature at the MAR.

As an example of local heating and convectional lifting, Figure 5-20 depicts a day on which the landmass of Florida was warmer than the surrounding Gulf of Mexico and Atlantic Ocean. Because the land gradually heats throughout the day and warms the air above it, convectional showers tend to form in the afternoon and early evening. The land becomes covered with clouds formed by this process. Towering cumulonimbus clouds are an expected summertime feature in the regions of North America affected by **mT** *Gulf/Atlantic* air masses, and to a lesser extent by the weaker **mT** *Pacific* air mass. Convectional precipitation also dominates along the intertropical convergence zone (ITCZ), over tropical islands, and anywhere that moist, unstable air is heated from below, or where the inflowing trade winds produce a dynamic convergence.

Orographic Lifting

The physical presence of a mountain acts as a topographic barrier to migrating air masses. **Orographic lifting** occurs when air is lifted upslope as it is pushed against a mountain. It cools adiabatically. Stable air that is forced upward in this manner may produce stratiform clouds, whereas unstable or con-

ditionally unstable air usually forms a line of cumulus and cumulonimbus clouds. An orographic barrier enhances convectional activity and causes additional lifting during the passage of weather fronts and cyclonic systems, thereby extracting more moisture from passing air masses. Figures 5-21 and 5-22 illustrate the operation of orographic lifting.

The wetter intercepting slope is termed the *windward slope*, as opposed to the drier far-side slope, known as the *leeward slope*. Moisture is condensed from the lifting air mass on the windward side of the mountain; on the leeward side the descending air mass is heated by compression, causing evaporation of any remaining water in the air. Thus, air can begin its ascent up a mountain warm and moist but finish its descent on the leeward slope hot and dry. In North America, *chinook winds* (called föhn or foehn winds in Europe) are the warm, downslope air flows characteristic of the leeward side of mountains. Such winds can bring a 20C° (36F°) jump in temperature and greatly reduced relative humidity.

The term **rain shadow** is applied to dry regions leeward of mountains. East of the Cascade Range, Sierra Nevada, and Rocky Mountains, such rain-shadow patterns predominate. In fact, the precipitation pattern of windward and leeward slopes persists worldwide, as confirmed by the precipitation maps for North America in Chapter 6 and the world shown in Chapter 7.

FIGURE 5-21
Orographic barrier and precipitation patterns—unstable conditions assumed. Warm, moist air is forced upward against a mountain range, producing adiabatic cooling and eventual saturation. On the leeward slopes, compressional heating as the air travels downward leads to hot, relatively dry air in the rain shadow of the mountain.

FIGURE 5-22
Orographic lifting along the Pacific Coast. Can you identify the level at which the dew point was reached? [Photo by author.]

Frontal Lifting

The leading edge of an advancing air mass is called its **front**. That term was applied by Vilhelm Bjerknes (1862–1951) and a team of meteorologists working in Norway during World War I, because it seemed to them that migrating air-mass "armies" were doing battle along their fronts. A front is a place of atmospheric discontinuity, a narrow zone that is the contact between two air masses of considerable difference in temperature, pressure, humidity, wind direction and speed, and cloud development. The leading edge of

"Mountains Set Precipitation Records"

Orographic lifting is limited in areal extent to locations where a topographic barrier exists, thus making it the least dominant lifting mechanism. However, mountain ranges are the most consistent precipitation-inducing mechanisms. Both the greatest average annual precipitation and maximum annual precipitation on Earth occur on the windward slopes of mountains that intercept moist tropical trade winds.

The world's greatest average annual precipitation occurs in the United States on the windward slopes of Mount Waialeale, on the island of Kauai, Hawaii, which rises 1569 m (5147 ft) above sea level. Rainfall averages 1234 cm (486 in. or 40.5 ft) a year (records for 1941-1992). In contrast, the rain-shadow side of Kauai receives only 50 cm (20 in.) of rain annually. If no islands existed at this location, this portion of the Pacific Ocean would receive only an average 63.5 cm (25 in.) of precipitation a year.

Another place receiving world-record precipitation is Cherrapunji, India, 1313 m (4309 ft) above sea level at 25° N latitude, in the Assam Hills south of the Himalayas. It is located just north of Bangladesh, where tropical cyclones, torrential rains, and flooding in 1970 and 1991 killed hundreds of thousands of people. Because of the summer monsoons that pour in from the Indian Ocean and the Bay of Bengal, Cherrapunji has received 930 cm (366 in. or 30.5 ft) of rainfall in *one month* and a total of 2647 cm (1042 in. or 86.8 ft) in *one year*! Not surprisingly, Cherrapunji is the all-time precipitation record holder for a single year and for every other time interval from 15 days to 2 years. The average annual precipitation there is 1143 cm (450 in., 37.5 ft), placing it second only to Mount Waialeale.

---

a cold air mass is a **cold front**, whereas the leading edge of a warm air mass is a **warm front**.

Cold Front. On weather maps, such as the one illustrated in Figure 5-28, a cold front is identified by a line marked with triangular spikes pointing in the direction of frontal movement along an advancing **cP** air mass. The steep face of the cold air mass suggests its ground-hugging nature, caused by its density and uniform physical character (Figure 5-23).

Warmer, moist air in advance of the cold front is lifted upward abruptly and is subjected to the same adiabatic rates and factors of stability/instability that pertain to all lifting air parcels. A day or two ahead of the cold front's passage, high cirrus clouds appear, telling observers that a lifting mechanism is on the way. The cold front's advance is marked by a wind shift, temperature drop, and lowering barometric pressure. Air pressure reaches a local low as the line of most intense lifting passes, usually just ahead of the front itself. Clouds may build up along the cold front into characteristic cumulonimbus types and may appear as an advancing wall of clouds. Precipitation usually is hard, containing large droplets, and can be accompanied by hail, lightning, and thunder. The aftermath usually brings northerly winds in the Northern Hemisphere (southerly winds in the Southern Hemisphere), lower temperatures, increasing air pressure from the cooler, denser air, and broken cloud cover.

A fast-advancing cold front can cause violent lifting and create a zone slightly ahead of the front

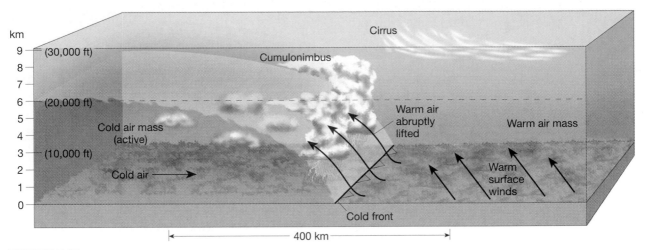

FIGURE 5-23

A typical cold front. Warm moist air is forced to lift abruptly by denser advancing cold air. As the air lifts, it cools by expansion at the DAR, cooling to the dew-point temperature as it rises to a level of active condensation and cloud formation. Cumulonimbus clouds may produce large rain drops, heavy showers, lightning and thunder, and hail.

called a *squall line*. Along a squall line such as the one in the Gulf of Mexico shown in Figure 5-24, wind patterns are turbulent and wildly changing and precipitation is intense. The well-defined wall of clouds in the photo rises abruptly, with new thunderstorms forming along the distinct front. Tornadoes also may develop along such a squall line.

Warm Front. A warm front is denoted on weather maps by a line marked with semicircles facing the direction of frontal movement (Figure 5-28). The leading edge of an advancing warm air mass is unable to displace cooler, passive air, which is more dense. Instead, the warm air tends to push the cooler, underlying air into a characteristic wedge shape, with the warmer air sliding up over the cooler air. Thus, in the cooler-air region, a temperature inversion is present, sometimes causing poor air drainage and stagnation.

Figure 5-25 illustrates a typical warm front in which **mT** air is gently lifted, leading to stratiform cloud development and characteristic nimbostratus clouds and drizzly precipitation. Associated with the warm front are the high cirrus and cirrostratus clouds that announce the advancing frontal system. Closer

FIGURE 5-24

Squall line in the Gulf of Mexico with clouds rising to 16,800 m (55,120 ft). The passage of such a frontal system over land often produces strong winds and possibly tornadoes. [Space Shuttle photo from NASA.]

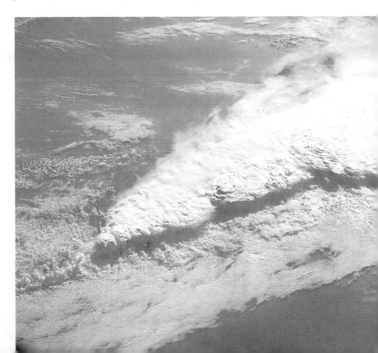

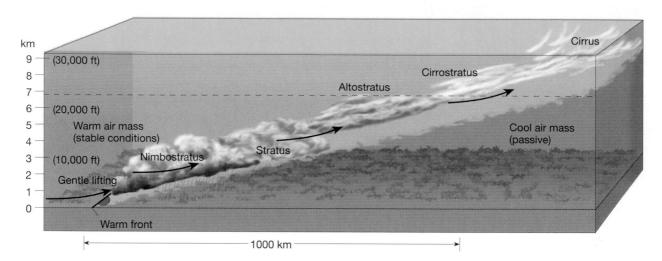

FIGURE 5-25

A typical warm front. Note the sequence of cloud development as the warm front approaches. Warm air slides upward over a wedge of cooler, passive air near the ground. Gentle lifting of the warm, moist air produces stratus and nimbostratus clouds and drizzly rain showers in contrast to the more dramatic precipitation associated with the passage of a cold front.

to the front, the clouds lower and thicken to alto-stratus, and then to stratus within several hundred kilometers of the front.

Midlatitude Cyclonic Systems

The conflict between contrasting air masses along fronts leads to conditions appropriate for the development of a midlatitude **wave cyclone**, a migrating center of low pressure, with converging, ascending air, spiraling inward counterclockwise in the Northern Hemisphere (or converging clockwise in the Southern Hemisphere). This cyclonic motion is generated by the pressure gradient force, the Coriolis force, and surface friction, discussed in Chapter 4.

Wave cyclones form a dominant type of weather pattern in the middle and higher latitudes of both the Northern and Southern Hemispheres and act as a catalyst for air-mass conflict. Such a *midlatitude cyclone*, or extratropical cyclone, can be born along the polar front, particularly in the region of the Icelandic and Aleutian subpolar low-pressure cells in the Northern Hemisphere (see Figures 4-15 and 4-18).

Development and strengthening of a midlatitude wave cyclone is known as *cyclogenesis*. In addition to the polar front, certain other areas are associated with wave cyclone development and intensification: the eastern slope of the Rockies and other north-south mountain barriers, the Gulf Coast, and the east coasts of North America and Asia.

Life Cycle of a Midlatitude Cyclone

Figure 5-26 shows the birth, maturity, and death of a typical midlatitude cyclone in four stages. On the average, a midlatitude cyclone takes 3–10 days to progress through these stages and cross the North American continent.

Along the polar front, cold and warm air masses converge and conflict. The polar front forms a discontinuity of temperature, moisture, and winds that establishes potentially unstable conditions. However, for a wave cyclone to form along the polar front, a point of air *convergence* at the surface must be matched by a compensating area of air *divergence* aloft. Even a slight disturbance along the polar front, perhaps a small change in the path of the jet stream, can initiate the converging, ascend-

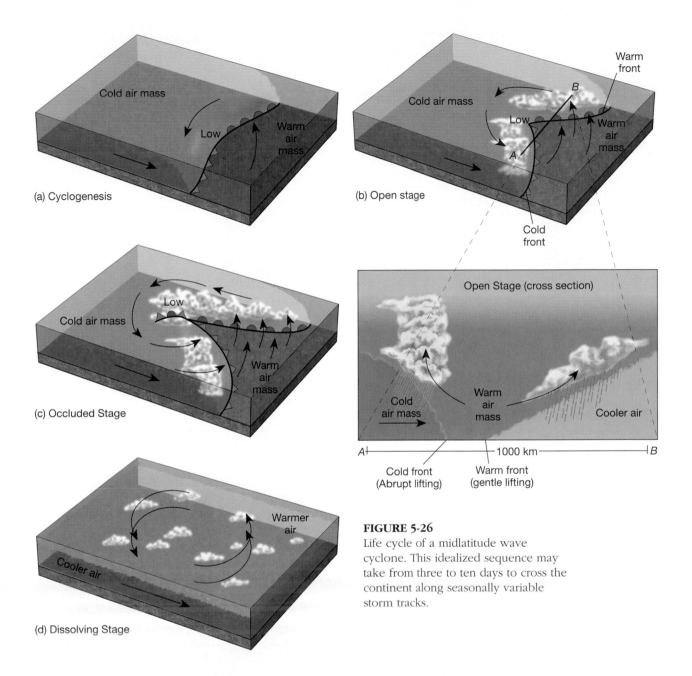

(a) Cyclogenesis

(b) Open stage

(c) Occluded Stage

(d) Dissolving Stage

Open Stage (cross section)

Cold front (Abrupt lifting)   Warm front (gentle lifting)

FIGURE 5-26
Life cycle of a midlatitude wave cyclone. This idealized sequence may take from three to ten days to cross the continent along seasonally variable storm tracks.

ing motion of a surface low-pressure system (Figure 5-26a).

As the midlatitude cyclone matures, the counterclockwise flow (in the Northern Hemisphere) draws the cold air mass from the north and west and the warm air mass from the south. A characteristic open stage forms as shown in Figure 5-26b and in the enlarged cross section.

These cyclonic storms—1600 km (1000 mi) wide—and their attendant air masses move across the continent along storm tracks, which shift latitudinally with the Sun, crossing North America far-

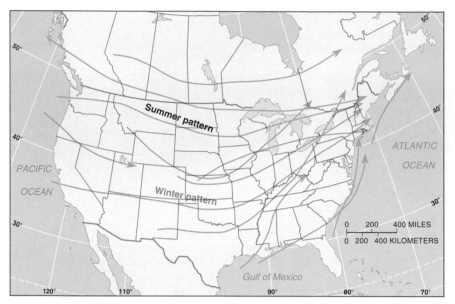

FIGURE 5-27
Cyclonic storm tracks for North America vary seasonally.

ther to the north in summer and farther to the south in winter (Figure 5-27). As the storm tracks begin to shift northward in the spring, **cP** and **mT** air masses are brought into their clearest conflict. This is the time of strongest frontal activity, with associated thunderstorms and tornadoes. Storm tracks follow the path of upper-air winds and jet streams which direct storm systems across the continent.

Because the **cP** air mass is more homogeneous in temperature and pressure than the **mT** air mass, the cold front leads a denser, more unified mass, and therefore moves faster than the warm front. Cold fronts can average 40 kmph (25 mph), whereas warm fronts average 16–24 kmph (10–15 mph). Thus, a cold front may overrun the cyclonic warm front, producing an **occluded front** within the overall cyclonic circulation (Figure 5-26c). Precipitation may be moderate to heavy initially and then taper as the warmer air wedge is lifted higher by the advancing cold air mass.

The final dissolving stage of the midlatitude cyclone occurs when its lifting mechanism is completely cut off from the warm air mass, which was its source of energy and moisture (Figure 5-26d). Remnants of the cyclonic system then dissipate in the atmosphere.

Daily Weather Map and the Midlatitude Cyclone

The four-stage sequence in Figure 5-26 illustrates the evolution in an ideal midlatitude cyclone model, actually the pattern of cyclonic passage over North America demonstrates a great variety of shape, form, and duration. However, we can apply this model along with our understanding of warm and cold fronts to an actual weather-map for April 2, 1988 (Figure 5-28).

On the April 2 map you can identify the temperatures reported by various stations, the patterns created, and the location of warm and cold fronts. The distribution of air pressure is defined by the use of *isobars,* the lines connecting points of equal pressure on the map. Although the low-pressure center identified is not intense, it is well defined over eastern Nebraska, with winds following in a counterclockwise, cyclonic path around the low pressure. The map and satellite image also illustrate the pattern of precipitation. The **mT** air mass, advancing northward from the Gulf/Atlantic source region, provides the moisture for afternoon and evening thundershowers. Note the plot of the storm track (dashed line) shown on the map.

To improve data gathering and weather forecasting, a new series of weather satellites is being developed.

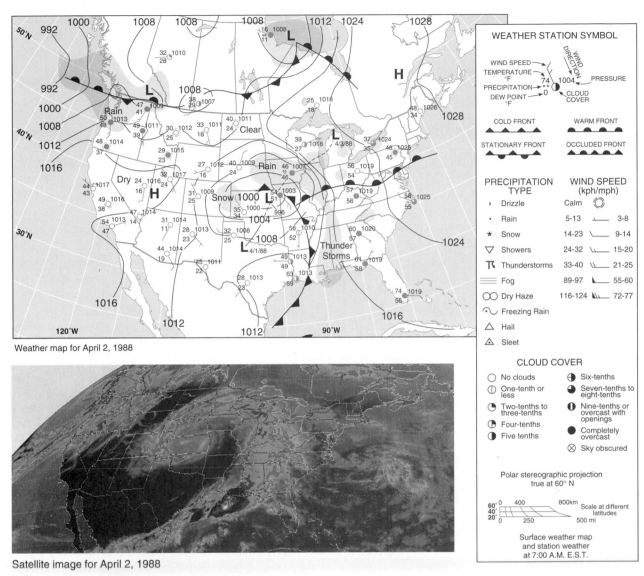

Weather map for April 2, 1988

Satellite image for April 2, 1988

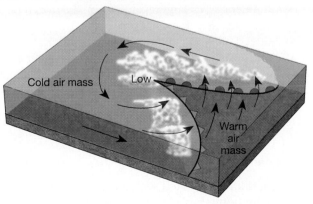

FIGURE 5-28

Weather map, *GOES-7* satellite image for April 2, 1988, and model of a midlatitude cyclonic system. The standard weather symbols used are identified and explained. Note the track of the center of low pressure from April 1 through April 3, beginning in Texas and ending in northern Michigan (dotted line). [National Weather Service and National Oceanic and Atmospheric Administration, National Climatic Center, Satellite Data Services Division.]

The first of these is *GOES-8* (*Geostationary Operational Environmental Satellite*), launched April 13, 1994. It is parked 36,000 km (22,400 mi) above Earth's surface at 75° west longitude. The images produced exceed in detail and content those from satellites previously used.

Violent Weather

Weather provides a continuous reminder of nature's power. The flow of energy across the latitudes can at times set into motion violent and potentially destructive conditions. Thunderstorms, tornadoes, and tropical cyclones are such forms of violent weather. We now consider each.

Thunderstorms

Tremendous energy is liberated by the condensation of large quantities of water vapor. This process locally heats the air causing violent updrafts and downdrafts as rising parcels of air pull surrounding air into the column and as the frictional drag of raindrops pulls air toward the ground. As a result, giant cumulonimbus clouds can create dramatic weather moments—squall lines of heavy precipitation, lightning, thunder, hail, blustery winds, and tornadoes. Thunderstorms may develop within an air mass, in a line along a front (particularly a cold front), or where mountain slopes cause orographic lifting.

Thousands of thunderstorms occur on Earth at any given moment. Equatorial regions and the ITCZ experience many of them, as exemplified by the city of Kampala in Uganda, East Africa (north of Lake Victoria), which sits virtually on the equator and averages 242 days a year of thunderstorms—a record. Figure 5-29 shows the distribution of annual days with thunderstorms across the United States and Canada. You can see that, in North America, most thunderstorms occur in areas dominated by **mT** air masses.

Lightning and Thunder. An estimated 8 million lightning strikes occur each day on Earth (see the chapter opening photo). **Lightning** refers to flashes of light caused by enormous electrical discharges—tens to hundreds of millions of volts—which briefly superheat the air to temperatures of 15,000° to 30,000°C (27,000°–54,000°F). The violent expansion of this abruptly heated air sends shock waves through the atmosphere—the sonic bangs known as **thunder**. The greater the distance a lightning stroke travels, the longer the thunder echoes. Lightning at great distance from the observer may not be accompanied by thunder and is called *heat lightning*.

Lightning is created by a buildup of electrical energy between areas within a cumulonimbus cloud or between the cloud and the ground, with sufficient electrical potential to overcome the high electrical resistance of the atmosphere and leap from one surface to the other—a giant spark!

Lightning poses a hazard to aircraft and to people, animals, trees, and structures. Individuals should acknowledge that certain precautions are mandatory when the threat of a lightning discharge is near. Nearly 200 deaths and thousands of injuries occur each year in the United States and Canada. When such lightning is imminent, agencies such as the National Weather Service issue severe storm warnings and caution people to remain indoors.

Hail. Hail is common in the United States and Canada, although somewhat infrequent at any given location, hitting a specific location perhaps every one or two years in the highest frequency areas. Annual hail damage in the United States tops $750 million. **Hail** generally forms within a cumulonimbus cloud when a raindrop circulates repeatedly above and below the freezing level in the cloud, adding layers of ice until its weight no longer can be supported by the circulation in the cloud.

Pea-sized and marble-sized hail are common, although hail the size of golf balls and baseballs also is reported at least once or twice a year somewhere in North America. For larger hail to form, the frozen pellets must stay aloft for longer periods. The largest authenticated hailstone in the world landed in Kansas in 1970; it measured 14 cm (5.6 in.) in diameter and weighed 758 grams (1.67 pounds)! The pattern of hail occurrence across the United States and Canada is similar to that of thunderstorms shown in Figure 5-29, dropping to zero approximately by the 60th parallel.

Tornadoes

The updrafts associated with a cold-front squall line and cumulonimbus cloud development appear on

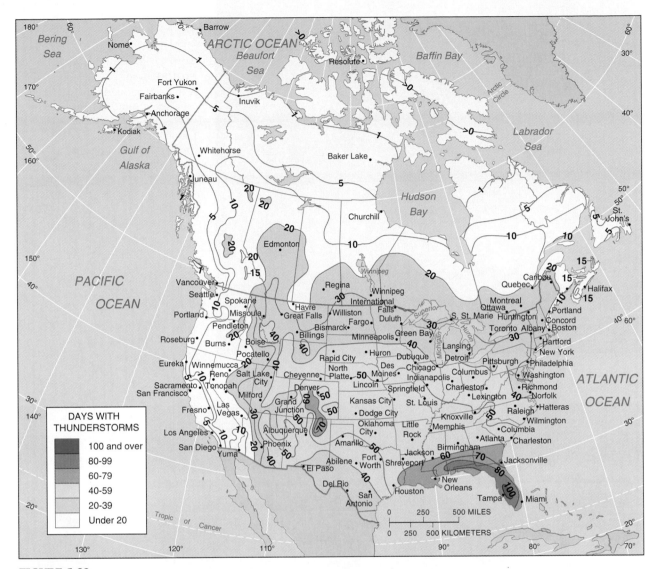

FIGURE 5-29

Average annual number of days with thunderstorms. [Courtesy of National Weather Service and Map Series 3, *Climatic Atlas of Canada,* Atmospheric Environment Service.]

satellite images as pulsing bubbles of clouds. According to one hypothesis, a spinning, cyclonic, rising column of mid-troposphere-level air forms a **mesocyclone**. A mesocyclone can range up to 10 km (6 mi) in diameter and rotate over thousands of meters vertically within the parent cloud. As a mesocyclone extends vertically and contracts horizontally, wind speeds accelerate in an inward vortex (much as ice skaters accelerate while spinning by pulling their arms in closer to their bodies). A well-developed mesocyclone most certainly will produce heavy rain, large hail, blustery winds, and lightning; some mature mesocyclones will generate tornado activity.

As more moisture-laden air is drawn up into the circulation of a mesocyclone, more energy is liberated, and the rotation of air becomes more rapid. The narrower the mesocyclone, the faster the spin of converging parcels of air being sucked into the rotation. The swirl of the mesocyclone itself is visible, as are smaller, dark-gray **funnel clouds** that pulse from the bottom side of the parent cloud. The terror of this stage of development is the lowering of a funnel cloud to Earth—a **tornado** (Figure 5-30).

FIGURE 5-30
A fully developed tornado locks onto the ground. [Photo courtesy of National Severe Storms Laboratory, National Oceanic and Atmospheric Administration.]

FIGURE 5-31
Aerial view of tornado devastation to homes and lives in Edmond, Oklahoma. [Photo by Chris Johns/Allstock.]

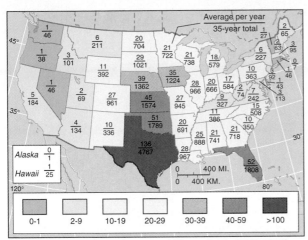

(a)   Average tornadoes per year and 35-year totals (1959-1993)

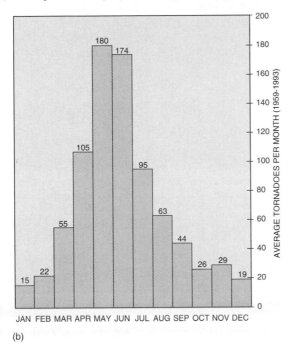

(b)

FIGURE 5-32
(a) Average tornadoes per year/total tornadoes for the past 35 years (1959 to 1993); (b) average tornadoes per month, 1959-1993 (U.S. tornado data). During an average year, at least 830 tornadoes strike throughout the United States and Canada. [Data courtesy of John Halmstad, National Severe Storm Forecast Center, National Weather Service.]

A tornado can range from a few meters to a few hundred meters in diameter and can last anywhere from a few moments to tens of minutes. Pressures inside a tornado usually are about 10% less than those in the surrounding air. The in-rushing convergence created by such a horizontal pressure gradient causes severe wind speeds that can exceed 485 kmph (300 mph), with at least a third of all tornadoes exceeding 182 kmph (113 mph). The damage produced by such wind is catastrophic (Figure 5-31). (See News Report #4, next page.)

All 50 states have experienced tornadoes, as have all the Canadian provinces and territories. May and June are the peak months, as you see in Figure 5-32b. A small number of tornadoes are reported on other continents each year, but North America receives the greatest share because its latitudinal position and shape permit contrasting air masses access to one another.

When tornado circulation occurs over water, a **waterspout** forms, and surface water is drawn up into the funnel some 3–5 m (10–16 ft). The rest of the waterspout funnel is made visible by the rapid condensation of water vapor.

Tropical Cyclones

A powerful manifestation of Earth's energy and moisture systems is the **tropical cyclone**, which originates entirely within tropical air masses. The tropics extend from the Tropic of Cancer at 23.5° N to the Tropic of Capricorn at 23.5° S, encompassing the equatorial zone between 10° N and 10° S. Approximately 80 tropical cyclones occur annually worldwide. Table 5-2 presents the criteria for classification of tropical cyclones based on wind speed.

The cyclonic systems forming in the tropics are quite different from midlatitude cyclones because the air of the tropics is essentially homogeneous, with no fronts or conflicting air masses. In addition, the warm air and warm seas ensure an abundant supply of water vapor and thus the necessary latent heat to fuel these storms.

Meteorologists now think that the cyclonic motion begins with slow-moving easterly waves of low pressure in the trade wind belt of the tropics, such as the Caribbean area. It is along the eastern (leeward) side of these migrating troughs of low pressure, a place of convergence and rainfall, that tropical cyclones are formed. Surface air flow then spins in toward the low-pressure area (convergence), ascends, and flows outward aloft. This important divergence aloft acts as a chimney, pulling more moisture-laden air into the developing system.

"Recent Tornado Records Are a Call for Action"

In the United States, 28,928 tornadoes were recorded in the 35 years from 1959 through 1993, with 755 of these causing 2684 deaths. In addition, these tornadoes resulted in over 55,000 injuries and damages over $16.7 billion. The yearly average of $475 million in damage is rising at about $48 million a year.

Interestingly, 1992 set a record for the most tornadoes in the United States in a single year, reaching an all-time high of 1297. In June alone, 421 occurred. The year 1993 ranked second with 1173 tornadoes; 1990 ranked third with 1133 tornadoes; 1991 ranked fourth with 1132. The annual average tornado occurrences for the preceding ten years was only 827. The reason for this marked increase in tornado frequency in North America is being researched, with explanations ranging from global climate change to better reporting by a larger population.

It is not enough to account statistically for past tornadoes; the need is for accurate forecasting. The National Severe Storms Forecast Center in Kansas City, Missouri, is the U. S. government's headquarters for such activities. Modernization of the National Weather Service includes increasing use of *Doppler radar* that enables scientists to detect the specific flow of moisture droplets in mesocyclones and allows a warning period of 30 minutes to 1 hour in areas at risk. A severe storm watch sometimes is possible a few hours in advance. But at other times, this most intense phenomenon of the atmosphere can strike with destructive suddenness and little warning.

Table 5-2

Tropical Cyclone Classification

Designation	Winds	Features
Tropical disturbance	Variable, low	Definite area of surface low pressure; patches of clouds
Tropical depression	Up to 34 knots (63 kmph, 39 mph)	Gale force, organizing circulation, light to moderate rain
Tropical storm	35 to 63 knots (63–118 kmph, 39–73 mph)	Closed isobars, definite circular organization, heavy rain, assigned a name
Hurricane (Atlantic and E. Pacific) Typhoon (W. Pacific) Cyclone (Indian Ocean, Australia)	Greater than 64 knots (119 kmph, 74 mph)	Circular, closed isobars; heavy rain, storm surges, tornadoes in right-front quadrant

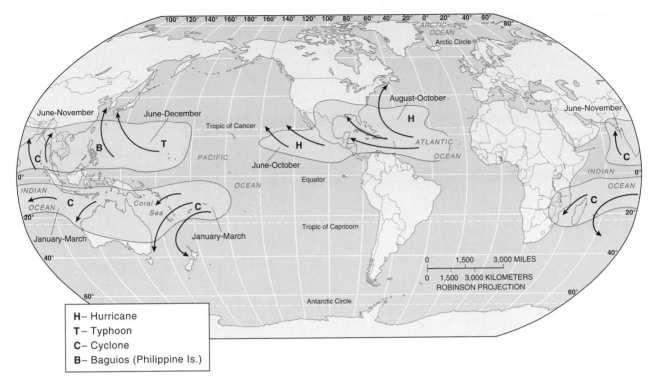

FIGURE 5-33

Worldwide pattern of the most intense tropical cyclones. These are typical tropical storm tracks with principal months of occurrence and regional names. For the North Atlantic region from 1871 to 1986, 934 tropical cyclones (storms and hurricanes) developed. The peak day in this Atlantic region (assuming a 9-day moving average calculation) is September 10, with 60 cyclones for the 116-year period.

The map in Figure 5-33 shows tropical cyclone formation areas, some of the characteristic storm tracks traveled, and indicates the range of months during which tropical cyclones are most likely to appear. As the map shows, tropical cyclones tend to occur when the equatorial low-pressure trough is farthest from the equator, that is, during the months that follow the summer solstice in each hemisphere. About 10% of all tropical disturbances have the right ingredients to intensify into a full-fledged **hurricane** or **typhoon**.

Hurricanes and Typhoons. Tropical cyclones are potentially the most destructive storms experienced by humans, costing an average of several hundred million dollars in property damage and thousands of lives each year. The tropical cyclone that struck Bangladesh in 1970 killed an estimated 300,000 people and the one in 1991 claimed over 200,000. In the United States, death tolls are much lower, but still significant: the Galveston, Texas, hurricane of 1900 killed 6000; Hurricane Audry (1957), 400; Hurricane Gilbert (1988),

318; Hurricane Camille (1969), 256; and Hurricane Agnes (1972), 117. FYI Report 5-1, "1992–The Year of Hurricanes," discusses the recent hurricanes.

Fully organized tropical cyclones possess an intriguing physical appearance. They range in diameter from a compact 160 km (100 mi), to 960 km (600 mi), to some western Pacific typhoons that reach 1300–1600 km (800–1000 mi). Vertically, these storms dominate the full height of the troposphere. The inward-spiraling clouds form dense *rain bands*, with a central area designated the *eye*, around which a thunderstorm cloud called the *eyewall* swirls, producing the area of most intense precipitation. The eye remains an enigma, for in the midst of devastating winds and torrential rains the eye has quiet, warm air with even a glimpse of blue sky or stars possible.

Hurricane Gilbert (September 1988) illustrates these general characteristics. For a Western Hemisphere hurricane, Gilbert achieved the record size (1600 km or 1000 mi in diameter) and the record low barometric pressure (888 mb or 26.22 in. of mercury). In addi-

(a)

FIGURE 5-34
Hurricane Gilbert, September 13, 1988: (a) *GOES-7* satellite image; (b) side-view radar image (Side-Looking Airborne Radar–*SLAR*) taken from an aircraft flying through the central portion of the storm. Rain bands of greater cloud density are false-colored in yellows and reds. The clear sky in the central eye is dramatically portrayed in both the satellite and radar images. [National Oceanic and Atmospheric Administration and the National Hurricane Center, Miami.]

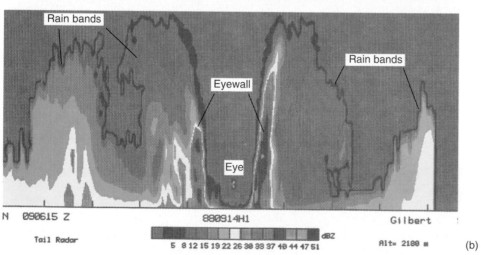

(b)

"Benefits from Hurricanes?"

According to the weather records of the past 50 years, about one-third of all hurricanes making landfall in the United States have created beneficial conditions inland in terms of the precipitation contributed to local water budgets. Dry soil conditions and low reservoirs are recharged by the rains, streams are flushed of pollution, and local water tables are restored. The tragic loss of life must not be forgotten, of course. In terms of overall effect, hurricanes should be viewed as normal and natural meteorological events that certainly have destructive potential but also contribute to the precipitation regimes of the southern and eastern United States.

One of the most devastating hurricanes to strike the United States coastline this century was Hurricane Camille. Ironically, it was significant not only for the disaster it brought (256 dead, $1.5 billion damage), but also for the moisture benefit it delivered and the drought it abated in Mississippi, Tennessee, and Kentucky. Camille's monetary benefits inland outweighed damage costs by 2 to 1, it is estimated.

---

tion, it sustained winds of 298 kmph (185 mph) with peak winds exceeding 320 kmph (200 mph). Figure 5-34 portrays Gilbert from overhead and in cross section. From these images you can see the vast counterclockwise circulation, the central eye, eyewall, rain bands, and overall hurricane appearance.

The entire storm travels at 16–40 kmph (10–25 mph). When it moves ashore, or makes *landfall*, it pushes *storm surges* of seawater inland. The strongest winds of a tropical cyclone are usually recorded in its right-front quadrant (relative to the storm's directional path), where dozens of fully developed tornadoes may be embedded at the time of landfall. For example, Hurricane Camille in 1969 had up to 100 tornadoes embedded in that quadrant.

Typhoons form in the vast region of the Pacific, where the annual frequency of tropical cyclones is greatest. The magnitude and intensity of typhoons generally exceed those of hurricanes. In fact, the lowest sea-level pressure recorded on Earth was 870 mb (25.69 in.) in the center of Typhoon Tip, in October 1979, 837 km (520 mi) northwest of Guam (17° N 138° E). The Geostationary Meteorological Satellite (GMS), operated by the Japan Weather Association, provides images for tracking these storms.

SUMMARY—Atmospheric Water and Weather

The next time it rains where you live, pause and reflect on the journey each of those water molecules has made. Water molecules came from within Earth over billions of years, in a process called **outgassing**, and began endless cycling through the hydrologic system of evaporation-condensation-precipitation. Water is the most common compound on the surface of Earth and it possesses unusual heat characteristics. Part of Earth's uniqueness is that its water naturally exists in all three states—solid, liquid, and gas; this is not so for any other planet in the Solar System. Liquid water covers about 71% of Earth. Approximately 97% of it is salty seawater, and the remaining 3% is freshwater—most of it frozen.

In many ways **water vapor** is the most important gas in the atmosphere. The transfer of massive

1992–The Year of Hurricanes

The period between August 24 and September 11, 1992, displayed nature's power and capacity to cause personal, societal, and economic tragedy. These 19 days saw Hurricane Andrew (Florida and Louisiana), Hurricane Iniki (Kauai, Hawaii), and Typhoon Omar (Guam) strike with record-breaking fury. Figure 1 presents three AVHRR images for August 24, 25, and 26, showing Andrew as it moved across Florida, westward across the Gulf of Mexico, and onshore in Louisiana. The central eye is clearly visible, as are the tight rainbands marking the rapid inward-spiraling, counterclockwise wind flow.

Andrew caused the third greatest dollar loss from any natural disaster in history as it swept across Florida and Louisiana—property damage exceeded $20 billion. (Andrew is topped in dollar loss only by the 1993 Midwest floods and the 1994 Northridge earthquake in Los Angeles, California, which exceeded $30 billion in damage each.) Sustained winds were 225 kmph (140 mph), with gusts to 282+ kmph (175+ mph). Studies recently completed by meteorologist Theodore Fujita estimated that winds in the eyewall reached 320 kmph (200 mph) in small vortices. (See the opening photograph for Chapter 4 for a sense of this wind devastation.)

The tragedy from Andrew is that the storm destroyed or seriously damaged 70,000 homes and left 200,000 people homeless between Miami and the

(a)

(b)

(c)

FIGURE 1
NOAA-10 and *11* AVHRR images of Hurricane Andrew, 1992: (a) August 24 (4:41 P.M. EDT), (b) August 25 (9:28 A.M. EDT) and (c) August 26 (4:17 P.M. EDT). Hurricane Andrew moved westward from Florida, across the Gulf of Mexico, and onshore in Louisiana. Note the time of day and direction of sunlight as it highlights the features of the hurricane. [Images from National Oceanographic and Atmospheric Administration as supplied by the EROS Data Center.]

Florida Keys. A year later 60,000 people were still homeless and reconstruction was progressing slowly. Approximately 8% of the agricultural industry in Florida's Dade County (Miami region) was destroyed outright—exceeding $1 billion in lost sales.

The remnants of Hurricane Andrew continued northward, reaching Quebec and eastern Canada by August 28th. Locally, heavy rains were produced as the remnants of Andrew interacted with weather fronts moving through the area. In Quebec many local records were broken for 24-hour precipitation totals.

The Everglades, the coral reefs north of Key Largo, some 10,000 acres of mangrove wetlands, and coastal southern pine forests all took significant hits from Andrew. Natural recovery will take years, but it does offer opportunities for study. Initial scientific assessments judge the Everglades to be quite resilient because the region's ecosystems evolved along with periodic hurricanes. An important fact is that storms damage human structures more than they damage natural systems. Remember this perspective: *threats from urbanization, agriculture, and water diversion represent a greater ongoing crisis to the Everglades than do these great storms!*

New buildings, apartments, and governmental offices are opening right next to still-visible rubble and bare foundation pads. Unfortunately, careful hazard planning to guide the settlement of these high-risk areas has never been policy. Thoughtful hazard perception is rarely practiced by public or private decision makers, whether along coastal lowlands, river floodplains, or earthquake fault zones.

The consequence is that our entire society bears the financial cost of poor perception, whether in Florida, California, or the Midwest, not to mention those victims who directly shoulder the physical, emotional, and economic hardship of the event. This recurrent, yet avoidable cycle—construction, devastation, reconstruction, devastation—was reinforced in *Fortune* magazine over 25 years ago after the devastation of Hurricane Camille:

> Before long the beachfront is expected to bristle with new motels, apartments, houses, condominiums, and office buildings. Gulf Coast businessmen, incurably optimistic, doubt there will ever be another hurricane like Camille, and even if there is, they vow, the Gulf Coast will rebuild bigger and better after that.*

---

*Fortune, October 1969, p. 62.

amounts of energy, powering the atmospheric and oceanic circulations and the weather, is accomplished through the absorption and release of **latent heat**. The measurement of water vapor in the air is expressed as **humidity** and **relative humidity**. Saturated air and the presence of **condensation nuclei** create conditions that produce *condensation* and eventual **cloud** formation, which occurs in a variety of classes and types. **Fog** is a cloud that occurs at the surface.

The **stability** of the lower atmosphere determines whether active lifting of air parcels will occur. A simple comparison of the **dry adiabatic rate (DAR)** and **moist adiabatic rate (MAR)** in a vertically moving parcel of air with that of the **environmental lapse rate** in the surrounding air determines the atmosphere's stability.

The interaction of **air masses** of differing temperature and moisture characteristics is the essence of weather phenomena. Unstable conditions create lifting air masses that cool, saturate, and condense to form clouds. These masses can be lifted by convection heating, topographic barriers, and frontal contrasts. Midlatitude wave cyclones organize air masses in conflicting patterns as they migrate across the continents and oceans from west to east. The violent power of some weather phenomena is displayed in thunderstorms, lightning, tornadoes, hurricanes, and typhoons.

Four record-breaking **tornado** years in the United States, powerful **hurricanes** and **typhoons**, occurrences of persistent drought, and dramatic storms brought weather to the headlines in the 1990s. Hurricane, typhoon, and flood damage in the United States exceeded $60 billion during 1992–1993.

Weather is the short-term condition of the atmosphere, compared to **climate**, which reflects long-term atmospheric conditions. **Meteorology** is the scientific study of the atmosphere. The spatial implications of atmospheric science and its relationship to human activities strongly link meteorology to physical geography. Analyzing and understanding patterns of wind, pressure, temperature, and moisture conditions portrayed on daily weather maps is key to weather forecasting. As forecasting and the public's perception of weather-related hazards improves, loss of life and property could stabilize rather than continue to increase with each event.

KEY TERMS

adiabatic
air mass
cloud
  cirrus
  cumulonimbus
  cumulus
  nimbostratus
  stratocumulus
  stratus
condensation nuclei
dew-point temperature
dry adiabatic rate (DAR)
environmental lapse rate
eustasy
fog
  advection fog
  evaporation fog
  radiation fog
  upslope fog
  valley fog
front
  cold front
  occluded front
  warm front
hail
humidity
  relative humidity
  specific humidity

hurricane
lapse rate
  normal lapse rate
  environmental lapse rate
latent heat
latent heat of condensation
latent heat of evaporation
lightning
mesocyclone
meteorology
moist adiabatic rate (MAR)
moisture droplet
orographic lifting
outgassing
phase change
rain shadow
saturated
stability
storm surge
sublimation
thunder
tornado
tropical cyclone
typhoon
vapor pressure
waterspout
wave cyclone
weather

REVIEW QUESTIONS

1. If the quantity of water on Earth has been quite constant in volume for at least 2 billion years, how can sea level have fluctuated? Explain.
2. Describe the locations of Earth's water, both oceanic and fresh. What is the largest repository of freshwater at this time? In what ways is this significant to modern society?
3. In what three states does water exist on Earth? How can it exist in three different states? Why is this important?
4. What happens to the physical structure of water as it cools below 4°C (39°F)? What in the significance of this?
5. What is latent heat? How is it involved in the phase changes of water? What amounts of energy are involved?
6. Where does the bulk of evaporation and precipitation on Earth take place?
7. Define relative humidity. What is meant by saturation and dew-point temperature?
8. Using several different measures of humidity in the air, derive relative humidity values (vapor pressure/saturation vapor pressure; specific humidity/maximum specific humidity).
9. How do the two instruments described in this chapter measure relative humidity?
10. Specifically, what is a cloud? Describe the droplets that form a cloud.
11. What are the basic forms of clouds? Describe how the basic cloud forms vary with altitude.
12. What type of cloud is fog? List and define the principal types of fog.
13. Differentiate between stability and instability relative to a parcel of air rising vertically in the atmosphere.
14. Why is there a difference between the dry adiabatic rate (DAR) and the moist adiabatic rate (MAR)?
15. How does a source region influence the type of air mass that forms over it? Give specific examples of each basic classification.
16. Explain why it is necessary for an air mass to rise if there is to be precipitation. What are the three main lifting mechanisms described in this chapter?
17. Differentiate between a cold front and a warm front.
18. What is meant by cyclogenesis? Where does it occur and why?
19. Diagram a midlatitude cyclonic storm during its open stage. Label each of the components in your illustration, and add arrows to indicate wind patterns in the system.
20. What constitutes a thunderstorm? What type of cloud is involved? What type of air masses would you expect in an area of thunderstorms in North America?
21. Lightning and thunder are powerful phenomena in nature. Briefly describe how they develop.
22. Evaluate the pattern of tornado activity in the United States from Figure 5-32. What generalizations can you make about the distribution and timing of tornadoes?
23. What are the different classifications for tropical cyclones? List the various names used worldwide for hurricanes.
24. What factors contributed to the incredible damage figures produced by Hurricane Andrew?

Spillway from Oroville Dam, California, at full discharge. [*Photo by author.*]

6

WATER
RESOURCES

THE HYDROLOGIC CYCLE
  A Hydrologic Cycle Model
THE WATER-BALANCE CONCEPT
  The Water-Balance Equation
  Three Examples of Water Balances
  Water Management: An Example
GROUNDWATER RESOURCES
  Groundwater Description
  Groundwater Utilization
  Pollution of the Groundwater Resource
DISTRIBUTION OF STREAMS
OUR WATER SUPPLY
  Daily Water Budget
  World Water Economy
GLOBAL OCEANS AND SEAS
SUMMARY
FYI REPORT 6-1 OGALLALA AQUIFER OVERDRAFT
FYI REPORT 6-2 ACID DEPOSITION: A BLIGHT ON THE
  LANDSCAPE

The availability of water is controlled by atmospheric processes and climate patterns. Because water is not always naturally available when and where it is needed, humans must rearrange water resources. The maintenance of a houseplant, the distribution of local water supplies, an irrigation program on a farm, the rearrangement of river flows—all involve aspects of *water-resource management*.

We begin this chapter by following the movement of water through the Earth-atmosphere system, using the hydrologic cycle model. Applying this concept we look at the water balance, which is an accounting of water for a specific area. Groundwater is an important water resource and is examined next. The groundwater resource is closely tied to surface-water budgets and generally is an abused resource.

We also consider the water we withdraw and consume from available resources, in terms of both quantity and quality. For many countries, water availability looms as a key environmental issue in the near future, for others, it is already a problem. The chapter concludes with a glimpse at the greatest repository of water on Earth—the oceans.

The Hydrologic Cycle

Vast currents of water, water vapor, ice, and energy are flowing about us continuously in an elaborate, open global plumbing system. Together they form the **hydrologic cycle**, which has operated for billions of years, from the lower atmosphere down to several kilometers beneath Earth's surface. The cycle involves the circulation and transformation of water throughout Earth's atmosphere, hydrosphere, lithosphere, and biosphere.

Modern study of the hydrologic cycle involves, computer modeling, direct observation, and remote sensing. A better understanding of the hydrologic cycle is central to understanding water resources and global climate change. Progress has proven difficult owing to the inherent chaos of such a nonlinear system as the hydrologic cycle.

A Hydrologic Cycle Model

Figure 6-1 is a simplified model of this complex system. Let's use the ocean as a starting point for our discussion. More than 97% of Earth's water is in the ocean, and here most evaporation and precipitation occurs. We can trace 86% of all evaporation to the ocean. The other 14% comes from the land, including water moving from the soil into plant roots and passing through their leaves (a process called *transpiration*, described later in this chapter).

In Figure 6-1, you can see that of the ocean's evaporated 86%, 66% combines with 12% advected from the land to produce 78% of all precipitation that falls back into the ocean. The remaining 20% of moisture evaporated from the ocean, plus 2% of land-derived moisture, produces the 22% of all precipitation that falls over land. Clearly, the bulk of continental precipitation comes from the oceanic portion of the cycle.

Precipitation that reaches Earth's surface follows two basic pathways—overland flow and movement into the ground. Water soaks in through **infiltration**, or penetration of the soil surface. It then permeates soil or rock through a vertical movement called **percolation**.

The atmospheric advection of water vapor from sea to land and land to sea at the top of Figure 6-1 appears to be unbalanced: 20% moving inland but only 12% moving out to sea. However, this exchange is balanced by 8% as runoff that flows from land to sea. Most of this runoff—about 95%—comes from surface waters washing across land as overland flow and streamflow. Only 5% of this runoff is attributable to slow-moving subsurface groundwater. These percentages indicate that the small amount of water in rivers and streams is very dynamic, whereas the large quantity of subsurface groundwater is sluggish in comparison and represents only a small portion of total runoff.

The residence time of water in any part of the hydrologic cycle determines its relative importance in affecting Earth's climates. The short time spent by water in transit through the atmosphere (10-day average) is reflected in temporary fluctuations in re-

FIGURE 6-1

The hydrologic cycle model with percentages and directional arrows denoting flow paths. Global average values are shown as percentages; note that all evaporation (86% + 14% = 100%) equals all precipitation (78% + 22% = 100%).

gional weather patterns. Long residence times, such as the 3000–10,000 years in deep-ocean circulations, groundwater aquifers, and glacial ice, act to moderate temperatures and climates. These slower parts of the cycle work as a "system memory"; heat is stored and released to buffer change.

To observe and describe the cycle and its related energy budgets, scientists established the Global Energy and Water Cycle Experiment (GEWEX) in 1988. This is part of the World Climate Research Program and represents a central focus in climate-change studies. Now that you are acquainted with the hydrologic cycle and surface water, let's examine the concept of the water balance as a method of assessing water resources.

The Water-Balance Concept

A *water balance* can be established for any area of Earth's surface—a continent, nation, region, or field—by calculating the total precipitation input and

the total water output. The water-balance is a portrait of the hydrologic cycle at a specific site or area for any period of time, including estimation of streamflow. Charles Warren Thornthwaite (1899–1963), a geographer pioneering in applied water-resource analysis, worked with others to develop a water-balance methodology. They related water-balance concepts to geographical problems, especially to accurately determining irrigation quantity and timing.

The Water-Balance Equation

Figure 6-1 portrays the general circulation of the hydrologic cycle, whereas the water balance allows us to examine specific aspects of the cycle at Earth's surface. To understand the method we must first examine the terms and concepts in the water-balance equation as presented in Figure 6-2. Refer to the equation in the figure as you read through the next series of headings.

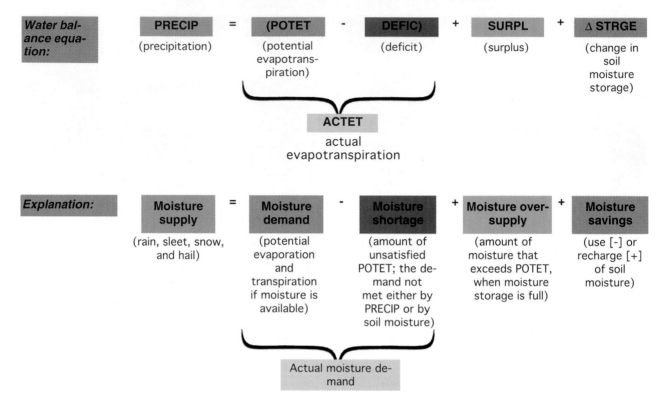

FIGURE 6-2

The water-balance equation explained. The outputs (components to the right) are an accounting of the precipitation input.

Precipitation. The supply of moisture to Earth's surface is **precipitation** (PRECIP)—rain, sleet, snow, and hail (recall from the hydrologic cycle that 22% of Earth's PRECIP falls on land). In some climates, additional deposits of water at Earth's surface occur in the form of dew and fog. Precipitation is measured with the **rain gauge**, which receives precipitation through an open top so that water can be measured by depth, weight, or volume (Figure 6-3a).

According to the World Meteorological Organization, more than 40,000 weather monitoring stations are operating worldwide, with over 100,000 places measuring precipitation. Precipitation patterns over the United States and Canada are shown in Figure 6-4. A map displaying the pattern of world precipitation is included in the next chapter as Figure 7-2.

Potential Evapotranspiration. **Evaporation** is the net movement of free water molecules away from

a wet surface into air that is less than saturated. **Transpiration** in plants is the outward movement of water through small openings (stomata) in the underside of leaves. Both evaporation and transpiration are water balance outputs and directly respond to temperature and humidity. Transpiration rates are partially controlled by the plants themselves as they conserve or release water with control cells around the stomata. Transpired quantities can be significant: on a hot day, a single tree may transpire hundreds of liters of water. Evaporation and transpiration are combined into one term—**evapotranspiration** (14% of evaporation and transpiration occurs from land and plants).

Potential evapotranspiration (POTET) is the water that would evaporate and transpire under *optimum* moisture conditions (adequate precipitation and full soil moisture capacity). Filling a bowl with water and letting it evaporate illustrates this concept: when the bowl becomes dry, is there still an evap-

188

oration demand? A demand of course is present, even when the bowl is dry. Similarly, the amount of water that would evaporate and/or transpire if water always was available is the POTET, or the amount that would evaporate from the bowl if it constantly was supplied with water.

Determining POTET. The easiest method for measuring POTET employs an **evaporation pan**, or evaporimeter (Figure 6-3b). As evaporation occurs, water in measured amounts is automatically replaced in the pan. Mesh screens over the pan protect against overmeasurement created by wind, which accelerates evaporation.

A *lysimeter* is a relatively elaborate device for measuring POTET. It is a buried tank, open at the surface and approximately a cubic meter in size. Using an actual portion of a field, a lysimeter isolates soil, subsoil, and plants so that the moisture moving through the sampled area can be measured. Lysimeters and evaporation pans are limited somewhat in availability across North America, but provide a data base with which to develop ways of calculating POTET. Other methods of estimating POTET based on meteorological data are widely used and can be easily implemented for regional applications.

Figure 6-5 presents POTET values derived by the Thornthwaite method for the United States and Canada. Please compare this POTET (demand) map with Figure 6-4, which is the PRECIP (supply) map

for the same region. The relationship between PRECIP supplies and POTET demands determines the remaining components of the water-balance equa-

(a)

(b)

FIGURE 6-3
(a) A rain gauge installation. A standard rain gauge is cylindrical, 20.3 cm (8.0 in.) in diameter and 58 cm (23 in.) deep. Modern styles include the tipping-bucket and weighing-type gauges. Water is guided by a funnel into a measuring tube. The device is designed to minimize evaporation, which could lead to inaccurate readings. (b) Standard evaporation pan. [Photos by author.]

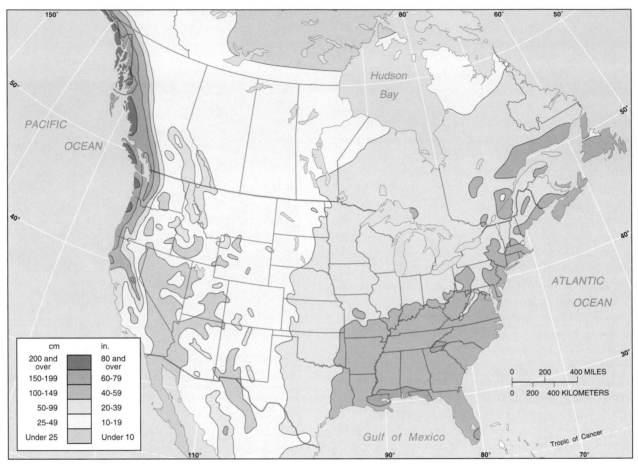

FIGURE 6-4
Annual precipitation (water supply) in the United States and Canada. [Adapted from U.S.
National Weather Service, U.S. Department of Agriculture, and Environment Canada.]

tion. From the two maps can you identify regions
where PRECIP is higher than POTET? (Eastern U.S.)
Or where POTET is higher than PRECIP? (Southwestern U.S.) Where you live, is the water demand
usually met by the precipitation supply? Or, does
your area experience a natural shortage? One way
to assess this is to note whether people use permanent or temporary sprinklers for lawns and gardens
during summer months?

Deficit.   POTET can be satisfied by PRECIP or by
moisture derived from the soil, or it can be artificially
met through irrigation. If these sources are inadequate, the location experiences a moisture shortage.
This unsatisfied POTET is **deficit** (DEFIC). By subtracting DEFIC from POTET, we determine the **ac-**

tual evapotranspiration, or ACTET, that takes
place. Under ideal conditions POTET and ACTET are
about the same, so that plants do not experience a
water shortage; prolonged deficits would lead to drought.

Surplus.   If POTET is satisfied and soil moisture is
full, then additional water input becomes **surplus**
(SURPL), or water oversupply. This excess water
may sit on the surface, or flow through the soil to
groundwater storage. Surplus water that flows across
the surface toward stream channels is termed *overland flow*. Such flow combines with other precipitation and subsurface flows into river channels to
make up the **total runoff** from the area—in
essence the *surface water resource*. Because streamflow, or runoff, is generated mostly from surplus

water, the water-balance approach is a useful tool for indirectly estimating streamflow.

Soil Moisture Storage.   Soil moisture storage is a "savings account" of water that can receive deposits and allow withdrawals as conditions change in the water balance. Soil moisture storage (STRGE) refers to the amount of water that is stored in the soil and is accessible to plant roots. (Soil moisture is discussed in more detail in Chapter 15.)

The moisture retained in the soil comprises two classes of water, only one of which is accessible to plants. The inaccessible class is *hygroscopic water,* the thin molecular layer that is tightly bound to each

soil particle by the natural hydrogen bonding of water molecules (Figure 6-6, left). Such water exists even in the desert, but is not available to meet POTET demands. Soil is said to be at the **wilting point** when all that is left in the soil is unextractable water; the plants wilt and eventually die after a prolonged period of such moisture stress. Unfortunately, most of us have watched a house plant go through this process.

The soil moisture that is generally accessible to plant roots is **capillary water**, held in the soil by surface tension and hydrogen-bonding between the water and the soil (Figure 6-6, middle). Almost all capillary water that remains in the soil is **available**

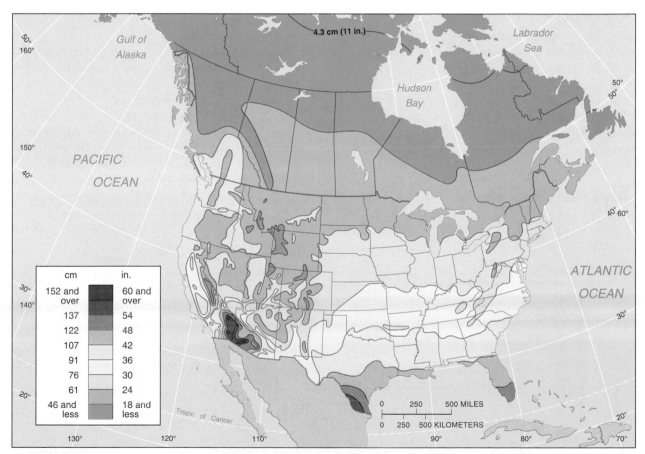

FIGURE 6-5
Potential evapotranspiration (water demand) for the United States and Canada. [From Charles W. Thornthwaite, 1948, "An Approach Toward a Rational Classification of Climate," *Geographical Review* 38, p. 64. Adapted by permission from the American Geographical Society. Canadian data adapted from Marie Sanderson, "The Climates of Canada According to the New Thornthwaite Classification," *Scientific Agriculture* 28 (1948): 501-517.]

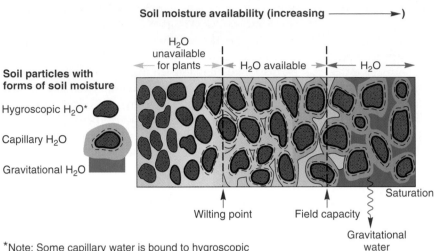

Soil moisture availability (increasing ⟶)

H₂O unavailable for plants ⟷ H₂O available ⟷ H₂O ⟶

Soil particles with forms of soil moisture

Hygroscopic H₂O*

Capillary H₂O

Gravitational H₂O

Wilting point   Field capacity   Saturation

Gravitational water

*Note: Some capillary water is bound to hygroscopic water on soil particle and is also unavailable.

FIGURE 6-6

Classes of soil moisture: hygroscopic, capillary, available, and gravitational. [After Donald Steila, *The Geography of Soils*, © 1976, p. 45. Reprinted by permission of Prentice Hall, Inc., Englewood Cliffs, NJ.]

water in soil moisture storage and is removable for POTET demands through the action of plant roots and surface evaporation. After water drains from the larger pore spaces, the available water remaining for plants is termed **field capacity**, or storage capacity. Field capacity is specific to each soil type and is an amount that can be determined by soil surveys.

When soil becomes saturated after rainfall or snowmelt, some soil water surplus becomes **gravitational water**. This excess water percolates downward under the influence of gravity from the shallower capillary zone to the deeper groundwater zone (Figure 6-6, right).

Figure 6-7 is a simplified graph of soil water capacity, showing the relationship of soil texture to soil moisture content. Various plant types send roots to different depths and therefore are exposed to varying amounts of soil moisture. For example, shallow-rooted crops such as spinach, beans, and carrots send roots down about 65 cm (25 in.) in a silt-loam soil, whereas deep-rooted crops such as alfalfa and shrubs exceed a depth of 125 cm (50 in.) in such a soil. A soil blend that maximizes available water is best for supplying plant water needs. Based on Figure 6-7, can you determine the soil texture with the greatest quantity of available water?

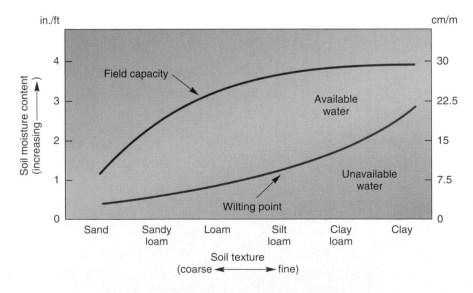

in./ft   cm/m

Soil moisture content (increasing ⟶)

Field capacity

Available water

Unavailable water

Wilting point

Sand   Sandy loam   Loam   Silt loam   Clay loam   Clay

Soil texture
(coarse ⟷ fine)

FIGURE 6-7

The relation between soil moisture availability and soil texture. [After U.S. Department of Agriculture, *1955 Yearbook of Agriculture–Water*, p. 120.]

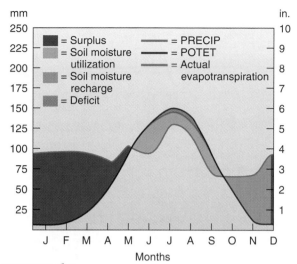

FIGURE 6-8
Annual average water balance components graphed for Kingsport, Tennessee.

Plant roots exert a tensional force on soil moisture to absorb it. As water transpires through the leaves, a pressure gradient is established throughout the plant, thus "pulling" water into the plant roots. When soil moisture is at field capacity, plants obtain water with less effort. As the soil water is reduced by **soil moisture utilization**, the plants must exert greater effort to extract the same amount of moisture. As a result, even though a small amount of water may remain in the soil, plants may be unable to exert enough pressure to utilize it. The *unsatisfied demand* that results is a *deficit*. Avoiding a deficit and reducing plant growth inefficiencies are the goals of irrigation, for the harder plants must work to get water, the less their yield and growth will be.

Rainwater, snowmelt, and irrigation water provide **soil moisture recharge**. The texture and structure of the soil dictate the rate of soil moisture recharge. Remember, soil moisture storage is a water savings account with withdrawals (utilization) and deposits (recharge). (Chapter 15 presents more on soil texture, structure, and moisture characteristics.)

Three Examples of Water Balances

Using all of these concepts, we now can apply the water-balance equation to several specific locales to observe how it works. We look at Kingsport, Tennessee, Ottawa, Ontario, and Phoenix, Arizona.

Kingsport is at 36.6° N, 82.5° W, with an elevation of 390 m (1280 ft), and has maintained weather records for more than half a century. Its water-balance data appear on the graph in Figure 6-8.

Figure 6-8 shows average water supply (PRECIP) and demand (POTET) for a year. We can see that Kingsport experiences a net supply from January through May and from October through December. Soil moisture estimates are at field capacity from January through May, with the excess supply providing a surplus each month. However, the warm days and months from June through September result in net demands for water. In June the net moisture demand begins, most of which is withdrawn from soil moisture storage.

July, August, and September feature soil moisture utilization so that by the end of September shallow-rooted soil moisture is near the wilting point. During October, however, net supply is generated and is credited as soil moisture recharge. Recharge occurs again in November and December, bringing the soil moisture back up to field capacity by the end of the year.

Obviously, not all stations experience the surplus moisture patterns of this humid-continental city and surrounding region; different climatic regimes experience different interactions among their water-balance components. Also, medium- and deep-rooted plants produce different results. Figure 6-9 presents water-balance graphs for the moist and cooler climate of Ottawa, Ontario, Canada, and the drier and warmer climate of Phoenix, Arizona.

Water Balance and Water Resources

Because water is distributed unevenly over space and time, most water resource management projects attempt to rearrange water either geographically, from one place to another, or across the calendar, from one month to another. Substantial human effort has been expended to override natural water-balance regimes and recurring drought. In North America, several large water-management projects are already in operation: the Bonneville Power Authority in Washington State, the Tennessee Valley Authority in the Southeast, the California Water Project, the Central Arizona Project, and the Churchill Falls and Nelson River Projects of Manitoba. Controversy and battles among conflicting interests invariably accompany water projects of this magnitude, especially because most ideal sites for multipurpose hydroelectric projects already are taken.

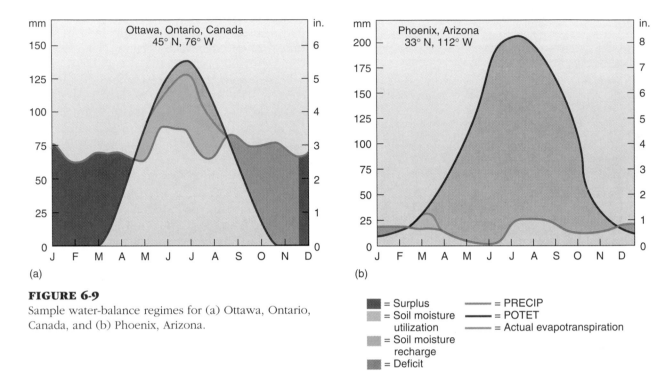

(a)

(b)

FIGURE 6-9
Sample water-balance regimes for (a) Ottawa, Ontario, Canada, and (b) Phoenix, Arizona.

■ = Surplus
▨ = Soil moisture utilization
▨ = Soil moisture recharge
▨ = Deficit

─── = PRECIP
━━━ = POTET
─── = Actual evapotranspiration

FIGURE 6-10
Snowy Mountains irrigation water makes possible the opening of new lands in Australia's eastern interior. [Photo by Snowy Mountains Hydroelectric Authority.]

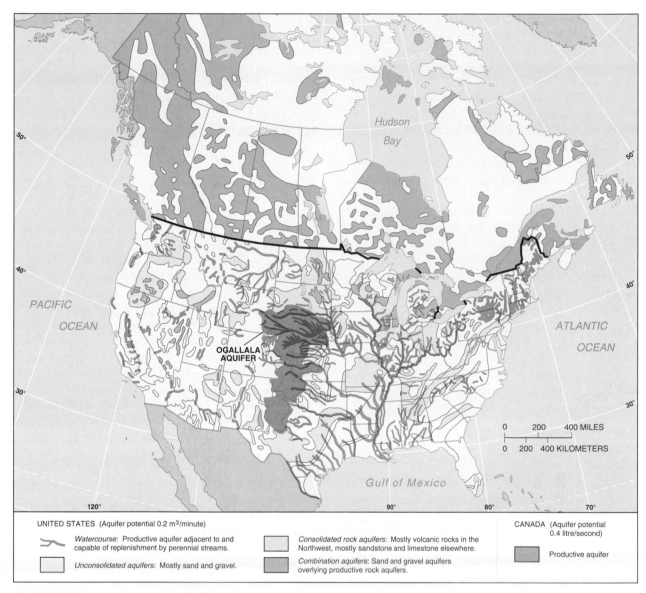

FIGURE 6-11

Groundwater resource potential for the United States and Canada. Highlighted areas of the United States are underlain by productive aquifers capable of yielding freshwater to wells. [Courtesy of Water Resources Council for the United States and the *Inquiry on Federal Water Policy* for Canada.]

A particularly ambitious regional water project is the Snowy Mountains Hydroelectric Authority Scheme of Australia, where water-balance variations created the need to relocate water. In the Snowy Mountains, part of the Great Dividing Range in extreme southeastern Australia, precipitation ranges from 100 to 200 cm (40–80 in.) a year, whereas interior Australia receives under 50 cm (20 in.), and drops to less than 25 cm (10 in.) farther inland (Figure 7-2). POTET values are greater throughout the Australian interior, creating water deficits, and less in the higher elevations of the Snowy Mountains, creating water surpluses.

The Snowy Mountains Scheme was begun in the 1950s and completed over 20 years later. The plan was to take surplus water that flowed down the Snowy River eastward to the Tasman Sea and reverse the flow to support newly irrigated farmland in the interior of New South Wales and Victoria. Through some of the longest tunnels ever built, vast pumping systems, numerous reservoirs, and power plants, this plan was completed and even expanded to include many tributary streams and drainage systems high in the mountains. The westward-flowing rivers—the Murray, Tumut, and Murrumbidgee—now have increased flow, and as a result, new acreage is in production in what was dry outback, formerly served only by wells drawing on meager groundwater (Figure 6-10).

Groundwater Resources

Groundwater is a part of the hydrologic cycle that lies beneath the surface but is tied to surface supplies. Groundwater is the largest potential source of freshwater in the hydrologic cycle—larger than all surface lakes and streams combined. Between Earth's land surface and a depth of 4 km (13,100 ft) worldwide, some

22 percent of Earth's freshwater resides. Despite this volume and its obvious importance, groundwater is widely abused. In many areas it is polluted or consumed in quantities beyond natural recharge rates.

About 50% of the U.S. population derives a portion of its freshwater from groundwater sources. Between 1950 and 1990, the annual groundwater withdrawal increased 160%. Figure 6-11 shows potential groundwater resources in the United States and Canada.

Groundwater Description

Figure 6-12 illustrates the following discussion. Groundwater is fed by surplus water, which percolates downward from the zone of capillary water as gravitational water. This excess surface water moves through the **zone of aeration** where soil and rock are less than saturated, and some pore spaces contain air. Eventually water reaches an area of collected subsurface water known as the **zone of saturation**, where the pores contain only water. Like a sponge made of sand, gravel, and rock, this zone stores water within its structure, filling all available pores and voids.

The **porosity** of this layer is dependent on the arrangement, size, and shape of individual compo-

FIGURE 6-12
Subsurface groundwater characteristics, processes, water flows, and human interactions. Begin at the far left and work your way across this integrative illustration.

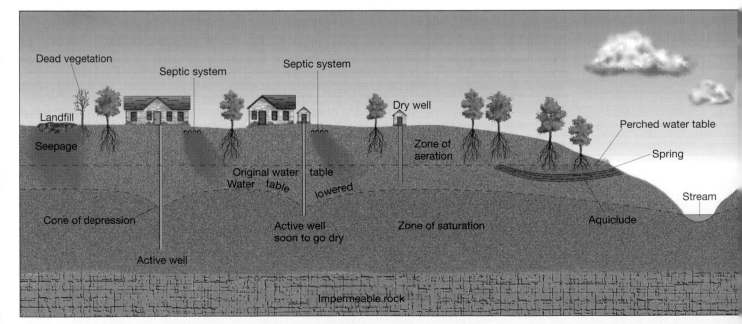

nent particles, the cement between them, and their degree of compaction. Subsurface structures are referred to as **permeable** or **impermeable**, depending on whether they permit or obstruct water flows.

An **aquifer** is a rock layer that is permeable to groundwater flow in usable amounts. An **aquiclude** is a body of rock that does not conduct water in usable amounts. The zone of saturation is an **unconfined aquifer**, a water-bearing layer that is not confined by an overlying impermeable layer. The upper limit of the water that collects in the zone of saturation is called the **water table**; it is the contact surface between the zones of saturation and aeration. (See these components on the left section in Figure 6-12.)

Groundwater movement is controlled by the slope of the water table, which generally follows the contours of the land surface (Figure 6-12). Water wells must penetrate the water table to optimize their potential flow. Where the water table intersects the surface, it creates springs or feeds lakes or river beds (Figure 6-13). Ultimately, groundwater may enter stream channels and flow as surface water (near the center in Figure 6-12). During dry periods the water table can act to sustain river flows.

A confined aquifer differs from an unconfined aquifer in several important ways. A **confined aquifer** is bounded above and below by impermeable layers of rock or sediment, while an unconfined

aquifer is not (Figure 6-12, right). In addition, the two aquifers differ in **aquifer recharge area**, which is the surface area where water enters an aquifer to recharge it. For an unconfined aquifer, that area generally extends above the entire aquifer to the ground surface. But in the case of a confined aquifer, the recharge area is far more restricted (area to the right of the industrial pollution site in Figure 6-12). Once these recharge areas are identified and mapped, the appropriate government body should zone them to prohibit pollution discharges, septic and sewage system installations, or hazardous-material dumping. Of course, an informed population would insist on such action.

Aquifers also differ in water pressure. A well drilled into an unconfined aquifer (Figure 6-12, left) must be pumped to make the water level rise above the water table. In contrast, the water in a confined aquifer is under the pressure of its own weight, creating a pressure level called the *piezometric surface,* which actually can be above ground level (to the right in the figure). Under this condition, **artesian water**, or groundwater confined under pressure, may rise up in wells and even flow out at the surface if the head of the well is below the piezometric surface. In other wells, however, pressure may be inadequate, and the artesian water must be pumped the remaining distance to the surface.

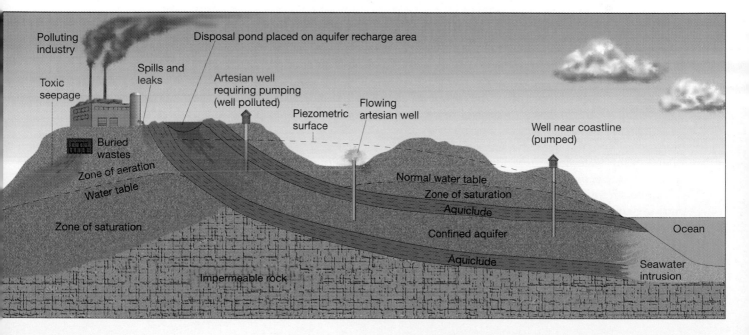

FIGURE 6-13
An active spring, evidence of groundwater and a
probable exposed aquifer. [Photo by author.]

Groundwater Utilization

As water is pumped from a well, the adjacent water
table within an unconfined aquifer will **drawdown**,
or become lower, if the rate of pumping exceeds the
horizontal flow of water in the aquifer. The resultant
shape of the water table around the well is called a
cone of depression (Figure 6-12, left).

Aquifers frequently are pumped beyond their flow
and recharge capacities, resulting in *groundwater
mining*. Large tracts experience chronic groundwater overdrafts in the Midwest, West, lower Mississippi
Valley, Florida, and the Palouse region of eastern
Washington State. In many places the water table or
artesian water level has declined more than 12 m (40

ft). Groundwater mining is of special concern today
in the Ogallala aquifer, which is the topic of FYI Report 6-1.

Another possible effect of water removal from an
aquifer is that the rock layer loses its internal support—the water in the pore spaces between the rock
grains—and overlying rock may crush the aquifer,
resulting in land subsidence and lower surface elevations. For example, within an 80 km (50 mi) ra-
dius of Houston, Texas, land has subsided more than
3 m (10 ft). In the Fresno area of the San Joaquin
Valley in central California, after years of intensive
pumping of groundwater for irrigation, land levels
have dropped almost 10 m (32 ft) because of water
removal and soil compaction. Unfortunately, collapsed aquifers may not be rechargeable even if surplus gravitational water becomes available, because
pore spaces may be permanently closed. Aquifers
also are destroyed by surface mining ("strip mining")
for coal in eastern Texas, northeastern Louisiana,
Wyoming, Arizona, and other locales.

An additional problem arises when aquifers are
overpumped near the ocean. In such a circumstance
the groundwater contact between freshwater and
seawater often migrates inland. As a result, wells
may become contaminated with saltwater, and the
aquifer may become useless as a freshwater source
(Figure 6-12 illustrates this *seawater intrusion* on the
far right margin).

Pollution of the Groundwater Resource

When surface water is polluted, groundwater also
becomes contaminated because it is recharged from
surface-water supplies. Groundwater migrates slowly
compared to surface water. Surface water flows
rapidly and flushes pollution downstream, but sluggish groundwater, once contaminated, remains polluted virtually forever.

Pollution can enter groundwater from industrial-
waste injection wells, septic tank outflows, seep-
age from hazardous-waste disposal sites, industrial
toxic-waste dumps, residues of agricultural pesti-
cides, herbicides, fertilizers, and residential and
urban wastes in landfills. As an example, there are
an estimated 10,000 underground gasoline storage
tanks suspected of leaking at active and abandoned
gas (filling) stations. Thus, pollution can come ei-

Ogallala Aquifer Overdraft

The largest known aquifer in the world lies below a six-state area extending from southern South Dakota and Nebraska to Texas, a region known as the American High Plains (Figure 1). The sand and gravel of the Ogallala aquifer (High Plains aquifer) were charged with meltwaters from retreating glaciers for several hundred thousand years. However, for the past 100 years, Ogallala groundwater has been heavily mined. The mining has been especially intense since World War II, when center-pivot irrigation devices were introduced. Their arms sweep in circles, spreading water over large fields of wheat, sorghums, cotton, corn, and about 40% of the grain fed to cattle in the United States.

The Ogallala aquifer is used to irrigate an area representing about one-fifth of all U.S. cropland: some 5.7 million hectares (14 million acres) are provided with water from 150,000 wells. In this process, water is being removed from the aquifer at 26 billion m³ (21 million acrefeet; see inside back cover for common water conversions) a year, an increase of more than 300% since 1950. In the past four decades the water table in the aquifer has dropped more than 30 m (100 ft), and throughout the 1980s it has averaged a 2 m (6 ft) drop *per year* because of extensive groundwater mining. U.S. Geological Survey scientists estimate that recharging the recoverable portions of the Ogallala aquifer (those portions that have not been crushed or subsided) would take at least 1000 years if water mining stopped today!

The aquifer is most imperiled in Texas, where the saturated layer is shallower and some 75,000 wells are in operation. The declining water table in Floyd County, Texas, illustrates the overdraft problem (Figure 2). Between 1952 and 1982, pumping costs in Floyd County

FIGURE 1

Map of the Ogallala aquifer. [After D. E. Kromm and S. E. White, "Interstate Groundwater Management Preference Differences: The Ogallala Region," *Journal of Geography* 86, no. 1 (Jan.- Feb. 1987): p. 5.]

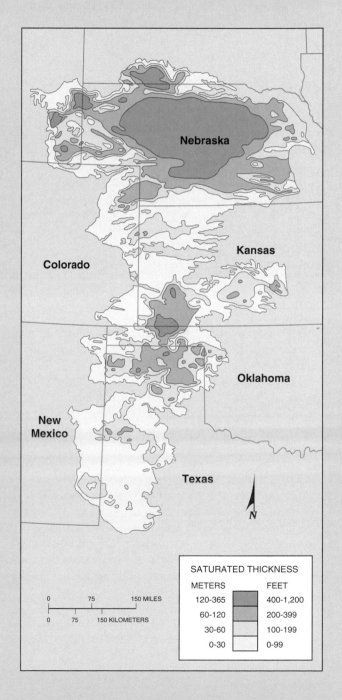

SATURATED THICKNESS

METERS		FEET
120-365		400-1,200
60-120		200-399
30-60		100-199
0-30		0-99

0   75   150 MILES

0   75   150 KILOMETERS

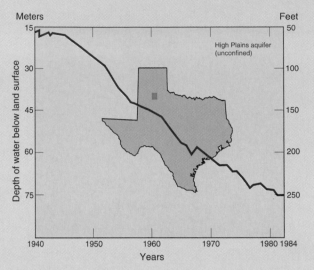

FIGURE 2

Water levels in Floyd County, Texas, wells. [After John E. Schefter, *Declining Ground-Water Levels and Pumping Costs: Floyd County, Texas*, U.S. Geological Survey Water Supply Paper No. 2275, Washington, DC: Government Printing Office, 1985, p. 114.]

have increased 221% (adjusted for increases in the price index for crops), even though irrigated acreage dropped during that same time by almost 15%.

Obviously, billions of dollars of agricultural activity cannot be abruptly halted, but neither can profligate water mining continue. This issue raises tough questions involving the management of cropland, reliance on extensive irrigation, the demand to produce exports, and the need to assess the importance of certain crops that are in chronic oversupply, as well as various crop subsidies. Present agricultural practices, if continued, will result in the loss of about half the total Ogallala aquifer resource, and a two-thirds loss in the Texas portion alone, by the year 2020.

ther from a point source (about 35% does so) or from a largegeneral area (a nonpoint source, 65%), and it can spread over a great distance, as illustrated in Figure 6-12.

Despite widespread publicity about groundwater pollution in the United States, Canada, and Europe, few new protective laws have been established. Laws such as the U.S. Safe Drinking Water Act (1974) and the U.S. Resource Conservation and Recovery Act (1976, 1984) either neglect or do not explicitly direct action for groundwater protection.

Groundwater contamination to some degree is reported in every state. In 1988 the General Accounting Office (GAO) completed a nationwide survey of groundwater protection standards. It found that only 26 states have any groundwater standards. In the face of government inaction, serious contamination of groundwater continues nationwide.

> One characteristic is the practical irreversibility of groundwater pollution, causing the cost of clean-up to be prohibitively high. . . . It is a questionable ethical practice to impose the potential risks associated with groundwater contamination on future generations when steps can be taken today to prevent further contamination.*

Distribution of Streams

Streams represent only a tiny fraction of all water, the smallest volume of any of the freshwater categories we have discussed. Yet, streams are the portion of the hydrologic cycle on which we most depend, representing four-fifths of all the water we use. Whether streams are perennial (constantly flowing) or intermittent, the total runoff that moves through them comes from surplus surface-water

*James Tripp, "Groundwater Protection Strategies," *Groundwater Pollution, Environmental and Legal Problems*. Washington, DC: American Association for the Advancement of Science 1984, p. 137.

"Water in the Middle East: Running on Empty"

The Persian Gulf states soon may run out of freshwater. Their vast groundwater resource is being overpumped to such an extent that saltwater is filling aquifers tens of kilometers inland. By the year 2000, groundwater on the Arabian Peninsula may be undrinkable. Imagine having the hottest issue in the Middle East become water, not oil!

Remedies for groundwater overutilization are neither easy nor cheap. In the Persian Gulf area, additional freshwater is being obtained through desalinization plants along the coasts. These plants remove salt from water. In fact, approximately 60% of the world's 4000 desalinization plants are presently operating in Saudi Arabia and other Persian Gulf States. A very costly water pipeline that will carry water overland from Turkey is being seriously considered to import 6 million m^3 a day (1.68 billion gallons a day) through two branching pipes—one toward Israel, the other through the Arabian Peninsula. The pipeline would run 1500 km (930 mi, the equivalent distance from New York City to St. Louis). Estimates predict the cost of water piped through the "Peace Pipeline" to be about the same as that of the desalinated water.

To lessen the crisis, end-use water patterns could be altered, including improvements in traditional agricultural practices and adding greater efficiency to urban usage—all to help reduce the demand for groundwater. Aquifers and rivers could be shared, but at present no negotiated accords exist for this purpose in the Middle East. The 1991 Persian Gulf War posed a grave threat to desalinization facilities in Saudi Arabia, and many in Iraq and Kuwait were damaged or destroyed. Future political and military unrest may well be related to water resources.

---

runoff, subsurface drainage, and groundwater (Figure 6-14). Because of their rapid renewal, as compared to lake water or groundwater, streams are the best single measure of available water supply.

A stream's flow rate is termed its **discharge**, usually expressed as the volume of water that passes a given point per unit of time. In volume of runoff, the Western Hemisphere exceeds the Eastern Hemisphere, primarily because the Amazon River (Figure 6-15) carries a far greater volume than any other, and because four of the world's highest-discharge river systems are in the Western Hemisphere (Mississippi-Missouri, Orinoco, St. Lawrence, and Mackenzie river systems). The Atlantic Ocean receives about 1.5 times more runoff than the Pacific Ocean. Table 6-1 lists the outflow locations, lengths, and discharge values for the world's largest rivers.

Streamflows generally increase downstream because the area being drained increases. However, the discharge of a stream that originates in a humid region and subsequently flows through an arid (dry) region usually decreases with distance because of high POTET rates. Such a stream is called an **exotic stream** and is exemplified by the Nile River (Table 6-1). This great river, Earth's longest, drains much of northeastern Africa. As it courses through the deserts of Sudan and Egypt, evaporation and water withdrawals for agriculture increase. By the time it empties into the Mediterranean Sea, the Nile's flow dwindles so much that it ranks only thirty-third in volume worldwide.

In the United States, the Colorado River flow also decreases with distance and, in fact, no longer reaches its mouth in the Gulf of California! The exotic Colorado River is depleted not only by passage across dry, desert lands but by upstream removal of water for agriculture and municipal uses, which is the subject of an FYI Report in Chapter 12.

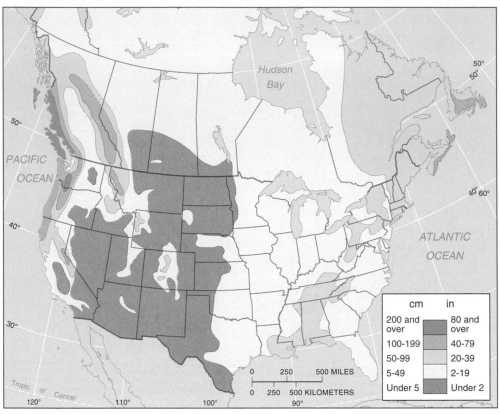

FIGURE 6-14
Distribution of runoff in the United States and Canada. [Data courtesy of Water Resources Council for the United States and *Currents of Change* by Environment Canada.]

FIGURE 6-15
The Amazon River discharges a fifth of all freshwater that enters the world's oceans through its 160 km (100 mi) wide mouth. Millions of tons of sediments are derived from a drainage basin as large as the Australian continent. This astronuat's view is to the northeast out over large islands of sediment where the river's discharge leaves the mouth and flows into the Atlantic Ocean. [Space Shuttle photograph from NASA.]

Table 6-1

Largest Rivers on Earth Ranked by Discharge Volume

Rank by Volume	Average Discharge at Mouth in Thousands of cms (cfs)	River (with Tributaries)	Outflow/Location	Length km (mi)	Rank by length
1	212.5 (7500)	Amazon (Ucayali, Tambo, Ene, Apurimac)	Atlantic Ocean/Amapá-Pará, Brazil	6570 (4080)	2
2	79.3 (2800)	La Plata (Paraná)	Atlantic Ocean/Argentina	3945 (2450)	16
3	39.7 (1400)	Congo, also known as the Zaire (Lualaba)	Atlantic Ocean/Angola, Zaire	4630 (2880)	10
4	38.5 (1360)	Ganges (Brahmaputra)	Bay of Bengal/India	2898 (1800)	23
5	21.8 (770)	Yangtze (Chàng Chiang)	East China Sea/Kiangsu, China	5980 (3720)	4
6	17.4 (614)	Yenisey (Angara, Selenga or Selenge, Ider)	Yenisey Gulf of Kara Sea/Siberia	5870 (3650)	5
7	17.3 (611)	Mississippi (Missouri, Ohio, Tennessee, Jefferson, Beaverhead, Red Rock)	Gulf of Mexico/Louisiana	6020 (3740)	3
8	17.0 (600)	Orinoco	Atlantic Ocean/Venezuela	2737 (1700)	27
9	15.5 (547)	Lena	Laptev Sea/Siberia	4400 (2730)	11
10	14.2 (500)	St. Lawrence	Gulf of St. Lawrence/Canada and U.S.	3060 (1900)	21
33	2.83 (100)	Nile (Kagera, Ruvuvu, Luvironza)	Mediterranean Sea/Egypt	6690 (4160)	1

In other regions, stream drainage that does not flow into the ocean, but instead outflows through evaporation or subsurface gravitational flow, terminates in areas of **internal drainage**. Portions of Asia, Africa, Australia, Mexico, and the United States have regions with such internal drainage patterns. Internal drainage dominates a region of the western United States as shown on maps in Figures 11-1 and 12-16a.

Estimates of available water supplies for the coninents are presented in Table 6-2. Population figures are included to give an idea of the demand for the resource. Comparing the 182 mm (7.1 in.) mean annual runoff in the United States (48 states) to the 583 mm (23.0 in.) in South America indicates the runoff variability that occurs worldwide under varying climatic regimes.

A worsening pollution problem is the increasing level of acid compounds introduced into streams and the environment. FYI Report 6-2 presents an introduction to this blight.

Our Water Supply

Water is critical for survival. Just to produce our food requires enormous quantities of water. Yet the quality and quantity of our supply often is taken for granted. For large-scale assessments, water managers speak in terms of *millions of gallons per day (MGD)* or *billions of gallons per day (BGD)* or *billions of liters per day*. In the western United States, where irrigated agriculture is so important, the measure frequently used is acre-feet per year; one acre-foot contains 325,872 gallons (43,560 ft³; 1234 m³; 1,233,429 liters).

Table 6-2

Estimates of Available Global Water Supply

	Land Area in Thousands of km² (mi²)	Mean Annual Discharge (Water Supply) in km³ per yr (BGD)	Mean Annual Runoff in mm (in.)	Population in Millions for	
				1995	2000
Africa	30,600 (11,800)	4,220 (3,060)	139 (5.5)	700	877
Asia	44,600 (17,200)	13,200 (9,540)	296 (12.0)	3400	3544
Australia-Oceania	8,420 (3,250)	1,960 (1,420)	245 (9.6)	28	30.4
Europe	9,770 (3,770)	3,150 (2,280)	323 (13.0)	728	733
North America (Canada, Mexico, U.S.)	22,100 (8,510)	5,960 (4,310)	286 (11.0)	382	397
United States (48 states)	7,820 (3,020)	1,620 (1,231)	182 (7.1)	—	—
United States (50 states)	9,360 (3,620)	2,340 (1,700)	250 (9.9)	261	268
South America	17,800 (6,880)	10,400 (7,510)	583 (23.0)	311	360
Global (excluding Antarctica)	134,000 (51,600)	38,900 (28,100)	290 (11.0)	5610	6127

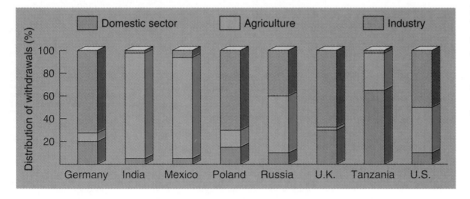

FIGURE 6-16

Water withdrawal by sector for selected countries. [After Gilbert R. White and others, 1982, *Resources and Needs: Assessment of the World Water Situation*, U.N. Water Conference, 1977.]

The U.S. water supply is derived from surface and groundwater sources that are fed by an average daily precipitation of 4244 BGD. That sum is based on an average annual precipitation value of 76.2 cm (30 in.) divided evenly among the 48 contiguous states (excluding Alaska and Hawaii).

The 4244 BGD U.S. water supply is not evenly distributed across the country or during the calendar year. Abundant water supplies in New England result in only about 1% of renewable available water being consumed each year, whereas in the dry Colorado River basin mentioned earlier, the discharge is completely consumed.

Daily Water Budget

The daily U.S. water balance has two general outputs: 71% ACTET (nonirrigated land) and 29% sur-

plus. About 71% (3013 BGD) of the daily supply passes through nonirrigated land—farm lands and pastures, forest and browse (young leaves, twigs, and shoots), and noneconomic (noncrop) vegetation. Only 29% (1231 BGD) is surplus runoff, available for withdrawal and consumption. The estimated U.S. withdrawal of water for the late-1990s is 625 BGD, a sizable increase from the 450 BGD withdrawn in the late-1970s.

The three main flows of this withdrawn water are *irrigation* (34%), *industry* (57%), and *municipalities* (9%). In contrast, Canada uses only 7% of its withdrawn water for irrigation, and 84% for industry. Figure 6-16 compares several nations in their use of withdrawn water and graphically illustrates the differences between developed and developing countries.

Consumptive uses are those that remove water from the stream at some point, without returning it

F.Y.I. Report 6-2

Acid Deposition: A Blight on the Landscape

Acid deposition is a major environmental problem in some areas of the world. Deposition is both dry (dust or aerosols) and wet (acid rain or snow). In addition, winds can carry the acid-producing chemicals many kilometers from their sources before they settle on the landscape, entering streams and lakes as runoff and groundwater flows. This inflow of acid threatens essential water resources.

Acid deposition is causally linked to declining fish populations in the northeastern United States, southeastern Canada, Sweden, and Norway; to widespread forest damage in these same places and in Germany; and generally to damage to buildings, sculptures, and historical artifacts. Corrective action has been delayed because of the complexity of the problem and a lot of politics.

1981   1989

FIGURE 1
The harm done to forests and crops by acid deposition is well established, especially in Europe. A heavily industrialized area of Poland was scanned by a *Landsat* satellite in 1981 and again in 1989 for a GIS-type comparison of the forest cover. False coloration helps in the analysis: brighter colors on the 1981 image (left) depict healthier vegetation, whereas the predominately blue patterns on the 1989 image denote over 50% of the forests in decline and death. [Image courtesy of Earth Satellite Corporation, Rockville, MD.]

The acidity of precipitation is measured on the *pH scale*. A pH of 7.0 is neutral (neither acidic nor basic), values below 7.0 are increasingly acidic, and values above 7.0 are increasingly basic or alkaline. (A pH scale for soil acidity and alkalinity is portrayed in Chapter 15, Figure 15-6.) Natural precipitation averages a pH of 5.65. The normal pH range for precipitation is 5.3–6.0. Thus, normal precipitation is always slightly acidic.

Certain anthropogenic gases are converted in the atmosphere into acids that are removed by wet and dry deposition. Most commonly involved are nitrogen and sulfur oxides (NO_x and SO_x) released in the combustion of fossil fuels, which can produce nitric acid (HNO_3) and sulfuric acid (H_2SO_4) in the atmosphere. Figures 2-27 and 2-29 depict the patterns of NO_x and SO_x emissions and wet deposition in the United States and Canada. Precipitation as acidic as pH 2.0 has fallen in the eastern United States, Scandinavia, and Europe. By comparison, vinegar and lemon juice register slightly less than pH 3.0. Aquatic plant and animal life perishes when lakes drop below pH 4.8. More than 50,000 lakes and some 96,500 km (60,000 mi) of streams in the United States and Canada are at a pH level below normal (i.e., below 5.3), with several hundred lakes incapable of supporting any aquatic life.

Regional-scale decline in forest cover is significant due to the rearrangement of soil nutrients and death of soil microorganisms. Europe is experiencing a more advanced impact on their forests than is evidenced elsewhere, principally due to the burning of coal and the density of industrial activity. In Germany, up to 50% of the forests are dead or damaged; in Switzerland 30% are afflicted. The percentage of forest loss is higher in the nations of eastern Europe.

Earth Satellite Corporation analyzed a small area of Poland that is heavily industrialized. The region produces and burns about 98% of Poland's coal. The two *Landsat* images allow you to do a GIS-type comparison of changes between 1981 and 1989 (Figure 1). Over 50% of Poland's forests are dead or dying. A high correlation exists between these devastated areas and acid rain–producing industrial activity.

In the United States, trees at higher elevations in the Appalachians are being injured by acid-laden cloud cover. In New England, some stands of spruce are as much as 75% affected, as evidenced through analysis of tree-growth rings, which become narrower in adverse growing years. Another possible indicator of forest damage is the reduction by almost half of the annual production of United States and Canadian maple sugar.

The National Academy of Sciences (NAS) and the National Research Council have identified causal relationships between fossil fuel combustion and acidic deposition, citing evidence that is "overwhelming". In spite of this consensus, politics and special interests have halted preventive measures. Government estimates place damage in the United States, Canada, and in Europe at over $50 billion annually. Because wind and weather patterns are international, efforts at reducing acidic deposition also must be international in scope. The NAS stated that a reduction of 50% in combustion emissions would reduce acid deposition by 50%.

farther downstream. *Withdrawal* refers to water that is removed from the supply, used for various purposes, and then returned to the supply. Withdrawn water is an opportunity to extend the resource through reuse and recycling. Often, however, withdrawn water is returned after being contaminated with chemical pollutants, waste (from industry, agriculture, and municipalities), or heat (power plants and industry). Contaminated or not, because this water is returned to the stream, it becomes a part of all water systems downstream. Thus, for example, the entire Missouri-Ohio-Mississippi River drainage is of immediate concern to residents of New Orleans, the last city to withdraw municipal water from the polluted river.

World Water Economy

When precipitation is budgeted, the limits of the water resource becomes apparent. How can we satisfy the growing demand for water? Water supplies available per person decline as population increases, thus world population growth since 1970 has reduced per capita water supplies by a third. Also, pollution limits the water-resource base, so that even before *quantity* constraints are felt, *quality* problems may limit the health and growth of a region.

Internationally, the world's water future is very complex. About one-fourth of the annual renewable water worldwide will be actively utilized by the year 2000. By that time, one-half billion people will be

"Personal Water Use"

On an individual level, statistics suggest that urban dwellers directly use an average 680 liters (180 gallons) of water per person per day, whereas rural populations average only 265 liters (70 gallons) per person per day. However, each of us indirectly accounts for a much greater water demand because of our use of food and products produced with water. For example, growing 77g (2.7 oz) of broccoli requires 42 L (11 gal) of water; producing 250 ml (8 oz) of milk consumes 182 L (48 gal) of water; 28 g (1 oz) of cheese, 212 L (56 gal); and 1 egg, 238 L (63 gal).

For the United States, dividing the total water withdrawal (625 BGD) by a population of 266 million people (estimate for 1996) yields a per capita direct and indirect use of 8930 liters (2350 gallons) of water per day! *Clearly, the average American lifestyle depends on enormous quantities of water and thus is vulnerable to shortfalls or quality problems.*

---

depending on the polluted Ganges River alone. Furthermore, of the 200 major river basins in the world (basins where rivers drain into an ocean, lake, or inland sea), 148 are shared by two nations, and 52 are shared by from three to ten nations. Clearly, water issues are international in scope, yet we continue toward a water crisis without a concept of a *world water economy* as a frame of reference.

Global Oceans and Seas

The ocean, Earth's greatest repository of water and one of the last great scientific frontiers, is of great interest to geographers. Oceans and seas are important for food, minerals, transportation, and for climate control through interactions with the atmosphere. From a geographical point of view, ocean and land surfaces are not evenly distributed. In fact, from certain perspectives, Earth appears to have a distinct "oceanic hemisphere" and a distinct "land hemisphere" (Figure 6-17). The present distribution of the world's oceans and major seas is illustrated in Figure 6-18. Specific surface areas, volumes, and depths are given in Table 6-3. (Names and locations of major ocean currents are shown in Figure 4-23.)

FIGURE 6-17

Two perspectives that divide Earth's surface into (a) an ocean hemisphere and (b) a land hemisphere.

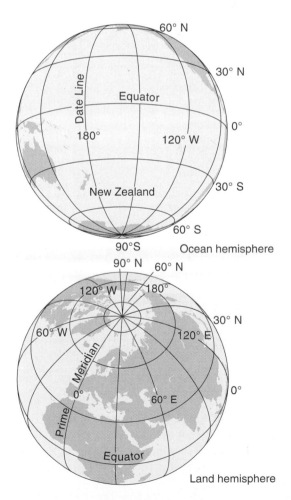

207

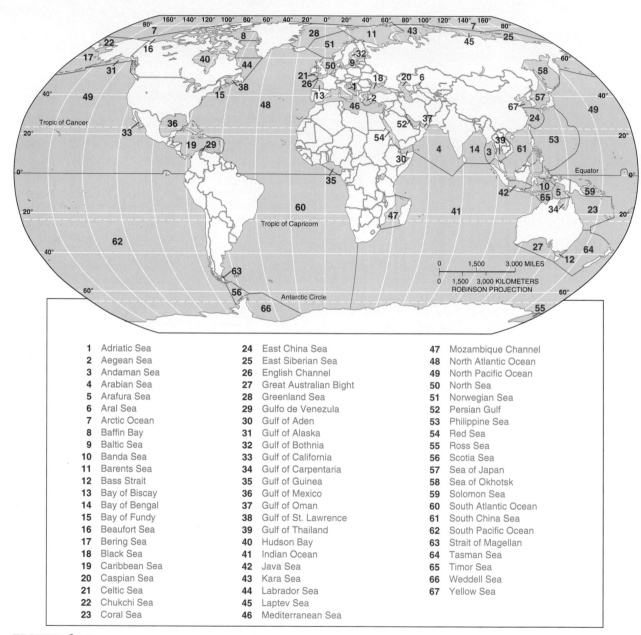

FIGURE 6-18
Principal oceans and seas of the world.

1	Adriatic Sea	24	East China Sea	47	Mozambique Channel
2	Aegean Sea	25	East Siberian Sea	48	North Atlantic Ocean
3	Andaman Sea	26	English Channel	49	North Pacific Ocean
4	Arabian Sea	27	Great Australian Bight	50	North Sea
5	Arafura Sea	28	Greenland Sea	51	Norwegian Sea
6	Aral Sea	29	Gulfo de Venezuela	52	Persian Gulf
7	Arctic Ocean	30	Gulf of Aden	53	Philippine Sea
8	Baffin Bay	31	Gulf of Alaska	54	Red Sea
9	Baltic Sea	32	Gulf of Bothnia	55	Ross Sea
10	Banda Sea	33	Gulf of California	56	Scotia Sea
11	Barents Sea	34	Gulf of Carpentaria	57	Sea of Japan
12	Bass Strait	35	Gulf of Guinea	58	Sea of Okhotsk
13	Bay of Biscay	36	Gulf of Mexico	59	Solomon Sea
14	Bay of Bengal	37	Gulf of Oman	60	South Atlantic Ocean
15	Bay of Fundy	38	Gulf of St. Lawrence	61	South China Sea
16	Beaufort Sea	39	Gulf of Thailand	62	South Pacific Ocean
17	Bering Sea	40	Hudson Bay	63	Strait of Magellan
18	Black Sea	41	Indian Ocean	64	Tasman Sea
19	Caribbean Sea	42	Java Sea	65	Timor Sea
20	Caspian Sea	43	Kara Sea	66	Weddell Sea
21	Celtic Sea	44	Labrador Sea	67	Yellow Sea
22	Chukchi Sea	45	Laptev Sea		
23	Coral Sea	46	Mediterranean Sea		

Salinity and Composition

Water often is called the "universal solvent," dissolving at least 57 of the 92 natural elements. In fact, most natural elements and the compounds they form are found in the seas as dissolved solids.

Thus, seawater is a solution, and the concentration of dissolved solids is called **salinity**.

The oceans remain a remarkably homogeneous mixture, and the ratio of individual salts does not change, despite minor fluctuations in overall salinity. In 1874 the British ship *HMS Challenger* sailed

Table 6-3

Area, Volume, and Depth of Earth's Oceans

Ocean	% of Earth's Ocean Area	*Area in km² (mi²)	*Volume in km³ (mi³)	Mean Depth of Main Basin in m (ft)	Deepest Point in m (ft)	
Pacific	48%	179,670 (69,370)	724,330 (173,700)	4280 (14,040)	Mariana Trench	11,033 (36,198)
Atlantic	28%	106,450 (41,100)	355,280 (85,200)	3930 (12,890)	Puerto Rico Trench	8605 (28,224)
Indian	20%	74,930 (28,930)	292,310 (70,100)	3960 (12,900)	Java Trench	7125 (23,376)
Arctic	4%	14,090 (5440)	17,100 (4100)	1205 (3950)	Eurasian Basin	5450 (17,876)

*Data in thousands (,000); includes all marginal seas.

around the world, taking surface and depth measurements and collecting samples of seawater. Analyses of those samples demonstrated the uniform composition of seawater. Ocean chemistry is a result of complex exchanges among seawater, the atmosphere, the influx of minerals, bottom sediments, and living organisms. In addition, significant amounts of mineral-rich water flow from Earth's crust into the ocean through *hydrothermal* vents in the ocean floor ("black smokers") and from mid-ocean rifts. The uniformity of seawater results from complementary chemical reactions and continuous mixing.

Seven elements comprise more than 99% of the dissolved solids in seawater: chlorine (Cl), sodium (Na), magnesium (Mg), sulfur (S), calcium (Ca), potassium (K), and bromine (Br). Seawater also contains dissolved gases (such as carbon dioxide, nitrogen, and oxygen), solid and dissolved organic matter, and a multitude of trace elements. Only sodium chloride (common table salt), magnesium, and a little bromine are commercially extracted in any significant amount. Future mining of minerals from the seafloor is technically feasible, although it remains uneconomical.

There are several ways to express salinity in seawater. Here are examples, using the worldwide average value:

- 3.5% (% = parts per hundred)
- 35,000 ppm (parts per million)
- 35,000 mg per liter
- 35 g/kg
- 35‰ (‰ = parts per thousand, the most common notation)

Salinity worldwide normally varies between 34‰ and 37‰; variations are attributable to atmospheric conditions above the water and to the quantity of freshwater inflows. In equatorial water, precipitation is high throughout the year, diluting salinity values to slightly lower than average (34.5‰). In subtropical oceans—where evaporation rates are high due to the influence of hot, dry, subtropical high-pressure cells—salinity is more concentrated, increasing to 36.5‰.

The term *brine* is applied to water that exceeds the average of 35‰ salinity, whereas *brackish* applies to water that is less than 35‰. Generally, oceans are lower in salinity near landmasses because of river discharges and runoff. Extreme examples include the Baltic Sea (north of Poland and Germany) and the Gulf of Bothnia (between Sweden and Finland), which average 10‰ or less salinity because of heavy freshwater runoff and low evaporation rates. On the other hand, the Sargasso Sea, within the North Atlantic subtropical gyre, averages 38‰. The Persian Gulf has a salinity of 40‰ as a result of high evaporation rates in an almost-enclosed basin. Deep pockets near the floor of the Red Sea register a very salty 225‰.

Salinity remains in the ocean, but the water recycles endlessly through the hydrologic cycle, driven by energy from the Sun. The water you drink today may have water molecules in it that not long ago were in the Pacific Ocean, the Yangtze River, or maybe they were groundwater in Sweden, or perhaps were airborne in the clouds over Peru!

SUMMARY—Water Resources

The water you use daily began endless flows through the **hydrologic cycle** billions of years ago—evaporation, condensation, precipitation, runoff, and streamflow. Sustainable management of this resource is of paramount importance to society. Scientists use the *water-balance method* to compare the supply of **precipitation** (PRECIP) to the natural demands of **potential evapotranspiration** (POTET) for any area of land—a yard, farm, or region—or for a global assessment. The uneven geographical distribution and seasonal variation of water supply and demand often require intervention if civilization is to succeed in an area for we are affected by patterns of **surplus**, **deficit**, and **soil moisture storage**. The bulk of all water is in the salty oceans. Our water supply consists of surplus water that generates groundwater and surface **total runoff**. Freshwater is only a tiny fraction of Earth's water and the major portion of this is frozen in glaciers.

Groundwater is not an independent water source; it is fed by surface surpluses. In our present era, *groundwater mining* is occurring beyond natural recharge rates in many areas. Pollution of groundwater is being reported to a greater extent than ever before. Once contaminated, an **aquifer** remains permanently ruined (in the human timeframe), making this an issue of national significance because of society's dependence on groundwater.

The streams that drain the landscape often constitute our immediate water supply. Americans in the 48 contiguous states withdraw approximately one-half of the available surplus runoff for irrigation, industry, and municipal uses. Water-resource planning on a regional and global scale, using water-balance principles, is an important consideration if future societies everywhere are to have enough water of adequate quality.

The last great earthbound physical frontier is in water—Earth's oceans and seas. The blend of elements in seawater forms its **salinity**, which varies in concentration in response to different climates and inputs of stream discharge from the continents.

KEY TERMS

aquiclude
aquifer
   confined aquifer
   aquifer recharge area
   unconfined aquifer
artesian water
available water
capillary water
cone of depression
deficit
discharge
drawdown
evaporation
evaporation pan
evapotranspiration
   actual evapotranspiration (ACTET)
   potential evapotranspiration (POTET)
exotic stream
field capacity
gravitational water
hydrological cycle

impermeable
infiltration
internal drainage
percolation
permeable
porosity
precipitation
rain gauge
salinity
soil moisture storage
soil moisture recharge
soil moisture utilization
surplus
total runoff
transpiration
water
water table
wilting point
zone of aeration
zone of saturation

REVIEW QUESTIONS

1. What are the components that make up the water-balance concept? How does the concept relate to the hydrologic cycle model?
2. Write out the water-balance equation and include a definition of each term beneath each component.
3. Explain how to derive actual evapotranspiration in the water-balance equation.
4. What is potential evapotranspiration? How do we go about measuring this potential rate? What method did Thornthwaite use to determine this value?
5. Explain the operation of soil moisture storage, soil moisture utilization, and soil moisture recharge. Include discussion of field capacity, capillary water, and wilting point concepts.
6. What does PRECIP – POTET tell us about the availability of water resources?
7. In terms of water balance and water management, explain the logic behind the Snowy Mountains Scheme in southeastern Australia.
8. Are groundwater resources independent of surface supplies or interrelated with them? Explain your answer.
9. Make a simple sketch of the subsurface environment, labeling zones of aeration and saturation and the water table in an unconfined aquifer. Next, add a confined aquifer to the sketch.
10. At what point does groundwater utilization become groundwater mining? Use the Ogallala aquifer example to explain your answer.
11. List some sources of groundwater pollution. Can a contaminated aquifer be cleaned up easily? Explain.
12. What are the five largest rivers on Earth in terms of discharge? Relate these to the weather patterns in each area. Can these flows be explained through an analysis of regional water budgets?
13. Characterize water use in the United States. Compare use between average urban and rural dwellers.
14. Describe the ways in which Earth's oceans and seas are a resource.

Hawaiian landscape, Kauai, Hawaii. [*Photo by A. R. Christopherson.*]

7

EARTH'S CLIMATES

CLIMATE SYSTEM COMPONENTS
  Temperature and Precipitation
CLASSIFICATION OF CLIMATIC
  REGIONS
  The Köppen Classification System
  Tropical A Climates
  Mesothermal C Climates
  Microthermal D Climates
  Polar E Climates
  Dry Arid and Semiarid B Climates
FUTURE CLIMATE CHANGE
  Developing Climate Models
  Global Warming
  Warming Indications and the Future
  Consequences of Climatic Warming
  Global Cooling
SUMMARY
FYI REPORT 7-1 THE EL NIÑO PHENOMENON

Earth experiences an almost infinite variety of weather. This variability, when considered along with the average conditions at a place over time, constitutes climate. Climates are so diverse that no two places on Earth's surface experience exactly the same climatic conditions, although general similarities permit grouping and classification for regional studies.

Early climatologists faced the challenge of identifying patterns as a basis for climate classification. Currently at the forefront of scientific effort by climatologists is the development of computer models that can simulate complex interactions in the atmosphere and hydrosphere. Vast linkages exist in the Earth-atmosphere system: for example, strong monsoonal rains in West Africa correlate with the development of intense Atlantic hurricanes; an El Niño in the Pacific is tied to drought-breaking rains in California, a drought in Australia, and floods in Louisiana; the eruption of a volcano in the Philippines temporarily lowers global temperatures; and human-made pollution is affecting climate patterns.

A contemporary emphasis for climatologists is the complex change occurring in global patterns of climate. Prediction of such change and its implications for society is a principal goal of contemporary climatology.

Climate System Components

Our study of climate synthesizes the energy-atmosphere phenomena and water-weather phenomena discussed in the first six chapters of this text. The condition of the atmosphere at any given place and time is called *weather*. The consistent behavior of weather over time is a region's **climate**. Climate includes not just the averages but the extremes experienced in a place, for a climate may have temperatures that average above freezing every month, yet still threaten severe frost problems for agricultural activities. Think of climate as dynamic rather than static.

Climatology, the study of climate, analyzes long-term weather patterns over time or space. One type of climatic analysis discerns areas of similar weather statistics and groups them into **climatic regions** that contain characteristic weather patterns. Synoptic analysis of weather elements—that is, the weather over a region at a specific time—is like a snapshot of prevailing conditions. Weather measurements and observations, gathered simultaneously from different points within a region, are plotted on maps and compared to identify climate types. Climate classification is an effort to formalize these observed patterns.

Weather components that combine to produce Earth's climates include patterns of insolation, temperature, humidity, atmospheric pressure and winds, seasonal distribution and amounts of precipitation, types of weather disturbances, and cloud coverage. Climates may be humid with distinct seasons, dry with consistent warmth, moist and cool—almost any combination is possible. There are places where it rains more than 20 cm (8 in.) each month and average monthly temperatures remain higher than 27°C (80°F) throughout the year. Other places may go without rain for 10 years at a time. Expected climatic patterns often are disrupted, as is the case with the recurring El Niño phenomenon, discussed in FYI Report 7-1.

Climates form an essential basis for *ecosystems,* the natural, self-regulating communities formed by plants and animals in their nonliving environment. On land, the basic climatic regions determine to a large extent the location of the world's major ecosystems. These regions, called *biomes,* include forest, grassland, savanna, tundra, and desert (Figure 7-1). Plant, soil, and animal communities are associated with these biomes.

Because climate cycles through periodic change and is never really stable, these ecosystems should be thought of as being in a constant state of adaptation and response. The present global climatic warming trend probably is producing such changes in plant and animal distributions at this time. Climatologists speculate that changes in climate and natural vegetation in the next 50 years could exceed the total of all changes since the peak of the last ice-age episodes, some 18,000 years ago.

The El Niño Phenomenon

Climate is defined as the consistent behavior of weather over time. However, average weather conditions may include extremes that depart from what is considered normal. The El Niño-Southern Oscillation phenomenon in the Pacific Ocean is such a disruption of expected climate patterns.

Normally, as shown in Figure 4-23, the region off the west coast of South America is dominated by the northward-flowing current called the Peru Current. These cold waters move toward the equator and join the westward movement of the South Equatorial Current. The Peru Current is part of the overall counterclockwise circulation that normally guides the winds and surface ocean currents around the subtropical high-pressure cell dominating the eastern subtropical Pacific in the Southern Hemisphere (visible on the world pressure maps in Figure 4-13). As a result, Guayaquil, Ecuador, normally receives 91.4 cm (36 in.) of precipitation per year, with high pressures dominant, whereas islands in the western Pacific and the maritime landmasses of the Indonesian archipelago receive over 254 cm (100 in.), with low pressures dominant.

Occasionally, and for as-yet unexplained reasons, pressure patterns alter and shift from their usual locations, thus affecting surface ocean currents and weather on both sides of the Pacific. Unusually high pressure develops in the western Pacific and lower pressure in the eastern Pacific. This regional change is an indication of large-scale ocean-atmosphere interactions. Trade winds normally moving from east to west (*a* in figure) weaken and can be replaced by an eastward (west-to-east) flow (*b* in figure). Sea-surface temperatures of the central and eastern Pacific then rise above normal, sometimes becoming more than 8C° (14F°) warmer, replacing the normally cold, upwelling, nutrient-rich water along Peru's coastline. Such ocean-surface warming, a "warm pool," may extend to the International Date Line.

People fishing the waters off Peru have come to expect a slight warming of surface waters that lasts for a few weeks at irregular intervals. The name El Niño ("the boy child") is used by Peruvians because these episodes periodically occur around the traditional December celebration time of Christ's birth (although they have occurred as early as spring and summer).

The shifting of atmospheric pressure and wind patterns across the Pacific, known as the Southern Oscillation, both initiates and supports El Niño (Figure 1). Thus, the oceanic *El Niño/Southern Oscillation (ENSO)* designation is derived. National Oceanographic and Atmospheric Administration (NOAA) scientists speculate that ENSO events historically occurred in 1877–1878, 1884, 1891, 1899–1900, 1925, 1931, and 1941–1942, among other times. And, they are certain that such ENSO events occurred in 1953, 1957–1958, 1965, 1969–1970, 1972–1973, 1976–1977, 1986–1987, 1991–1992, and 1993, with the most intense period of this century during 1982–1983. The expected interval for occurrence is 3–5 years, but may range from 2 to 12 years.

The 1982–1983 El Niño strengthened by an extremely strong Southern Oscillation, led scientists to further clarification of complex global interconnections among pressure patterns in the Pacific, sea-surface temperatures, occurrences of drought in some places, excessive rainfall in others, and the disruption of fisheries and wildlife. Related effects occurred into the midlatitudes: droughts in South Africa, southern India, Australia, and the Philippines; strong hurricanes in the Pacific, including Tahiti and French Polynesia; and flooding in the southwestern United States and mountain states, Bolivia, Cuba, Ecuador, and Peru. The Colorado River flooding shown in the FYI Report in Chapter 12 was in large part attributable to the 1982–1983 El Niño.

Estimates place the overall damage at more than $8 billion worldwide for that one event. It appears that the climate of one location is related to climates elsewhere—although it should be no surprise that Earth operates as a vast integrated system. "It is fascinating that what happens in one area can affect the whole world. As to why

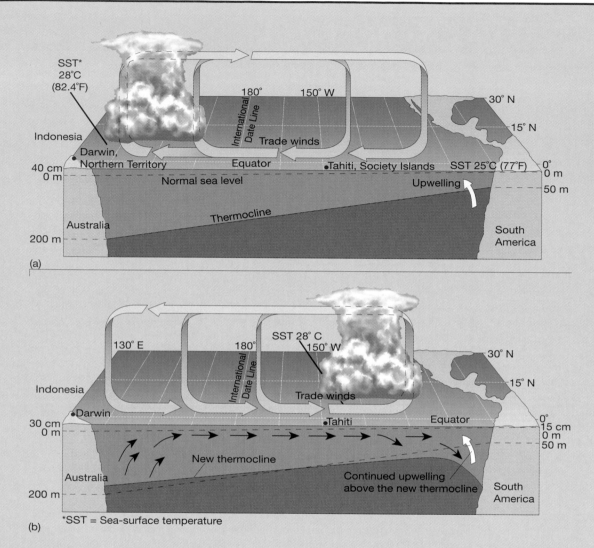

FIGURE 1

(a) Normal and (b) reversed pressure, wind, and current patterns across the Pacific Ocean. Scientists have linked these pressure and related wind changes to occurrences of El Niños. Note the change in the depth of the thermocline, a zone of strong temperature gradient, along South American coast during an El Niño. The thermocline is deeper and warmer water is dominant near the surface. [Adapted after C. S. Ramage, "El Niño, June 1986, p. 76." Copyright © 1986 by Scientific American, Inc.]

this happens, that's the question of the century. Scientists are trying to make order out of chaos."*

*Alan Strong of NOAA's National Environmental Satellite, Data, and Information Service, quoted by Barry Siegel in "El Niño: Global Climatic Patterns Become Chaotic," *Los Angeles Times,* August 17, 1983.

A six-year dry spell in the western United States was ended in early 1993 by record rain and snowfall produced by lingering effects from the 1992 El Niño. These unpredictable changes in expected climatic patterns present a challenge to science to decipher the Earth-atmosphere system. Understanding will involve many principles of physical geography and will require a global-scale spatial perspective.

FIGURE 7-1
A northern coniferous forest biome, indicative of a cool, moist climate at this
elevation in the Canadian Rockies. [Photo by author.]

Temperature and Precipitation

Before we look at climatic regions, let's briefly review key concepts from the first six chapters. Uneven insolation over Earth's surface, varying with
latitude, is the energy input for the climate system
(Chapter 2). Daylength and temperature patterns
vary daily and seasonally. The principal controls of
temperature are latitude, altitude, land-water heating differences, and the amount and duration of
cloud cover. The pattern of world temperatures and
annual temperature ranges is discussed in Chapter 3
and portrayed in Figures 3-15, 3-17, and 3-18.

The hydrologic cycle provides the basic means of
transferring energy and mass through Earth's climate
system. The moisture input to climate is precipitation in the form of rain, sleet, snow, and hail. Figure 6-4 portrays the distribution of mean annual

precipitation in North America. Figure 7-2 shows the
worldwide distribution of precipitation and identifies several patterns, such as the way precipitation
decreases from western Europe inland toward the
heart of Asia, however high precipitation values
dominate southern Asia where the landscape receives moisture-laden monsoonal winds in the summer. Southern South America reflects a pattern of
wet western (windward) slopes and dry eastern (leeward) slopes.

Most of Earth's deserts and regions of permanent
drought are located in lands dominated by subtropical high-pressure cells, with bordering lands grading
to grasslands and to forests as precipitation increases.
The most consistently wet places on Earth straddle
the equator in the Amazon region of South America,
the Zaire region of Africa, and the Indonesian and
Southeast Asian area, all of which are influenced by

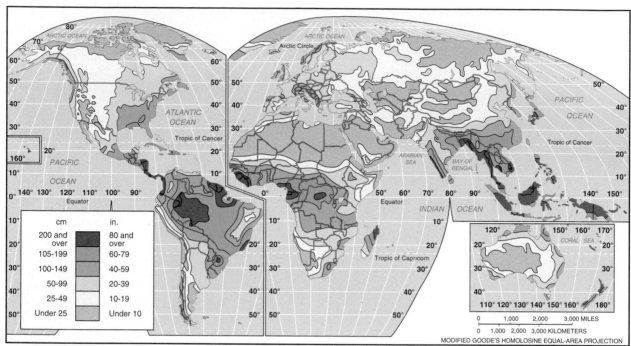

FIGURE 7-2
Worldwide average annual precipitation.

equatorial low pressure and the intertropical convergence zone (Figure 4-15).

Simply comparing the two principal climatic components—temperature and precipitation—reveals important relationships (Figure 7-3). These patterns of temperature and precipitation, along with other weather factors, provide the key to climate classification. Here is a handy list of maps for your reference as you study Earth's climates:

Figure	Subject
2-10	Average daily net radiation
3-1	Average annual solar radiation receipt
3-15, -17, -18	Global average temperatures
4-13	Average global barometric pressure
4-23	Major ocean currents
5-18	North American air masses
6-4	Annual precipitation–U.S. and Canada
7-2	Annual precipitation–global

Classification of Climatic Regions

The ancient Greeks simplified their view of world climates into three zones: the "torrid zone" referred to warmer areas south of the Mediterranean, the "frigid zone" was to the north, and the area where they lived was labeled the "temperate zone," which they considered the optimum climate. Through exploration and satellite observations, Earth's myriad climatic variations are revealed today. Note that climate cannot actually be observed and really does not exist at any particular moment. Climate is therefore a conceptual *statistical construction* from measured weather elements according to the assumptions of the classification scheme employed.

Classification is the process of ordering or grouping data or phenomena in related classes. Such generalizations are important organizational tools in science and are especially useful for the spatial analysis of climatic regions. A climate classification based on *causative* factors—for example, the interaction of air masses—is called a **genetic classifi-**

FIGURE 7-3
Temperature and precipitation schematic showing climatic relations.

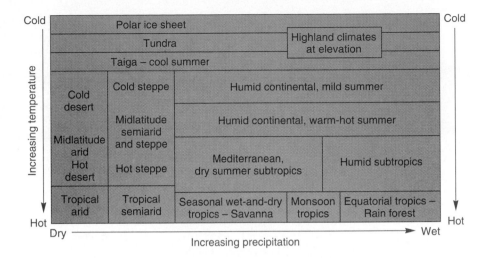

cation. A climate classification based on *statistics* or other data is an **empirical classification**. Climate classifications based on temperature and precipitation data are examples of empirical classifications.

Many climate classifications have been proposed over the years, based upon a variety of assumptions. Some are genetic, explaining climates based upon net radiation, thermal regimes, or air-mass dominance over a region. Others are empirical, and describe climates as observed. One empirical classification system, published by C. W. Thornthwaite, identified moisture regions using aspects of the water-balance approach (discussed in Chapter 6) and vegetation types. Another empirical classification system, and the one we describe here, is the Köppen classification system.

The Köppen Classification System

The Köppen classification system, widely used for its ease of comprehension, was designed by Wladimir Köppen (1846–1940), a German climatologist and botanist. After years of development and research his first wall map showing world climates, coauthored with his student Rudolph Geiger, was introduced in 1928 and widely adopted. Köppen continued to refine this system until his death.

Classification Criteria. The basis of any empirical classification system is the choice of criteria used to draw lines between categories. **Köppen-Geiger**

climate classification uses *average monthly temperatures, average monthly precipitation,* and *total annual precipitation* to devise its spatial categories and boundaries. But we must remember that boundaries really are transition zones of gradual change. The trends and overall patterns of boundary lines are more important than their precise placement, especially with the small scales generally used for world maps.

The modified Köppen-Geiger system has its drawbacks. It does not consider winds, temperature extremes, precipitation intensity, amount of sunshine, cloud cover, or net radiation. Yet the system is important because its *correlations with the actual world are reasonable* and the *input data are standardized and readily available*. For our purposes of general understanding, a modified Köppen-Geiger classification is useful.

A simplified Köppen-Geiger classification system is presented on the world map in the inside-front cover of this text. The cover opens so that the climate map can be referred to as you read this chapter. You may want to cross reference the schematic in Figure 7-3 with this climate map. Maps of each classification are in Figure 7-4.

Köppen's Climatic Designations. The Köppen system uses capital letters (A, B, C, D, E, H) to designate climatic categories from the equator to the poles. Five of the climate classifications are based upon thermal criteria:

THERMAL CLASSIFICATIONS

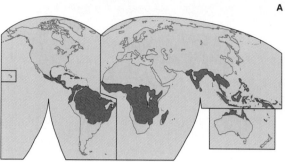

A World distribution of A climates.

A   Tropical Climates

Consistently warm with all months averaging above 18°C (64.4°F); annual PRECIP exceeds POTET.

Af - Tropical rain forest:
  f = All months receive PRECIP in excess of 6 cm (2.4 in.).

Am - Tropical monsoon:
  m = A marked short dry season with 1 or more months receiving less than 6 cm (2.4 in.) PRECIP; an otherwise excessively wet rainy season. ITCZ 6-12 months dominant.

Aw - Tropical savanna:
  w = Summer wet season, winter dry season; ITCZ dominant 6 months or less, winter water-balance deficits.

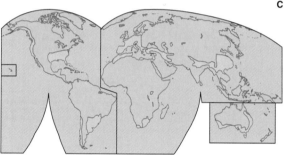

C World distribution of C climates.

C   Mesothermal Climates

Warmest month above 10°C (50°F); coldest month above 0°C (32°F) but below 18°C (64.4°F); seasonal climates.

Cfa, Cwa - Humid subtropical:
  a = Hot summer; warmest month above 22°C (71.6°F).
  f = Year-round PRECIP.
  w = Winter drought, summer wettest month 10 times more PRECIP than driest winter month.

Cfb, Cfc - Marine west coast:
  Mild-to-cool summer.
  f = Receives year-round PRECIP
  b = Warmest month below 22°C (71.6°F) with 4 months above 10°C.
  c = 1 to 3 months above 10°C.

Csa, Csb - Mediterranean summer dry:
  s = Pronounced summer drought with 70% of PRECIP in winter.
  a = Hot summer with warmest month above 22°C (71.6°F).
  b = Mild summer; warmest month below 22°C.

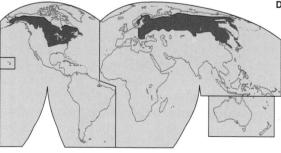

D World distribution of D climates.

D   Microthermal Climates

Warmest month above 10°C (50°F); coldest month below 0°C (32°F); cool temperate-to-cold conditions; snow climates. In Southern Hemisphere, only in highland climates.

Dfa, Dwa - Humid continental:
  a = Hot summer; warmest month above 22°C (71.6°F).
  f = Year-round PRECIP.
  w = Winter drought.

Dfb, Dwb - Humid continental:
  b = Mild summer; warmest month below 22°C (71.6°F).
  f = Year-round PRECIP.
  w = Winter drought.

Dfc, Dwc, Dwd - Subarctic:
  Cool summers, cold winters.
  f = Year-round PRECIP.
  w = Winter drought.
  c = 1 to 4 months above 10°C.
  d = Coldest month below –38°C (–36.4°F), in Siberia only.

FIGURE 7-4
Köppen-Geiger world climate classification system with locator maps for each of the principal climatic regions.

E Polar Climates

Warmest month below 10°C (50°F); always cold; ice climates.

EF - Ice cap:
Warmest month below 0°C (32°F); PRECIP exceeds very small POTET demand; the polar regions.

ET - Tundra:
Warmest month between 0-10°C (32-50°F); PRECIP exceeds small POTET demand; snow cover 8-10 months.

EM - Polar marine:
All months above −7°C (20°F), warmest month above 0°C; annual temperature range <17°C (30°F).

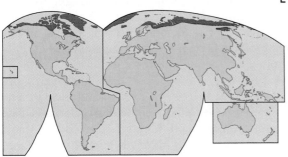

E World distribution of E climates.

MOISTURE CLASSIFICATION

B Dry Arid and Semiarid Climates

POTET exceeds PRECIP in all **B** climates. Subdivisions based on PRECIP timing and amount and mean annual temperature. Boundaries determined by formulas and graphs.

Earth's arid climates.
BWh - Hot low-latitude desert
BWk - Cold midlatitude desert

BW = PRECIP less than 1/2 POTET.
 h = Mean annual temperature >18°C (64.4°F).
 k = Mean annual temperature <18°C.

Earth's semiarid climates.
BSh - Hot low-latitude steppe
BSk - Cold midlatitude steppe

BS = PRECIP more than 1/2 POTET but not equal to it.
 h = Mean annual temperature >18°C (64.4°F).
 k = Mean annual temperature <18°C.

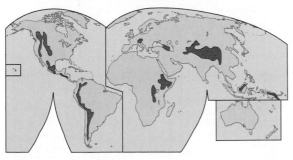

B World distribution of B climates.

B - Climate determinations: formulas in metric units [in English units]

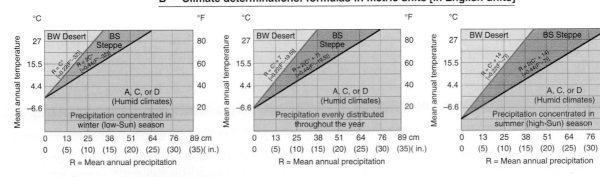

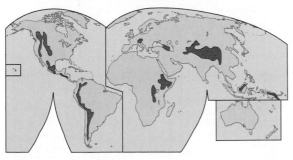

H World distribution of H climates.

A Tropical (equatorial regions)

C Mesothermal (Mediterranean, humid subtropical, marine west coast regions)

D Microthermal (humid continental, subarctic regions)

E Polar (polar regions)

H Highland (compared to lowlands at the same latitude, highlands have lower temperatures—recall the normal lapse rate—and more efficient precipitation due to lower POTET demand)

Only one climate classification is based on moisture as well:

B Dry (deserts and steppes).

Figure 7-4 shows the criteria for each letter symbol within that climate type. Additional lowercase letters are used to signify temperature and moisture conditions. For example, in an *Af tropical rain forest climate,* the *A* tells us that the average coolest month is above 18°C (64.4°F, average for the month), and the *f* indicates that the weather is constantly wet, with the driest month receiving at least 6 cm (2.4 in.) of precipitation. (The designation *f* is from the German *feucht,* for moist.) As you can see in the map of *A* climates the *Af tropical rain forest* climate dominates along the equator and equatorial rain forest.

In a *Dfa* climate, the *D* means that the average warmest month is above 10°C (50°F), with at least one month falling below 0°C (32°F); the *f* says that at least 3 cm (1.2 in.) of precipitation falls during every month; and the *a* indicates a warmest summer month averaging above 22°C (71.6°F). Thus, a *Dfa* climate is a humid-continental, hot-summer climate in the microthermal category. Using Figure 7-4 and the weather data for your city or town, see if you can determine the Köppen classification for where you live. (Also refer to the larger-scale climate map.)

Let's not get lost in the alphabet with this system, for it is meant to help you understand climate through a simplified set of criteria. Always say the descriptive name of the climate as well as its symbol; for example: *Csa Mediterranean summer dry, ET tundra,* or *BSk cold midlatitude steppe.*

Global Climate Patterns. In the following sections, **climographs** are presented for cities that ex-

emplify particular climates. These climographs show monthly temperature and precipitation, location coordinates, annual temperature range, total annual precipitation, annual hours of sunshine (as an indication of cloudiness), the local population, and a location map.

Discussions of soils, vegetation, and major terrestrial biomes that fully integrate these global climate patterns are presented in Chapters 15 and 16. Table 16-1 integrates all this information and will enhance your understanding of this chapter, so please refer to it along with the climate map inside the front cover as you read the text.

Tropical A Climates

Tropical *A* climates are the most extensive, occupying about 36% of Earth's surface, including both ocean and land areas (Figure 7-4). The *A* climate classification extends along all equatorial latitudes, straddling the tropics from about 20° N to 20° S and stretching as far north as the tip of Florida and south-central Mexico, central India, and Southeast Asia. The key temperature criterion for an *A* climate is that the coolest average month must be warmer than 18°C (64.4°F), making these climates truly winterless. The consistent daylength and high Sun altitude (angle above the horizon) throughout the year generate this warmth.

Tropical Rain Forest Climates (Af). Subdivisions of the *A* climates are based upon the distribution of precipitation during the year. Thus, in addition to consistent warmth, an *Af tropical rain forest* climate is constantly moist, with no month recording less than 6 cm (2.4 in.) of precipitation. Indeed, most stations in *Af tropical rain forest* climates receive in excess of 250 cm (100 in.) of rainfall a year. Not surprisingly, the water balances in these regions exhibit enormous water surpluses, creating the world's largest stream discharges in the Amazon and Congo (Zaire) rivers.

The source of this abundant precipitation is convectional thunderstorms triggered by local heating, tending to peak each day from midafternoon to late evening inland and earlier in the day where marine influence is strong. This precipitation pattern coincides with the location of the intertropical convergence zone (ITCZ) discussed in Chapter 4. The ITCZ

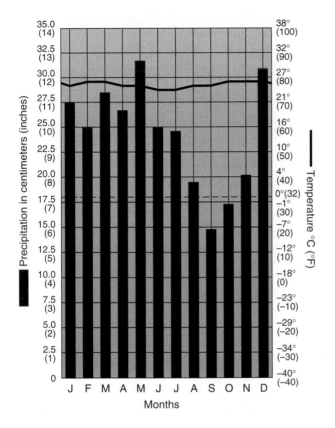

Station: Uaupés, Brazil **Af**   **Elevation:** 86 m (282.2 ft)
Lat/long: 0°08' S, 67°05' W   **Population:** 7500
Avg. Ann. Temp.: 25°C (77°F)   **Ann. Temp. Range:**
Total Annual Precipitation: 2C° (3.6F°)
          291.7 cm (114.8 in.)   **Ann. Hrs. of Sunshine:** 2018

FIGURE 7-5
Climograph for Uaupés, Brazil (Af climate).

shifts northward and southward with the high-summer Sun throughout the year but influences *Af tropical rain forest* regions during all 12 months.

The world climate map shows that the only interruption in the equatorial extent of *Af tropical rain forest* climate occurs in the highlands of the Andes and in East Africa, where high elevations produce lower temperatures. Mount Kilimanjaro, at 5895 m (19,340 ft), is 4° south of the equator but has permanent snow near its summit. Köp-pen placed such sites in the *H highland* climate designation.

The high rainfall of *Af tropical rain forest* climates sustains lush evergreen broadleaf tree growth, producing Earth's equatorial and tropical rain forests. Their leaf canopy is so dense that light does not easily reach the forest floor, leaving the ground surface in low light and sparse plant cover. A major ecological issue facing the world society today is the rampant deforestation of these rain forests, discussed in more detail in Chapter 16.

The high temperatures of this climate contribute to high bacterial action in the soil so that organic ma-

terial is quickly broken down, and certain minerals and nutrients are washed away by the heavy precipitation. The resulting soils are somewhat sterile and unable to support intensive agricultural activity without the addition of fertilizers.

Uaupés, Brazil (Figure 7-5), is characteristic of the *Af tropical rain forest* classification. On the climograph, you can see that the month with the lowest precipitation receives nearly 15 cm (6 in.), and the annual range of temperature is barely 2C° (3.6F°). In all *Af tropical rain forest* climates, the diurnal temperature range exceeds the annual average minimum-maximum range; in fact, day-night temperatures can range more than 11C° (20F°), which is more than five times the annual average range. Other characteristic *Af tropical rain forest* stations include Iquitos, Peru; Kiribati, Nigeria; Mbandaka, Zaire; Jakarta, Indonesia; and Singapore.

Tropical Monsoon Climates (Am). Tropical *A* climates that have a dry season—one or more months with rain of less than 6 cm (2.4 in.)—are candidates for this classification. The migration of the ITCZ affects these areas from 6 to 12 months of the year, with the dry season occurring when the convergence zone is not overhead. *Am tropical monsoon* climates lie principally along coastal areas within the *Af tropical rain forest* climatic realm and experience seasonal variation of winds and precipitation. Evergreen trees grade into thorn forests on the drier margins near the adjoining savanna climates. Typical *Am tropical monsoon* stations include Freetown, Sierra Leone (west coast of Africa); Columbo, Sri Lanka; and Cairns, Australia.

Tropical Savanna Climates (Aw). The world climate map shows the pattern of *Aw tropical savanna* climates on either margin of the *Af tropical rain forest* climates to the north and south. *Aw tropical savanna* climates occur in Africa, South America, India, Southeast Asia, and the northern portion of Australia. The ITCZ dominates *Aw tropical savanna* climates for approximately six months or less of the year as it

shifts with the high Sun. Summers are wetter than winters in *Aw tropical savanna* climates because of the influence of convectional rains that accompany the shifting ITCZ. The *w* designation signifies this winter-dry condition, when rains have left the area and high pressures dominate. As a result, POTET rates exceed PRECIP in winter, and deficits in the water balance result. *Aw tropical savanna* vegetation is characterized by scattered drought-resistant trees throughout dominant grasslands (Figure 7-6).

Mérida, Mexico, on the northern edge of the Yucatán Peninsula, is a characteristic *Aw tropical savanna* station (Figure 7-7). Temperatures in Mérida are consistent with the pattern of tropical *A* climates, although the comparative winter dryness offers a few months of lower humidity. The summer rain, associated with thunderstorms, is sometimes augmented by tropical cyclones. In September of 1988, Hurricane Gilbert presented an extreme example of this moisture source, devastating most of Mérida with rain and strong winds.

FIGURE 7-6
Characteristic tropical savanna landscape in Kenya with plants adapted to seasonally dry water budgets. [Photo by Stephen J. Krasemann/DRK Photo.]

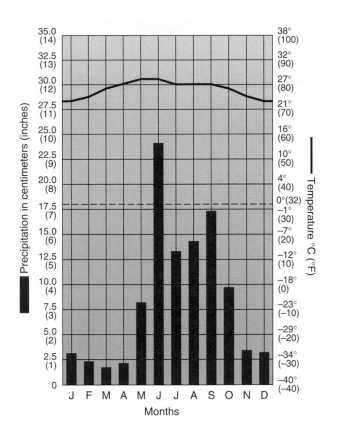

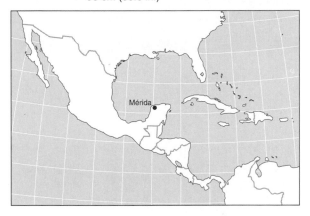

Station: Mérida, Mexico   **Aw**   **Elevation:** 22 m (72.2 ft)
Lat/long: 20°59' N, 89°39' W   **Population:** 400,000
Avg. Ann. Temp.: 26°C (78.8°F)   **Ann. Temp. Range:** 5C°(9F°)
Total Annual Precipitation: **Ann. Hrs. of Sunshine:** 2379
      93 cm (36.6 in.)

FIGURE 7-7
Climograph for Mérida, Mexico (Aw climate).

Mesothermal C Climates

The mesothermal *C* climates occupy the second-largest percentage of Earth's land-and-sea surface, totaling about 27%. However, they rank only fourth when land area alone is considered. Together, *A* and *C* climates dominate over half of Earth's oceans and about one-third of its land area. The largest percentage of the world's population—approximately 55%—resides in the *C* climates.

The word mesothermal ("middle temperature") suggests warm and temperate conditions, with the coldest month averaging below 18°C (64.4°F) but with all months averaging above 0°C (32°F). The mesothermal *C* climates, and nearby portions of the microthermal *D* climates, are regions of great weather variability, for these are the latitudes of greatest air-mass conflict. The *C* climatic region marks the beginning of true seasonality; contrasts in temperature are evidenced by vegetation, soils, and human lifestyle adaptations. Subdivisions of the *C* classification are based on precipitation variability as given in Figure 7-4.

Humid Subtropical Hot Summer Climates (Cfa, Cwa).

The humid subtropical hot summer climates, with a warmest month above 22°C (71.6°F), form two principal types. The *Cf* type is moist all year, with the *f* indicating that all months receive precipitation above 3 cm (1.2 in.); the *Cw* type has a pronounced winter-dry period, designated by a *w*.

The *Cfa humid subtropical hot summer* climate is influenced during the summer by the maritime tropical air masses that are generated over warm coastal waters off eastern coasts. Note on the climate map the presence of *Cfa* climates in the southeastern United States, eastern South America, southeastern Australia and eastern Asia. The warm, moist, unstable air forms convectional showers over land. In fall, winter, and spring, maritime tropical and continental polar air masses interact, generating frontal activity and frequent midlatitude cyclonic storms. Overall, precipitation averages 100–200 cm (40–80 in.) per year. *Cfa humid subtropical hot summer* climates include New Orleans; Nashville; Atlanta; Baltimore; Tokyo; Rosario, Argentina; and Sydney, Australia.

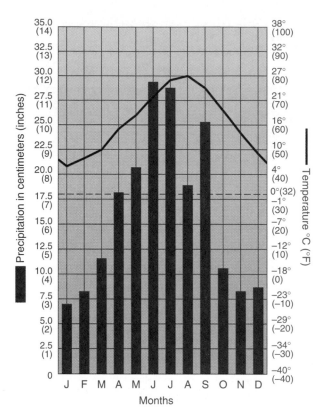

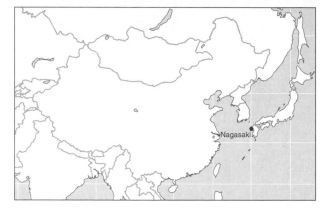

Station: Nagasaki, Japan   **Cfa**   **Elevation:** 27 m (88.6 ft)
Lat/long: 32°44' N, 129°52' E   **Population:** 449,000
Avg. Ann. Temp.: 16°C (60.8°F)   **Ann. Temp. Range:**
Total Annual Precipitation: 21C° (37.8F°)
                   195.7 cm (77 in.)   **Ann. Hrs. of Sunshine:** 2131

FIGURE 7-8
Climograph for Nagasaki, Japan (Cfa climate), and the landscape of the Nagasaki region. [Photo by Noboru Komine/Photo Researchers.]

Nagasaki, Japan (Figure 7-8), is a characteristic station with no month receiving less than 7.3 cm (3.0 in.) of precipitation. Unlike *Cfa humid subtropical hot summer* stations in the United States, however, Nagasaki's summer precipitation is quite a bit higher than its winter precipitation because of the effects of the Asian monsoon.

Cwa humid subtropical winter drought climates are related to the winter-dry, seasonal pulse of the monsoons and extend poleward from the *Aw tropical savanna* climates. Köppen identified the wettest *Cwa* summer month as receiving 10 times more precipitation than the driest winter month. The *Cwa* regions affected by monsoonal rains hold precipitation records.

A representative station of this humid subtropical wet summer and dry winter regime is Ch'engtu, China. Figure 7-9 demonstrates the strong correlation between precipitation and the high Sun of summer. Other examples of *Cwa humid subtropical winter drought* climates include Allahabad, India; Lashio, Myanmar (Burma); Hong Kong; Tsinan, China; and Bandeirantes, Brazil.

The habitability of the *Cfa humid subtropical hot summer* and *Cwa humid subtropical winter drought* climates and their ability to sustain populations are borne out by the concentrations in north-central India, the bulk of China's 1.2 billion people, and the many who live in climatically similar portions of the United States.

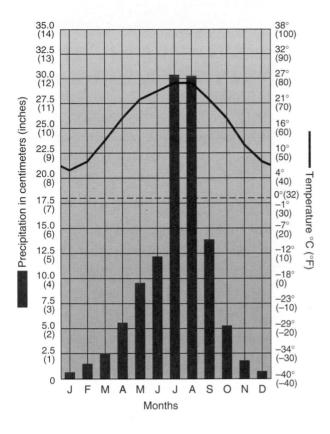

Station: Ch'engtu, China **Cwa**   **Elevation:** 498 m (1633.9 ft)
Lat/long: 30°40' N, 104°04' E   **Population:** 2,260,000
Avg. Ann. Temp.: 17°C (62.6°F)   **Ann. Temp. Range:**
Total Annual Precipitation: 20C° (36F°)
            114.6 cm (45.1 in.) **Ann. Hrs. of Sunshine:** 1058

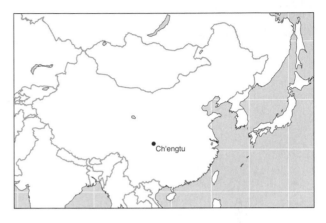

FIGURE 7-9
Climograph for Ch'engtu, China (Cwa climate).

Marine West Coast Climates (Cfb, Cfc). Certainly the dominant climates of Europe and other middle-to-high-latitude west coasts are the marine west coast climates, featuring mild winters and cool summers. The *b* indicates that the warmest summer month is below 22°C (71.6°F), with at least four months averaging above 10°C (50°F); the *c* indicates that only one to three months are above 10°C; again, the *f* denotes consistent moisture throughout the year. Chapter 3 discusses the moderating effects of a marine location on temperature patterns, an effect you can see clearly as you locate the *Cfb marine west coast* climates on the climate map.

Such climates demonstrate an unusual mildness for their latitude. These marine west coast climates extend all along the coastal margins of the Aleutians in the North Pacific, covering the southern third of Iceland in the North Atlantic, as well as Scandinavia and the British Isles. Unlike those in Europe, the climates in Canada, Alaska, Chile, and Australia are backed by mountains and remain restricted to coastal environs. Representative marine west coast stations include Seattle; Vancouver; Valdivia, Chile;

Hobart, Tasmania; Dublin, Ireland; Greenwich and London; Bordeaux, France; and Stuttgart, Germany.

The cooler *Cfc* version of the marine west coast regime occurs at higher latitudes and elevations in Alaska, Iceland, and Scotland. The climograph for Vancouver demonstrates the moderate temperature patterns and the annual temperature range for a *Cfb marine west coast* station (Figure 7-10).

The dominant air mass in *Cfb* and *Cfc marine west coast* climates is the maritime polar, which is cool, moist, and unstable (see Figure 5-18). Weather systems forming along the polar front access these regions throughout the year, making weather quite unpredictable. Coastal fog, totaling a month or two of days each year, is a part of the moderating marine influence. Frosts are possible and tend to shorten the growing season.

An interesting anomaly occurs in the eastern United States. In portions of the Appalachian highlands, increased elevation moderates summer temperatures in the *Cfa humid subtropical hot summer* classification, producing a *Cfb marine west coast* designation (inside front-cover map). Vegetation simi-

"*Cwa* Climate Region Sets Precipitation Records"

Cherrapunji, India, in the Assam Hills south of the Himalayas, is the all-time precipitation record holder for a single year and for every other time interval from 15 days to 2 years. This region of *Cwa* climate is located just north of Bangladesh, where tropical cyclones, torrential rains, and flooding in 1970, 1988, and 1991 killed hundreds of thousands of people.

Because of the summer monsoons that pour in from the Indian Ocean and the Bay of Bengal, Cherrapunji has received 930 cm (30.5 ft) of rainfall in one month and a total of 2647 cm (86.8 ft) in one year. In that location the contrast between the dry and wet monsoons is most severe, ranging from dry winds in the winter to these torrential rains and floods in the summer.

---

larities between the Appalachians and the Pacific Northwest are quite noticeable, enticing many emigrants who relocate from the East to settle in these climatically familiar environments in the West.

Mediterranean Dry Summer Climates (Csa, Csb).

Across the planet during summer months, shifting cells of subtropical high pressure block moisture-bearing winds from adjacent regions. As an example, in summer the continental tropical air mass over the Sahara in Africa shifts northward over the Mediterranean region and blocks maritime air masses and cyclonic systems (see world pressure maps in Figure 4-13). This shifting of stable, warm-to-hot, dry air over an area in summer and away from these regions in the winter creates a pronounced dry-summer and wet-winter pattern. The designation *s* specifies that at least 70% of annual precipitation occurs during the winter months.

Cool offshore currents (the California Current, Canary Current, Peru Current, Benguela Current, and West Australian Current, Figure 4-23) produce stability in overlying maritime tropical air masses along west coasts, poleward of subtropical high pressure. The world climate map shows *Csa* and *Csb* Mediter-

ranean dry summer climates along the western margins of North America, central Chile, and the southwestern tip of Africa, as well as across southern Australia and the Mediterranean basin—the climate's namesake region.

Figure 7-11 compares the climographs of Mediterranean dry summer cities Sevilla, Spain *(Csa)* and San Francisco *(Csb)*. Coastal maritime effects moderate San Francisco's summer so that the warmest month falls below 22°C (71.6°F). Along most west coasts, the warm, moist air overlying cool ocean water produces frequent summer fog.

The Mediterranean dry-summer climate brings water balance deficits in summer. Winter precipitation recharges soil moisture, but water utilization usually exhausts soil moisture by late spring. Thus, local water balances must be augmented by irrigation for large-scale agriculture, although some subtropical fruits, nuts, and vegetables are uniquely suited to this warm-to-hot, dry-summer condition. Natural vegetation is a hard-leafed, drought-resistant variety known locally in the western United States as *chaparral*.

A few representative *Csa* stations include Sacramento; Rome; Adelaide and Perth, Australia; Izmir,

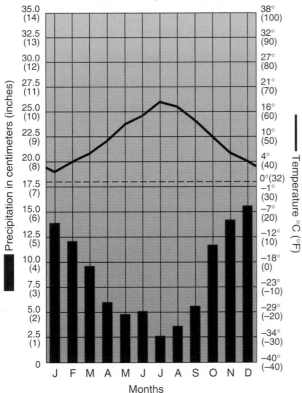

Station: Vancouver, British Columbia   **Elevation:** sea level
Lat/long: 49°11' N, 123°10' W   **Population:** 431,000
Avg. Ann. Temp.: 10°C (50°F)   **Cfb**   **Ann. Temp. Range:**
Total Annual Precipitation: 16C°(28.8F°)
              104.8 cm (41.3 in.)   **Ann. Hrs. of Sunshine:**
                                                          1723

FIGURE 7-10

Climograph of Vancouver, British Columbia. The natural vegetation of nearby Cape Scott, on Vancouver Island, is representative of the marine west coast climate (Cfb climate). [Photo by Michael Collier.]

Turkey; and Tunis, Tunisia. Coastal *Csb* stations include Santa Monica, California; Portland, Oregon; Valparaíso, Chile; and Lisbon, Portugal.

Microthermal D Climates

Humid microthermal climates experience a long winter season with some summer warmth. Microthermal signifies conditions that are cool temperate to cold, with at least one month averaging below 0°C (32°F) and at least one month above 10°C (50°F). Approximately 21% of Earth's land surface is influenced by *D* climates, equaling about 7%

of Earth's total surface. These climates occur poleward of *C* climates and experience severe ranges of temperature related to continentality and airmass conflicts.

Because the Southern Hemisphere lacks substantial landmasses, *D* climates do not develop there except in highland areas (see the global climate map). The tertiary letters *a, b,* and *c* (hot, warm, and cool, respectively) that are used with *D* climates carry the same meaning that they had with *C* climates. The *D* climates range from a *Dfa* in Chicago to the formidable extremes of a *Dwd* in the region of Verkhoyansk, Siberia.

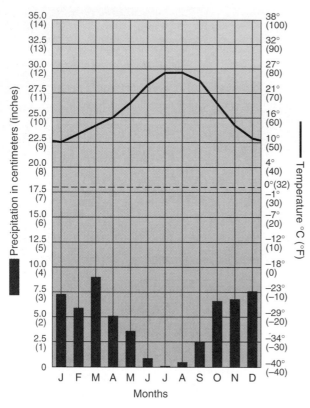

Station: Sevilla, Spain
Lat/long: 37°22' N, 6°00' W
Avg. Ann. Temp.:   **Csa**
     18°C (64.4°F)
Total Annual Precipitation:
     55.9 cm (22 in.)

Elevation: 13 m (42.6 ft)
Population: 651,000
Ann. Temp. Range:
     16C° (28.8F°)
Ann. Hrs. of Sunshine:
     2862

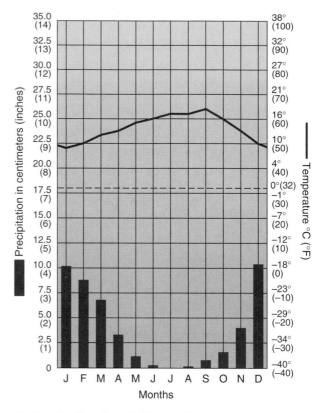

Station: San Francisco, California
Lat/long: 37°37' N, 122°23' W
Avg. Ann. Temp.:   **Csb**
     14°C (57.2°F)
Total Annual Precipitation:
     47.5 cm (18.7 in.)

Elevation: 5 m (16.4 ft)
Population: 750,000
Ann. Temp. Range:
     9C° (16.2F°)
Ann. Hrs. of Sunshine:
     2975

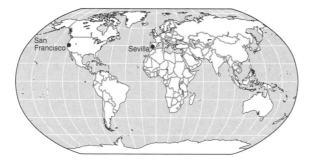

FIGURE 7-11
Climographs for (a) Sevilla, Spain (Csa climate), and (b) San Francisco (Csb climate).

Humid Continental Hot Summer Climates (Dfa, Dwa).

The humid continental hot summer climates are distinguished from each other by the distribution of precipitation during the year (*f, w*). The dry winter associated with the vast Asian landmass, specifically Siberia, is exclusively assigned *w* because it is dominated by an extremely dry and frigid anticyclone in winter. The dry monsoons of

southern and eastern Asia are produced in the winter months by this high-pressure system, with winds blowing outward to the Pacific and Indian oceans.

Both *Dfa* and *Dwa humid continental hot summer* climates are influenced by maritime tropical air masses in the summer. In North America, frequent frontal activity is possible between conflicting maritime tropical and continental polar air masses, especially in winter.

The climograph for New York City illustrates the *Dfa* climate (Figure 7-12). Other examples of *Dfa* climates include Chicago and Omaha, and Varna, Bulgaria. *Dwa* climates are typified by Beijing and Tsingtao in China.

Originally, forests covered the *Dfa humid continental hot summer* climatic region of the United States as far west as the Indiana-Illinois border. Beyond that approximate line, the tall-grass prairies extended westward to about the 98th meridian (98° W) and the approximate location of the 51 cm (20 in.) *isohyet* (line of equal precipitation), with the short-grass prairies beyond to the west.

The deep sod posed problems for the first emigrant settlers, as did the climate. However, native grasses soon were replaced with domesticated wheat and barley, and various eastern inventions (barbed wire, the self-scouring steel plow, well-drilling techniques, railroads, etc.) helped open the region further. In the United States today, the *Dfa humid continental hot summer* climatic region is the location of corn, soybean, hog, and cattle production. The soybean was first domesticated in China and is now widely grown in this region, exceeded only by corn and wheat production.

FIGURE 7-12
Climograph for New York City (Dfa climate).

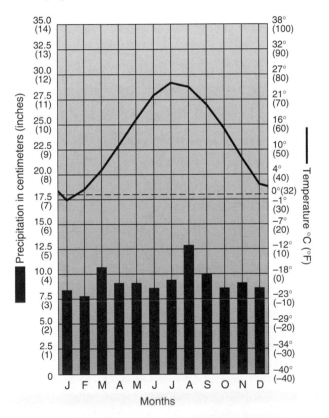

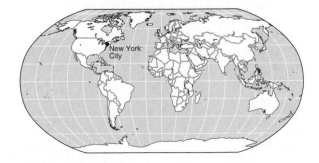

Station: New York, New York
Lat/long: 40°46' N, 74°01' W
Avg. Ann. Temp.:   **Dfa**
    13°C (55.4°F)
Total Annual Precipitation:
    112.3 cm (44.2 in.)
Elevation: 16 m (52.5 ft)
Population: 7,100,000
Ann. Temp. Range:
    24C° (43.2F°)
Ann. Hrs. of Sunshine:
    2564

Humid Continental Mild Summer Climates (Dfb, Dwb).

Soils are thinner and less fertile in these cooler *D* climates, yet agricultural activity is important and includes dairy cattle, poultry, flax, sunflowers, sugar beets, wheat, and potatoes. Frost-free periods range from fewer than 90 days in the north to as many as 225 days in the south. Overall, precipitation is lower than it is in the hot summer regions to the south; however, heavier snowfall is notable and important to soil moisture recharge when it melts. Various snow-capturing strategies have been used, including fences and tall stubble (plant stalks left standing after harvest) to create snow drifts and thus more moisture retention on the ground.

The dry-winter aspect of this cool summer climate occurs only in Asia, in a far-eastern area poleward of the winter-dry *C* climates. A representative *Dwb humid continental mild summer* climate is Vladivostok, Russia, usually one of only two ice-free ports in that nation. Characteristic *Dfb* stations are Duluth, Minnesota, and Moscow and Saint Petersburg in Russia. Figure 7-13 presents a climograph for Moscow, which is at 55° N, or about the same latitude as the southern shore of Hudson Bay in Canada.

Subarctic Climates (Dfc, Dwc, Dwd).

As we move farther poleward, seasonality becomes greater. The growing season, though short, is more intense during long summer days, with at least one to four

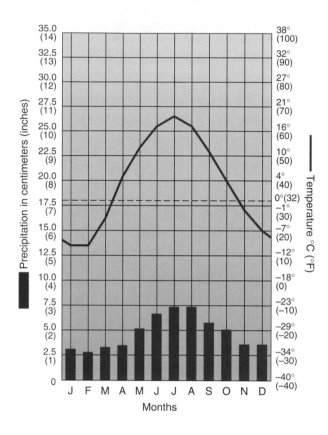

Station: Moscow, Russia   **Dfb**   **Elevation:** 156 m (511.8 ft)
Lat/long: 55°45' N, 37°34' E   **Population:** 9,900,000
Avg. Ann. Temp.: 4°C (39.2°F)   **Ann. Temp. Range:**
Total Annual Precipitation: 29C° (52.2F°)
        57.5 cm (22.6 in.)   **Ann. Hrs. of Sunshine:**
                1597

FIGURE 7-13
Climograph for Moscow, Russia (Dfb climate).

months averaging above 10°C (50°F). The world climate map illustrates the land covered by these three cold climates: vast stretches of Alaska, Canada, northern Scandinavia, and Russia. Discoveries of minerals and petroleum reserves have led to new interest in these regions.

Those subarctic areas that receive 25 cm (10 in.) or more of precipitation per year on the northern continental margins are covered by the so-called "snow forest" of fir, spruce, larch, and birch—the boreal forests of Canada and the taiga of Russia. These forests are a transition to the more northern open woodlands and to the tundra region of the far north. Soils are thin in these lands once scoured by glaciers, and precipitation is low. However, POTET is also low, so soils are generally moist and either partially or totally frozen beneath the surface, a phenomenon known as *permafrost.*

The Churchill, Manitoba climograph (Figure 7-14) shows average monthly temperatures below freezing for seven months of the year, during which time light snow cover and frozen ground persist. Churchill

is representative of the *Dfc subarctic* climate, with an annual temperature range of 40C° (72F°) and a low precipitation total of 44.3 cm (17.4 in.), spread fairly evenly over the year but reaching a maximum in late summer. High pressure dominates Churchill during its cold winter—this is the source area for the continental polar air mass. Other *Dfc subarctic* climates are found at Fairbanks and Eagle, Alaska; Dawson, Yukon; Fort Vermillion, Alberta; and Trondheim, Norway.

The *Dwc* and *Dwd subarctic* climates occur only within Russia. Köppen selected the tertiary letter *d* for the intense cold of Siberia and north-central and eastern Asia; it designates a coldest month with an average temperature lower than –38°C (–36.4°F).

A typical *Dwd subarctic* station is Verkhoyansk, Siberia (Figure 7-15). For four months of the year average temperatures fall below –34°C (–29.2°F). Verkhoyansk frequently reaches minimum winter temperatures that are lower than –68°C (–90°F). However, as pointed out in Chapter 3 (Figures 3-16 and 3-17), summer temperatures in the same area

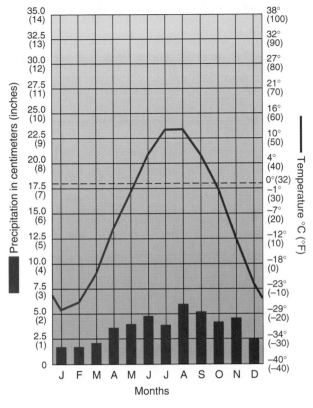

Station: Churchill, Manitoba   **Dfc**   **Elevation:** 35 m (114.8 ft)
Lat/long: 58°45' N, 94°04' W   **Population:** 1400
Avg. Ann. Temp.: −7°C (19.4°F)   **Ann. Temp. Range:**
Total Annual Precipitation: 40C° (72F°)
              44.3 cm (17.4 in.)   **Ann. Hrs. of Sunshine:**
                                                              1732

FIGURE 7-14
Climograph for Churchill, Manitoba (Dfc climate).

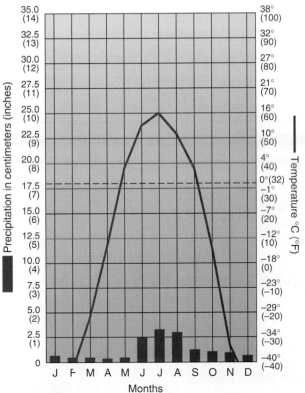

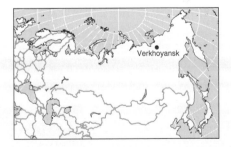

Station: Verkoyansk, Russia   **Dwd**   **Elevation:** 137 m (449.5 ft)
Lat/long: 67°33' N, 135°23' E   **Population:** 1400
Avg. Ann. Temp.: −15°C (5°F)   **Ann. Temp. Range:**
Total Annual Precipitation: 63C° (113.4F°)
              15.5 cm (6.1 in.)   **Ann. Hrs. of Sunshine:**
                                                    not available

FIGURE 7-15
Climograph for Verkhoyansk, Russia (Dwd climate).

produce the world's greatest annual temperature range from winter to summer, a remarkable 63C° (113F°) range. Winter lifestyles feature brittle metals and plastics, triple-thick window panes, and temperatures that render straight antifreeze a solid.

Polar E Climates

The polar climates—*ET tundra, EF ice cap,* and *EM polar marine*—cover about 19% of Earth's total surface. Poleward of the Arctic and Antarctic circles, daylength increases in summer until daylight becomes continuous, yet average monthly temperatures never rise above 10°C (50°F). Daylength, which in part determines the amount of insolation received, and low Sun altitude are the principal climatic factors in these frozen and barren regions.

In an *ET tundra* climate, the temperature of the warmest month is between 0°C and 10°C (32°F and 50°F), and PRECIP is in excess of a very small POTET demand. The land is under continuous snow cover for 8–10 months, but when the snow melts and spring arrives, numerous plants appear—stunted sedges, mosses, flowering plants, and lichens. Much of the area experiences permafrost. The tundra is also the summer home of mosquitoes of legend and black gnats. Like the *D* climates, *ET tundra* climates are strictly a Northern Hemisphere occurrence, except for elevated mountain locations in the South-

ern Hemisphere and a portion of the Antarctic Peninsula.

Most all of Antarctica falls within the *EF ice cap* climate, as does the North Pole, with all months averaging below freezing. Both regions are dominated by dry, frigid air masses, with vast expanses that never warm above freezing. The area of the North Pole is actually a sea covered by ice, whereas Antarctica is a substantial continental landmass covered by Earth's greatest ice sheet. For comparison, winter minimum air temperatures at the South Pole (July) frequently drop below that of frozen CO_2 or "dry ice" (−78°C; −109°F).

Antarctica is constantly snow covered but receives less than 8 cm (3 in.) of precipitation each year. Because of consistent winds, it is difficult to discern what is new snow and what is old snow being blown around. Antarctic ice is several kilometers thick and is the largest repository of freshwater on Earth. This ice represents a vast historical record of Earth's atmosphere, within which thousands of volcanic eruptions worldwide have deposited ash layers, and ancient combinations of atmospheric gases lie trapped in frozen bubbles (Figure 7-16). An analysis of an ice core taken from Antarctica is presented in FYI Report 14-1 in Chapter 14.

EM polar marine temperatures are more moderate than other polar climates in winter, with no month below −7°C (20°F), yet they are not as warm as those

FIGURE 7-16
The frozen Antarctic landscape in the daylight of midnight—Earth's largest repository of freshwater. [Photo by Wolfgang Kaehler.]

of *ET tundra* climates. Because of marine influences, annual temperature ranges do not exceed 17C° (30F°). *EM polar marine* climates, not shown on the climate map, occur along the Bering Sea, the tip of Greenland, northern Iceland, Norway, and in the Southern Hemisphere, generally over oceans between 50° and 60° S.

Dry Arid and Semiarid B Climates

The *B* climates are the only ones that Köppen classified according to the *amount and distribution of precipitation,* rather than temperature. These are the world's arid and semiarid regions—deserts and steppes possessing unique plants, animals, and physical features. The mountains, rock strata, long vistas, and the resilient struggle for life are all magnified by the dryness. Because of the sparseness of vegetation, the landscape is bare and exposed; POTET exceeds PRECIP in all parts of *B* climates, creating varying permanent water deficits. The severity of these deficits distinguishes the different subdivisions within this climatic group. Vegetation is typically *xerophytic:* drought resistant, waxy, hard leafed, and adapted to aridity and low transpiration loss (Figure 7-17). Along water courses, plants called *phreatophytes,* or "water-well plants," penetrate to great depths for the water they need.

The *B* climates occupy more than 35% of Earth's land area, clearly the most extensive climate over land. The world climate map reveals the pattern of Earth's dry climates, which cover broad regions between 15° and 30° N and S. In these areas the subtropical high-pressure cells predominate, with subsiding, stable air and low relative humidity. Under generally cloudless skies, these subtropical deserts extend to western continental margins, where cool, stabilizing ocean currents flow offshore and summer advection fogs form. The Atacama Desert of Chile, the Namib Desert of Namibia, the Western Sahara of Morocco, and the Australian Desert each lie adjacent to a coastline. Extension of these dry regions into higher latitudes is associated with rain shadows, produced by orographic lifting over western mountains in North America and Argentina. Interior Asia, far distant from any moisture-bearing air masses, is also included within the *B* climates.

The major subdivisions are the *BW deserts,* where PRECIP is *less* than one-half of POTET, and the *BS steppes,* where PRECIP is *more* than one-half of POTET. (The *W* designation is from the German *wuste,* for desert.) In an effort to better describe the dry climates, Köppen developed simple formulas to determine the usefulness of rainfall, based on the season in which it falls—whether it falls principally in the summer with a dry winter, or in the winter with a dry summer, or whether the rainfall is evenly distributed. Small graphs on Figure 7-4 demonstrate these rela-

FIGURE 7-17
Desert plants are particularly adapted to the harsh environment of the Mojave desert. [Photo by author.]

tionships. A tertiary lowercase letter is defined by the average annual temperature: temperatures above 18°C (64.4°F) are signified by an *h* for hot; those below 18°C are represented by a *k* for cold. (The *k* designation is from the German *kalt,* for cold.)

Hot Low-Latitude Desert Climates (BWh).

The *BWh hot low-latitude desert* climates are Earth's true tropical and subtropical deserts. They usually are concentrated on the western sides of continents, although Egypt, Somalia, and Saudi Arabia also fall within this classification. Rainfall, usually the result of local convectional showers during the summer season, is very unreliable in *BWh hot low-latitude desert* climates. Some regions receive near zero amounts, whereas others may receive up to 35 cm (14 in.). Representative *BWh hot low-latitude desert* stations include Yuma, Arizona; Aswan, Egypt; Lima, Peru; and Riyadh, Saudi Arabia (Figure 7-18).

Cold Midlatitude Desert Climates (BWk).

On the world climate map, this classification covers only a small area: countries along the southern border of Russia, the Gobi Desert, and Mongolia in Asia; the central third of Nevada and areas of the American Southwest, particularly at high elevations; and Patagonia in Argentina. Because of lower temperatures and lower POTET values, rainfall must be low for stations to meet the *BWk cold midlatitude desert* criteria; consequently, total average annual rainfall is only about 15 cm (6 in.). Albuquerque, New Mexico, with 20.7 cm (8.1 in.) of precipitation and an annual average temperature of 14°C (57.2°F), is representative of this climate type (Figure 7-19).

Hot Low-Latitude Steppe Climates (BSh).

These climates are generally located around the periphery of hot deserts, where shifting subtropical high-pressure cells create a distinct summer-dry and winter-wet precipitation pattern. This climate type is best identified around the periphery of the Sahara and in Iran, Afghanistan, and Kazakhstan. Along the southern margin of the Sahara lies the Sahel, a drought-tortured region of expanding desert condi-

Station: Riyadh, Saudi Arabia **BWh** **Elevation:** 609 m (1998 ft)
Lat/long: 24°42' N, 46°43' E
Avg. Ann. Temp.: 26°C (78.8°F)
Total Annual Precipitation:
        8.2 cm (3.2 in.)
Population: 1,308,000
Ann. Temp. Range:
        24C° (43.2F°)
Ann. Hrs. of Sunshine:
        not available

FIGURE 7-18
Climograph for Riyadh, Saudi Arabia (BWh climate).

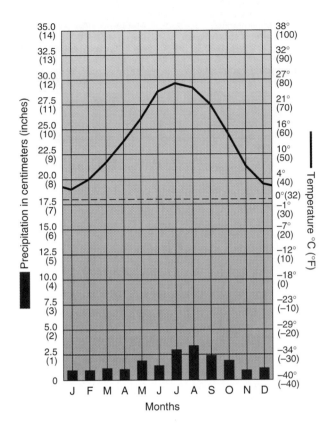

Station: Albuquerque, **BWk** New Mexico
Lat/long: 35°03' N, 106°37' W
Avg. Ann. Temp.: 14°C (57.2°F)
Total Annual Precipitation: 20.7 cm (8.1 in.)
Elevation: 1620 m (5315 ft)
Population: 370,000
Ann. Temp. Range: 24C° (43.2F°)
Ann. Hrs. of Sunshine: 3420

FIGURE 7-19
Climograph for Albuquerque, New Mexico (BWk climate).

tions. Human populations in this region have suffered great hardship as desert conditions have gradually expanded over their countries. Average annual precipitation in *BSh hot low-latitude steppe* areas is usually below 60 cm (23.6 in.). A few sample stations include Santiago, Chile; Kayes, Mali; Las Palmas, Canary Islands; Daly Waters, Australia; and Poona, India.

Cold Midlatitude Steppe Climates (BSk). The *BSk cold midlatitude steppe* climates occur poleward of about 30° latitude and the *BWk cold midlatitude desert* climates. Such midlatitude steppes are not generally found in the Southern Hemisphere. As with other *B* climate classifications, *BSk cold midlatitude steppe* climate rainfall variability is large and is undependable, ranging from 20 to 40 cm (7.9–15.7 in.). Not all rainfall is convectional, for cyclonic systems penetrate the continents; however, most produce little precipitation in this climate region. A sample of *BSk cold midlatitude steppe* stations includes Denver; Salt Lake City; Cheyenne, Wyoming;

Lethbridge, Alberta; Ulaanbaatar, Mongolia; and Semipalatinsk (Semey), Kazakhstan (Figure 7-20).

Future Climate Change

Significant climatic change has occurred on Earth in the past and most certainly will occur in the future. There is nothing society can do about long-term influences that cycle Earth through swings from ice ages to warmer periods. However, our global society must address possible short-term changes that are influencing global temperatures within the life span of present generations. A first step is to develop ways of computer modeling Earth's atmosphere.

Developing Climate Models

Imagine the tremendous complexity of building a mathematical model of all the climatic components over different time frames and at various scales! The challenge is to predict future climatic patterns in

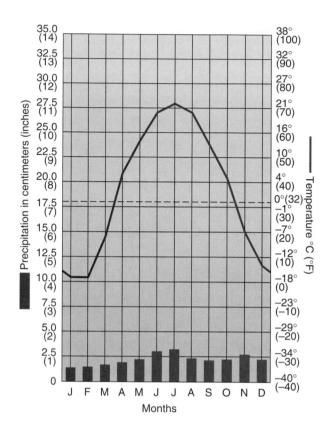

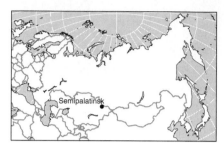

Station: Semipalatinsk, **BSk**
    (Semey), Kazakhstan
Lat/long: 50°21' N, 80°15' E
Avg. Ann. Temp.: 3°C (37.4°F)
Total Annual Precipitation:
    26.4 cm (10.4 in.)

Elevation: 206 m (675.9 ft)
Population: 330,000
Ann. Temp. Range:
    39C° (70.2F°)
Ann. Hrs. of Sunshine:
    not available

FIGURE 7-20
Climograph for Semipalatinsk, Kazakhstan (BSk climate).

what is essentially a nonlinear, chaotic natural system. Computers make possible the processing of global data with which to assess trends in average temperatures and sea level. At best their conclusions are only generalizations of reality; the complex ocean-atmosphere interface and the role of clouds still elude climatologists.

Climatic models can be only as accurate as the data input and assumptions built into them. Adequate data availability is a problem that is slowly being resolved. Thermometers have been in use since the late 1600s, but a significant worldwide density of readings has been available only from around the turn of this century. Today, approximately 1000 stations worldwide balloon-launch automated instrument packages twice a day. These provide temperature, humidity, and pressure measurements at varying altitudes. New land-based methods employing lasers to obtain these readings are gradually increasing the data base.

The most complex climate models, known as **general circulation models (GCMs)**, are essentially based on mathematical models that were originally established for forecasting weather. The further development of *coupled general circulation models* (CGCMs) that link atmospheric and oceanic components is critical to improving climate predictions.

At present, four complex GCMs (three-dimensional models) are in use in the United States and four in other countries: CCM—National Center for Atmospheric Research; GISS—Goddard Institute for Space Studies (NASA); GFDL—Geophysical Fluid Dynamics Laboratory (Princeton); OSU—Oregon State University; CCC—Canadian Climate Center; MRI—Meteorological Research Institute, Japan; CSIRO—Commonwealth Scientific and Industrial Research Organization, Australia; and UKMO—Meteorological Office, United Kingdom. As you can see, the effort to predict climate change is international.

Climatic sensitivity to doubling of CO_2 levels in the atmosphere is a comparative benchmark among GCMs. These GCMs do not predict specific temperatures but rather offer various scenarios of global warming. Interestingly, they tend to agree on climate-change predictions at the global scale, but vary in specific changes at the regional level.

Global Warming

Human activities are enhancing the greenhouse effect. There is little scientific doubt that air temperatures are the highest since recordings were begun in earnest more than 100 years ago. In terms of **paleoclimatology**, the science that studies past climates (discussed in Chapter 14), Earth is within 1C° of equaling the highest average temperature of the past 125,000 years (based on ice-core data). The rate of warming in the past 30 years exceeds any comparable period in the temperature record, according to NASA scientists, and are thought to be attributable to a buildup of greenhouse gases. Climatologists Richard Houghton and George Woodwell describe the present climatic condition:

> The world is warming. Climatic zones are shifting. Glaciers are melting. Sea level is rising. These are not hypothetical events from a science fiction movie; these changes and others are already taking place, and we expect them to accelerate over the next years as the amounts of carbon dioxide, methane, and other trace gases accumulating in the atmosphere through human activities increase.*

---

*Richard Houghton and George Woodwell, "Global Climate Change," *Scientific American,* April 1989, p. 36.

Scientists are attempting to determine the difference between forced fluctuations (human-caused) and unforced fluctuations (natural) as a key to predicting future climate trends. Because the gases generating temperature changes are human in origin, various management strategies are possible to reduce their input. Let's begin by examining the problem at its roots.

Carbon Dioxide and Global Warming. *Radiatively active gases* are atmospheric gases, such as carbon dioxide (CO_2), methane, CFCs, and water vapor, that absorb and radiate infrared wavelengths (heat). Carbon dioxide is the principal radiatively active gas causing Earth's natural greenhouse effect. CO_2 is transparent to light but opaque to the infrared wavelengths radiated by Earth, and thus its presence delays heat loss to space. While detained, that heat energy is absorbed and reradiated over and over producing warmth in the lower atmosphere. As concentrations of these heat-absorbing gases increase, more heat is maintained in the atmosphere and temperatures increase.

The Industrial Revolution, beginning in the mid-1700s, initiated a tremendous surge in the burning of fossil fuels. This, coupled with the destruction and inadequate replacement rate of forests, is causing atmospheric CO_2 levels to increase 7.3–9.1 billion metric tons (8–10 billion tons) per year (Figure 7-21).

FIGURE 7-21
Industrial landscapes exemplify the increasing output of CO_2 into the atmosphere. [Photo by author.]

Date	CO_2 Concentration (%)	Parts per Million
1825	0.021	210
1888	0.028	280
1970	0.032	320
1985	0.035	350
1995 (estimate)	0.037	370
2020 (estimate)	0.055	550
2050 (estimate)	0.060	600

Table 7-1

Lower Atmosphere Concentration of CO_2

CO_2 alone is thought to be responsible for almost 60% of the global warming trend.

Table 7-1 shows the increasing percentage of CO_2 in the lower atmosphere from 1825 to present and gives estimates for the future. CO_2 is presently increasing in concentration at the rate of 0.4% per year. This increase appears sufficient to override any natural climatic tendencies toward a cooling trend, as well as to produce possible unwanted global warming in the next century.

Figure 7-22 shows past and projected sources of excessive (non-natural) carbon dioxide by country or region, clearly identifying developing nations as the sector with the greatest probable growth in fossil-fuel consumption and new CO_2 production. However, national and corporate policies could alter this forecast by actively steering developing countries toward more appropriately scaled alternative energy sources (low temperature, renewable, labor intensive).

Especially important in the United States are policies to eliminate energy waste and inefficiency, both in terms of reducing carbon dioxide production and in setting a positive example. Yet relatively simple solutions are not being promoted, and throughout the 1980s the United States was uncooperative in worldwide efforts to halt this human-induced increase in CO_2. This occurred despite the fact that the United States presently produces 27% of the excess CO_2 annually! U.S. emissions instead are predicted to rise 15% over 1990 levels by the year 2000. The European Community and a majority of nations favor stabilizing emissions at 1990 levels by the year 2000.

The Office of Technology Assessment and the National Academies of Science and Engineering, in sep-

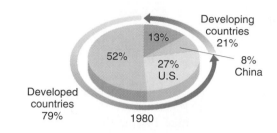

(a)

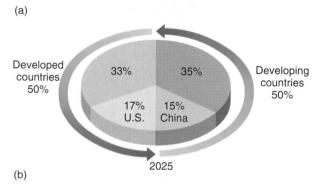

(b)

FIGURE 7-22

Countries and regions of origin (a) for excessive CO_2 in 1980 and (b) forecast for 2025.

arate assessments, concluded that the United States could hold to 1990 levels by the year 2015 at little or no additional cost. (See OTA's "Changing by Degrees: Steps to Reduce Greenhouse Gases," February 5, 1991, and "Policy Implications of Greenhouse Warming," by the Committee on Science, Engineering, and Public Policy of NAS, 1991.) The OTA report concluded that

> Some reductions may even be at a net savings if the proper policies are implemented. . . . The efficiency of practically every end use of energy can be improved relatively inexpensively.

Methane and Global Warming. Another radiatively active gas contributing to the overall greenhouse effect is **methane** (CH_4), which is increasing in concentration at about 1% per year, although much less in absolute amount than CO_2. Air bubbles in ice show that concentrations of methane in the past—500 to 27,000 years ago—were approximately 0.7 ppm, whereas current atmospheric concentrations are 1.7 ppm.

Methane is generated by organic processes, such as digestion and rotting in the absence of oxygen (anaerobic processes). About 50% of the excess methane being produced comes from bacterial action in the intestinal tracts of livestock and from underwater bacteria in rice paddies. Burning of vegetation causes another 20% of the excess, and bacterial action inside the digestive systems of increasing termite populations also is a significant source. Methane is now believed responsible for at least 12% of the total atmospheric warming, complementing the warming caused by the buildup of CO_2 and equaling about one-half the contribution of CFCs.

Chlorofluorocarbons (CFCs) and Global Warming. CFCs are thought to contribute about 25% of the global warming. CFCs absorb infrared in wavelengths missed by carbon dioxide and water vapor in the lower troposphere. As radiatively active gases, CFCs enhance the greenhouse effect, and also play a negative role in stratospheric ozone depletion.

Warming Indications and the Future

Global mean temperatures between 1980 and 1993 registered *nine of the warmest years in the history of instrumental measurement*. A global average temperature of 15.4°C (59.8°F) was reached in 1990. The eruption of Mount Pinatubo lowered temperatures in 1991; otherwise, that year would have placed first.

Comparing annual temperatures and 5-year mean temperatures gives a sense of overall trends. Figure 7-23 shows observed temperatures over the 114 years from 1880 through 1993. The overall pattern is clear—we are entering a period of record warmth. The 1990 Scientific Assessment of Climate Change, an international working group from 30 nations, reached a consensus that the warming will be socially significant, and that warming rates will accelerate in the near future. This agreement among scientists was echoed by the IPCC, a broad-based group of 380 scientists established by the United Nations Environment Programme and representing the efforts of 63 countries and 18 nongovernmental organizations. The 1990 IPCC report concluded that there is ". . .virtual unanimity among greenhouse experts that a warming is on the way and that the consequences will be serious."

IPCC affirmed their major conclusions in 1992 and 1994 updates: human-produced emissions of CO_2, CH_4, CFCs, and N_2O (nitrous oxide) are substantially increasing atmospheric concentrations of these gases. Also, they affirm that a warming trend is occurring. However, uncertainties remain as to magnitude, timing, and specific regional patterns. A minority of scientists persist as critics of the greenhouse warming model, and although no member of the IPCC agrees with these critics, all points of view must be considered.

There is a wide range of uncertainty in estimates of future temperature increase, from 1.5 to 4.5C° (2.7–8.1F°). According to the updated IPCC assessment, an average warming on the low side of forecasts of 2.5C° (4.5F°) is predicted for the timeframe 1990 to 2029. If carbon dioxide concentrations reach 550 ppm by the year 2020 (they are presently at 360 ppm), a temperature increase of 3C° (5.4F°) is forecast for the equatorial regions. This equatorial increase translates to a high-latitude increase of about 10C° (18F°).

The GCM operated by the Goddard Institute produced temperature projections for the years 2000 and 2029 using a middle case scenario (Figure 7-24). The maps predict a distinct warming pattern.

Relative to warming and the oceans, during the decade of the 1980s, average ocean surface temperatures rose at an annual rate of 0.11C° (0.2F°). Of course, any change in ocean temperature has a profound effect on weather and, indirectly, on agriculture and soil moisture. Even the warming of the ocean itself will contribute about 25% of sea-level rise due to thermal expansion of the water.

The Scripps Institution of Oceanography, in La Jolla, California, has kept ocean temperature records since 1916. From 1950 to the present, they registered a 0.8C° (1.44F°) warming of the upper 100 m (330 ft). Significant increases are being recorded to depths of over 300 m (1000 ft) as ocean temperature records are set.

Consequences of Climatic Warming

The consequences of uncontrolled atmospheric warming are complex. Regional climate responses are expected as temperature, precipitation, soil moisture, and air-mass characteristics change. Although the ability to accurately forecast such regional

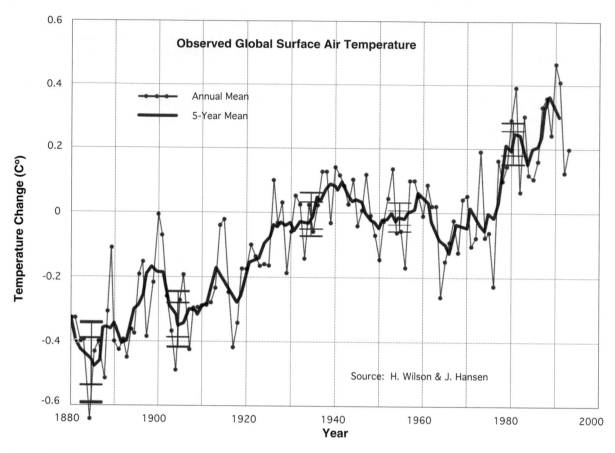

FIGURE 7-23
Global temperature trends from 1880 to 1993. The 0 baseline represents the 1951–1980
global average temperature. Note the cooling impact of the 1991 eruption of Mount
Pinatubo and the subsequent return to a warming trend. [Courtesy of Dr. James E.
Hansen and Helene Wilson, "Update of GISS Global Temperature Analysis Through
1993," NASA Goddard Institute for Space Studies, New York.]

changes is still evolving, some implications of warming are forecasted.

**Effects on World Food Supply and the Bios-
phere.** Modern single-crop agriculture is more delicate and susceptible to changes in temperature,
water demand and irrigation needs, and soil chemistry, than traditional multicrop agriculture. The
southern and central grain-producing areas of North
America are forecast to experience hot and dry
weather by the middle of the next century as a result
of higher temperatures. An increased probability of
extreme heat waves is forecast for these U.S. grain

regions. Available soil moisture is projected to be at
least 10% less throughout the midlatitudes in the
next 30 years. Scientists have considered possible
adaptations, such as changing crop varieties to late-
maturing, heat-resistant types, and adjustments in
soil-management practices, such as fertilizer applications and irrigation schedules.

Today, the United States and Canada sell wheat to
China and the Commonwealth of Independent States
(former USSR). If the foreign grain areas become
wetter and gain a longer frost-free period, grain sales
could shift in the other direction. Specifically, if U.S.
production decreases, the longer growing seasons,

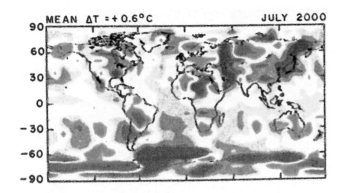

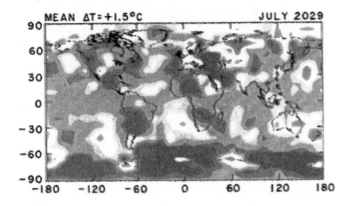

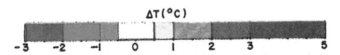

FIGURE 7-24
July temperature projections by the GISS general circulation model (GCM) for the years 2000 and 2029 for a middle-case scenario. [From J. Hansen et al., "Global Climate Changes as Forecast by Goddard Institute for Space Studies Three-Dimensional Model," *Journal of Geophysical Research* 43, no. 8 (August 1988): plate 6. Published by the American Geophysical Union.]

increased temperatures and thermal energy patterns, and possible increased precipitation at higher latitudes could benefit Canadian cropping patterns. The possibility exists that billions of dollars of agricultural losses in one region could be countered by billions of dollars of gain in another.

Studies completed at the universities of Toronto and Wisconsin–Madison suggest, with a degree of uncertainty, that greenhouse warming will increase the temperature of the Great Lakes, which in turn will be an advantage for various species of fish. On the other hand, a warming of small Canadian lakes might prove harmful to aquatic life.

Also, land-based animals will have to adapt to changing patterns of available forage, for animals are adapted to the climate zones in which they live. Increased temperatures already are causing thermal stress to some embryos as they reach their thermal limits. Particularly affected are amphibians whose embryos develop in shallow water. Warming of ocean water is endangering corals worldwide, and they appear to be deteriorating at record levels.

Melting Glaciers and Ice Sheets. Perhaps the most pervasive climatic effect of increased warming would be the rapid escalation of ice melt. The additional water, especially from continental ice masses that are on land, would raise sea level worldwide. Satellite remote-sensing technology allows monitoring of global sea-ice and continental-ice to track evolving trends. Worldwide measurements confirm that sea level has been gradually rising this century.

Scientists currently are studying the ice sheets of Greenland and Antarctica for possible changes in the operation of the hydrologic cycle, including snowlines and the rate at which icebergs break off (calve) into the sea. The key area being watched is the *West Antarctic ice sheet,* where the Ross Ice Shelf holds back vast grounded ice masses (ice presently located on land). The flow from several visible ice streams (channels of flowing ice) is accelerating from the ice sheet. Several interpretations for these changes are possible.

The possible sea-level rise must be expressed as a range of values that are under constant reassessment. Estimates of future sea level rise range from a low of 30–110 cm (1–4 ft) to a high of 6 m (20 ft). These levels could be further augmented by melting of alpine and mountain glaciers, which have experienced about a 30% decrease in overall ice mass during this century. In *Climate Change 1992* (p. 158), IPCC states:

> there is conclusive evidence for a worldwide recession of mountain glaciers. . . . This is among the clearest and best evidence for a change in energy balance at the Earth's surface since the end of the last century.

A quick survey of world coastlines shows that even moderate changes could bring disaster of unparalleled proportions. At stake are the river deltas, lowland coastal farming valleys, and low-lying mainland areas. In the worst possible scenario, a rise of 6 m (20 ft) would flood 20% of Florida; inundate the Mississippi floodplain as far inland as St. Louis; flood the Pampas of Argentina; drown Venice; submerge the Bahamas, The Netherlands, the Maldives; and more. All surviving coastal areas would contend with high water and high tides—but again this is with a worst possible scenario.

Solutions. The position is held by some government leaders that society has adapted to many problems in the past. Capital can be raised and technology harnessed to block the flood and guard the coastlines and deltas. Although such adaptive strategies may be needed, no substantive long-range planning is going forward. Why are we waiting? Why not take action now? Whatever the actual impact—from major financial and cultural irritation to outright catastrophe—solutions exist to reduce the severity of the hazard. In addition, individual conservation and a demand for energy efficiency remain cheap ways to stretch present energy resources, reduce imports, cut carbon dioxide emissions, and save money.

A cooperative global network of nations, under the United Nations Environment Programme (UNEP), participates in the World Weather Watch System and the Global Change Program to gather temperature, weather, and climatic information to assist climate modelers and policy makers. Many studies and participants at numerous world conferences have agreed that governmental policies can profoundly alter the greenhouse effect. The 1989 Cairo Compact and the comprehensive 1990, 1992, and 1994 IPCC reports are steps in the right direction.

One of the significant agreements that emerged from the 1992 United Nations Conference on Environment and Development (UNCED, the "Earth Summit" detailed in FYI Report 17-1) was the *U.N. Framework Convention on Climate Change*—a legally binding agreement signed by over 160 nations that is a first-ever attempt to evaluate and address global warming on an international scale. We are in a unique position to alter, perhaps slow, Earth's probable anthropogenic warming trend by controlling the production of certain radiatively active gases.

Global Cooling

In the 1980s and early 1990s, several volcanic eruptions also affected global temperatures. The 1982 El Chichón eruption in south-central Mexico sent a cloud of sulfuric acid and ash 25 km (15.5 mi) into the atmosphere. Within 21 days the cloud had circled the globe, increasing the atmospheric albedo and lowering temperatures by 0.5C° (0.9F°). Some of the finer particles remained in the stratosphere through 1985.

The largest eruption so far in this century began in June 1991. Mount Pinatubo in the Philippines caused significant global temperature effects by injecting a blanket of acid mist and ash into the atmosphere. This increased atmospheric albedo, which decreased insolation reaching the surface. The volcanic aerosols increased shortwave reflectance (atmospheric albedo) by 4.3 W/m², whereas heat absorption increased only 1.8 W/m², producing a net radiation reduction of 2.5 W/m². This change in net radiation lowered average temperatures in the Northern Hemisphere from 0.5C° to 1C° (0.9F° to 1.8F°), and about half that amount in the Southern Hemisphere.

Another potential cooling effect was postulated by Paul Crutzen (a Dutch atmospheric scientist with the Max Planck Institute of Germany) and John Birks (a chemist at the University of Colorado) in "The Atmosphere After a Nuclear War: Twilight at Noon." Their study, published in the Royal Swedish Academy of Sciences publication *Ambio* in June 1982, launched an avalanche of scientific analyses, assessments, and general confirmations (see Suggested Readings at the end of the text for additional references on this subject).

The **nuclear winter hypothesis**, as it has come to be known, now encompasses a whole range of ecological, biological, and climatic impacts associated with the detonation of a relatively small number of nuclear warheads within the biosphere. Resulting urban fires and fire storms would produce such a pall of insolation-obscuring smoke that surface heating would be drastically reduced. Temperatures could drop 25C° (45F°) and perhaps more in the midlatitudes. This potential global impact of modern warfare affects literally all aspects of Earth systems and therefore is appropriate to our study of physical geography.

SUMMARY—Earth's Climates

Climate is dynamic rather than static or inactive. **Climate** is a synthesis of **weather** phenomena at many scales, from planetary to local. The consistent behavior of weather over time is a region's climate. Climate includes not just the averages but the extremes experienced in a place. Various aspects of weather including temperature, precipitation, humidity, air masses, air pressure, and global wind systems contribute to the pattern of climates. Earth experiences a wide variety of climatic conditions that can be grouped by general similarities into **climatic regions**.

Climatology, the study of climate, analyzes long-term weather patterns over time or space. Climate **classification** is an effort to formalize these observed patterns—the process of ordering or grouping data or phenomena in related classes. A climate classification based on *causative* factors is called a **genetic classification.** A climate classification based on *statistics* or other data is an **empirical classification.** The basis of any empirical classification system is the choice of criteria used to draw lines between categories. The **Köppen-Geiger climate classification** uses *average monthly temperatures, average monthly precipitation*, and *total annual precipitation* to devise its spatial categories and boundaries.

A modified empirical climate classification system developed by Köppen provides a basis for describing the various distinct environments on Earth: tropical *A,* mesothermal *C,* microthermal *D,* polar *E,* and arid *B* climates. Highland climates are assigned an *H* to distinguish them from the others and to denote the conditions created by elevation. The main climatic groups are divided into subgroups, based upon temperature and the seasonal timing of precipitation.

Many activities of present-day society are producing climatic changes, particularly a global warming trend. The 1980s and early 1990s were dominated by the highest average annual temperatures experienced since the advent of instrumental measurements. There is little scientific doubt that this warming is related to the greenhouse effect, which is increasing due to addition of CO_2, methane, and CFCs to the atmosphere and that various control options need to be started in a cooperative global effort. Many of the actions that could reduce warming trends are cost effective and produce net benefits.

The Intergovernmental Panel on Climate Change (IPCC) predicted surface-temperature response to a doubling of CO_2 with a range of increase between 1.5C° and 4.5C° (2.7–8.1F°) for the next century. Climatologists are constructing and refining general circulation models (GCMs) to better forecast future climate patterns. People and their political institutions then may use this information to form policies aimed at reducing unwanted climate change.

KEY TERMS		
	classification	general circulation models (GCMs)
	climate	genetic classification
	climatic regions	Köppen-Geiger climate classification
	climatology	methane
	climographs	nuclear winter hypothesis
	empirical classification	paleoclimatology

REVIEW QUESTIONS

1. Define climate and compare it to weather. What is climatology?
2. What creates the climatic disruption associated with El Niño? What are some of the changes and effects that occur worldwide?
3. What are the differences between a genetic and an empirical classification system?
4. Describe Köppen's approach to climatic classification. What was the basis of the system he devised?
5. List and discuss each of the six principal climate designations. In which one of these general types do you live? Which classification is the only type associated with the distribution and amount of precipitation?
6. What is a climograph, and how is it used to display climatic information?
7. Which of the major climate types occupies the most land and ocean area on Earth?
8. Characterize tropical *A* climates in terms of temperature, moisture, and location.
9. Using Africa's tropical climates as an example, characterize the climates produced by the seasonal shifting of the ITCZ with the high Sun.
10. Mesothermal *C* climates occupy the second largest portion of Earth's entire surface. Describe their temperature, moisture, and precipitation characteristics.
11. How can a marine west coast climate type *(Cfb)* occur in the Appalachian region of the eastern United States? Explain.
12. What role do offshore currents play in the distribution of the *Csb* climate designation? What type of fog is formed in these regions?
13. Discuss the climatic designation for the coldest places on Earth outside the poles. What do each of the letters in the Köppen classification indicate?
14. In general terms, what are the differences among the four desert classifications? How are moisture and temperature distributions used to differentiate these subtypes?
15. Describe any scientific consensus that has been reached about climate change.
16. Characterize worldwide temperature patterns from 1880 to 1993. What have temperatures been like since 1980?
17. Describe the potential climatic effects of global warming on polar and high-latitude regions. What are the implications of these climatic changes for persons living at lower latitudes? The effects on agricultural activity?
18. What is the IPCC? UNCED?

PART 3

EARTH'S CHANGING LANDSCAPES

CHAPTER 8
The Dynamic Planet

CHAPTER 9
Earthquakes and Volcanoes

CHAPTER 10
Weathering, Karst Landscapes, and Mass Movement

CHAPTER 11
Rivers and Related Landforms

CHAPTER 12
Wind Processes and Desert Landscapes

CHAPTER 13
Coastal Processes and Landforms

CHAPTER 14
Glacial and Periglacial Landforms

Sunset in the Endicott Mountains, Brooks Range, Alaska, Gates of the Arctic National Park. [Photo by Tom Bean.]

Earth is a dynamic planet whose surface is actively shaped by physical agents of change. Part 3 is organized around two broad systems of these agents—endogenic and exogenic. The **endogenic system** (Chapters 8 and 9) encompasses internal processes that produce flows of heat and material from deep below the crust, powered by radioactive decay—*the solid realm of Earth*. Earth's surface responds by moving, warping, and breaking, sometimes in dramatic episodes of earthquakes and volcanic eruptions.

At the same time, the **exogenic system** (Chapters 10–14) involves external processes that set air, water, and ice into motion, powered by solar energy—*the fluid realm of Earth's environment*. These media are sculpting agents that carve, shape, and reduce the landscape. One such process—weathering—breaks up and dissolves the crust, making materials available for erosion, transport, and deposition by rivers, winds, wave action, and flowing glaciers. Thus, Earth's surface is the interface between two systems, one that builds the landscape and one that reduces it.

The dramatic ramparts of Cuernos del Paine, Torres del Paine National Park, Chile. [Photo by Wolfgang Kaehler, EPX.]

8

THE DYNAMIC PLANET

THE PACE OF CHANGE
EARTH'S STRUCTURE AND INTERNAL
   ENERGY
   Earth in Cross Section
GEOLOGIC CYCLE
   Rock Cycle
CONTINENTAL DRIFT AND PLATE
TECTONICS
   A Brief History
   Sea-floor Spreading and Production of New
     Crust
   Subduction of Crust
   The Formation and Breakup of Pangaea
PLATE TECTONICS
   Plate Boundaries
   Earthquakes and Volcanoes
   Hot Spots
SUMMARY

The twentieth century has been a time of great discovery about Earth's internal structure and dynamic crust, yet much remains undiscovered. This is a time of revolution in our understanding of how the present arrangement of continents and oceans evolved. One task of physical geography is to explain the spatial implications of all this new knowledge and its effect on Earth's surface and society.

Earth's interior is highly structured, with uneven heating generated by the radioactive decay of unstable elements. The results are irregular patterns of surface fractures, the occurrence of earthquakes and volcanic activity, and the formation of mountain ranges. This chapter describes Earth's endogenic system, for it is within Earth that we find the driving forces that affect the crust.

A new era of Earth-systems science is emerging, combining various sciences within the study of physical geography. Remember that geography is an *approach*, a way of looking at things over physical space, the *spatial science* that examines place, location, human-Earth relationships, the uniqueness of regions, and movement (Figure 1-1). The geographic essence of geology, geophysics, paleontology (fossil study), seismology (earthquake study), and geomorphology (landform study) are all integrated by geographers to produce an overall picture of Earth's surface environment and of specific physical phenomena that affect human society. Improved remote-sensing capabilities and the development of computer-based geographic information and global positioning systems are improving these efforts.

The Pace of Change

The **geologic time scale** is a summary timeline of all Earth history. Figure 8-1 reflects currently accepted names of time intervals that encompass Earth's history, from vast eons through briefer eras, periods, and epochs. The *sequence* in this scale is based upon the relative positions of rock strata above or below each other—*relative time*. An important general principle is that of *superposition*, which states that rock and sediment always are arranged with the youngest beds "superposed" near the top of a rock formation and

the oldest at the base; if they have not been disturbed. The *absolute ages* on the scale, determined by scientific methods such as dating by radioactive isotopes, are used to refine the time-scale sequence—*absolute time*. Some highlights of Earth's history are listed on the geologic time scale in the figure.

The guiding principle of Earth science is uniformitarianism, first proposed by James Hutton in his *Theory of the Earth* (1795) and later amplified by Charles Lyell in *Principles of Geology* (1830). **Uniformitarianism** assumes that *the same physical processes active in the environment today have been operating throughout geologic time.* "The present is the key to the past" describes this principle. Uniformitarianism is supported by evidence unfolding from modern exploration and from the landscape record of volcanic eruptions, earthquakes, and processes that shape the landscape.

In contrast, the principle of *catastrophism* fits the vastness of Earth's age and the complexity of its rocks into a shortened time span. Catastrophism holds that Earth is young and that mountains, canyons, and plains formed through catastrophic events that did not require eons of time. Because there is little physical evidence to support this idea, catastrophism is more appropriately considered a belief than a serious scientific hypothesis. Ancient landscapes, such as the Appalachians, represent a much broader time scale of events (Figure 8-2).

However, geologic time is punctuated by dramatic events, such as massive landslides, earthquakes, and volcanic episodes. Within the principle of uniformitarianism, these localized catastrophic events occur as small interruptions in the generally uniform processes that shape the slowly evolving landscape.

We now start our journey deep within the planet. A knowledge of Earth's internal structure and energy is key to understanding the surface.

Earth's Structure and Internal Energy

Along with the other planets and the Sun, Earth is thought to have condensed and congealed from a nebula of dust, gas, and icy comets about 4.6 billion years ago (Chapter 2). The oldest surface rock yet

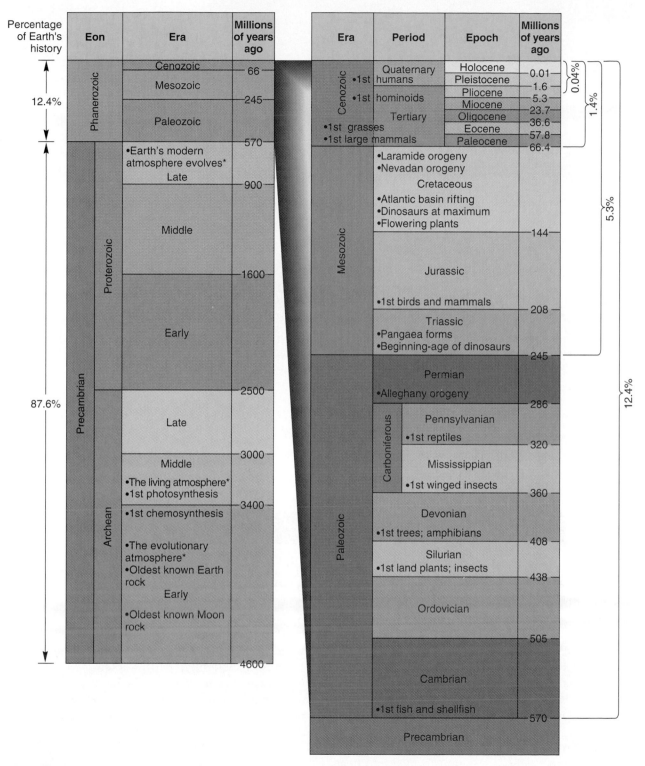

FIGURE 8-1

The geologic time scale is organized using absolute and relative dating methods. Absolute dates are determined through modern technological means, especially radioactive dating. In the column at left note that over 88% of geologic time occurred during the Precambrian. (*Approximate stages in development of Earth's atmosphere.) [Data from *1983 Geologic Time Scale*, Geological Society of America.]

FIGURE 8-2
Ancient rock formations of the
Appalachians, which began
forming during the Alleghany
Orogeny over 250 million years
ago. The rock outcrop is
quartzite, a metamorphic rock.
[Photo by Coco McCoy,
Rainbow.]

discovered on Earth (known as the Acasta Gneiss) lies in northwestern Canada and dates back 3.96 billion years. Scientists from Saint Louis University, the Canadian Geological Survey, and Australian National University certified the sample's age in 1989. Previously, rocks from Greenland held the record at 3.8 billion years old. These finds tell us something very significant: Earth was forming continental crust nearly 4 billion years ago.

Throughout Earth's formation process, heat was accumulating. As the protoplanet compacted into a smaller volume, energy was transformed into heat through compression. In addition, significant heat was trapped in Earth from the decay of unstable forms (isotopes) of uranium, thorium, potassium, and other radioactive elements.

Earth in Cross Section

As Earth solidified, heavier elements slowly gravitated toward the center, and lighter elements slowly welled upward to the surface, concentrating in the crust. Consequently, Earth's interior is arranged roughly in concentric layers, each one distinct either in chemical composition or temperature, with heat radiating outward from the center by conduction and then by physical convection in the more plastic levels nearer the surface.

Our knowledge of Earth's *internal differentiation* into these concentric layers has been acquired entirely through indirect evidence, because we are unable to drill more than a few kilometers into Earth. There are several physical properties of Earth mate-

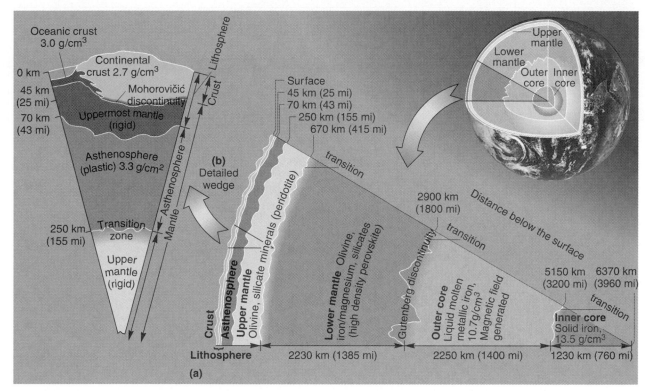

FIGURE 8-3

(a) Earth's interior from the central core to the crust is illustrated in cross section.
(b) The structure of the lithosphere and its relation to the asthenosphere is
detailed in the enlarged wedge.

rials that enable us to approximate the nature of the interior. For example, when an earthquake or underground nuclear test sends shock waves through the planet, the cooler areas, which generally are more rigid, transmit these **seismic waves** at a higher velocity than do the hotter areas.

Density also has an effect on seismic velocities, and fluid or plastic zones simply do not transmit some types of seismic waves; they absorb them. Some seismic waves are reflected as densities change; others are refracted, or bent, as they travel through Earth. Thus, the distinctive ways in which seismic waves pass through Earth and the time they take to travel between two surface points help *seismologists* deduce the structure of Earth's interior (Figure 8-3).

Figure 8-4 illustrates the dimensions of Earth's interior in comparison to surface distances in North America to give you a sense of size.

Earth's Core. A third of Earth's entire mass, but only a sixth of its volume, lies in its dense core. The **core** is differentiated into two regions—*inner core* and *outer core*—divided by a transition zone of several hundred kilometers (Figure 8-3). Estimated core temperatures range from 3000°C (5400°F) to as high as 6650°C (12,000°F). The inner core is thought to be solid iron. It remains solid even though it is at high temperature because of tremendous pressures. The iron is not pure, but probably is combined with silicon, and possibly oxygen and sulfur. The outer core is molten, metallic iron with lighter densities than the inner core.

Earth's Magnetism. The fluid outer core generates at least 90% of Earth's magnetic field and the magnetosphere that surrounds and protects Earth from the solar wind. One hypothesis describes circulation patterns in the outer core that are influenced by Earth's rotation. This circulation generates elec-

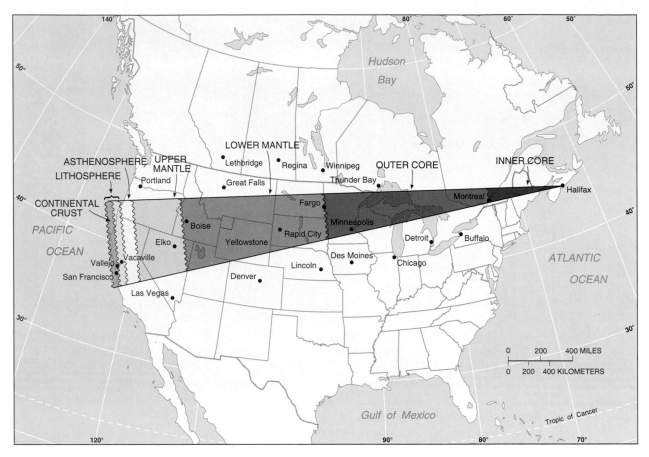

FIGURE 8-4
Surface map, using the distance from Halifax, Nova Scotia, to San Francisco to compare the distance from Earth's center to the surface. The continental crust thickness is the same as the distance between Vallejo (in the eastern portion of San Francisco Bay) and the city of San Francisco.

tric currents, which in turn induce the magnetic field. Think of the outer core flowing at several kilometers per year, over a million times faster than movement in the overlying mantle.

An intriguing feature of Earth's magnetic field is that its polarity sometimes fades to zero and then returns to full strength with north and south magnetic poles reversed! In the process, the field does not blink on and off but instead oscillates slowly to low intensity and then rapidly regains its full strength. This **magnetic reversal** has taken place nine times during the past 4 million years and hundreds of times over Earth's history. The average period of a magnetic reversal is 500,000 years; occurrences pos-

sibly vary from as short as several thousand years to as long as tens of millions of years.

There appears to be a trend in recent geologic time toward more frequent reversals with shorter intervals. When Earth is without polarity in its magnetic field, a random pattern of magnetism results. The effects of these low-intensity episodes on life are still speculative, but without a magnetosphere, the surface environment is unprotected from cosmic radiation and solar wind. Given present rates of magnetic field decay, we are perhaps 2000 years away from the next transition and reversal.

The reasons for these magnetic reversals are unknown; however, the spatial patterns they create at

Earth's surface are important diagnostic evidence in understanding the evolution of landmasses and the movement of the continents. When new iron-bearing rocks solidify from molten material at Earth's surface, the small magnetic particles in the rocks align according to the orientation of the magnetic poles at that time. This pattern is then locked in place as the rocks cool and harden. All across Earth, rocks bear this identical record of magnetic reversals. In addition, these magnetic patterns record the migration of the continents over time in relation to Earth's magnetic poles.

Earth's Mantle. A transition zone of several hundred kilometers divides the outer core from the mantle. Scientists at the California Institute of Technology analyzed more than 25,000 earthquakes and determined that this transition area is bumpy and uneven, with ragged peak-and-valleylike formations. Some of the motions in the mantle may be created by this rough texture at what is called the *Gutenberg discontinuity*. A *discontinuity* is a place where physical differences occur between adjoining regions in Earth's interior, such as between the outer core and lower mantle.

The **mantle** represents about 80% of Earth's total volume. It is rich in oxides and silicates of iron and magnesium (FeO, MgO, and SiO_2) and is more dense and tightly packed at depth, grading to lesser densities toward the surface. Of the mantle's volume, 50% is in the lower mantle, which is composed of rock that probably has never been at Earth's surface. The upper mantle is separated from the lower mantle by a broad transition zone of several hundred kilometers. The entire mantle experiences a gradual temperature increase with depth.

The upper mantle is divided into three fairly distinct layers. Outermost, just below the crust, is a high-velocity zone approximately 45–70 km (25–43 mi) thick, so-called because seismic waves are transmitted rapidly through this rigid, cooler layer. This *uppermost mantle*, along with the crust, makes up the *lithosphere.*

Below the lithosphere, from about 70 km down to 250 km (43–155 mi), is the **asthenosphere**, or plastic layer, from the Greek *asthenos,* meaning weak. It contains pockets of increased heat from radioactive decay and is susceptible to slow convective currents in these hotter (and therefore less dense) materials.

Because of this dynamic condition, the asthenosphere is the least rigid region of the mantle. About 10% of the asthenosphere is molten in asymmetrical patterns and hot spots. The resulting slow movement in this zone disturbs the overlying crust and creates tectonic activity—folding, faulting, and general deformation of surface rocks.

Below the asthenosphere resides the third layer, the rest of the upper mantle. Here the rocks are solid again, a zone of greater density, high seismic velocity, and increasing pressures.

Lithosphere and Crust. The lithosphere includes the entire **crust**, which represents only 0.01% of Earth's mass, and the upper mantle down to a depth of about 70 km (43 mi), with an imprecise transition marking the lower boundary. The boundary between the crust and the high-velocity portion of the lithospheric upper mantle is another discontinuity called the **Mohorovičić discontinuity**, or **Moho** for short, named for the Yugoslavian seismologist who determined that seismic waves change at this depth, owing to sharp contrasts of materials and densities.

Figure 8-3b illustrates the relationship of the crust to the rest of the lithosphere and the asthenosphere below. Crustal areas below mountain masses extend farther downward, perhaps to 50–60 km (31–37 mi), whereas the crust beneath continental interiors averages about 30 km (19 mi) in thickness and oceanic crust averages only 5 km (3 mi).

The composition, density, and texture of continental and oceanic crust are quite different. Continental crust is basically **granite**. It is crystalline and high in silica, aluminum, potassium, calcium, and sodium. (Sometimes continental crust is called *sial,* shorthand for *si*lica and *al*uminum.) Oceanic crust is **basalt**. It is granular and high in silica, magnesium, and iron. (Sometimes oceanic crust is called *sima,* shorthand for *si*lica and *ma*gnesium.)

The principle of buoyancy (that something less dense, like wood, floats in something denser, like water) and the principle of balance were developed into the important principle of **isostasy** to explain certain movements of Earth's crust. Think of

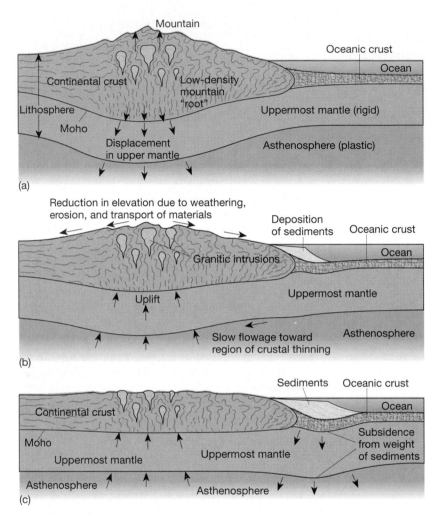

FIGURE 8-5

Isostatic adjustment of the crust. Earth's entire crust is in a constant state of compensating adjustment as suggested by these three sequential stages. In (a), the mountain mass slowly sinks, displacing mantle material. In (b), because of the loss of mass through erosion and transportation, the crust isostatically adjusts upward and sediments accumulate in the ocean. As the continent thins, (c) the heavy sediment load offshore begins to deform the lithosphere beneath the ocean.

Earth's outer crust as floating on the denser layers beneath, much as a boat floats on water. With a greater load (e.g., glaciers, sediment, mountains), the crust tends to ride lower in the asthenosphere. Without that load (e.g., when a glacier melts), the crust rides higher, in a recovery uplift known as an *isostatic rebound.* Thus, the entire crust is in a constant state of compensating adjustment, or *isostasy,* slowly rising and sinking in response to its own weight, and pushed and dragged about by currents in the asthenosphere (Figure 8-5).

Earth's crust is the outermost irregular shell that resides restlessly on a dynamic and diverse interior. Let us now examine the processes at work on this crust and the variety of rock types that compose the landscape.

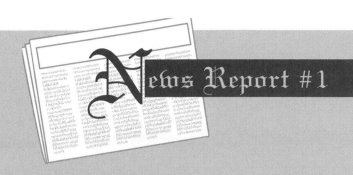

"Drilling the Crust to Record Depths"

Scientists wanting to sample mantle material directly have unsuccessfully tried for decades to penetrate Earth's crust to the Moho discontinuity. The longest-lasting deep-drilling attempt is on the northern Kola Peninsula in Russia, 250 km north of the Arctic Circle. Twenty years of high-technology drilling has produced a hole 12 km deep (7.5 mi or 39,400 ft). Presently, crystalline rocks 1.4 billion years old at 180°C (356°F) are being penetrated by diamond-tipped drills.

Oceanic crust is thinner than continental crust and is the object of several drilling attempts. The International Ocean Drilling Program (ODP), a cooperative effort directed by Texas A & M University, drilled a 2.5 km-deep hole in an oceanic rift near the Galápagos Islands from 1975 to 1993, with further drilling planned. An ocean floor of distinct layers of lava and deeper magma chambers were found. However, the Moho and Earth's mantle remain untapped.

FIGURE 1.
The modern drilling ship, the *JOIDES Resolution,* of the Ocean Drilling Program (at right). [Photo by ODP Texas A & M University.]

Geologic Cycle

Earth's crust is in an ongoing state of change, being formed, deformed, moved, and broken down by physical, chemical, and biological processes. The endogenic (internal) system is at work building landforms, while the exogenic (external) system wears them down. This vast give-and-take at the Earth-at- mosphere interface is called the **geologic cycle**. It is fueled by Earth's internal heat and by solar energy, influenced by the leveling force of Earth's gravity. Figure 8-6 illustrates the geologic cycle, combining many of the elements discussed in this text.

As you can see in the figure, the geologic cycle itself is composed of three principal cycles—hydrologic, rock, and tectonic. The *hydrologic cycle,* along

259

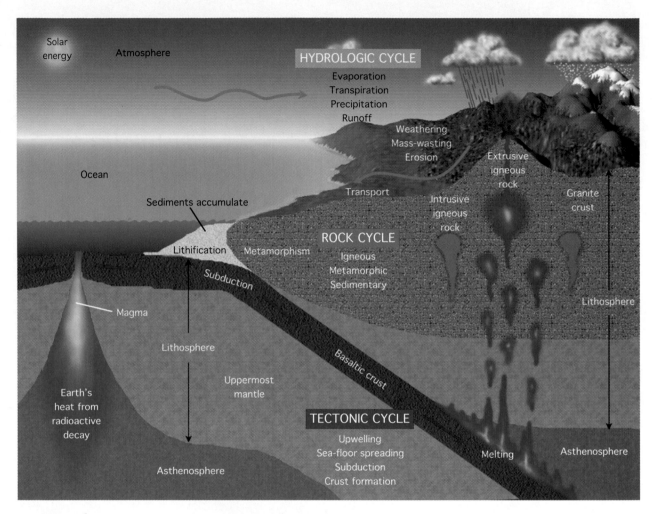

FIGURE 8-6
Geologic cycle, showing the interactive relationship of the rock cycle, tectonic
cycle, and hydrologic cycle, as endogenic (internal) and exogenic (external)
processes operate at or near Earth's surface.

the top of Figure 8-6, erodes, transports and deposits
Earth materials through the chemical and physical
action of water and ice. The *rock cycle* produces the
three basic rock types found in the crust—igneous,
metamorphic, and sedimentary. The *tectonic cycle*
brings heat energy and new material to the surface
and recycles material, creating movement and deformation of the crust.

Rock Cycle

Of Earth's crust, 99% is composed of only eight natural elements, and only two of these—oxygen and

silicon—account for 75.4% (Figure 8-7). Oxygen is
the most reactive gas in the lower atmosphere, readily combining with other elements. For this reason,
the percentage of oxygen is higher in the crust than
the 21% concentration in the atmosphere. The relatively large percentages of lightweight elements such
as silicon and aluminum in the crust are explained
by the internal differentiation process in which less
dense elements migrated toward the surface, as discussed earlier.

A **mineral** is an element or combination of elements that forms an inorganic natural compound. A
mineral can be described with a specific symbol or

chemical formula and possesses specific qualities, including a crystalline structure. One of the most widespread mineral families on Earth is the *silicates,* because silicon (Si) readily combines with oxygen and other elements. This mineral family includes quartz, feldspar, and clay minerals and numerous gemstones. Another important mineral family is the *carbonate* group, which features carbon in combination with oxygen and other elements such as calcium, magnesium, and potassium. Of the nearly 3000 minerals, only 20 are common, with just 10 of those making up 90% of the minerals in the crust.

A **rock** is an assemblage of minerals bound together (such as granite, a rock containing three minerals), or it may be a mass of a single mineral (such as rock salt). Literally thousands of rocks have been identified, all the result of three rock-forming processes: *igneous, sedimentary,* and *metamorphic.* Fig-ure 8-8 illustrates the interrelationships among these three processes in the **rock cycle**. The next three sections examine each process.

Igneous Processes. Rocks that solidify and crystallize from a molten state are called **igneous rocks**. They form from **magma**, which is molten rock beneath the surface (hence the name *igneous,* which means "fire-formed" in Latin). Magma is fluid, highly gaseous, and under tremendous pressure. It is either *intruded* into crustal rocks, known as *country rock,* or *extruded* onto the surface as **lava**.

The cooling history of the rock—how fast it cooled and how steadily the temperature dropped—determines its texture and degree of crystallization. These range from coarse-grained (slower cooling, with more time for larger crystals to form) to fine-grained or glassy (faster cooling). Most rocks in the

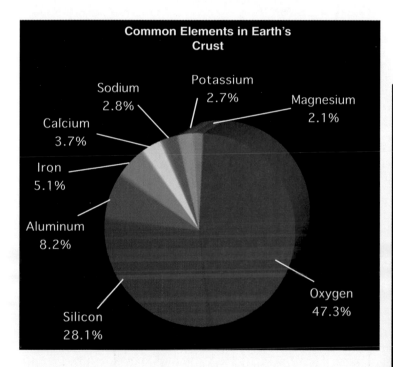

Common Elements in Earth's Crust

- Potassium 2.7%
- Sodium 2.8%
- Magnesium 2.1%
- Calcium 3.7%
- Iron 5.1%
- Aluminum 8.2%
- Oxygen 47.3%
- Silicon 28.1%

FIGURE 8-7

(a) Eight natural common elements comprise almost 99 percent of Earth's crust. (b) Quartz crystals (SiO_2) consist of Earth's two most abundant elements, silicon (Si) and oxygen (O). Quartz is one of the crust's most common minerals. [Photo from *Laboratory Manual in Physical Geology,* 3rd ed., Richard M. Busch, ed., p. 27. © 1993 Macmillan Publishing Company.]

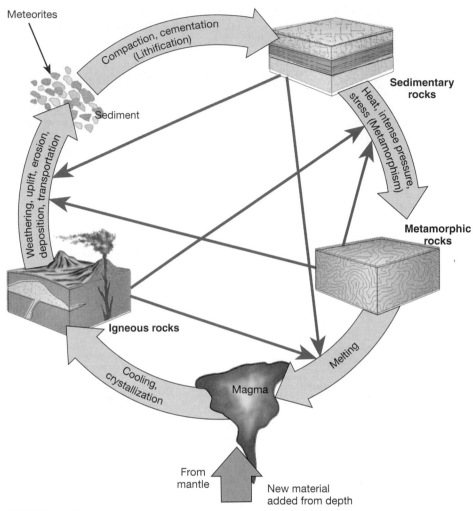

FIGURE 8-8
A rock-cycle schematic demonstrating the relationships among igneous, sedimentary, and metamorphic processes. The arrows indicate that each rock type can enter the cycle at various points and be transformed into other rock types. [Adapted by permission from *Laboratory Manual in Physical Geology*, 3d ed., Richard M. Busch, editor, p. 48. Copyright © 1993 by Macmillan Publishing Company.]

crust are igneous, although they frequently are covered by sedimentary rocks, soil, or the ocean. Figure 8-9 illustrates the variety of occurrences of igneous rocks, both on and beneath Earth's surface.

Intrusive igneous rock that cools slowly in the crust forms a **pluton**, a general term for any intrusive igneous rock body, regardless of size or shape. It is named after the Roman god of the underworld, Pluto. The largest pluton form is a **batholith** (Figure 8-9), defined as an irregular-shaped mass with a surface exposure greater than 100 km² (40 mi²) that

has invaded layers of crustal rocks. Batholiths form the mass of many large mountain ranges, for example, the Sierra Nevada batholith in California, the Idaho batholith, and the Coast Range batholith of Canada and extreme northern Washington State (Figure 8-10a). Smaller plutons are shown in Figure 8-9.

Basalt is the most common fine-grained extrusive igneous rock. It makes up the bulk of the ocean floor, comprising 71% of Earth's surface. It appears in lava flows such as those on the Galápagos Islands (Figure 8-10b).

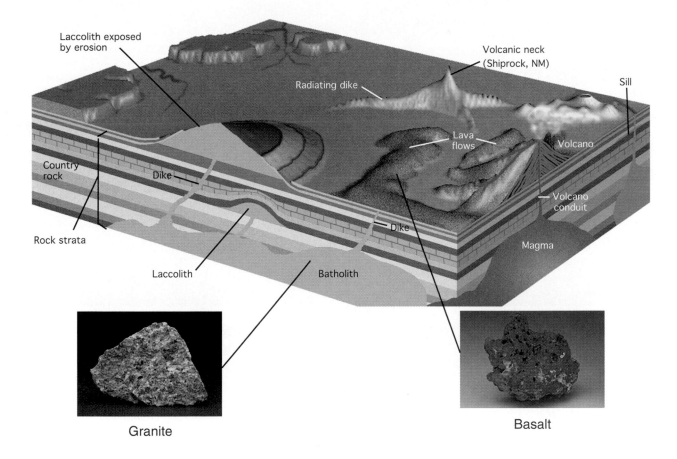

FIGURE 8-9
The variety of occurrences of igneous rocks, both intrusive (below the surface) and extrusive (on the surface). Inset photographs show samples of granite (intrusive) and basalt (extrusive). [Photos from *Laboratory Manual in Physical Geology,* 3rd ed., Richard M. Busch, editor, pp. 61, 63. Copyright © 1993 by Macmillan Publishing Company.]

FIGURE 8-10
(a) Exposed granites of the Sierra Nevada batholith; (b) basaltic lava flows on James Island of the Galápagos Islands, Ecuador. [(a) Photo by author; (b) photo by Galen Rowell.]

(a)   (b)

(a)

(b)

Sandstone

Limestone

FIGURE 8-11

Two types of sedimentary rock: (a) sandstone formation with sedimentary strata subjected to differential weathering; note weaker underlying siltstone, which erodes faster. (b) A limestone landscape, formed by chemical sedimentary processes, appears pitted and weathered in County Clare, Ireland. [(a) Photo by author; (b) photo by Tom Bean.]

Sedimentary Processes. Most sedimentary rocks are derived from existing rocks or from organic materials such as bone and shell. The exogenic processes of weathering and erosion provide the raw materials needed to form sedimentary rocks. Bits and pieces of former rocks—principally quartz, feldspar, and clay minerals—are eroded and then mechanically transported (by water, ice, wind, and gravity) to other sites where they are deposited. In addition, some minerals are dissolved into solution and form sedimentary deposits by precipitating from those solutions; this is an important process in the oceanic environment.

Various sedimentation forms are created in different environments—deserts, glaciers, beaches, tropics, and so on. Characteristically, sedimentary rocks are laid down by wind, water, and ice in horizontally layered beds. These layered strata are important records of past ages, and **stratigraphy** is the study of their sequence (superposition), thickness, and spatial distribution for clues to the age and origin of the rocks. Figure 8-11a is a sandstone sedimentary rock in a desert landscape. Note the various layers in the formation and how differently they resist erosional processes.

The cementation, compaction, and hardening of sediments into **sedimentary rocks** is called **lithification**. Various cements fuse rock particles together; lime ($CaCO_3$, or calcium carbonate) is the most common, followed by iron oxides (Fe_2O_3), and silica (SiO_2). Particles also can be united by drying (dehydration), heating, or chemical reactions. The two primary sources of sedimentary rocks are the mechanically transported bits and pieces of former rock, called *clastic sediments,* and the dissolved minerals in solution, called *chemical sediments.* Clastic sedimentary rocks, such as sandstone, are derived from weathered and fragmented rocks that are further worn in transport—everything from boulders to microscopic clay particles.

Chemical sedimentary rocks are not formed from physical pieces of broken rock, but instead are dissolved, transported in solution, and chemically precipitated out of solution (they are essentially nonclastic). The most common chemical sedimentary rock is **limestone**, which is lithified calcium carbonate, $CaCO_3$. A similar form is dolomite, which is lithified calcium-magnesium carbonate, $CaMg(CO_3)_2$. About 90% of all limestone comes from organic sources making it biochemical—derived from shell and bone produced by biological activity. Once formed, these rocks are vulnerable to chemical weathering which produces unique landforms, as discussed in the weathering section of Chapter 10 and as shown in Figure 8-11b.

Another sedimentary rock of organic origin is coal. **Coal** is really a biochemical rock, very rich in carbon. Plants incorporate carbon into their tissues by taking in carbon dioxide (CO_2) during photosynthesis, and it is this process that brings organic carbon into the rock cycle. During the Carboniferous period (360–286 million years ago) lush vegetation flourished and died in tropical swamps. Coal formed in these bogs through progressive stages.

Pieces of former plants can be easily recognized in *peat,* a brownish, compacted fibrous organic sediment—a first step in forming coal. Peat has been burned as a low-grade fuel for centuries, producing a low-heat energy output and a great deal of air pollution. Following more compaction and hardening, a soft, coal-like substance known as *lignite* forms. Further lithification leads to *bituminous coal,* the first true sedimentary rock in this sequence. Burial of coal deposits under sedimentary overburden results in the formation of coal seams, usually between beds of shale or sandstone. Where conditions are right, metamorphism of bituminous coal results in *anthracite coal,* a metamorphic rock that is extremely hard and therefore difficult to mine. More than half of U.S. electrical production is generated by coal-fired power plants principally using bituminous coal as the fuel.

Chemical sediments also form from inorganic sources when water evaporates and leaves behind a residue of salts. These **evaporites** may exist as common salts, such as gypsum or sodium chloride (table salt), to name only two, and often appear as flat, layered deposits across a dry landscape. This process is dramatically demonstrated in the pair of photographs in Figure 8-12, taken in Death Valley National Monument (now a National Park) one day and one month after a record 2.57 cm (1.01 in.) rainfall.

Chemical deposition also occurs in the water of natural hot springs from chemical reactions between minerals and oxygen; this is a sedimentary

FIGURE 8-12
A Death Valley landscape (a) one day after a record rainfall event when the valley was covered by several square kilometers of water yet only a few centimeters deep. (b) One month later the water had evaporated and the same valley is coated with evaporites (borated salts). [Photos by author.]

(a)

(b)

FIGURE 8-13
An example of an ancient metamorphic rock formation in Antarctica. [Photo courtesy of U.S. Geological Survey.]

process related to *hydrothermal activity*. A deposit of travertine, a form of calcium carbonate, at Mammoth Hot Springs in Yellowstone National Park is an example.

Metamorphic Processes. Any rock, either igneous or sedimentary, may be transformed into a **metamorphic rock**, by going through profound physical and/or chemical changes under pressure and increased temperature. (The name metamorphic comes from the Greek, meaning to "change form.") Metamorphic rocks generally are more compact than the original rock and therefore are harder and more resistant to weathering and erosion (Figure 8-13).

Several conditions can cause metamorphism, particularly when subsurface rock is subjected to high temperatures and high compressional stresses occurring over millions of years. Igneous rocks are compressed during collisions of portions of Earth's crust (described under "Plate Tectonics" later in this chapter). Or, igneous rocks may be sheared and stressed along earthquake fault zones. Sometimes they simply are crushed under great weight when a crustal area is thrust beneath other crust.

Another metamorphic condition occurs when sediments collect in broad depressions in Earth's crust and, because of their own weight, create enough pressure in the bottommost layers to transform the sediments into metamorphic rock. Also, molten magma rising within the crust may "cook" adjacent rock, a process called *contact metamorphism*.

Metamorphic rocks may be changed both physically and chemically from the original rocks. If the mineral structure demonstrates a particular alignment after metamorphism, the rock is *foliated* and some minerals may appear as wavy striations (streaks or lines) in the new rock. On the other hand, parent rock with a more homogeneous (evenly mixed) makeup may produce a *nonfoliated* rock. Table 8-1

Table 8-1
Metamorphic Rocks

Parent Rock	Metamorphic Equivalent	Texture
Shale (clay minerals)	Slate	Foliated
Granite, slate, shale	Gneiss	Foliated
Basalt, shale, periodite	Schist	Foliated
Limestone, dolomite	Marble	Nonfoliated
Sandstone	Quartzite	Nonfoliated

Examples

Folded gneiss
(foliated)

Marble
(nonfoliated)

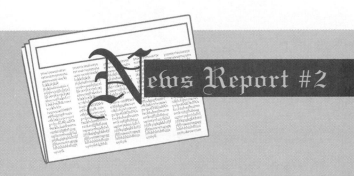

"Canyon Rocks Harder than Steel"

The ancient roots of mountains are composed predominantly of metamorphic rocks. Tourists visiting the Grand Canyon in Arizona witness this fact first hand. At the bottom of the inner gorge (the lower 455 m, or 15500 ft), a Precambrian metamorphic rock called the *Vishnu Schist* is over 2 billion years in age. A sample of the rock and a metal file allow visitors to test the rock's hardness against steel. Despite the fact that the schist is harder, the Colorado River has cut down into the uplifted Colorado Plateau, exposing and eroding these old rocks. Given enough time the processes of nature will wear away the canyon only to be replaced by new rock formations produced by Earth's internal processes.

lists some metamorphic rock types and their parent rocks. The two inset photographs demonstrate foliated and nonfoliated textures.

You have seen how the rock-forming processes yield the igneous, sedimentary, and metamorphic materials of Earth's crust. Next we look at how vast **tectonic processes** press, push, and drag portions of the crust in large-scale movements, causing the continents to "drift."

Continental Drift and Plate Tectonics

Have you ever looked at a world map and noticed that a few of the continental landmasses appear to have matching shapes like pieces of a jigsaw puzzle—particularly South America and Africa? The incredible reality is that the continental pieces indeed once were fitted together! Continental landmasses not only migrated to their present locations but continue to move at speeds up to 6 cm (2.4 in.) per year. We say that the continents are "adrift" because convection currents in the asthenosphere and upper mantle are dragging them around. The key point is that the arrangement of continents and oceans we see today is not permanent but is in a continuing state of change.

A continent such as North America is actually a collage of pieces and fragments of crust that have migrated to form the landscape we know. Through a historical chronology, let us trace the discoveries that produced this major revolution in science, called *plate tectonics*. The facts that the continents drift and the young seafloor spreads from rifts are revolutionary discoveries in the tectonic cycle that came to light only in this century.

A Brief History

As early mapping gained accuracy some observers noticed a symmetry. Abraham Ortelius (1527-1598) described the apparent fit of some continental coastlines in his *Thesaurus Geographicus* (1596). In 1620, the English philosopher Sir Francis Bacon noted gross similarities between the shapes of Africa and South America (although he did not suggest that they had drifted apart). Others wrote—unscientifically—about such apparent relationships, but it was not until much later that a valid explanation was proposed. In 1912, German geophysicist and meteorologist Alfred Wegener publicly presented in a lecture his idea that Earth's landmasses migrate. His book, *Origin of the Continents and Oceans,* appeared in

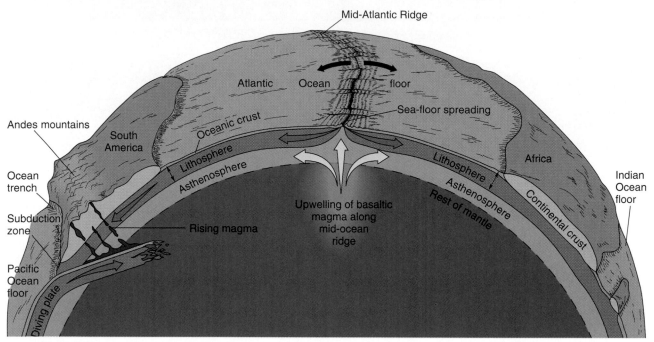

FIGURE 8-14

Sea-floor spreading, upwelling currents, subduction, and plate movements, shown in cross section. [After Peter J. Wyllie, *The Way The Earth Works*. Copyright © 1976, by John Wiley & Sons, Inc. Adapted by permission of John Wiley & Sons, Inc.]

1915. Wegener today is regarded as the father of this concept, which he called *continental drift*. But scientists at the time, knowing little of Earth's interior structure and bound by an inertial mindset as to how continents and mountains were formed, were unreceptive of Wegener's revolutionary proposal and, in a nonscientific spirit, rejected it outright.

Wegener thought that all landmasses were united in one supercontinent approximately 225 million years ago, during the Triassic period—see Figure 8-16b. (As you read this, it may be helpful to refer to the geologic time scale in Figure 8-1.) This one landmass he called **Pangaea**, meaning "all Earth." Although his initial model kept the landmasses together far too long and his idea about the driving mechanism was incorrect, Wegener's overall configuration of Pangaea was right.

To come up with his Pangaea fit, he began studying the research of others, specifically the geologic record, the fossil record, and the climatic record for the continents. He concluded that South America and Africa were related in many complex ways. He also concluded that the large midlatitude coal deposits, which stretch from North America to Europe to China, and date to the Permian and Carboniferous periods (245–360 million years ago), existed because these regions once were more equatorial in location and therefore covered by lush vegetation that became coal.

As modern scientific capabilities built the case for continental drift, the 1950s and 1960s saw a revival of interest in Wegener's concepts, and finally confirmation. Aided by an avalanche of discoveries, the theory today is nearly universally accepted as an accurate model of the way Earth's surface evolves, and virtually all Earth scientists accept the fact that continental masses move about in dramatic ways now referred to as plate tectonics.

Sea-floor Spreading and Production of New Crust

The key to establishing the fact of continental drift was a better understanding of the seafloor. The seafloor has a remarkable feature: an interconnected mountain chain (ridge) some 64,000 km (40,000 mi) in extent and averaging more than 1000 km (600 mi) in width. A striking view of this great undersea mountain chain opens Chapter 9. How did this mountain chain get there?

In the early 1960s, geophysicists Harry H. Hess and Robert S. Dietz proposed **sea-floor spreading** as the mechanism that builds this mountain chain and drives continental movement. Hess said that these submarine mountain ranges, called the **mid-ocean ridges**, were the direct result of upwelling flows of magma from hot areas in the upper mantle and asthenosphere and perhaps deeper sources. When mantle convection brings magma up to the crust, the crust is fractured and the magma spills out and cools to form new seafloor, building the ridges

and spreading laterally. Figure 8-14 suggests that the ocean floor is rifted and scarred along mid-ocean ridges, with arrows indicating the direction in which the seafloor spreads.

As new crust is generated and the seafloor spreads, the alignment of the magnetic field in force at the time dictates the alignment of magnetic particles in the cooling rock, creating a kind of magnetic tape recording in the seafloor. Each magnetic reversal and reorientation of Earth's polarity is recorded in the oceanic crust. Figure 8-15 illustrates just such a recording, from the Mid-Atlantic ridge near Iceland. Note the magnetic mirror images that develop on either side of the sea-floor rift as a result of these magnetic reversals recorded in the rock.

If you measure the same distance from the rift to either side, you find the same orientation of magnetic content in the rock. Thus, the periodic reversals of Earth's magnetic field have proven a valuable clue to understanding sea-floor spreading. In essence, scientists were provided with a global pattern of magnetic stripes to use in fitting together pieces of Earth's crust.

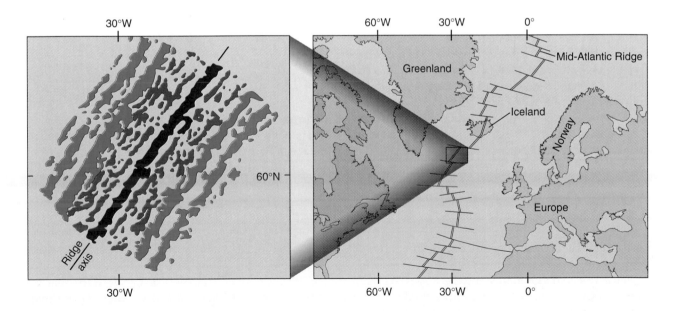

FIGURE 8-15

Magnetic reversals recorded in the seafloor south of Iceland along the Mid-Atlantic Ridge. [Magnetic reversals reprinted from *Deep-Sea Research* 13, J. R. Heirtzler, S. Le Pichon, and J. G. Baron. Copyright © 1966, Pergamon Press, p. 247.]

As the age of the seafloor was determined from these seafloor recordings of Earth's magnetic-field reversals and other measurements, the complex harmony of the two concepts of continental drift and sea-floor spreading became clearer. *The youngest crust anywhere on Earth is at the spreading centers of the mid-ocean ridges.* With increasing distance from these centers, Earth's surface gets increasingly older. The oldest seafloor is in the western Pacific near Japan, an area called the Pigafetta Basin.

Overall, the seafloor is relatively young—nowhere does it exceed 200 million years in age, remarkable when you remember that Earth's age is 4.6 billion years. The reason is that oceanic crust is not permanent—it is slowly plunging beneath continental crust along Earth's deep oceanic trenches. The discovery that the seafloor is so young demolished earlier geologic thinking that the oldest rocks would be found there.

Subduction of Crust

In contrast to the upwelling zones along the mid-ocean ridges are the areas of descending crust elsewhere. We must remember that the basaltic ocean crust has a greater average density (3.0 grams per cm³) than continental crust (2.7 grams cm³). As a result, when continental crust and oceanic crust slowly collide, the denser ocean floor will dive beneath the lighter continent. This forms a descending **subduction zone** (Figure 8-14). The world's **oceanic trenches** are subduction zones and are the deepest features on Earth's surface. The deepest is the Mariana Trench near Guam, which descends below sea level to −11,033 m or −36,198 ft.

The subducted portion of crust is dragged down into the mantle, where it remelts, is recycled, and eventually migrates back toward the surface through deep fissures and cracks in crustal rock (Figure 8-14, left). Volcanic mountains like the Andes in South America and the Cascade Range from northern California to the Canadian border form inland of these subduction zones. The fact that spreading ridges and subduction zones are areas of earthquake and volcanic activity provide important clues for the study of continental drift.

Sea-floor spreading, subduction, and mantle convection all are phenomena of plate tectonics, which by 1968 had become the all-encompassing term for the concepts that began with Wegener's original continental drift proposal. Using current scientific findings, let us go back and reconstruct the past, and Pangaea.

The Formation and Breakup of Pangaea

The supercontinent of Pangaea and its subsequent breakup into today's continents represent only the last 225 million years of Earth's 4.6 billion years, or only the most recent 1/23 of Earth's existence. During the other 22/23 of geologic time, other things were happening. The landmasses as we know them were barely recognizable.

Pre-Pangaea. Figure 8-16a shows the pre-Pangaea arrangement of 465 million years ago (during the Middle Ordovician period). The outlines of South America, Africa, India, Australia, and southern Europe appear within the continent called Gondwana in the Southern Hemisphere. What would become North America and Greenland is the inverted continent of Laurentia, which straddles the equator near the location of the present-day Fiji and Samoa island groups. The other land groups are identified.

Pangaea. Figure 8-16b shows an updated version of Wegener's Pangaea, 225–200 million years ago (Triassic–Jurassic periods). Areas of North America, North Africa, the Middle East, and Eurasia were near the equator and therefore were covered by plentiful vegetation. The landmasses focused near the South Pole and attached to the sides of Antarctica were covered by ice. Today, those portions of South America, southern Africa, India, Australia, and Antarctica bear the scars of this shared glacial blanket.

The India plate (the slab of Earth's crust that included modern India) appears larger on the map than the same area today because the Tibetan Plateau formed the northern portion of the plate. We also can see that Africa shared a common connection with both North and South America. Today, the Appalachian Mountains in the eastern United States and the Atlas Mountains of northwestern Africa reflect this common ancestry; they are, in fact, portions of the same mountain range. At the time of Figure 8-16b, the Atlantic Ocean did not exist. The only oceans were Panthalassa ("all seas"), which would become the Pacific, and the Tethys Sea (partly enclosed by the African and the Eurasian plates), which formed the Mediterranean Sea, with trapped portions becoming the present-day Caspian Sea.

Pangaea Breaks Up. Figure 8-16c shows the movement of plates that had occurred by 135 million years ago (the beginning of the Cretaceous period). New seafloor that had formed is highlighted. The active spreading center had rifted North America away from landmasses to the east, shaping the coast of Labrador. India was further along in its journey, with a spreading center to the south, and a subduction zone to the north where the leading edge of the India plate was diving beneath Eurasia. At that time a rift also began in what was to be the South Atlantic, and Antarctica and Australia were pulled into isolation from the other masses.

Modern Continents Take Shape. Figure 8-16d shows the arrangement 65 million years ago (beginning of the Tertiary period). Sea-floor spreading along the Mid-Atlantic Ridge had grown some 3000 km (almost 1900 mi) in 70 million years. Africa had moved northward about 10° in latitude, leaving Madagascar split from the mainland and opening up the Gulf of Aden. The rifting along what would be the Red Sea had begun. The India plate had moved three-fourths of the way to Asia, as Asia continued to rotate clockwise. Of all the major plates, India traveled the farthest. From this time until the present, according to Robert Dietz and John Holden, more than half of the ocean floor was renewed.

The Continents Today. Figure 8-16e shows modern geologic time (the late Cenozoic Era). The northern reaches of the India plate have underthrust the southern mass of Asia through subduction, forming the Himalayas in the upheaval created by the collision. Plate motions continue to this day and will proceed into the future.

Plate Tectonics

Plate tectonics is the comprehensive model for Wegener's continental drift proposal, sea-floor spreading, and other related aspects of this twentieth-century revolution in Earth sciences. *Tectonic,* from the Greek *tektonikós* meaning building or construction, refers to the deformation of Earth's crust as a result of internal forces, which can form various structures in the lithosphere. Tectonic processes include the upwelling of magma, plate movement, and subduction of crust, plus folds, faults, warps, fractures, earthquakes, and volcanic activity. As we discussed, the system is powered by density and temperature differences in an unevenly mixed mantle.

Earth's present crust is divided into at least 14 plates, of which about half are major and half are minor, in terms of size (Figure 8-17). These plates are composed of literally hundreds of smaller pieces and perhaps dozens of microplates that have migrated together. The arrows in the figure indicate the direction in which each plate is presently moving, and the length of the arrows suggests the rate of movement. The illustration of the seafloor that begins Chapter 9 is helpful in identifying the plate boundaries in the following discussion. Please correlate this map of the tectonic plates (Figure 8-17) with that illustration.

Plate Boundaries

The boundaries where plates meet clearly are dynamic places. The block diagram inserts in Figure 8-16e show the three general types of motion and interaction that occur along the boundary areas:

- *Divergent boundaries* (lower left) are characteristic of sea-floor spreading centers, where upwelling material from the mantle forms new seafloor, and crustal plates are spread apart. These are zones of tension. An example noted in the figure is the divergent boundary along the East Pacific rise, which gives birth to the Nazca plate and the Pacific plate. Figure 8-15 illustrates patterns of magnetic reversals on such a divergent plate boundary south of Iceland along the

FIGURE 8-16

The formation and breakup of Pangaea and the types of motions occurring at plate boundaries. [(a) from R. K. Bambach, Scotese, and Ziegler, "Before Pangaea: The Geography of the Paleozoic World," *American Scientist* 68 (1980): 26–38, reprinted by permission; (b) through (e) from Robert S. Dietz and John C. Holden, *Journal of Geophysical Research* 75, no. 26 (10 September 1970): 4939–56, copyright by the American Geophysical Union.]

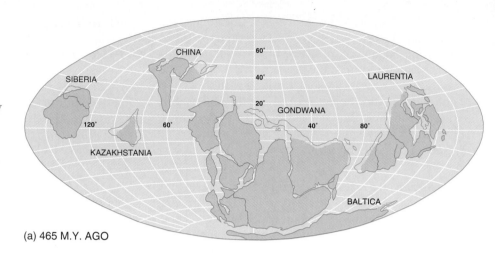

(a) 465 M.Y. AGO

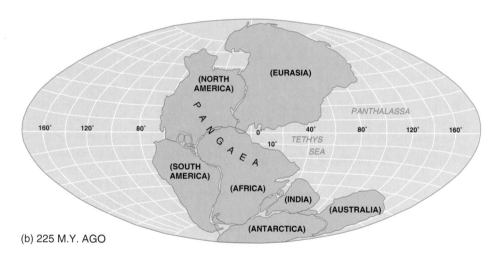

(b) 225 M.Y. AGO

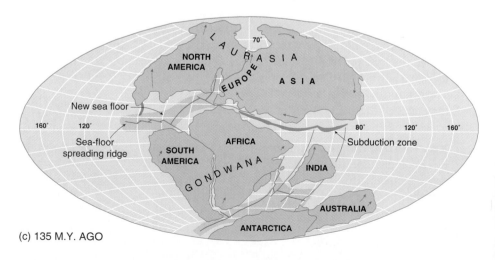

(c) 135 M.Y. AGO

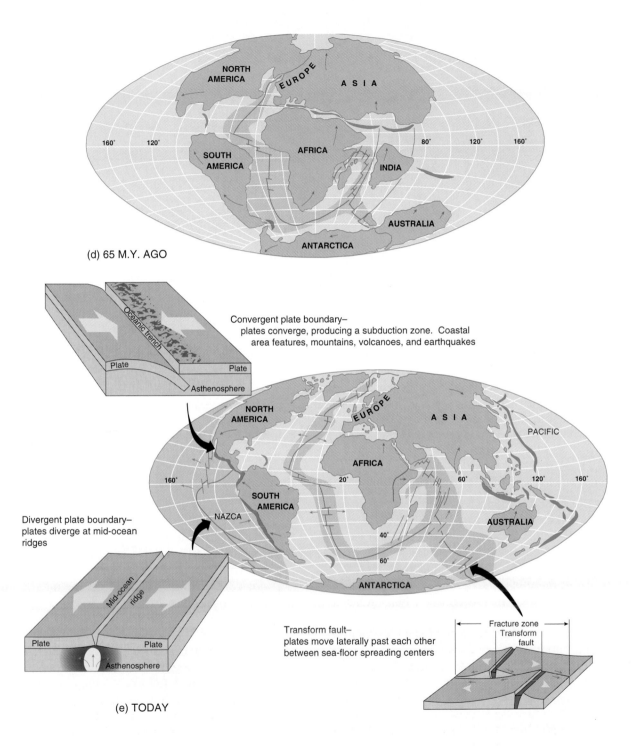

(d) 65 M.Y. AGO

Convergent plate boundary–
plates converge, producing a subduction zone. Coastal
area features, mountains, volcanoes, and earthquakes

Divergent plate boundary–
plates diverge at mid-ocean
ridges

Transform fault–
plates move laterally past each other
between sea-floor spreading centers

(e) TODAY

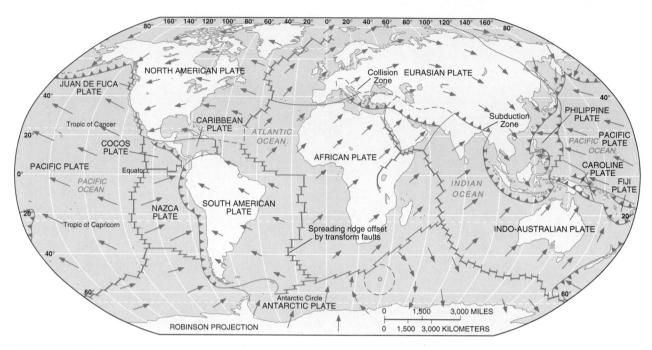

FIGURE 8-17

Earth's 14 major lithospheric plates and their motions. Each arrow represents 20 million years of movement, the longer arrows indicating that the Pacific and Nazca plates are moving more rapidly than the Atlantic plates. [Adapted from U. S. Geodynamics Committee and Sheldon Judson, Kenneth S. Deffeyes, and Robert B. Hargraves, *Physical Geology,* © 1976, p. 28. Reprinted by permission of Prentice Hall, Inc., Englewood Cliffs, NJ.]

Mid-Atlantic Ridge. While most divergent boundaries occur at mid-ocean ridges, there are a few within continents themselves. An example is the Great Rift Valley of East Africa, where continental crust is being rifted apart.

- *Convergent boundaries* (upper left) are characteristic of collision zones, where areas of continental and/or oceanic crust collide. These are zones of compression. Examples include the subduction zone off the west coast of South and Central America (noted in Figure 8-16e) and the area along the Japan and Aleutian trenches. Along the western edge of South America, the Nazca plate collides with and is subducted beneath the South American plate, creating the Andes Mountains chain and related volcanoes. The collision of India and Asia mentioned earlier is another example of a convergent boundary.

- *Transform boundaries* (lower right) occur where plates slide laterally past one another at right angles to a sea-floor spreading center, neither diverging nor converging, and usually with no volcanic eruptions. These are the right-angle fractures stretching across the mid-ocean ridge system worldwide (visible in Figure 8-17 and Chapter 9 opening illustration).

Across the entire ocean floor these boundaries are the location of *transform faults,* always parallel to the direction in which the plate is moving. You can see in the figures that mid-ocean ridges are not simple straight lines. When a mid-ocean rift begins, it opens at points of weakness in the crust. The fractures you see in the figure began as a series of offset breaks in the crust in each portion of the spreading center. As new material rises to the surface, building the mid-ocean ridges and spreading the plates, these offset areas slide past each other, in horizontal faulting motions. The famous San Andreas fault system in California, where continental crust has overridden a transform system, is related to this type of motion and is discussed in the next chapter.

Earthquakes and Volcanoes

Plate boundaries are the primary location of earthquake and volcanic activity, and the correlation of these phenomena is an important aspect of plate tectonics. Earthquakes and volcanic activity are discussed in more detail in the next chapter, but their general relationship to the tectonic plates is important to point out here. Earthquake zones and vol-

canic sites are identified in Figure 8-18. The "ring of fire" surrounding the Pacific basin, named for the frequent incidence of volcanoes, is most evident. The subducting edge of the Pacific plate thrusts deep into the crust and upper mantle, producing molten material that makes its way back toward the surface, causing active volcanoes along the Pacific Rim. Such processes occur similarly at plate boundaries throughout the world.

Hot Spots

A dramatic aspect of plate tectonics is the estimated 50–100 **hot spots** across Earth's surface. These are individual sites of upwelling material from the mantle,

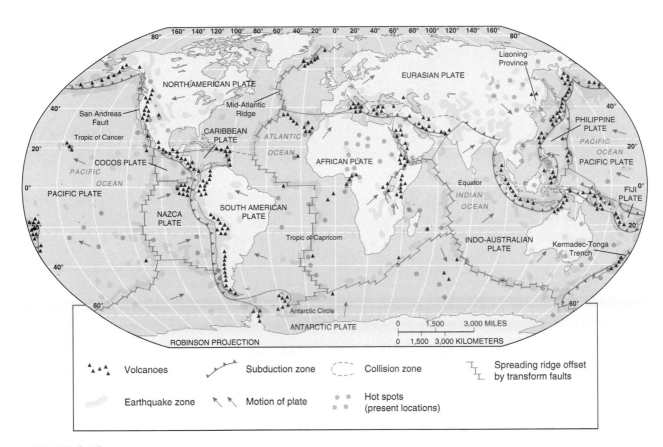

FIGURE 8-18

Earthquake and volcanic activity in relation to major tectonic plate boundaries, and principal hot spots. [Earthquake/volcano data from *Earthquakes* by Bruce A. Bolt. Copyright © 1988 W. H. Freeman and Company; reprinted with permission. Hot spots from "Hot Spots on the Earth's Surface," August 1976, p. 52, Kevin C. Burke and J. Tuzo Wilson. Copyright © 1976 by Scientific American, Inc. All rights reserved.]

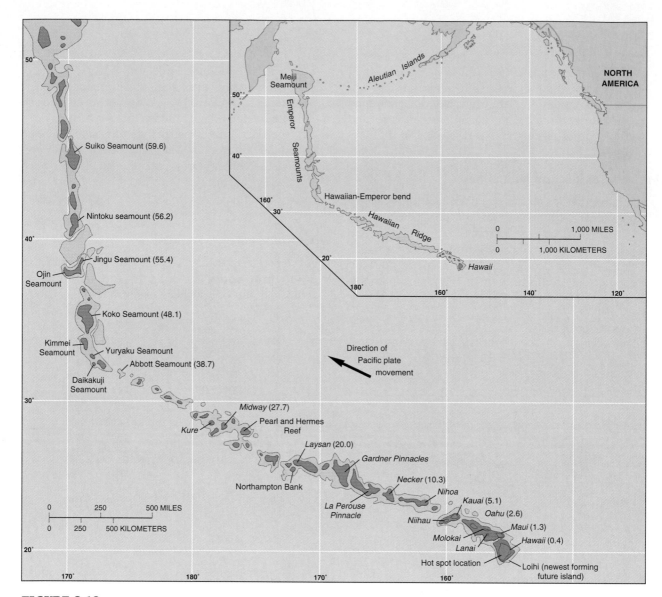

FIGURE 8-19

Hawaii and the linear volcanic chain of islands known as the Emperor Seamounts.
Ages of islands and seamounts in the chain are shown in millions of years. [After
David A. Clague, "Petrology and K-Ar (Potassium-Argon) Ages of Dredged
Volcanic Rocks from the Western Hawaiian Ridge and the Southern Emperor
Seamount Chain," *Geological Society of America Bulletin* 86 (1975): 991.]

noted on Figure 8-18. Hot spots occur beneath both
oceanic and continental crust and appear to be deeply
anchored in the mantle, tending to remain fixed beneath migrating plates. Unlike convectional currents in
the asthenosphere that principally drive plate motions,
scientists now think hot spots are initiated by upwelling plumes that are rooted below the 670 km
transition zone. Thus, the area of a plate that is above

a hot spot is locally heated for the brief geologic time
it is there (a few hundred thousand or million years).

An example of an isolated hot spot is the one that
has formed the Hawaiian-Emperor islands chain
(Figure 8-19). The Pacific plate has moved across this
hot, upward-erupting plume for almost 80 million
years, with the resulting string of volcanic islands
moving northwestward away from the hot spot.

"Yellowstone on the Move"

Yellowstone National Park, in northwestern Wyoming, is above a hot spot, and because the North American plate is moving westward, the hot spot is leaving a track across the western United States. Evidence such as basaltic flows is cast through Idaho, Oregon, and Washington.

Hydrothermal features such as Yellowstone and most linear volcanic chains have a hot spot at one end, where mantle material works its way up through the passing plate. The track left by these eruptions matches the direction of the plate. If the North American plate continues drifting westward, imagine the phenomenon of Yellowstone, including thermal springs, lava flows, and earthquake and volcanic activity, eventually migrating through the Dakotas!

Thus, the age of each island or seamount in the chain increases northwestward from the island of Hawaii, as you can see from the ages marked in the figure. The oldest island in the Hawaiian part of the chain is Kauai, approximately 5 million years old; it is weathered, eroded, and deeply etched with canyons and valleys.

The big island of Hawaii actually took less than 1 million years to build to its present stature. The youngest island in the chain is still a *seamount*, a submarine mountain that does not reach the surface. It rises 3350 m (11,000 ft) from the ocean floor but is still 975 m (3200 ft) beneath the ocean surface. Even though this new island will not experience the Sun for about 10,000 years, it is already named Loihi.

To the northwest of this active hot spot in Hawaii, the island of Midway rises as a part of the same system. From there the Emperor Seamounts stretch northwestward until they reach about 40 million years of age. At that point, this linear island chain shifts direction northward, evidently reflecting a change in the movement of the Pacific plate at that earlier time. At the northernmost extreme, the seamounts that formed about 80 million years ago are now approaching the Aleutian Trench, where they eventually will be subducted beneath the Eurasian plate.

Iceland is an example of an active hot spot sitting astride a mid-ocean ridge (Figure 8-15). It also is the best example of a segment of mid-ocean ridge rising above sea level. This hot spot has generated enough material to form Iceland and continues to cause eruptions from deep in the asthenosphere. As a result, Iceland is still growing in area and volume. The youngest rocks are near the center of Iceland, with rock age increasing toward the eastern and western coasts.

Take a moment to locate these features (e.g., mid-ocean ridges, subduction zones, Hawaiian and Icelandic hot spots) on the Chapter 9 opening illustration.

SUMMARY — The Dynamic Planet

Earth's surface is where the **endogenic** (internal) **system**, powered by heat energy from within the planet, and the **exogenic** (external) **system**, powered by insolation and influenced by gravity, work together to produce Earth's diverse landscape. The **geologic time scale** depicts the vast span of geologic time and the sequence of events that produced our planet as we know it today. The nature of the endogenic system lies hidden beneath Earth's crust and is studied by scientists indirectly through analysis of **seismic** (earthquake) **wave** behavior. The **geologic cycle** is a model of these internal and external interactions. It comprises three other cycles: the *hydrologic cycle, tectonic cycle*, and **rock cycle**. The rock cycle describes the three principal rock-forming processes—**igneous, sedimentary**, and **metamorphic**.

A knowledge of Earth's interior is important in understanding its magnetic field and the dynamics of crustal motion: **upwelling, sea-floor spreading**, and **subduction**. Polarity reversals in Earth's magnetism are recorded in cooling magma and sediments that contain iron minerals. The pattern of changing **magnetic reversals** frozen in rock helps scientists piece together the history of Earth's drifting crust. Earth's **crust** is broken into huge slabs or plates, each in motion in response to slow currents in the **mantle** below the crust. The present configuration of the oceans and the continents bears the imprint of these vast tectonic forces. Likewise, occurrences of often damaging earthquakes and volcanoes correlate with plate boundaries. The theory of **plate tectonics** has produced a twentieth-century revolution in Earth sciences.

KEY TERMS

asthenosphere	magnetic reversal
basalt	mantle
batholith	metamorphic rock
coal	mid-ocean ridge
core	mineral
crust	Mohorovičić discontinuity (Moho)
endogenic system	oceanic trench
evaporite	Pangaea
exogenic system	plate tectonics
geologic cycle	pluton
geologic time scale	rock
granite	rock cycle
hot spot	sea-floor spreading
igneous rock	sedimentary rock
isostasy	seismic wave
lava	stratigraphy
limestone	subduction zone
lithification	tectonic process
magma	uniformitarianism

REVIEW QUESTIONS

1. Define the endogenic and the exogenic systems. Describe their energy sources.
2. How is geologic time organized? What is the basis for the time scale? What era, period, and epoch are we living in?
3. Contrast uniformitarianism and catastrophism.
4. Make a simple sketch of Earth's interior, label each layer, and list its physical characteristics, temperature, composition, and range of size on your drawing.
5. What is the present thinking on how Earth generates its magnetic field? Is this field constant, or does it change? Explain the implications of your answer.
6. Describe the asthenosphere. Why and under what circumstances is it mobile? What are the consequences of this movement?
7. Define isostasy and isostatic rebound, and explain the crustal equilibrium concept.
8. What is a mineral? A mineral family? Name the most common minerals on Earth.
9. Describe igneous processes. What is the difference between intrusive and extrusive forms?
10. Briefly describe sedimentary processes and lithification. Give an example of a sedimentary rock.
11. What is metamorphism? Name some original parent rocks and their metamorphic equivalents.
12. Briefly review the history of continental drift, sea-floor spreading, and the all-inclusive plate tectonics theory. What was Alfred Wegener's role?
13. Define upwelling and describe related features associated with such action on the ocean floor. Define subduction and explain the process.
14. What was Pangaea? What happened to it during the past 225 million years?
15. Characterize the three types of plate boundaries and the probable conditions associated with each type.
16. Why is there a relationship between plate boundaries and volcanic and earthquake activity?

Floor of the Oceans, 1975, by Bruce C. Heezen and Marie Tharp. [Copyright © 1980 by Marie Tharp. Reproduced by permission of Marie Tharp, 1 Washington Ave., South Nyack, NY 10960.]

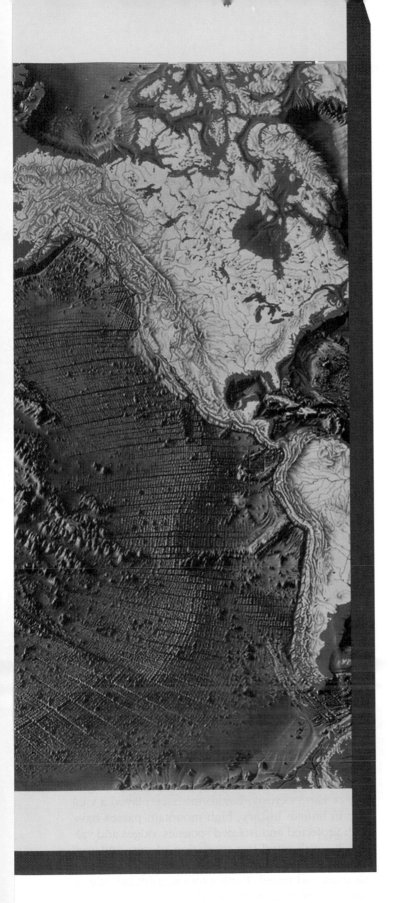

9

EARTHQUAKES AND VOLCANOES

THE OCEAN FLOOR

EARTH'S SURFACE RELIEF FEATURES
Crustal Orders of Relief
Earth's Topographic Regions

CRUSTAL FORMATION PROCESSES
Continental Shields
Building Continental Crust

CRUSTAL DEFORMATION PROCESSES
Folding   Faulting

OROGENESIS (MOUNTAIN BUILDING)
Types of Orogenies
The Appalachian Mountains
World Structural Regions

EARTHQUAKES
Earthquake Essentials
The Nature of Faulting
Earthquake Forecasting, Preparedness, and Planning

VOLCANISM
Locations of Volcanic Activity
Types of Volcanic Activity

SUMMARY
FYI REPORT 9-1: THE 1980 ERUPTION OF MOUNT SAINT
HELENS

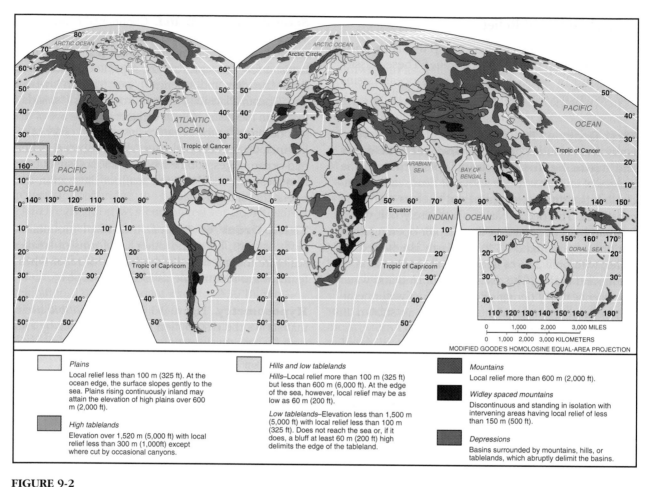

Plains
Local relief less than 100 m (325 ft). At the ocean edge, the surface slopes gently to the sea. Plains rising continuously inland may attain the elevation of high plains over 600 m (2,000 ft).

High tablelands
Elevation over 1,520 m (5,000 ft) with local relief less than 300 m (1,000ft) except where cut by occasional canyons.

Hills and low tablelands
Hills–Local relief more than 100 m (325 ft) but less than 600 m (6,000 ft). At the edge of the sea, however, local relief may be as low as 60 m (200 ft).

Low tablelands–Elevation less than 1,500 m (5,000 ft) with local relief less than 100 m (325 ft). Does not reach the sea or, if it does, a bluff at least 60 m (200 ft) high delimits the edge of the tableland.

Mountains
Local relief more than 600 m (2,000 ft).

Widely spaced mountains
Discontinuous and standing in isolation with intervening areas having local relief of less than 150 m (500 ft).

Depressions
Basins surrounded by mountains, hills, or tablelands, which abruptly delimit the basins.

FIGURE 9-2
Earth's topographic regions. [After Richard E. Murphy, "Landforms of the World," Map Supplement No. 9, *Annals of the Association of American Geographers* 58, no. 1 (March 1968). Adapted by permission.]

and elevation (the Greek *hypsos* means height). Relative to Earth's diameter of 12,756 km, the surface generally is of low relief; for comparison, Mount Everest is only 8.8 km (5.5 mi) above sea level and Mauna Loa in Hawaii is 10.0 km (6.2 mi) measured from the seafloor. The average elevation of Earth's solid surface is actually under water: −2070 m (−6790 ft) below mean sea level. The average elevation just for exposed land is +875 m (+2870 ft). For the ocean depths, the average elevation is −3800 m (−12,470 ft). From this you can see that, on the average, the oceans are much deeper than continental regions are high. Overall, the underwater ocean basins, ocean floor, and submarine mountain ranges form Earth's largest area.

Earth's Topographic Regions

The three orders of relief can be further generalized into six topographic regions: plains, high tablelands, hills and low tablelands, mountains, widely spaced mountains, and depressions (Figure 9-2). Each type of topography is defined by an arbitrary elevation or descriptive limit that is in common use (see legend in the figure). Four of the continents possess extensive *plains,* which are identified as areas with local relief of less than 100 m (325 ft) and slope angles of 5° or less. Some plains have high elevations of over 600 m (2000 ft); in the United States the high plains achieve elevations above 1220 m (4000 ft). The Colorado Plateau and Antarctica are notable *high table-*

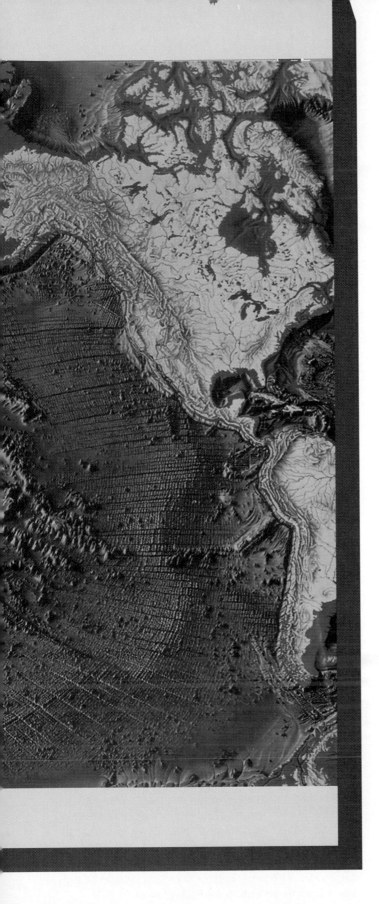

9

EARTHQUAKES AND VOLCANOES

THE OCEAN FLOOR

EARTH'S SURFACE RELIEF FEATURES
Crustal Orders of Relief
Earth's Topographic Regions

CRUSTAL FORMATION PROCESSES
Continental Shields
Building Continental Crust

CRUSTAL DEFORMATION PROCESSES
Folding   Faulting

OROGENESIS (MOUNTAIN BUILDING)
Types of Orogenies
The Appalachian Mountains
World Structural Regions

EARTHQUAKES
Earthquake Essentials
The Nature of Faulting
Earthquake Forecasting, Preparedness, and Planning

VOLCANISM
Locations of Volcanic Activity
Types of Volcanic Activity

SUMMARY
FYI REPORT 9-1: THE 1980 ERUPTION OF MOUNT SAINT
HELENS

Tectonic forces were brought home to hundreds of millions of television viewers as they watched the 1989 baseball World Series from Candlestick Park near San Francisco. Less than half an hour before the call to "play ball," a powerful earthquake rocked the region and turned sportscasters into newscasters and sports fans into disaster witnesses.

Four and one-half years later, at California State University in Northridge, the bookstore shelves were stocked, registration was completed for 26,000 students, and apartments were rented. Everything was in place for the start of classes when a magnitude 6.8 earthquake hit the university and the Los Angeles metropolitan region with tremendous ground acceleration. Damage to date exceeds $30 billion—$350 million to the campus alone.

Such tectonic activity has repeatedly deformed, recycled, and reshaped Earth's crust during its 4.6-billion-year existence. The principal tectonic and volcanic zones lie along plate boundaries. The arrangement of continents and oceans, the origin of mountain ranges, topography, and the locations of earthquake and volcanic activity are all the result of dynamic Earth processes.

The Ocean Floor

We begin our look at earthquakes and volcanoes where many occur unseen: on the ocean floor. The illustration that opens this chapter is a striking representation of Earth with its blanket of water removed. The ocean floor is revealed to us through decades of direct and indirect observation. Careful examination of this portrayal provides a helpful review of concepts learned in the previous chapter, laying the foundation for this and subsequent chapters. The scarred ocean floor is clearly visible, its sea-floor spreading centers marked by over 64,000 km (40,000 mi) of oceanic ridges (Earth's longest mountain chain), its subduction zones indicated by deep oceanic trenches, and its transform faults slicing at right angles through the oceanic ridges.

On the floor of the Indian Ocean you can locate the wide track along which the India plate traveled northward to its collision with the Eurasian plate. Vast sediment patterns in the Indian Ocean, south of the Indus and Ganges rivers on either side of India mark centuries of discharge of these former soils. Sediments derived from the Himalayan Range blanket the floor of the Bay of Bengal (south of Bangladesh) to a depth of 20 km (12.4 mi). The principle of isostasy can be examined in central and west-central Greenland, where the weight of ice has pressed portions of the land far below sea level.

In the area of the Hawaiian Islands, the hot spot track is marked by a chain of islands and seamounts which you can follow along the Pacific plate from Hawaii to the Aleutians. You can find Iceland's position on the Mid-Atlantic Ridge. Subduction zones south and east of Alaska and Japan as well as along the western coast of South and Central America are quite visible as dark trenches. In addition, you can follow the East Pacific Rise (an ocean ridge) northward as it trends beneath the west coast of the North American plate, disappearing under earthquake-prone California. The continents, submerged continental shelves and slopes, and the expanse of the sediment-covered abyssal plain are all identifiable on this illustration.

See if you can correlate this sea-floor illustration with the maps of plate boundaries and earthquake/volcano activity shown in Figures 8-17 and 8-18.

Earth's Surface Relief Features

Relief refers to vertical elevation differences in the landscape. Examples include, the low relief of Iowa and Nebraska, medium relief in foothills along mountain ranges, and high relief in the Rockies and Himalayas. The undulating form of Earth's surface is called **topography**, the feature portrayed so effectively on topographic maps. The relief and topography of Earth's crustal landforms have played a vital role in human history: high mountain passes have both protected and isolated societies; ridges and valleys have dictated transportation routes; and vast plains have encouraged better methods of communication and travel. Earth's topography has stimulated human invention and adaptation.

Crustal Orders of Relief

For convenience, geographers group the landscape's topography into three *orders of relief.* These classify landscapes by scale, from enormous ocean basins and continents down to local hills and valleys.

First Order of Relief. The broadest category of landforms includes huge **continental platforms** and **ocean basins**. Continental platforms are the masses of crust that reside above or near sea level, including the undersea continental shelves along the coastlines. The ocean basins are entirely below sea level and are portrayed in the chapter-opening illustration. Approximately 71% of Earth is covered by water, with only 29% of its surface appearing as continents and islands. The distribution of land and water in evidence today demonstrates a distinct water hemisphere and continental hemisphere, as shown in Chapter 6 (see Figure 6-17).

Second Order of Relief. Continental features that are classified in the second order of relief include continental masses, mountain masses, plains, and lowlands. A few examples are the Alps, Rocky Mountains (both Canadian and American), west Siberian lowland, and Tibetan Plateau. The great rock cores ("shields") that form the heart of each continental mass are of this second order. In the ocean basins, second order of relief includes continental rises, slopes, abyssal plains, mid-ocean ridges, submarine canyons, and subduction trenches—all visible in the seafloor illustration that opens this chapter.

Third Order of Relief. Individual mountains, cliffs, valleys, hills, and other landforms of smaller scale are included in the third order of relief. These features are identifiable as local landscapes.

Hyposometry. Figure 9-1 is a *hypsographic curve,* that shows the distribution of Earth's surface by area

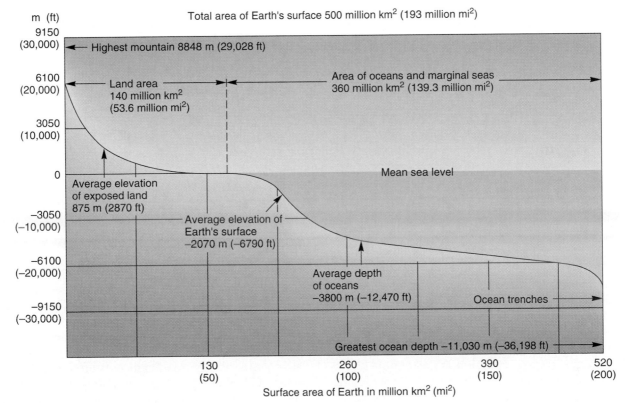

FIGURE 9-1
Hypsographic curve of Earth's surface, charting elevation as related to mean sea level. From the highest point above sea level (Mount Everest) to the deepest oceanic trench (Mariana Trench) Earth's overall relief is almost 20 km (12.5 mi).

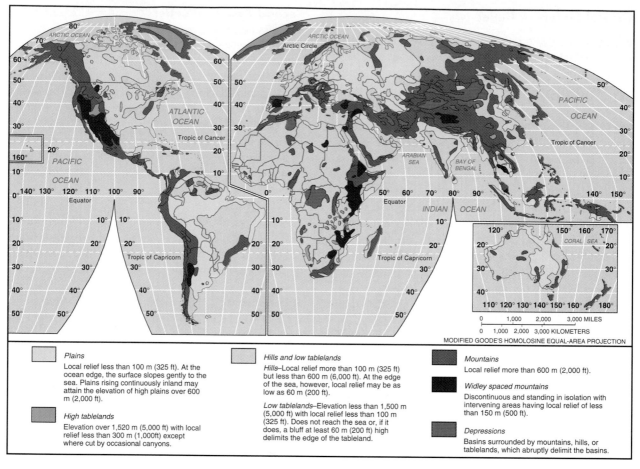

FIGURE 9-2
Earth's topographic regions. [After Richard E. Murphy, "Landforms of the World,"
Map Supplement No. 9, *Annals of the Association of American Geographers* 58, no.
1 (March 1968). Adapted by permission.]

and elevation (the Greek *hypsos* means height). Relative to Earth's diameter of 12,756 km, the surface generally is of low relief; for comparison, Mount Everest is only 8.8 km (5.5 mi) above sea level and Mauna Loa in Hawaii is 10.0 km (6.2 mi) measured from the seafloor. The average elevation of Earth's solid surface is actually under water: −2070 m (−6790 ft) below mean sea level. The average elevation just for exposed land is +875 m (+2870 ft). For the ocean depths, the average elevation is −3800 m (−12,470 ft). From this you can see that, on the average, the oceans are much deeper than continental regions are high. Overall, the underwater ocean basins, ocean floor, and submarine mountain ranges form Earth's largest area.

Earth's Topographic Regions

The three orders of relief can be further generalized into six topographic regions: plains, high tablelands, hills and low tablelands, mountains, widely spaced mountains, and depressions (Figure 9-2). Each type of topography is defined by an arbitrary elevation or descriptive limit that is in common use (see legend in the figure). Four of the continents possess extensive *plains,* which are identified as areas with local relief of less than 100 m (325 ft) and slope angles of 5° or less. Some plains have high elevations of over 600 m (2000 ft); in the United States the high plains achieve elevations above 1220 m (4000 ft). The Colorado Plateau and Antarctica are notable *high table-*

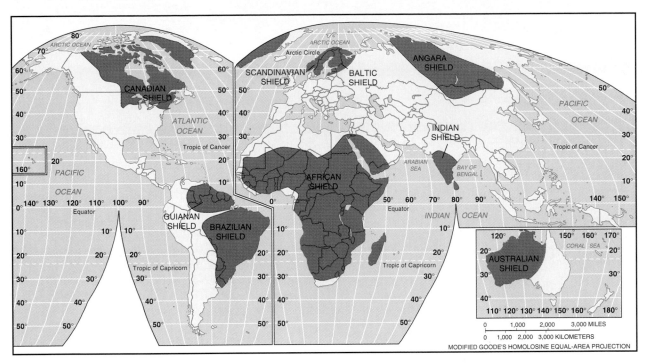

FIGURE 9-3
Portions of major continental shields that have been exposed by erosion.
[After Richard E. Murphy, "Landforms of the World," Map Supplement No. 9,
Annals of the Association of American Geographers 58, no. 1 (March 1968).
Adapted by permission.]

lands, with elevations exceeding 1520 m (5000 ft).
Africa is dominated by *hills and low tablelands.*

Mountain ranges are characterized by local relief exceeding 600 m (2000 ft) and appear on each continent.
Earth's relief and topography are undergoing constant
change as a result of processes that form crust.

Crustal Formation Processes

How did Earth's continental crust form? What gives
rise to the three orders of relief just discussed? Ultimately, the answer is tectonic activity, driven by our
planet's internal energy.

Tectonic activity produces continental crust that is
quite varied. Nonetheless, continental crust generally can be thought of in three categories, all of
which are discussed in this chapter: (1) residual
mountains and continental cores ("shields") that are
inactive remnants of ancient tectonic activity; (2) tectonic mountains and landforms, produced by active

folding and faulting movements that deform the
crust; and (3) volcanic features, formed by the surface accumulation of molten rock from eruptions of
subsurface materials. Thus, several distinct processes
operate in concert to produce the continental crust.

Continental Shields

All continents have a nucleus of crystalline rock on
which the continent "grows" with additions of other
crust and sediments. The nucleus is the *craton,* or
heartland region, of the continental crust. Cratons
generally have been eroded to a low elevation and
relief and are old (most exceed two billion years in
age, but all are Precambrian, or older than 570 million years). Portions of cratons are covered with layers of sedimentary rock that are quite stable over
time. An example of such a stable *platform* is the region that stretches from the Rockies to the Appalachians and northward into central Canada. A
region where a craton is exposed at the surface is

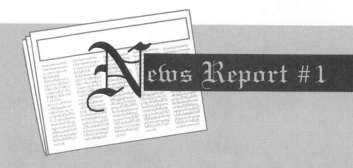

"James Michener On Terranes and Tectonics in *Alaska*"

Author James Michener is known for weaving physical geography and geology into his historical novels. In his 1988 book, *Alaska*, Michener tells of the construction of the continent through successive additions of displaced terranes, pasted onto the shield heartland of North America. Michener writes:

But the immediate task is to understand how this trivial ancestral [continental] nucleus could aggregate to itself the many additional segments of rocky land which would ultimately unite to comprise the Alaska we know. Like a spider waiting to grab any passing fly, the nucleus remained passive but did accept any passing terranes—those unified agglomerations of rock considerable in size and adventurous in motion—that wandered within reach. . . . And the great plates of Earth's crust never rest. . . .*

In the region surrounding the Pacific, such accreted terranes are significant. At least 25% of the growth of western North America can be attributed to the collection of terranes since the early Jurassic period (190 million years ago). Without these accreted terranes, much of Alaska, British Columbia, Washington, Oregon, and California would not exist today.

---

* James A. Michener, *Alaska* (New York: Random House, 1988), p. 5 and 9.

---

called a **continental shield**. Figure 9-3 shows the principal areas of exposed shields.

Building Continental Crust

Continental crust results from a complex process that involves sea-floor spreading and formation of oceanic crust, its subduction, remelting, and subsequent rise of magma as summarized in Figure 8-14. To understand this process, begin with the magma that originates in the asthenosphere and wells up along the mid-ocean ridges. It is less than 50% silica and is rich in iron and magnesium. This material rises at the spreading centers, cools to form ocean floor, spreads outward, and collides with continental crust. Being denser, the oceanic crust plunges back into the mantle, and remelts. This magma then rises and cools, forming more continent.

As the subducting oceanic plate works its way under a continental plate, it takes with it sediment and trapped water, melting and incorporating vari-

ous elements from the crust into the mixture. As a result, the magma (generally called a "melt") that migrates upward from a subducted plate contains 50–75% silica, is high in aluminum, and has a high-viscosity (thick) texture. Note that this is quite different in composition from the magma that arose directly from the asthenosphere at spreading centers to form new sea floor.

Bodies of the silica-rich magma may reach the surface in explosive volcanic eruptions, or they may stop short and become subsurface intrusive bodies in the crust, cooling slowly to form crystalline plutons such as batholiths (see Figures 8-9 and 8-10).

Terranes. A surprising discovery is that each of Earth's major plates actually is a collage of many crustal pieces acquired from a variety of sources. Accretion, or accumulation, has occurred as crustal fragments of ocean floor, curved chains (or arcs) of volcanic islands, and other pieces of continental crust have been forced against the edges of continental

FIGURE 9-4
General types of stress and resulting strain.

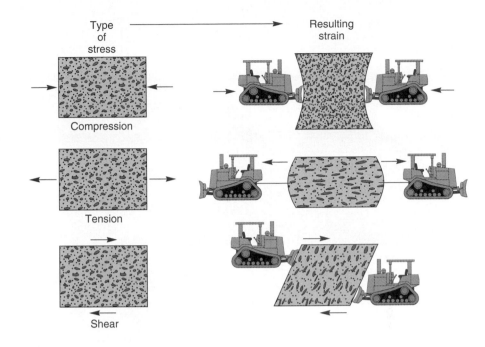

Type of stress → Resulting strain

Compression

Tension

Shear

shields and platforms. These migrating crustal pieces, which have become attached to the plates, are called **terranes** (not to be confused with "terrain," which refers to the topography of a tract of land).

Displaced terranes have histories different from those of the continents that capture them, are usually framed by fractures, and are distinguished in rock composition and structure from their new continental homes.

The Appalachian Mountains, extending from Alabama to the Maritime Provinces of Canada, possess bits of land once attached to ancient portions of Europe, Africa, South America, Antarctica, and various oceanic islands. These discoveries, barely a decade old, demonstrate how continents are assembled. (See News Report #1.)

Crustal Deformation Processes

Rocks, whether they be igneous, sedimentary, or metamorphic, can be subjected to powerful *stress* by tectonic forces, gravity, and the weight of overlying rocks. There are three types of stress: *compression* (shortening), *tension* (stretching), and *shearing* (stress when two pieces slide past each other), as shown in Figure 9-4. The *strain* that results from

these stresses (how the rocks respond) is expressed as *folding* (bending) or *faulting* (breaking). The patterns created by these processes are evident in the landforms we see today.

Folding

When layered flat strata are subjected to compressional forces, they are bent and deformed (Figure 9-5). Convergent plate boundaries worldwide intensely compress rocks, deforming them in a process known as **folding**. If we take pieces of thick fabric, stack them flat on a table, and then slowly push on opposite ends of the stack, the cloth layers will bend and form similar folds (Figure 9-5a). If we then draw a line down the center axis of a resulting ridge and trough, we are able to see how the names of the folds are assigned. Along the ridge of a fold, layers *slope downward away from the axis,* resulting in an **anticline**. In the trough of a fold, however, layers *slope downward toward the axis,* producing a **syncline** (Figure 9-5a, left).

If the axis of either type of fold is not "level" (horizontal, or parallel to Earth's surface), the layers then *plunge* (dip down) at an angle. Knowledge of how folds are angled to Earth's surface and where they are located is important for resource recovery. Pe-

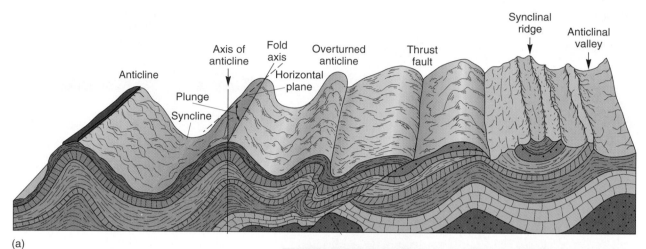

(a)

(b)

FIGURE 9-5
(a) Folded landscape and the basic types of fold structures; (b) a roadcut exposes a *synclinal ridge* near Hancock, Maryland. [Photo by John Thrasher.]

troleum, for example, collects in the upper portions of folds in permeable rock layers such as sandstone.

A residual ridge may form within a syncline, due to different rock strata of greater resistance to weathering processes (a "synclinal ridge," Figure 9-5a, right). An interstate highway roadcut dramatically exposes such a synclinal ridge in western Maryland (Figure 9-5b). Compressional forces often push folds far enough that they actually overturn on their own strata ("overturned anticline," near center of figure). Further stress eventually fractures the rock strata along distinct lines, and some overturned folds are thrust up, causing a considerable shortening of the original strata ("thrust fault"). The Canadian Rocky Mountains and the Appalachian Mountains illustrate well the complexity of the resulting folded landscape.

Satellites let us view many of these structures from an orbital perspective, as in Figure 9-6. Northwest of the Strait of Hormuz, just north of the Persian Gulf, are the Zagros Mountains of Iran. In the satellite image, anticlines form the parallel ridges; active weathering and erosion processes expose underlying strata.

Earth's continental crust also is subjected to broad warping actions, that produce an up-and-down bending of strata that are too expansive to be considered folding. Such forces can be the result of mantle convection, isostatic adjustments, and/or swelling from an underlying hot spot. Warping features range from small, individual, foldlike structures called *basins* and *domes* (Figure 9-7) up to regional features the size of the Ozark Mountain complex in

FIGURE 9-6

Southern Zagros Mountains in the Zagros crush zone between the Arabian and Eurasian plates. Originally, this area was a dispersed terrane (migrating crustal piece) that had separated from the Eurasian plate. However, the collision produced by the northward push of the Arabian block is now shoving this terrane back into Eurasia and forming the folded mountains shown. [NASA image.]

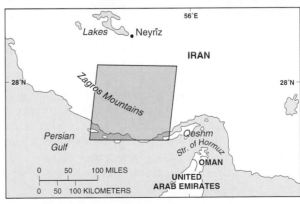

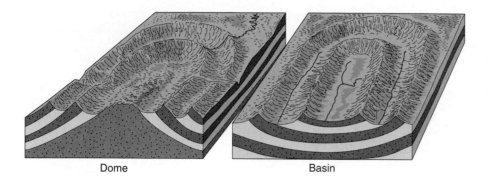

Dome   Basin

FIGURE 9-7
An upwarped dome and a structural basin. [After Joseph E. Van Riper, *Man's Physical World,* p. 436, copyright 1971 by McGraw-Hill.]

Arkansas and Missouri, the Colorado Plateau in the West, or the Black Hills of South Dakota.

The isostatic rebound of the Hudson Bay region represents a broad upwarping that will eventually lead to draining of the bay. As the last ice age ended and the glaciers melted, a great weight was lifted off this region of Canada. In response, the area gradually is rising in elevation—*isostatic rebound* as shown in Figure 8-5.

Faulting

When rock strata are strained beyond their ability to remain a solid unit, they fracture. Rocks on either side of the fracture are displaced relative to the other side in a process known as **faulting**. Thus, *fault zones* are areas of crustal movement. At the moment of fracture the fault line shifts and a sharp release of energy occurs, called an **earthquake** or *quake* (Figure 9-8).

The fracture surface along which the two sides of a fault move is called the *fault plane.* The tilt and orientation of the fault plane provides the basis for naming the three basic types of faults. A **normal fault**, or tension fault occurs when rocks are pulled apart. They move vertically along an inclined fault plane so that one "block" of rock ends up lower than the other (Figure 9-8a). A cliff formed by faulting is commonly called a *fault scarp,* or *escarpment.*

Compressional forces associated with converging plates force rocks to move *upward* along the fault plane. This is called a **reverse fault**, or compression fault (Figure 9-8b). On the surface it appears similar in form to a normal fault. If the fault plane forms a low angle relative to the horizontal, the fault is termed a **thrust fault**, or overthrust fault, indicating that the overlying block has shifted far over the underlying block (Figure 9-5a, "thrust fault"). Place one hand palm-down on the back of the other and move them past each other—this is the motion of a low-angle thrust fault with one side pushing over the other. In the Alps, three to five such overthrusts are in evidence from the compressional forces of the ongoing African plate collision with the Eurasian plate. Beneath the Los Angeles basin, such overthrust faults produce a high risk of earthquakes, and caused the 1971 Sylmar earthquake and the 1994 Northridge earthquake.

If movement along a fault plane is horizontal, as produced by a *transform fault,* it is called a **strike-slip fault**, or transcurrent fault (Figure 9-8c). The movement is described as *right-lateral* or *left-lateral,* depending on the motion perceived when you observe movement on one side of the fault relative to the other side.

Although strike-slip faults do not produce cliffs (scarps), they can create linear *rift valleys,* as is the case with the San Andreas fault system of California (Figure 9-9). The rift is clearly visible in the photograph where the edges of the North American and Pacific plates are grinding past one another as a result of transform faults associated with a former seafloor spreading center. In its westward drift, the North American plate rode over portions of this spreading center (later shown in Figure 9-14). Consequently, the San Andreas system is described as a series of *transform* (associated with former spreading center), *strike-slip* (horizontal motion), and *right-lateral* (one side moving to the right hand of the other) faults.

In the interior western United States, the Basin and Range Province experienced tensional forces caused by uplifting and thinning of the crust, which cracked the surface to form aligned pairs of normal faults and a distinctive landscape (please refer to Chapter 12,

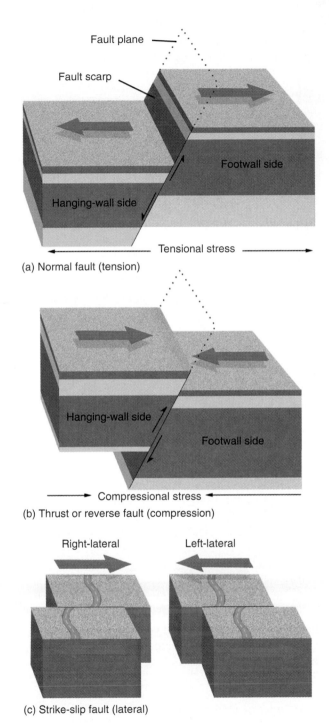

Fault plane

Fault scarp

Footwall side

Hanging-wall side

Tensional stress

(a) Normal fault (tension)

Hanging-wall side

Footwall side

Compressional stress

(b) Thrust or reverse fault (compression)

Right-lateral   Left-lateral

(c) Strike-slip fault (lateral)

FIGURE 9-8

Basic types of faults: (a) normal fault (tension); (b) reverse fault (compression); and (c) strike-slip fault (transcurrent or horizontal). In England, when coal miners worked along a reverse fault (b), they would stand on the lower side (footwall) and hang their lanterns on the upper side (hanging wall), giving rise to these terms.

Figure 12-17). The term **horst** is applied to upward-faulted blocks; **graben** refers to downward-faulted blocks. Examples of horst and graben landscapes include the Great Rift Valley of East Africa (associated with crustal spreading), which extends northward to the Red Sea, and the Rhine graben through which the Rhine River flows in Europe.

We have explained the factors that work to produce crust and discussed the tectonic forces that bend, warp, and break it. Now let's look at specific mountain-building processes.

Orogensis (Mountian Building)

Orogenesis literally is the birth of mountains (*oros* comes from the Greek for "mountain"). An *orogeny* is a mountain-building episode, occurring over millions of years, that thickens continental crust. It can occur through large-scale deformation and uplift of the crust. It also may include the capture of migrating terranes and cementation of them to the continental margins, or the intrusion of granitic magmas to form plutons. These granite masses often are exposed by erosion following uplift. Uplift is the final act of the orogenic cycle. Earth's major chains of folded and faulted mountains, called *orogens,* correlate remarkably with the plate tectonics model.

No orogeny is a simple event; many involve previous developmental stages dating back millions of years, and the processes are ongoing today. Major orogens include the Rocky Mountains (Laramide orogeny, 40–80 million years ago); the Sierra Nevada (Sierra Nevadan orogeny, 35 million years ago, with older batholithic intrusions dating back 130–160 million years); the Appalachians and the Ridge and Valley Province of the eastern U.S. (Alleghany orogeny, 250–300 million years ago, preceded by at least two earlier orogenies); and the Alps of Europe (Alpine orogeny, 20–120 million years ago and continuing to the present, with many earlier episodes).

FIGURE 9-9
The San Andreas fault in the eastern margin of the central Coast Ranges of California. This view is toward the north, so the Pacific plate is on the left (the west side). The Pacific plate is moving northward relative to the North American plate on the right (the east side). [Photo by Randall Marrett.]

The Sierra Nevada of California and the Grand Tetons of Wyoming are examples of later stages of mountain-building orogenesis. Each is a **tilted fault-block** mountain range, in which a normal fault on one side of the range has produced a tilted landscape of dramatic relief (Figure 9-10). Slowly cooling magma intruded into those blocks and formed granitic cores of coarsely crystalline rock. After tremendous tectonic uplift and the removal of overlying material through weathering, erosion, and transport, those granitic masses are now exposed in each mountain range. In some areas, the overlying material originally covered these batholiths by more than 7500 m (25,000 ft).

Types of Orogenies

Figure 9-11 illustrates convergent plate-collision patterns associated with orogenesis. Shown in Figure 9-11a is *oceanic plate–continental plate collision* orogenesis. This is occurring along the Pacific coast of the Americas and has formed the Andes, the Sierra of Central America, the Rockies, and other western mountains. We see folded sedimentary formations and intrusions of magma forming granitic plutons at the heart of these mountains. Their buildup was augmented by accretion of displaced terranes, which were cemented during their collision with the continental mass. Also, note the associated volcanic activity inland from the subduction zone.

Shown in Figure 9-11b is the *oceanic plate–oceanic plate collision,* where two portions of oceanic crust collide. Such collisions can pro-duce either simple volcanic island arcs or more complex arcs like Japan that include deformation and metamorphism of rocks, and granitic intrusions. These processes have formed the chains of island arcs and volcanoes that continue from the southwestern Pacific to the western Pacific, the Philippines, the Kurils, and through portions of the Aleutians.

Both the *oceanic–continental* and *oceanic– oceanic collision* types are active around the Pacific Rim. Both are thermal in nature, because the diving plate melts and migrates back toward the surface as magma. This region around the Pacific is known as the **circum-Pacific belt** or, more popularly, the *ring of fire* for its many volcanoes.

Shown in Figure 9-11c is the *continental plate–continental plate collision,* which occurs when two large

FIGURE 9-10
Example of a tilted fault block: the Teton Range in Wyoming. [Photo by Galen Rowell.]

continental masses collide. Here the orogenesis is quite mechanical; large masses of continental crust are subjected to intense folding, overthrusting, faulting, and uplifting. Deformation of shallow and deep marine sediments and basaltic oceanic crust is produced by crushing as the plates converge.

As mentioned earlier, the collision of India with the Eurasian landmass produced the Himalayan Mountains. That collision is estimated to have shortened the overall continental crust by as much as 1000 km (620 mi), and to have produced sequences of thrust faults at depths of 40 km (25 mi). The disruption created by that collision has reached far under China, and frequent earthquakes there signal the continuation of this collision. The Himalayas feature the tallest above-sea-level mountains on Earth, including Mount Everest at 8848 m (29,028 ft) elevation.

As orogenic belts increase in elevation, weathering and erosion processes work to reduce the mountains. The mountain mass is in continual isostatic adjustment as it builds and wears away (see Figure 8-5).

The Appalachian Mountains

The old, eroded, fold-and-thrust belt of the eastern United States contrasts with the younger mountains of the western portions of North America. As noted, the *Alleghany orogeny* followed at least two earlier orogenic cycles of uplift and the accretion of several captured terranes. (In Europe this is called the Hercynian orogeny.) The original material for the Appalachians and Ridge and Valley Province resulted from the collisions that pro-duced Pangaea. In fact, the Atlas Mountains of northwest Africa were originally connected to the Appalachians, but have rafted apart, embedded in their own tectonic plates. Similarly, folded and faulted rock structures in the British Isles and Greenland demonstrate a past relationship with the Appalachians.

The linear folds of the Appalachian system are well displayed on the satellite image, topographic map, and digitized relief map in Figure 9-12. These dissected ridges are cut across by rivers, forming *water gaps* that greatly influenced migration and settlement patterns during the early days of the United States. The initial flow of people, goods, and ideas was guided by this topography.

World Structural Regions

Figure 9-13 defines seven essential structural regions, highlights Earth's two major continental mountain systems, and allows interesting comparison with the

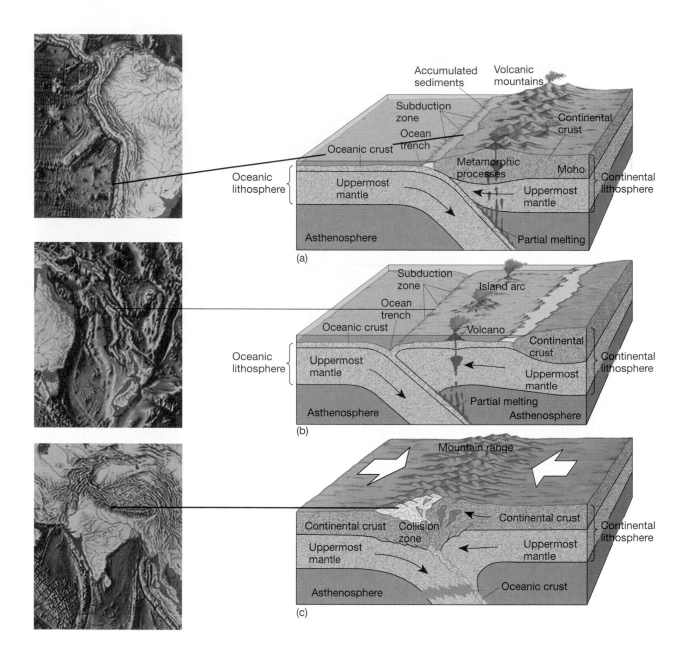

FIGURE 9-11

Types of plate convergence: (a) oceanic–continental (example: Nazca plate-South American plate collision and subduction); (b) oceanic–oceanic (example: New Hebrides Trench near Vanuatu, 16° S, 168° W); and, (c) continental–continental (example: India plate and Eurasian landmass collision and resulting Himalayan Mountains). [Inset illustrations derived from *Floor of the Oceans* © 1980 by Marie Tharp. Reproduced by permission of Marie Tharp, 1 Washington Ave., South Nyack, NY 10960.]

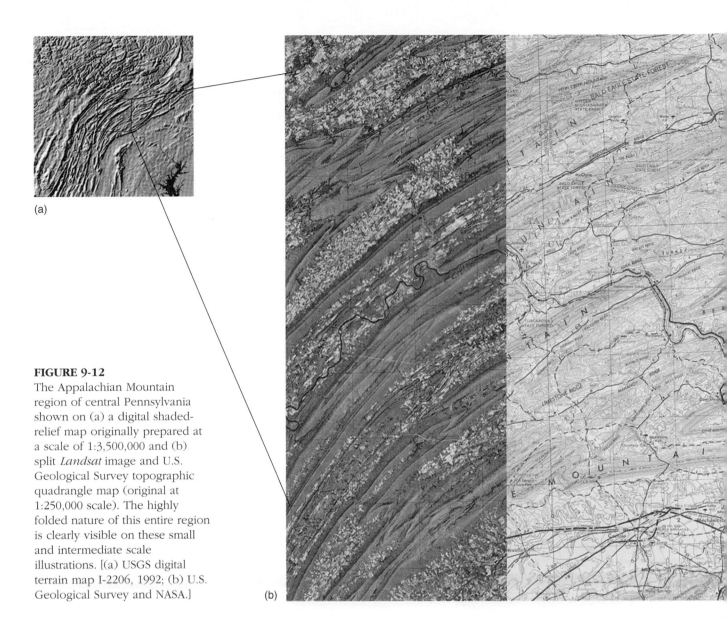

(a)

(b)

FIGURE 9-12
The Appalachian Mountain region of central Pennsylvania shown on (a) a digital shaded-relief map originally prepared at a scale of 1:3,500,000 and (b) split *Landsat* image and U.S. Geological Survey topographic quadrangle map (original at 1:250,000 scale). The highly folded nature of this entire region is clearly visible on these small and intermediate scale illustrations. [(a) USGS digital terrain map I-2206, 1992; (b) U.S. Geological Survey and NASA.]

chapter-opening illustration of the ocean floor. Looking at the distribution of world structural regions helps summarize the information presented about the three rock-forming processes (igneous, sedimentary, metamorphic), plate tectonics, landform construction, and orogenesis.

The map reveals two large mountain chains. The relatively young mountains along the western margins of the North and South American plates stretch from the tip of Tierra del Fuego to the massive peaks of Alaska, forming the **Cordilleran system**. The mountains of southern Asia, China, and northern India continue in a belt through the upper Middle East to Europe and the European Alps, constituting the **Eurasian-Himalayan system**.

Earthquakes

Crustal plates do not move smoothly past one another. Instead, stress builds strain in the rocks along the plate boundaries until the sides release and lurch into new

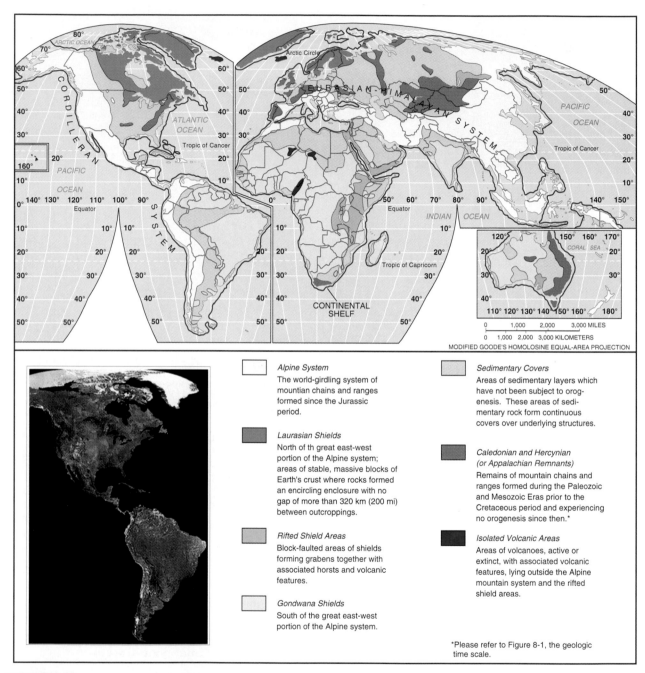

Alpine System
The world-girdling system of mountian chains and ranges formed since the Jurassic period.

Laurasian Shields
North of th great east-west portion of the Alpine system; areas of stable, massive blocks of Earth's crust where rocks formed an encircling enclosure with no gap of more than 320 km (200 mi) between outcroppings.

Rifted Shield Areas
Block-faulted areas of shields forming grabens together with associated horsts and volcanic features.

Gondwana Shields
South of the great east-west portion of the Alpine system.

Sedimentary Covers
Areas of sedimentary layers which have not been subject to orogenesis. These areas of sedimentary rock form continuous covers over underlying structures.

Caledonian and Hercynian (or Appalachian Remnants)
Remains of mountain chains and ranges formed during the Paleozoic and Mesozoic Eras prior to the Cretaceous period and experiencing no orogenesis since then.*

Isolated Volcanic Areas
Areas of volcanoes, active or extinct, with associated volcanic features, lying outside the Alpine mountain system and the rifted shield areas.

*Please refer to Figure 8-1, the geologic time scale.

FIGURE 9-13
World structural regions and major mountain systems. Because each structural region in this figure includes related landforms adjacent to the central feature of the region, some of the regions appear larger than the structures themselves. Structural regions in the Western Hemisphere are visible on this composite *Landsat* image inset. [After Richard E. Murphy, "Landforms of the World," Map Supplement No. 9, *Annals of the Association of American Geographers* 58, no. 1 (March 1968). Adapted by permission. Inset image from EROS Data Center and the National Geographic Society.]

positions. Devastating earthquakes can result from these rapid shifts of crust. In the Liaoning Province of northeastern China, ominous indications of tectonic activity began in 1970. Foreboding symptoms included land uplift and tilting, increased numbers of minor tremors, and changes in the region's magnetic field— all of this after almost 120 years of quiet.

These precursors of tectonic events continued for almost five years before Chinese scientists took the bold step of forecasting an earthquake. Finally, on February 4, 1975, at 2:00 P.M., some 3 million people were evacuated in what turned out to be a timely manner; the quake struck at 7:36 P.M., within the predicted time frame. Ninety percent of the buildings in the city of Haicheng were destroyed, but thousands of lives were saved, and success was proclaimed— an earthquake had been forecasted and preparatory action taken, for the first time in history.

Only 17 months later, at Tangshan in the northeastern province of Hebei (Hopei), an earthquake occurred on July 28, 1976, without warning. No preliminary activity occurred from which a forecast could be prepared. Consequently, this quake killed a quarter of a million people! It also destroyed 95% of the buildings and 80% of the industrial structures and severely damaged more than half the bridges and highways. The jolt was strong enough to throw people against the ceilings of their homes. An old, undetected fault had ruptured and offset 1.5 m (5 ft) along an 8 km (5 mi) stretch through Tangshan, devastating large areas just 145 km (90 mi) southeast of

Beijing, China's capital city. How is it possible that these two tectonic events produced such different human consequences?

Earthquake Essentials

Earthquakes associated with faulting are referred to as *tectonic earthquakes*. Their vibrations are transmitted as waves of energy throughout Earth's interior and are detected with a **seismograph**, an instrument that records vibrations in the crust. Earthquakes are rated on two different scales, an intensity scale and a magnitude scale.

An *intensity scale* is useful in classifying and describing damage to terrain and structures following an earthquake. Earthquake intensity is rated on the arbitrary *Mercalli scale,* a Roman numeral scale from I to XII representing "barely felt" to "catastrophic total destruction." It was designed in 1902 and modified in 1931 to be more applicable to conditions in North America (Table 9-1).

Earthquake *magnitude* is estimated according to a system designed by Charles Richter in 1935. First, the amplitude of seismic waves is recorded on a seismograph located at least 100 km (62 mi) from the center of origin of the quake (epicenter). That measurement is then charted on the **Richter scale** (Table 9-1). The scale is open ended and logarithmic; that is, each whole number on the scale represents a 10-fold increase in the measured wave amplitude. Translated into energy, each whole num-

Table 9-1
Earthquake Characteristics and
Frequency Expected Each Year

Characteristics in Populated Areas	Approximate Intensity (modified Mercalli scale)	Approximate Magnitude (Richter scale)	Number per Year
Nearly total damage	XII	>8.0	1 every few years
Great damage	X–XI	7–7.9	18
Considerable-to-serious damage to buildings; railroad tracks bent	VII–IX	6–6.9	120
Felt-by-all, with slight damage to buildings	V–VII	5-5.9	800
Felt-by-some to felt-by-many	III–IV	4–4.9	6200
Not felt, but recorded	I–II	2-3.9	500,000

SOURCE: Charles F. Richter, *Elementary Seismology*, Freeman, 1958, and others.

Table 9-2
A Sampling of Significant Earthquakes

Year	Date	Location	Deaths	Intensity	Magnitude
1556	Jan. 23	Shensi, China	830,000	—	—
1737	Oct. 11	Calcutta, India	300,000	—	—
1812	Feb. 7	New Madrid, Missouri	Several	XI–XII	—
1857	Jan. 9	Fort Tejon, California	—	X–XI	—
1870	Oct. 21	Montreal to Quebec, Canada	—	IX	—
1886	Aug. 31	Charleston, South Carolina	—	IX	6.7
1906	Apr. 18	San Francisco, California	3000	XI	8.25
1923	Sept. 1	Kwanto, Japan	143,000	XII	8.2
1939	Dec. 27	Erzincan, Turkey	40,000	XII	8.0
1964	Mar. 28	Southern Alaska	131	X–XII	8.6
1970	May 31	Northern Peru	66,000	—	7.8
1971	Feb. 9	San Fernando, California	65	VII–IX	6.5
1972	Dec. 23	Managua, Nicaragua	5000	X–XII	6.2
1976	Jul. 28	Tangshan, China	250,000	XI–XII	7.6
1978	Sept. 16	Iran	25,000	X–XII	7.7
1985	Sept. 19	Mexico City, Mexico	7000	IX–XII	8.1
1988	Dec. 7	Armenia–Turkey border	30,000	XII	6.9
1989	Oct. 17	Loma Prieta (near Santa Cruz, California)	67	VII–IX	7.1
1991	Oct. 20	Uttar Pradesh, India	1700	IX–XI	6.1
1994	Jan. 17	Northridge(Reseda), California	66	VII–IX	6.6

ber demonstrates a 31.5-fold increase in the amount of energy released. Thus, a 3.0 on the Richter scale represents 31.5 times more energy than a 2.0 and 992 times more energy than a 1.0. It is difficult to imagine the power released by the Tangshan quake, which was rated a 7.8 on the Richter scale.

Table 9-1 presents the characteristics and frequency of earthquakes that are expected annually worldwide. Table 9-2 lists a sampling of significant earthquakes. Note each earthquake's location, magnitude, and intensity rating (if available), and total loss of life (if known).

The subsurface area along a fault plane, where the motion of seismic waves is initiated, is called the *focus,* or hypocenter (see Figure 9-15). The area at the surface directly above is the **epicenter**. Shock waves produced by an earthquake radiate outward from both the focus and epicenter areas. An *aftershock* may occur after the main shock, sharing the same general area of the epicenter. A *foreshock* also is possible, preceding the main shock. (Before the June 1992 Landers earthquake in southern California, at least two dozen foreshocks occurred along that portion of the fault.) Usually, the greater the distance from an epicenter, the less severe the shock

that is experienced. At a distance of 40 km (25 mi), only 1/10 of the full effect of an earthquake normally is felt. However, if the ground is unstable, distant effects can be magnified, as they were in Mexico in 1985 and in San Francisco in 1989.

The Nature of Faulting

We earlier described specific types of faults and faulting motions. How a fault actually breaks is still under investigation, but the basic process is described by the **elastic-rebound theory**. Generally, two sides along a fault appear to be locked by friction, resisting any movement despite the powerful forces acting on adjoining pieces of crust. This stress continues to build strain along the fault surfaces, storing elastic energy like a wound-up spring. When movement finally does occur as the strain buildup exceeds the frictional lock, energy is released abruptly, returning both sides of the fault to a condition of less strain.

The San Francisco Earthquakes. In 1906 an earthquake devastated San Francisco, a city of 400,000 people at the time. Movement along the San Andreas fault was evident over 435 km (270 mi), and this

"Damage Strikes at Distance from Epicenters"

Normally, the greater the distance from an earthquake's epicenter, the less severe the shock. However, two recent earthquakes proved to be an exception to this rule. Scientists think they know why.

Mexico City, currently the world's second-largest city, is positioned on soft, moist sediments of an ancient lake bed.

On September 19, 1985, during rush-hour traffic, two major earthquakes (8.1 and 7.6 on the Richter scale) struck 400 km (250 mi) southwest of Mexico City. The epicenter was on the seafloor off Mexico and Central America. Despite the distance, seismic waves arrived and set the old lake bed beneath Mexico City in motion, magnifying the shock waves by more

than 500%. As a result, 250 buildings collapsed, some 8000 structures experienced damage, and about 7000 people perished.

The damage caused by the Loma Prieta earthquake in 1989 in the San Francisco Bay Area, where houses and freeways are built on fill and bay mud, was also at some distance from the epicenter. Like the situa-

tion in Mexico City, unstable soils magnified the effects of a distant quake. Scientists now think that, in these events, seismic waves were reflected off the crust-mantle boundary (Moho discontinuity) traveled from the focal point of the quake, bounced off the boundary zone, and back to the surface, where they caused severe damage.

---

prompted intensive research to discover the nature of faulting. (Realize that this occurred six years before Wegener's continental drift hypothesis was proposed.) The elastic-rebound theory developed as a result.

The San Andreas fault system in California provides a good example of the evolution of a spreading center overridden by an advancing continental plate (Figure 9-14). As shown in the figure, the East Pacific Rise developed as a spreading center with associated transform faults (a) while the North American plate was progressing westward after the breakup of Pangaea. Forces then shifted the transform faults toward a northwest-southeast alignment along a weaving axis (b). Finally, the western margin of North America overrode those shifting transform faults (c). In *relative terms*, the motion along the fault is right-lateral, whereas in *absolute terms* the North American plate is still moving westward.

The earthquake that disrupted the 1989 World Series, mentioned at the outset of this chapter, involved

a portion of the San Andreas fault approximately 16 km (9.9 mi) east of Santa Cruz and 95 km (59 mi) south of San Francisco (Figure 9-15). The quake occurred at 5:04 P.M. Pacific Daylight Time and registered 7.1 on the Richter scale. A fault had ruptured at a focus unusually deep for the San Andreas system—more than 18 km (11.5 mi) below the surface.

The fault plane suggested in Figure 9-15 shows the two plates moving approximately 2 m (6 ft) past each other deep below the surface, with the Pacific plate thrusting 1.3 m (4.3 ft) upward. This is an unusual motion for the San Andreas fault, especially when compared to the apparent motion in 1906, and perhaps is a clue that this fault is more complex than previously thought. In only 15 seconds, more than 2 km (1.2 mi) of freeway overpass and a section of the San Francisco–Oakland Bay Bridge collapsed, $8 billion in damage was generated, 14,000 people were displaced from their homes, 4000 were injured, and 67 people were killed.

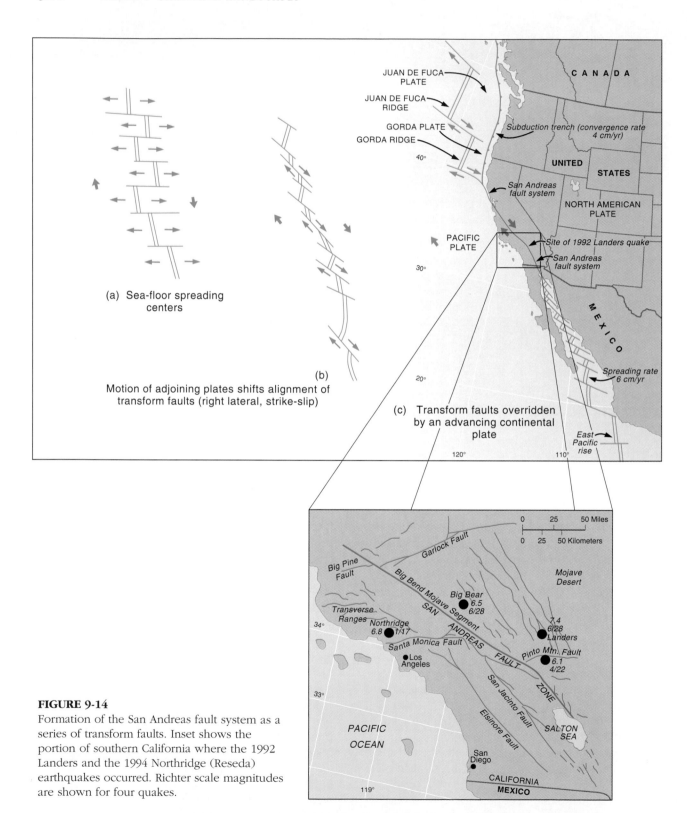

(a) Sea-floor spreading
centers

(b)
Motion of adjoining plates shifts alignment of
transform faults (right lateral, strike-slip)

(c) Transform faults overridden
by an advancing continental
plate

FIGURE 9-14
Formation of the San Andreas fault system as a
series of transform faults. Inset shows the
portion of southern California where the 1992
Landers and the 1994 Northridge (Reseda)
earthquakes occurred. Richter scale magnitudes
are shown for four quakes.

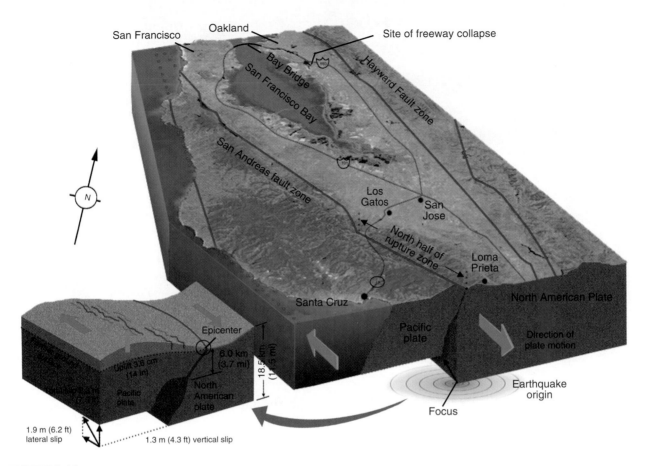

FIGURE 9-15
The 1989 Loma Prieta, California, earthquake fault-plane diagram of the lateral and
thrust movement at depth for this earthquake. [(a) After "The Loma Prieta
Earthquake of October 17, 1989," U.S. Geological Survey pamphlet by Peter J.
Ward and Robert A. Page. Washington, DC: November 1989, p. 1.]

The Southern California Earthquakes. Since
the mid-1980s the Coachella Valley area of California, east of Los Angeles, experienced seven earthquakes greater than 6 on the Richter scale. On April
22, 1992, a point near Joshua Tree, California, was
hit by a 6.1 magnitude quake. Many small aftershocks rippled along the faultline.

Two months later a 7.5-magnitude earthquake
rocked the relatively unpopulated area near Landers,
California, just north of the epicenters for the earlier
quakes (see the inset map on Figure 9-14c). This
event was the single largest quake in California in

30 years. However, because of the remote location,
both injury and damage were low.

Three hours passed; then a 6.5 magnitude earthquake hit near Big Bear City 60 km (37 mi) to the
northwest of Landers (see inset map). Aftershocks
were almost continuous among these epicenters. For
yet unexplained reasons, related earthquakes were
recorded over the next few weeks in Mammoth
Lakes 645 km (400 mi) to the north, at Mount Shasta
in northern California, in southern Nevada and Utah,
and 1810 km (1125 mi) distant in Yellowstone National Park, Wyoming!

FIGURE 9-16
A multi-level parking facility on the California State University–Northridge campus collapsed dramatically in the instant of the earthquake. In addition, damage to 20,000 apartment units across the metropolitan region left thousands homeless after the Northridge (Reseda) earthquake that occurred at 4:31 A.M., January 17, 1994. [Photo by author.]

Students at California State University–Northridge, in the San Fernando Valley of southern California, need no reminder of the power of earthquakes. Their campus was near the epicenter of the most devastating earthquake in U.S. history in terms of property damage. The January 17, 1994, Northridge (Reseda) earthquake, a Richter 6.8, and over 9200 aftershocks (as of September 1994), caused approximately $350 million damage to campus buildings two weeks before the beginning of spring classes (Figure 9-16). Amazingly, the semester began just three weeks late in 450 temporary trailers; a graduation was still held in May!

Across the Los Angeles region, tens of thousands of apartment units and homes were destroyed or damaged by the extreme ground acceleration (objects hurled upward against the pull of gravity). A deeply buried thrust fault caused the quake, focused 18 km (11 mi) beneath the San Fernando Valley (epicenter is shown on Figure 9-14 inset). The Richter scale ratings proved inadequate in expressing the force exerted by such vertical ground motion. Scientists think that southern California will be affected by more of these thrust-fault actions as strain continues to build along plate boundaries and the San Andreas system.

Earthquake Forecasting, Preparedness, and Planning

A major challenge for seismologists is how to predict the specific time and place for a quake. One approach is to examine the history of each plate boundary and determine the frequency of earthquakes in the past. Seismologists then construct maps that provide an estimate of expected earthquake activity. Areas that are quiet and overdue for an earthquake are termed *seismic gaps*; such an area forms a gap in the earthquake occurrence record and is therefore a place that possesses accumulated strain.

The areas around San Francisco and northeast of Los Angeles represent such gaps where the fault system appears to be locked by friction and is accumulating strain. The 1989 Loma Prieta earthquake was predicted in 1988 by the U.S. Geological Survey as having a 30% chance of occurring with a 6.5 magnitude, within 30 years. The actual quake filled a portion of the seismic gap in that region.

Actual implementation of an action plan to reduce deaths, injuries, and property damage from earthquakes is very difficult. The political environment adds complexity; sadly, the impact of an accurate earthquake prediction may be viewed as a negative threat to a region's economy. If we examine the potential socioeconomic impact of earthquake prediction on an urban community, it is difficult to imagine a chamber of commerce, bank, real estate agent, tax assessor, or politician who would privately welcome such a prediction. Long-range planning is a complex subject.

A valid and applicable generalization seems to be that *humans are unable or unwilling to perceive hazards in a familiar environment*. Such an axiom of human behavior certainly helps explain why large populations continue to live and work in earthquake-prone settings in developed countries. Similar questions also can be raised about populations in areas vulnerable to other disasters.

Volcanism

We are reminded of Earth's internal energy by the recent violent eruptions of Mount Pinatubo and Mount Mayon (Philippines, 1991 and 1993 respectively), Mount Unzen (Japan, 1991), Mount Hudson (Chile, 1992), Mount Spurr and Mount Redoubt (Alaska, 1992), and Galeras volcano (Colombia, 1993). A **volcano** forms at the end of a central vent or conduit that rises from the asthenosphere through the crust into a volcanic mountain. A **crater**, or circular surface depression, usually forms at the summit.

Magma rises and collects in a magma chamber deep below the volcano until conditions are right for an eruption. This subsurface magma produces tremendous heat and in some areas it boils groundwater, as seen in the thermal features of Yellowstone National Park. The steam given off is a potential energy source if it is accessible. Such **geothermal energy** has provided heating and electricity for more than 90 years in Iceland, New Zealand, and Italy. North of San Francisco, some 1300 megawatts of electrical capacity are generated by turbines that are spun by steam from geothermal wells.

The magma that actually issues from the volcano is termed **lava** (molten rock). Lava, gases, and **tephra** (pulverized rock and clastic materials ejected violently during an eruption) pass through the vent to the surface and build a volcanic landform. Lava can occur in many different textures and forms, which accounts for the varied behavior of volcanoes

and the different landforms they build. In this section we will look at five volcanic landforms and their origins: cinder cones, calderas, composite volcanoes, shield volcanoes, and plateau basalts.

A **cinder cone** is a small cone-shaped hill usually less than 450 m (1500 ft) high, with a truncated top formed from cinders that accumulate during moderately explosive eruptions. Cinder cones are made of tephra and scoria (cindery rock full of air bubbles, or vesicular), as exemplified by Paricutín in southwestern Mexico. Another distinctive landform is a large basin-shaped depression called a **caldera** (Spanish for "kettle"). It forms when summit material on a volcanic mountain collapses inward after an eruption or other loss of magma, forming a caldera that may fill with rainwater such as Crater Lake in southern Oregon.

Over 1300 volcanoes exist on Earth; fewer than 600 are active (have had at least one eruption in recorded history). North America has about 70 volcanoes (mostly inactive) along the western margin of the continent. Mount Saint Helens in Washington State is a famous example. In an average year about 50 volcanoes erupt worldwide, varying from modest activity to major explosions. Eruptions in remote locations and at depths on the seafloor go largely unnoticed, but an occasional eruption of great magnitude near a population center makes headlines.

Volcanoes produce some benefits. These include materials that contribute to fertile soils, as in Hawaii; conditions that can provide geothermal energy; and even new real estate added to Iceland, Japan, Hawaii, and elsewhere.

Locations of Volcanic Activity

The location of volcanic mountains on Earth is a function of plate tectonics and hot-spot activity. Volcanic activity occurs in three areas:

1. Along subduction boundaries at continental plate–oceanic plate convergence (Mount Saint Helens) or oceanic plate–oceanic plate convergence (Philippines and Japan).

2. Along sea-floor spreading centers on the ocean floor (Iceland) and areas of rifting on continental plates (the rift zone in east Africa).

3. At hot spots, where individual plumes of magma rise through the crust (Hawaii).

FIGURE 9-17
Principal mechanisms for each type of volcanic activity. [After U.S. Geological Survey, *The Dynamic Planet,* 1989.]

Figure 9-17 illustrates these three types of volcanic activity, which you can compare to the active volcano sites and plate boundaries shown in Figure 8-18. Figures 9-18 to 9-20 are included with Figure 9-17 to illustrate various aspects of volcanic activity.

Types of Volcanic Activity

The variety of forms among volcanoes makes them hard to classify; most fall in transition areas between one type and another. Even during a single eruption, a volcano may behave in several different ways. The primary factors in determining an eruption type are (1) the chemistry of its magma, which is related to the magma's source, and (2) its viscosity. Viscosity is magma's thickness (resistance to flow, or degree of fluidity); it ranges from low viscosity (very fluid) to high viscosity (thick and flowing slowly). We will consider two types of eruptions—effusive and explosive—and the characteristic landforms they build.

Effusive Eruptions. These are the relatively gentle eruptions that produce enormous volumes of lava annually on the seafloor and in places like Hawaii and Iceland. Direct eruptions from the asthenosphere produce a low-viscosity magma which is very fluid and cools to form a dark, basaltic rock (less than 50% silica and rich in iron and magnesium). Gases readily escape from this magma because of its texture, causing an **effusive eruption** that pours out on the surface, with relatively small explosions and little tephra. However, dramatic fountains of basaltic lava sometimes shoot upward, powered by jets of rapidly expanding gases (Figure 9-20).

An effusive eruption may come from a single vent or from a flank eruption through side vents in surrounding slopes. If such vents form a linear opening, they are called *fissures;* these sometimes create a dramatic curtain of fire during eruptions. In Iceland, active fissures are spread throughout the plateau landscape. In Hawaii, rift zones capable of erupting tend to converge on the central crater, or vent. The interior of such a crater is often a sunken

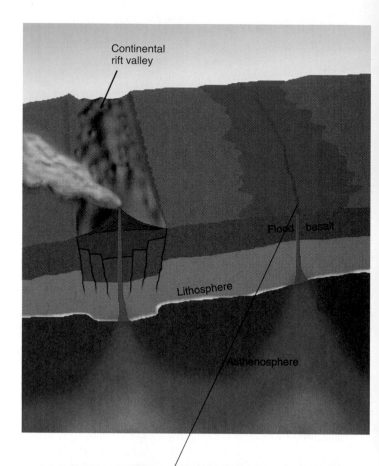

FIGURE 9-18
Two expressions of volcanic activity. In the foreground are plateau basalts characteristic of the Columbia Plateau in Oregon. Mount Hood in the background, is a composite volcano, part of the Cascade Range volcanic mountain chain, south of Mount Saint Helens. [Photo by author.]

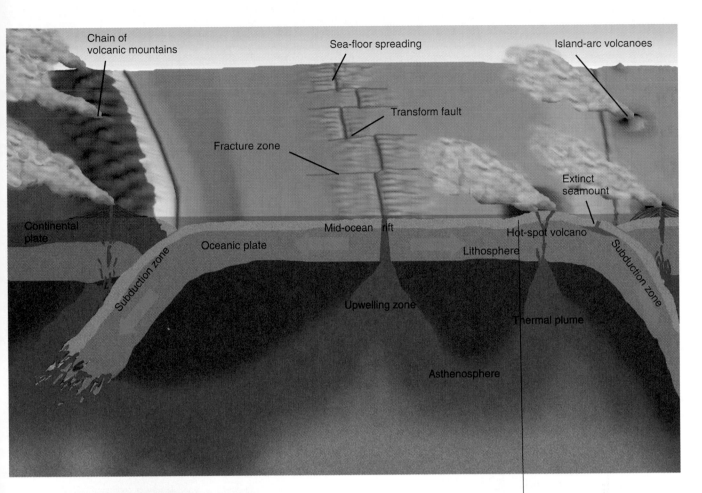

Chain of
volcanic mountains

Sea-floor spreading

Island-arc volcanoes

Transform fault

Fracture zone

Extinct
seamount

Continental
plate

Mid-ocean   rift

Hot-spot volcano

Subduction zone

Oceanic plate

Lithosphere

Subduction zone

Upwelling zone

Thermal plume

Asthenosphere

FIGURE 9-19
Basalt from Hawaii (left) and dacite from
Mount Saint Helens (right). [Photograph by
J. D. Griggs/U.S. Geological Survey.]

FIGURE 9-20
Kilauea fountain eruption from East Rift
spatter cone in Hawaii. [Photo by Kepa
Maly/USGS.]

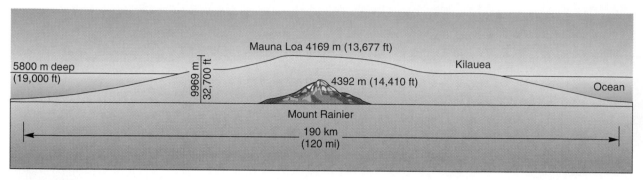

FIGURE 9-21

Comparison of Mauna Loa in Hawaii and Mount Rainier in Washington state. Their different profiles signify their different origins. [After U.S. Geological Survey, *Eruption of Hawaiian Volcanoes,* Washington, DC: Government Printing Office, 1986.]

caldera, which may fill with low-viscosity magma during an eruption, forming a molten lake, which may then overflow lava downslope.

On the island of Hawaii the continuing Kilauea eruption represents the longest eruptive episode in Hawaiian history, active since 1823! During 1989–1990, lava flows from Kilauea actually consumed several visitor buildings in the Hawaii Volcanoes National Park and homes in Kalapana, a nearby town. Kilauea has produced more lava than any other vent on Earth during recorded history.

A typical mountain landform built from effusive eruptions is gently sloped, gradually rising from the surrounding landscape to a summit crater. The shape is similar in outline to a shield of armor lying face up on the ground, and therefore is called a **shield volcano**. The shield shape and size of Mauna Loa in Hawaii is distinctive when compared to Mount Rainier in Washington, which is a different type of volcano and the largest in the Cascade Range (Figure 9-21). The height of the shield is the result of successive eruptions, flowing one on top of another. Mauna Loa is one of five shield volcanoes that make up the island of Hawaii, and it has taken at least 1 million years to accumulate its mass. Mauna Kea is slightly taller, but Mauna Loa is the most massive single mountain on Earth.

In rifting areas, effusive eruptions send material out through elongated fissures, forming extensive sheets on the surface. The Columbia Plateau of the northwestern United States represents the eruption

of **plateau basalts**, or *flood basalts* (Figure 9-18).

These extensive sheet eruptions also episodically flow from the mid-ocean ridges, forming new seafloor in the process. Worldwide, the volume of material produced in effusive fissure eruptions along spreading centers is far greater than that of the continental eruptions. Iceland continues to form in this effusive manner, with upwelling basaltic flows creating new Icelandic real estate.

Explosive Eruptions. Volcanic activity along subduction zones produces the well-known explosive volcanoes. Magma produced by the melting of subducted oceanic plate and other materials is thicker (more viscous) than magma from effusive volcanoes; it is 50–75% silica and high in aluminum. Consequently, it tends to block the magma conduit inside the volcano, trapping and compressing gases, causing pressure to build and setting the stage for an **explosive eruption**. This magma forms a lighter rock at the surface as illustrated in the comparison in Figure 9-19.

The term **composite volcano** is used to describe explosively formed mountains. (They are sometimes called *stratovolcanoes,* but because shield volcanoes also can exhibit a stratified structure, *composite* is the preferred term.) Composite volcanoes tend to have steep sides; they are more conical than shield volcanoes. Their alternating layers of lava and ash and cone shape gives them their name: *composite cones* (Figure 9-22). If a single summit vent remains stationary, a remarkable symmetry may develop as

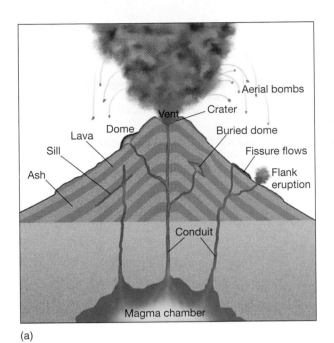

(a)

(b)

FIGURE 9-22

(a) A typical composite volcano with its cone-shaped form. (b) An eruption of
Mount Shishaldin, Alaska, a composite volcano. [Photo courtesy of AeroMap U.S.,
Inc., Anchorage, Alaska.]

the mountain grows in size, as demonstrated by
Mount Orizaba in Mexico, Mount Shishaldin in
Alaska, Mount Fuji in Japan, the pre-1980-eruption
shape of Mount Saint Helens in Washington, and
Mount Mayon in the Philippines.

As the magma in a composite volcano forms plugs
near the surface, blocked-off passages build tremendous pressure, keeping the trapped gases compressed and liquified. When the blockage can no
longer hold back this pressurized inferno, explosions
equivalent to megatons of TNT blast the tops and
sides off these mountains. Much less lava is produced than in shield eruptions but larger amounts
of tephra, which includes volcanic ash (<2 mm in
diameter), dust, cinders, lapilli (up to 32 mm in diameter), scoria (volcanic slag), pumice, and *pyroclastics* (explosively ejected rock pieces). Table 9-3
presents some notable composite volcano eruptions.
FYI Report 9-1 details the much-publicized 1980
eruption of Mount Saint Helens.

Unlike the volcanoes in Hawaii Volcanoes National Park, where tourists gather at observation plat-

forms to watch the relatively calm effusive eruptions,
composite volcanoes do not invite close inspection
and can explode with little warning:

- In A.D. 79, Mount Vesuvius buried the city of
  Pompeii, Italy, in three days of tephra and py-
  roclastic eruptions, even though the volcano
  had been dormant for centuries.

- In 1883 in Indonesia, an entire group of islands
  centered on the volcano Krakatoa (between the
  big islands of Java and Sumatra) was obliterated
  in two days of explosions. The ash cloud
  reached into the mesosphere, and the sounds of
  that August blast were heard in central Australia,
  the Philippines, and even 4800 km (3000 mi)
  away in the Indian Ocean! (The sound took
  some 4 hours to reach that listener, who noted
  the noise in his ship's log.) What was not blown
  into the sky fell back into the collapsed remains
  of a large caldera 275 m (900 ft) under the sea.

- In 1902 on the Caribbean island of Martinique,
  the beautiful port city of Saint-Pierre was dev-

Table 9-3

Composite Volcano Eruptions

Date	Name or Location	Number of Deaths	Amount extruded (mostly pyroclastics) in km³ (mi³)
Prehistoric	Yellowstone, Wyoming	Unknown	2400 (576)
4600 B.C.	Mount Mazama (Crater Lake, Oregon)	Unknown	50–70 (12–17)
1900 B.C.	Mount Saint Helens	Unknown	4 (0.95)
A.D. 79	Mount Vesuvius, Italy	20,000	3 (0.7)
1815	Tambora, Indonesia	66,000	80–100 (19–24)
1883	Krakatoa, Indonesia	36,000	18 (4.3)
1902	Mont Pelée, Martinique	29,000	Unknown
1912	Mount Katmai, Alaska	Unknown	12 (2.9)
1980	Mount Saint Helens	54	4 (0.95)
1985	Nevado del Ruiz, Colombia	23,000	1 (0.24)
1991	Mount Unzen, Japan	10	2 (0.5)
1991	Mount Pinatubo, Philippines	800	12 (3.0)
1992	Mount Redoubt, Alaska	0	Current
1993	Mount Mayon, Philippines	—	Current

astated in a few minutes by an eruption of Mont Pelée. That eruption featured a lateral blast of incandescent ash and super-hot gases, known as a "glowing cloud" or *nuée ardente* in French. Despite months of rumbling, ash, hot mud-flows, and minor bursts from the mountain, politicians discouraged citizens from fleeing, and all but two of 30,000 people in the town perished.

- In June 1991, following 600 years of dormancy, Mount Pinatubo in the Philippines erupted. The summit of the 1460 m (4795 ft) volcano exploded, devastating many surrounding villages and permanently closing Clark Air Force Base, operated by the United States (refer to Figure 1-5). Fortunately, local scientists assisted by the U.S. Geological Survey accurately predicted the eruption, allowing evacuation of the surrounding countryside and saving thousands of lives. Although volcanoes are regional events, their spatial implications are great, for their debris can spread worldwide. The single volcanic eruption of Mount Pinatubo was significant to the global environment and energy budget, as discussed in Chapters 2, 3, 4, and 7—a summary of these impacts:

- 15 to 20 million tons of ash and sulfuric acid mist were blasted into the atmosphere, concentrating at 16 to 25 km (10–15.5 mi) altitude;
- 12 km³ (3.0 mi³) of material were ejected and extruded in the eruption;
- 60 days after the eruption about 42% of the globe was affected (from 20° S to 30° N) by the thin, spreading aerosol cloud;
- colorful twilight and dawn skies were observed worldwide;
- an increase in atmospheric albedo of 1.5% (4.3 W/m²) occurred;
- an increase in the atmospheric absorption of insolation followed (2.5 W/m²);
- a decrease in net radiation at the surface (2.5 W/m²) and a lowering of Northern Hemisphere average temperatures of 0.5C° (0.9F°) were measured;
- atmospheric scientists and volcanologists were able to study the eruption aftermath using orbiting sensors aboard satellites and general circulation model simulations on computers.

In this era of Earth systems science volcanic eruptions and earthquakes are studied from a unique global perspective. Earth is measured and observed as a vast integrated system—a dynamic planet!

The 1980 Eruption of Mount Saint Helens

Probably the most studied and photographed composite volcano on Earth is Mount Saint Helens, located 70 km (45 mi) northeast of Portland, Oregon, and 130 km (80 mi) south of the Tacoma-Seattle area of Washington. Mount Saint Helens is the youngest and most active of the Cascade Range of volcanoes, which form a line from Mount Lassen in California to Mount Baker in Washington. The Cascade Range is the product of the Juan de Fuca sea-floor spreading center off the coast of northern California, Oregon, Washington, and British Columbia and the plate subduction that occurs offshore, as identified in Figure 9-14.

Associated events began on March 20, 1980, with a sharp earthquake at 3:48 P.M. registering 4.1 on the Richter scale. The first eruptive outburst occurred one week later, beginning with a 4.5 Richter quake and continuing with a thick black plume of ash and the development of an initial crater about the size of a football field.

April 1 marked the first harmonic tremor, or volcanic earthquake, from which it was inferred that magma was on the move within the mountain. *Harmonic tremors* are slow, steady vibrations, unlike the sharp releases of energy associated with earthquakes and faulting. A growing bulge on the north side of the mountain indicated the direction of the magma flow within the volcano. A bulge represents the greatest risk from a composite volcano, for it could mean a potential lateral burst across the landscape.

On Sunday, May 18, at 8:27 A.M., the area north of the mountain was rocked by a 5.0 Richter quake, the strongest to that date. The mountain, with its 245 m (800 ft) bulge, was shaken, but nothing else happened. Then a second quake (5.1) five minutes later loosened the bulge, and launched the eruption. David Johnson, a volcanologist with the U.S. Geological Survey, was only 8 km (5 mi) from the mountain, servicing monitoring instruments, when he saw

FIGURE 1
Scorched earth and tree blowdown area, covering some 39,000 hectares (95,000 acres). [Photo by author.]

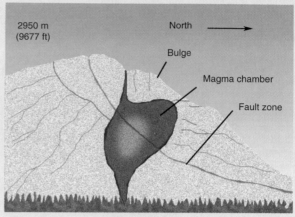

2950 m
(9677 ft)

North →

Bulge

Magma chamber

Fault zone

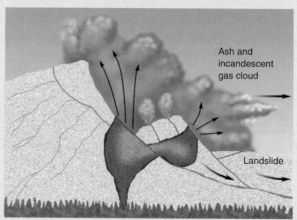

Ash and
incandescent
gas cloud

Landslide

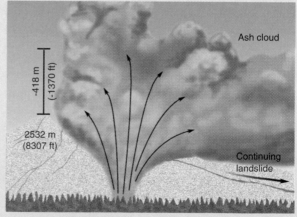

–418 m
(–1370 ft)

2532 m
(8307 ft)

Ash cloud

Continuing
landslide

it begin. He radioed headquarters in Vancouver, Washington, saying, "Vancouver, Vancouver, this is it!" but he perished in the eruption that followed.

As the contents of the mountain exploded, a surge of hot gas (about 300°C or 570°F), steam-filled ash, tephra, and a nuée ardente moved northward, hugging the ground and traveling at speeds up to 400 kmph (250 mph) for a distance of 28 km (17 mi) (Figure 1).

The slumping north face of the mountain produced the greatest landslide witnessed in recorded history; about 2.75 km³ (0.67 mi³) of rock, ice, and trapped air, all fluidized with steam, tumbled down at speeds approaching 250 kmph (155 mph). Landslide materials did not come to rest until they reached 21 km (13 mi) into the valley, blanketing the forest, covering a lake, and filling the rivers below. A series of photographs, taken at 10-second intervals from the east looking west, records this sequence (Figure 2). The eruption continued with intensity for nine hours, first clearing out old rock from the throat of the volcano and then blasting new material.

As destructive as such eruptions are, they also are constructive, for this is the way in which a volcano builds height. Before the eruption, Mount Saint Helens was 2950 m (9677 ft) tall; the eruption blew away 418 m (1370 ft). But today, Mount Saint Helens is growing by building a lava dome within the crater. The thick lava rapidly and repeatedly plugs and breaks in a series of dome eruptions that may continue for several decades. The dome already is over 300 m (1000 ft) high, so that a new mountain is being born from the eruption of the old. An ever-resilient ecology is recovering as plants and animals alike reclaim the devastated area.

SUMMARY—Tectonics and Volcanism

Earth's surface is dramatically shaped by tectonic forces generated within the planet. Significant scientific and technological breakthroughs give scientists the ability to analyze tectonic and volcanic activity. The continents are formed as a result of upwelling material from below generated by plate tectonics and enlarged through accretion of dispersed **terranes**. The crust is deformed by **folding**, broad warping, and **faulting**, which produce characteristic landforms. Earth's landscape features a great variety of **relief** and **topography**. **Orogenesis**, the process of mountain building, has produced Earth's major mountain belts—the **Cordilleran system** of the Americas and the **Eurasian-Himalayan system** that stretches from the Alps in Europe to the Himalayas in southern Asia.

Earthquakes, abrupt shifts along faults, generally occur along plate boundaries and cause disasters that affect the world's population. More is being learned all the time about the nature of faulting: stress and the buildup of strain, irregularities along fault plane surfaces, the way faults rupture, and the relationship among active faults. Earthquakes have claimed many lives and caused a great deal of damage through the centuries, so earthquake prediction and improved planning are active concerns of **seismology**, the study of earthquake waves and Earth's interior. The greatest civil disaster in U.S. history was caused by an earthquake—over $30 billion by the January, 1994, Northridge (Reseda) earthquake in California.

Volcanoes offer direct evidence of the makeup of the asthenosphere and uppermost mantle. Dramatic examples of volcanic activity have highlighted the first few years of the 1990s. Volcanoes are of two general types, based on the chemistry and the viscosity of the magma involved—*effusive* (like Kilauea in Hawaii) that build **shield volcanoes**; and *explosive* (like Mount Pinatubo in the Philippines) that build a **composite volcanoes**. Volcanic activity has produced some destructive moments in history, while constantly creating new seafloor, land, and soils.

KEY TERMS

anticline

caldera

cinder cone

circum-Pacific belt

continental platforms

continental shield

Cordilleran system

crater

earthquake

effusive eruption

elastic-rebound theory

epicenter

Eurasian-Himalayan system

explosive eruption

faulting

    normal fault

    reverse fault

    strike-slip fault

    thrust fault

folding

geothermal energy

graben

horst

lava

ocean basins

orogenesis

plateau basalts

relief

Richter scale

seismograph

syncline

tephra

terranes

tilted fault block

topography

volcano

    composite volcano

    shield volcano

REVIEW QUESTIONS

1. How does the ocean floor map (chapter-opening illustration) bear the imprint of the principles of plate tectonics? Describe and explain.
2. What is a craton? Relate this structure to continental shields and platforms and describe these regions in North America.
3. Diagram a folded landscape in cross section, and identify the features of the folding.
4. Distinguish between relief and topography. Give an example of each.
5. Define the four basic types of faults. How do these relate to earthquakes and seismic activity?
6. Define orogenesis. What is meant by the birth of mountain chains?
7. Identify and detail Earth's two largest mountain chains.
8. How are plate boundaries related to episodes of mountain building? Do different types of plate boundaries produce differing orogenic episodes?
9. Describe the differences between the two earthquakes that occurred in China in 1975 and 1976.
10. Differentiate between the Mercalli and Richter scales. How are these used to describe an earthquake? Did these scales prove adequate in describing the vertical ground acceleration in the earthquake in Los Angeles in 1994? Explain.
11. What is the relationship between an epicenter and the focus of an earthquake?
12. Describe the San Andreas fault and its relationship to sea-floor spreading movements.
13. What do you see as the biggest barrier to effective earthquake prediction?
14. What is a volcano?
15. Where do you expect to find volcanic activity in the world? Why?
16. Compare effusive and explosive eruptions. Why are they different? What types of landforms are produced by each type?
17. Describe several recent volcanic eruptions.

Alabama Hills and Inyo Mountains of eastern California. [Photo by author.]

10

WEATHERING, KARST LANDSCAPES, AND MASS MOVEMENT

LANDMASS DENUDATION
  Base Level of Streams
  Dynamic Equilibrium Approach to Landforms
WEATHERING PROCESSES
  Physical Weathering Processes
  Chemical Weathering Processes
KARST TOPOGRAPHY AND
  LANDSCAPES
  Caves and Caverns
MASS MOVEMENT PROCESSES
  Mass Movement Mechanics
  Classes of Mass Movements
  Human-Induced Mass Movements
SUMMARY

One of the benefits you receive from a physical geography course is an enhanced appreciation of the scenery. Whether you go on foot, or by car, train, or plane, travel is an opportunity to experience Earth's varied landscapes and witness the active processes that produce them. We now begin a five-chapter examination of various processes at work on the landscape. This chapter examines weathering and mass movement of the lithosphere. The other four chapters look at river systems, wind-influenced landscapes, coastal processes and landforms, and glaciated regions. Whether you enjoy time along a river, love the desert or coastline, or live in a place where glaciers passed, you will find something of interest in these chapters.

We discussed the internal processes of our planet and how the landforms of the crust are produced. However, as the landscape is formed, a variety of external processes simultaneously operate to wear it down. Perhaps you have noticed highways in mountainous and cold climates that appear rough and broken. Roads that experience freezing weather seem to pop up in chunks each winter and develop potholes. Or, maybe you have seen older marble structures such as tombstones, etched and dissolved by rainwater. Physical and chemical weathering processes similar to these are important to the overall reduction of the landscape and the release of essential minerals from bedrock.

Mass movement of surface materials rearranges landforms, providing often dramatic reminders of the power of nature. A news broadcast may bring you word of an avalanche in Colombia, mudflows in Los Angeles, a landslide in Turkey, or hot mud flows surging down the slopes of a volcano in Japan. The processes that produce such mass movement are continually operating in and on the landscape.

Landmass Denudation

Geomorphology is the science of landforms— their origin, evolution, form, and spatial distribution. It is an important aspect of physical geography. **Denudation** is any process that wears away or rearranges landforms. The principal denudation processes affecting surface materials include *weathering, mass movement, erosion, transportation,* and *deposition,* as produced by the agents of moving water, air, waves, ice, and the pull of gravity.

Base Level of Streams

The idea that Earth is billions of years old and continually changes is modern. Centuries ago it was popularly held that Earth was less than 7000 years old and that when change occurred it was in response to cataclysmic events. The theory of uniformitarianism ("the present is the key to the past") and the contrasting philosophy of catastrophism are discussed in Chapter 8. Acceptance of the vast span of geologic time led people to look anew at the process of change on Earth.

American geologist and ethnologist John Wesley Powell (1834–1902) was an early director of the U.S. Geological Survey, explorer of the Colorado River, and a pioneer in understanding the landscape (Figure 10-1). In 1875 he put forward the idea of **base level**, or a level below which a stream cannot erode its valley further. The hypothetical *ultimate base level* is sea level. You can imagine base level as a surface extending inland from sea level, inclined gently upward, under the continents (Figure 10-2). Ideally, this is the lowest practical level for all denudation processes.

Powell also recognized that not every landscape degraded down to sea level, for other base levels seemed to be in operation. A *local base level,* or a temporary one, may control the lower limit of local streams. That local base level might be a river, a lake, a hard and resistant rock structure, or a human-made dam (Figure 10-2). In arid landscapes, with their intermittent precipitation, local base level is determined by valleys, plains, or other low points.

Ideally, endogenic processes build and create *initial* landscapes, whereas exogenic processes work toward low relief, an ultimate condition of little change,

FIGURE 10-1
John Wesley Powell developed his base level concept during extensive exploration
of the West. He first ventured down the Colorado River in 1869 in heavy oak
boats, drifting by Dead Horse Point and the bend in the river shown in the photo.
[Photo by author.]

and the stability of *sequential* landscapes. Various theories have been proposed to model these denudation
processes and account for the appearance of the landscape at different developmental stages, not only for
humid temperate climates, but also for coastal, arid
desert, and equatorial landscapes.

A *dynamic equilibrium*, a balance among force,
form, and process, is the preferred model used by
many contemporary geomorphologists. Slope and
landform stability are considered a function of the resistance of rock materials to the attack of denudation
processes.

FIGURE 10-2
Ultimate and local base level
concepts.

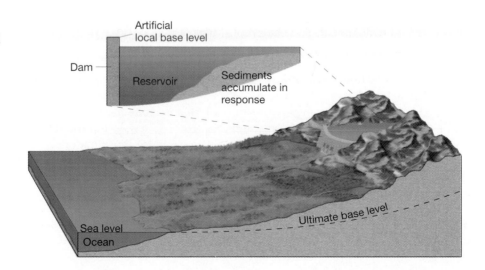

Dynamic Equilibrium Approach to Landforms

A landscape is an open system, with highly variable inputs of energy and materials: uplift creates the *potential energy of position* above base level and therefore a disequilibrium between relief and energy. The Sun provides radiant energy that is converted into *heat energy.* The hydrologic cycle imparts *kinetic energy* through mechanical motion. *Chemical energy* is made available from the atmosphere and various reactions within the crust.

As physical factors fluctuate in an area, the surface constantly responds in search of equilibrium. Every change produces compensating actions and reactions. This balancing act between tectonic uplift and reduction by weathering and erosion, between the resistance of rocks and the ceaseless attack of weathering and erosion, is summarized in the **dynamic equilibrium model**.

According to current thinking, landscapes in a dynamic equilibrium show ongoing adaptations to the ever-changing conditions of rock structure, climate, local relief, and elevation. Endogenic events, such as faulting, or a lava flow, or exogenic events, such as a heavy rainfall, or a forest fire, may provide new sets of relationships for the landscape. Following such destabilizing events, a landform system arrives at a **geomorphic threshold**—the point at which there is enough energy to overcome the resistance against movement. At this threshold the system breaks through to a new set of equilibrium relationships as the landform and its slopes enter a period of adjustment and realignment. The pattern over time is a sequence of (a) equilibrium stability, (b) a destabilizing event, (c) a period of adjustment, and (d) development of a new and different condition of equilibrium stability.

Slow, continuous-change events, such as soil development and erosion, tend to maintain an approximate equilibrium condition. Although less frequent, dramatic events such as a major landslide or dam collapse require longer recovery times before an equilibrium is reestablished.

Slopes. Material loosened by weathering is susceptible to erosion and transportation. However, if it is to move downslope, the forces of erosion must overcome other forces: friction, inertia (the resistance to movement), and the cohesion of particles to each other (Figure 10-3a). If the angle is steep enough for gravity to overcome frictional forces, or if the impact of raindrops, or moving animals, or even wind, dislodges material, then erosion of particles and transport downslope can occur.

Slopes, or hillslopes, are curved, inclined surfaces that form the boundaries of landforms. Figure 10-3b illustrates essential slope forms that may vary with conditions of rock structure and climate. Slopes generally feature an upper *waxing slope* near the top ("waxing" here means increasing). This convex surface curves downward and grades into the *free face* below. The presence of a free face indicates an outcrop of resistant rock that forms a steep scarp or cliff.

Downslope from the free face is a *debris slope,* that receives rock fragments and materials from above. In humid climates constantly moving water carries away material as it arrives, whereas in arid climates, graded debris slopes persist and accumulate. A debris slope grades into a *waning slope,* a concave surface along the base of the slope that forms a *pediment,* or broad, gently sloping erosional surface.

Slopes are open systems and seek an *angle of equilibrium* among the forces described here. Conflicting forces work together on slopes to establish an optimum compromise incline that balances these forces. A geomorphic threshold (change point) is reached when any of the conditions in the balance is altered. All the forces on the slope then compensate by adjusting to a new dynamic equilibrium.

The recently disturbed hillslope in Figure 10-4 is in the midst of compensating adjustment. The disequilibrium was created by the failure of saturated slopes, which a day earlier were under water.

Slopes, then, are shaped by the relationship between rates of weathering and breakup of slope materials and the rates of mass movement and erosion of those materials. A slope is thought of as stable if its strength is greater than these denudation processes, and unstable if materials are weaker than these processes.

Now, with the concepts of base level, landmass denudation, and slope development in mind, let's examine specific processes that operate to reduce landforms.

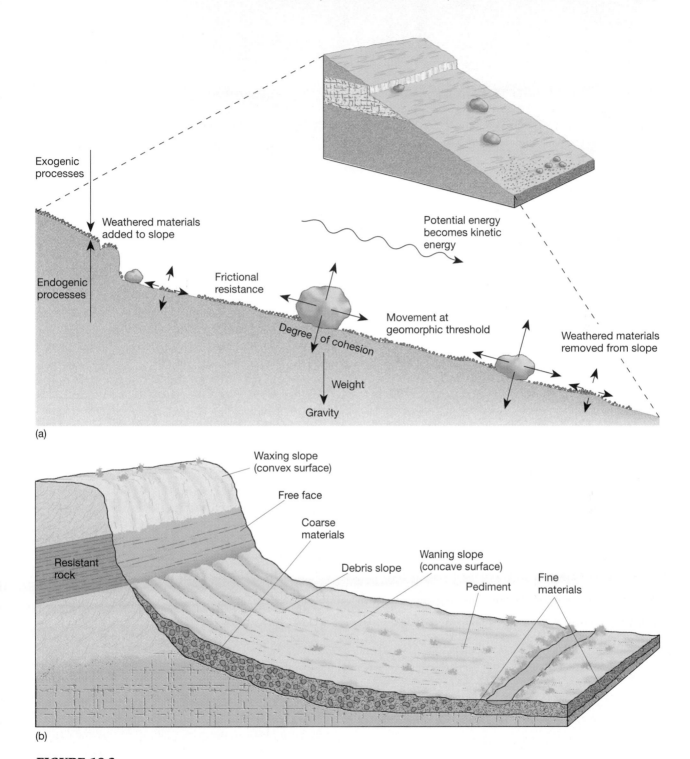

FIGURE 10-3

(a) Forces (noted by arrows) act on materials along an inclined slope; (b) the principal elements of slope form.

FIGURE 10-4
A slope in disequilibrium. Unstable, saturated soils gave way, leaving the roadbed destroyed and the guardrail suspended in air. The slope was underwater when a dam break downstream emptied the reservoir and exposed the slope. [Photo by author.]

Weathering Processes

Rocks at or just below Earth's surface are exposed to both physical and chemical weathering processes. **Weathering** encompasses a group of processes by which surface and subsurface rock disintegrates into mineral particles or dissolves in water. Weathering does not transport the weathered materials; it simply generates them for transport by the agents of water, wind, waves, and ice—all influenced by gravity. In most areas, the upper surface of bedrock undergoes continual weathering to broken-up rock called **regolith**. Loose surface material comes from further weathering of regolith and from transported and deposited regolith (Figure 10-5a). In some areas regolith may be missing or undeveloped, thus exposing an outcrop of unweathered bedrock.

Bedrock is the *parent rock* from which weathered regolith and soils develop. The bedrock is traceable through similarities in composition. For example, the sand and soils in Figure 10-5b derive their color and character from the parent rock in the background. Fragmented material, known as **sediment**, combines with weathered rock to form the **parent material** from which soil evolves.

Jointing in rock is important for weathering processes. **Joints** are fractures or separations in rock. The presence of these usually plane (flat) surfaces greatly increases the surface area of rock exposed to weathering.

Weathering is greatly influenced by the character of the bedrock: hard or soft, soluble or insoluble, broken or unbroken. Important controls on weathering rates are climatic elements, including the amount of precipitation, overall temperature patterns, and any freeze-thaw cycles. Other significant factors are the presence and position of the water table and the orientation of each exposed slope (which controls exposure to Sun, wind, and precipitation). Slopes facing away from the Sun's rays (north-facing slopes in the Northern Hemisphere) tend to be more moist and vegetated than those slopes that receive direct sunlight.

There is a definite relationship between climate (annual precipitation and temperature) and physical (mechanical) and chemical weathering processes: physical weathering dominates in drier, cooler climates, whereas chemical weathering dominates in wetter, warmer climates. Extreme dryness reduces most weathering to a minimum, as is experienced in desert climates *(BW low-latitude arid desert)*. In the hot, wet, tropical and equatorial rain-forest climates *(Af tropical rain forest)*, most rocks weather rapidly, and the weathering extends deep below the surface.

In climates where freezing is common another form of weathering is the freeze-thaw action that alternately expands the water in rocks (freezing) and then contracts it (thawing), creating forces great enough to mechanically split the rocks.

Vegetation is also a factor in weathering. Although vegetative cover can protect rock, it also can provide organic acids from the partial decay of organic matter and thus hasten chemical weathering. Plant roots can enter crevices and break up a rock, exerting enough pressure to drive rock segments apart, thereby exposing greater surface area to other weathering

FIGURE 10-5
(a) A cross section of a typical hillside. (b) These coral-pink dunes in southern Utah derive their color from the parent materials in the background. [Photo by author.]

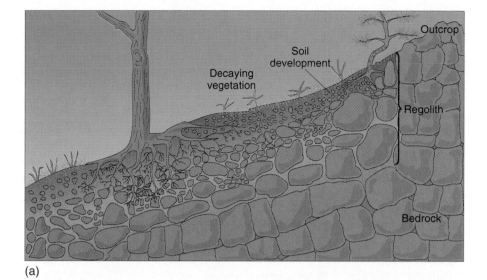

(a)

(b)

processes. You may have observed how tree roots can heave (lift) the sections of a sidewalk or driveway and crack the cement. The differing resistance of rock, coupled with these variations in the intensity of weathering, result in **differential weathering** as shown in Figure 8-11a.

Physical Weathering Processes

When rock is broken and disintegrated without any chemical alteration, the process is called **physical weathering**, or mechanical weathering. By break-

ing up rock, physical weathering greatly increases the surface area on which chemical weathering may operate. In the complexity of nature, physical and chemical weathering usually operate together in a combination of processes. We separate them here for the convenience of discussion.

Crystallization. Especially in arid climates, dry weather draws moisture to the surface of rocks. As the water evaporates, crystals form from dissolved minerals. As this process continues over time and the crystals grow and enlarge, they exert a force great enough to spread apart individual mineral grains and begin breaking up the rock. Such *crystallization,* or *salt-crystal growth,* is a form of physical weathering.

In the Colorado Plateau of the Southwest, deep indentations develop in sandstone cliffs, especially above impervious layers, where salty water slowly flows out of the rock strata. Crystallization then loosens the sand grains, and subsequent erosion and transportation by water and wind complete the sculpturing process. Over 1000 years ago, Native Americans built entire villages in these weathered niches at several locations, including Mesa Verde in Colorado and Arizona's Canyon de Chelly (pronounced "canyon duh shay," Figure 10-6).

Hydration. Another process involving water, but little chemical change, is **hydration**. It is a physical weathering process in which water is absorbed by a mineral. This addition of water initiates swelling and stress within the rock, mechanically forcing grains apart as the constituents expand. Hydration can lead to granular disintegration and further susceptibility of the rock to chemical weathering, especially by oxidation and carbonation.

Frost Action. Water expands as much as 9% of its volume as it freezes (see Chapter 5, "Ice, the Solid Phase"). This expansion creates a powerful mechanical force called **frost action**, or freeze-thaw action, that can exceed the tensional strength of rock. Freezing actions are important in humid microthermal climates *(Df humid continental)* and in subarctic and polar regimes *(subarctic Dfc and Dw climates and E polar climate).*

Repeated cycles of water freezing and thawing break rock segments apart. The work of ice begins in small openings, gradually expanding until rocks are cleaved (split). Figure 10-7 shows this action on

FIGURE 10-6
Cliff dwelling site in Canyon de Chelly, Arizona, used by the Anasazi people until about 900 years ago. [Photo by author.]

FIGURE 10-7
Physical weathering along joints in rock produces discrete blocks in the backcountry of Canyonlands National Park in Utah. [Photo by author.]

blocks of rock, causing *joint-block separation* along existing joints and fractures in the rock. This weathering action, called *frost-wedging,* pushes portions of the rock apart. Cracking and breaking can be in any shape, depending on the rock structure. The softer supporting rock underneath the slabs in the photo already has weathered physically—an example of differential weathering.

Several cultures have used this principle of frost action to quarry rock. Pioneers in the early American West drilled holes in rock, poured water in them, and then plugged them. During the cold winter months, expanding ice broke off large blocks along lines determined by the hole patterns. In spring the blocks were hauled to cities for construction. Frost action also produces unwanted fractures that damage pavement and split water pipes.

Spring can be a risky time to venture into mountainous terrain. As ice melts with the rising temperatures, newly fractured rock pieces fall without warning and may even start rock slides. The falling rock pieces may shatter on impact—another form of physical weathering (Figure 10-8).

Pressure-Release Jointing. To review from Chapter 8, rising magma that is deeply buried and subjected to high pressures forms intrusive igneous rocks called plutons, which cool slowly and produce coarse-grained, crystalline granitic rocks (see illustra-

tion of a pluton in Figure 8-9). As the landscape is subjected to uplift, the weathering regolith that covers such a pluton can be eroded and transported away, eventually exposing the pluton as a mountainous batholith.

As the tremendous weight of the overburden is removed from the granite, the pressure of deep burial is relieved. The batholith responds with a slow but enormous heave, and in a process known as *pressure-release jointing,* layer after layer of rock peels off in curved slabs or plates, thinner at the top of the rock structure and thicker at the sides. As these slabs weather, they slip off in a process called **sheeting** (Figure 10-9a). This *exfoliation process* creates arch-shaped and dome-shaped features on the exposed landscape, forming an **exfoliation dome** (Figure 10-9b). Such domes probably are the largest single weathering features on Earth.

Chemical Weathering Processes

Chemical weathering refers to actual decomposition and decay of the constituent minerals in rock due to chemical alteration of those minerals, always in the presence of water. The chemical breakdown becomes more intense as both temperature and precipitation increase. Although individual minerals vary in susceptibility, no rock-forming minerals are completely unresponsive to chemical weathering. A familiar ex-

ample of chemical weathering is the eating away of cathedral facades and the etching of tombstones caused by increasingly acid precipitation.

Spheroidal weathering is a chemical process that occurs when water penetrates joints and fractures in rock and dissolves the rock's cementing materials. The resulting rounded edges and corners resemble exfoliation but do not result from pressure-release jointing (Figure 10-10). This type of weathering and rock decay in turn opens spaces for frost action.

A boulder can be attacked from all sides, shedding spherical shells of decayed rock like the layers of an onion. As the rock breaks down, more and more surface area is exposed for further weathering. This type of weathering is visible in the photograph from the Alabama Hills that opens this chapter; note the spheroidal rock formations. (See News Report #1.)

Hydrolysis. When minerals chemically combine with water, the process is called **hydrolysis**. Compared to hydration, a physical process in which water is simply absorbed, the hydrolysis process involves active participation of water in chemical reactions to produce different minerals.

FIGURE 10-9
(a) Exfoliation processes loosen slabs of granite, freeing them for further weathering and downslope movement. (b) Exfoliated plates of rock are visible in layers, characteristic of dome formations in granites. View is from the east side of Half Dome in Yosemite National Park, California. [Photos by author.]

(a)

(b)

FIGURE 10-10
Chemical weathering processes act on the joints in granite, leading to spheroidal weathering and eventually to a rounding of individual rock segments. The two boulders are erratics left by retreating ice; they too are being attacked by these processes. [Photo by author.]

When weaker minerals in rock are changed by hydrolysis, the interlocking crystal network in the rock fails, and *granular disintegration* takes place. Such disintegrating granite may appear etched, corroded, and softened. By-products of this chemical weathering of granite include clay and silica. As clay is formed from the minerals in granite, quartz (SiO_2) particles are left behind. The resistant quartz may wash downstream, eventually becoming sand on some distant beach.

Oxidation. An example of chemical weathering occurs when the oxygen dissolved in water oxidizes (combines with) certain metallic elements to form oxides. This is a chemical weathering process known as **oxidation**. Perhaps most familiar is the "rusting" of iron in rocks or soil that produces a reddish-brown stain of iron oxide (Fe_2O_3). Such iron-bearing rocks then weather easily because these oxides are physically weaker than the original iron-bearing mineral.

Carbonation and Solution. The simplest form of chemical weathering occurs when a mineral dissolves into **solution**—for example, when sodium chloride (common table salt) dissolves in water. Water is called the *universal solvent* because it is capable of dissolv-

ing at least 57 of the natural elements and many of their compounds. It readily dissolves carbon dioxide, thereby yielding precipitation containing carbonic acid (H_2CO_3). This acid is strong enough to react with many minerals, especially limestone, in a process called **carbonation**. Carbonation simply means reactions whereby *carbon* combines with minerals. Such chemical weathering transforms minerals containing calcium, magnesium, potassium, and sodium into *carbonates*.

When rainwater attacks formations of *limestone* (which is calcium carbonate, $CaCO_3$), the constituent minerals are dissolved and wash away with the rainwater. Walk through an old cemetery and you can observe the carbonation of marble, a metamorphic form of limestone. Weathered limestone and marble, whether in tombstones or in rock formations, appear pitted and weathered wherever adequate water is available for carbonation. In this era of human-induced increases of acid precipitation, carbonation processes are greatly enhanced (see FYI Report 6-2, "Acid Deposition—A Blight on the Landscape").

Entire landscapes composed of limestone are dominated by the chemical weathering process of carbonation. These are the regions of karst topography, which we look at next.

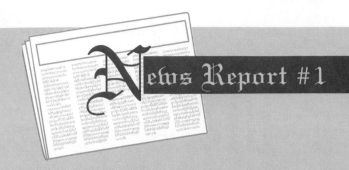

"Alabama Hills—A Popular Movie Location"

The spheroidal weathering and rock formations in the Alabama Hills have appeared in more movies and commercials than any star. The rugged landscape, just east of Mount Whitney and west of Death Valley, has provided scenic backdrops for films as diverse as *How the West Was Won*, *Gunga Din*, *Bad Day at Black Rock*, and *Tremors*. The Alabama Hills appears in many current television car commercials.

Rounded granite joint-blocks of chemically and physically weathered outcrops appear in a variety of rock forms throughout the Alabama Hills. In the chapter-opening photograph for Chapter 10 you can imagine a chase scene raising clouds of dust!

Karst Topography and Landscapes

Limestone is so abundant on Earth that many landscapes are composed of it. These areas are quite susceptible to chemical weathering. Such weathering creates a specific landscape of pitted and bumpy surface topography, poor surface drainage, and well-developed channels underground. Remarkable labyrinths of underworld caverns also may develop. These are the hallmarks of **karst topography**, originally named for the Krš Plateau in Yugoslavia where these processes were first studied. Approximately 15% of Earth's land area has some developed karst, with outstanding examples found in southern China, Japan, Puerto Rico, Cuba, the Yucatán of Mexico, Kentucky, Indiana, New Mexico, and Florida.

For a limestone landscape to develop into karst topography, several important prerequisites must exist. The limestone formation must contain 80% or more calcium carbonate. Complex patterns of joints in the otherwise impermeable limestone are needed for water to concentrate its solution activity and to form routes to subsurface drainage channels. Local surface relief also is important. Karst landscapes do display certain characteristic forms (Figure 10-11).

The weathering of limestone landscapes creates many **sinkholes**, which form circular depressions. If these collapse through the roof of an underground cavern, the term *doline* is also used, and the landscape is referred to as *doline karst*. A gently rolling limestone plain might be pockmarked with sinkholes and dolines 2 to 100 m (7–330 ft) deep and 10 to 1000 m (33–3300 ft) wide.

Through solution and collapse, dolines may coalesce to form *uvalas,* or valley sinks, which are depressions up to several kilometers in diameter sometimes referred to as a *karst valley*. In Florida, several sinkholes have made the news because lowered water tables have caused their collapse into underground solution caves, taking with them homes, businesses, and even new cars from an auto dealership. One such sinkhole collapsed in a Florida suburban area in 1993 (Figure 10-11b). A complex landscape in which dolines intersect is called a *cockpit karst*. The dolines can be symmetrically shaped in certain circumstances; one at Arecibo, Puerto Rico, is perfectly shaped for a radio telescope installation (Figure 10-12).

Another type of karst topography forms in the wet tropics, where deeply jointed, thick limestone beds are weathered into gorges, with isolated resistant blocks

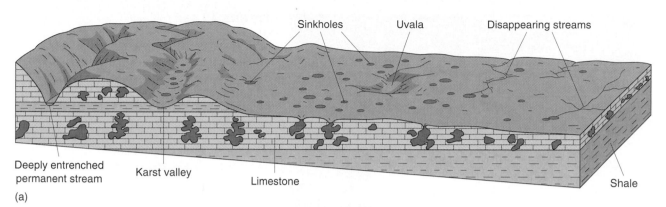

FIGURE 10-11

(a) Idealized features of a portion of karst topography in southern Indiana; (b) Florida, sinkhole, formed in 1993, 50 m (160 ft) in diameter. [(a) From William D. Thornbury, *Principles of Geomorphology;* illustration by William J. Wayne, p. 326. Copyright © 1954, by John Wiley and Sons, Inc. Reprinted by permission. (b) Photo by J. Watson, © Sygma.]

FIGURE 10-12
Cockpit karst topography near
Arecibo, Puerto Rico, provides a
natural depression for the dish
antenna of a giant radio
telescope. The Arecibo
Observatory is part of the
National Astronomy and
Ionosphere Center, which is
operated by Cornell University
under contract with the National
Science Foundation. [Photo
courtesy of Cornell University.]

left standing. These resistant cones and towers are most remarkable in several areas of China, where an otherwise lower-level plain is interrupted by *tower karst* up to 200 m (660 ft) high (Figure 10-13).

Caves and Caverns

Most caves are formed in limestone rock, because it is so easily dissolved by carbonation. The largest limestone caverns in the United States are Mammoth Cave in Kentucky, Carlsbad Caverns in New Mexico, and Lehman Cave in Nevada. Limestone formations in which the Carlsbad Caverns formed were themselves deposited 200 million years ago when the region was covered by shallow seas. Regional uplifts associated with the Rockies (the Laramide orogeny

40–80 million years ago) then brought the region above sea level, leading to active cave formation (Figure 10-14).

Caves generally form just beneath the water table, where a later lowering of the water level exposes them to further development. The *dripstones* that form under such conditions are produced as water containing dissolved minerals slowly drips from the cave ceiling. Calcium carbonate precipitates out of the evaporating solution, literally a molecular layer at a time, and accumulates at a point below on the cave floor. Depositional features called *stalactites* grow down from the ceiling and *stalagmites* build up from the floor, sometimes growing until they connect and form a continuous *column* (Figures 10-14 and 10-15). A dramatic subterranean world is thus created.

FIGURE 10-13
Tower karst of the Kwangsi Province, Yangzhou, China. [Photo by Wolfgang Kaehler.]

FIGURE 10-14
Carlsbad Caverns in New Mexico, where a series of underground caverns includes rooms over 1200 m (4000 ft) long and 190 m (625 ft) wide. Although a national park since 1930, unexplored portions remain. [Photo by Scott T. Smith.]

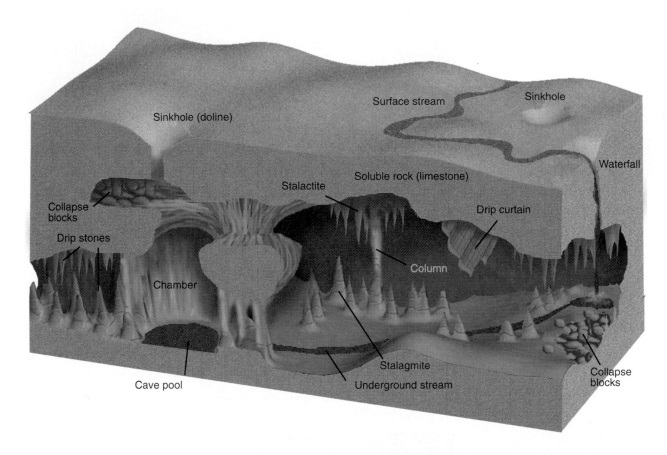

FIGURE 10-15
Formation of underground caves and related forms in limestone. [Inset photo of sinkhole in pasture by Thomas M. Scott, Florida Geological Survey.]

Mass Movement Processes

On November 13, 1985, at 11 P.M., after a year of earthquakes, 10 months of harmonic tremors, a growing bulge on its northeast flank, and 2 months of small summit eruptions, Nevado del Ruiz in central Colombia, South America, violently erupted in a lateral ex-

plosion. The volcano, northernmost of two dozen dormant peaks in the Cordilleran Central of Colombia, had erupted six times during the past 3000 years, killing 1000 people during its last eruption in 1845.

But on this night, the familiar tephra, lava, and the blast itself were not the problem. The hot eruption melted about 10% of the ice on the mountain's snowy

"Amateurs Make Cave Discoveries"

The exploration and scientific study of caves is called *speleology*. Although professional investigations involve physical and biological scientists, amateur cavers, or "spelunkers," have made many important discoveries.

Cave habitats are unique. They are nearly closed, self-contained ecosystems with simple food chains and great stability. In total darkness, bacteria synthesize inorganic elements and produce organic compounds that form the basis of food chains that sustain

many types of cave life, including algae, small invertebrates, amphibians, and fish among others.

The mystery, intrigue, and excitement of caves lies in the variety of dark passageways, enormous chambers that narrow to tiny crawl spaces,

strange formations, and underwater worlds only accessed by cave-diving. Many of the major caves were first discovered by private property owners and amateur adventurers, a fact that keeps this popular science/sport very much alive.

---

peak, liquefying mud and volcanic ash, sending a hot mudflow downslope. Such a flow is called a *lahar,* an Indonesian word referring to flows of volcanic origin. This lahar moved rapidly down the Lagunilla River toward the villages below. The wall of mud was at least 40 m (130 ft) high as it approached Armero, a regional center with a population of 25,000. The city slept as the lahar buried its homes and citizens: 23,000 people were killed there and in other afflicted river valleys, thousands were injured, and 60,000 were left homeless. The debris flow from the volcano is today a permanent grave for its victims.

Although less devastating to human life, the mudflow generated by the eruption of Mount Saint Helens also was a lahar. Not all mass movements are this destructive, but such processes are important in the overall denudation of the landscape.

Mass Movement Mechanics

Physical and chemical weathering processes create an overall weakening of surface rock, which makes it more susceptible to the pull of gravity. The term

mass movement applies to all downward movements of materials propelled and controlled by gravity, such as the lahar just described. These movements can range from dry to wet, slow to fast, small to large, and from free-falling to gradual or intermittent in motion. The term mass movement is sometimes used interchangeably with *mass wasting*. To combine the concepts, we can say that mass movement of material works to waste slopes and provide raw material for erosion, transportation, and deposition.

The Role of Slopes. All mass movements occur on slopes. If we try to pile dry sand on the beach, the grains will flow downward until an equilibrium is achieved. The steepness of the slope that results when loose sand comes to rest depends on the size and texture of the grains; this is called the *angle of repose*. This angle represents a balance of driving and resisting forces (gravity and friction) and commonly ranges between 33° and 37° (measured from a horizontal plane).

The driving force in mass movements is gravity, working in conjunction with the weight, size, and shape of the surface material, the degree to which the slope is oversteepened, and the amount and form of moisture available—whether frozen or fluid. The greater the slope angle, the more susceptible the surface material is to mass movement when gravity overcomes friction.

If the rock strata in a slope are underlain by clays, shales, and mudstones, all of which are highly susceptible to hydration (physical swelling in response to the presence of water), the rock strata will tend toward mass movement because less driving force energy is required. However, if the rock strata are such that material is held back from slipping, then more driving force energy may be required, such as that generated by an earthquake.

In the Madison River Canyon near West Yellowstone, Montana, on the Wyoming border, a blockade of dolomite (a magnesium-rich carbonate rock) had held back a deeply weathered and *oversteepened slope* (40–60° slope angle) for untold centuries (white area in Figure 10-16). Then, shortly after midnight on August 17, 1959, an earthquake measuring 7.5 on the Richter scale broke the dolomite structure along the foot of the slope. This released 32 million m³ (1.13 billion ft³) of mass, which moved downslope at 95 kmph (60 mph), creating gale force winds through the canyon. The material continued more than 120 m (390 ft) up the opposite canyon slope, entombing campers beneath about 80 m (260 ft) of rock and debris. These events convey a dramatic example of the role of slopes and tectonic forces in creating mass land movements.

The mass of material also effectively dammed the Madison River and created a new lake, dubbed Quake Lake. The landslide debris dam established a

FIGURE 10-16

Cross section of the geologic structure of the Madison River Canyon at the location of the landslide triggered by the 1959 earthquake: (1) prequake slope contour, (2) weathered dolomite that failed, (3) direction of landslide, and (4) landslide debris blocking the canyon and damming the Madison River. [After Jarvis B. Hadley, *Landslides and Related Phenomena Accompanying the Hebgen Lake Earthquake of 17 August 1959,* U.S. Geological Survey Professional Paper 435-K, p. 115. Inset photo by Randall Christopherson.]

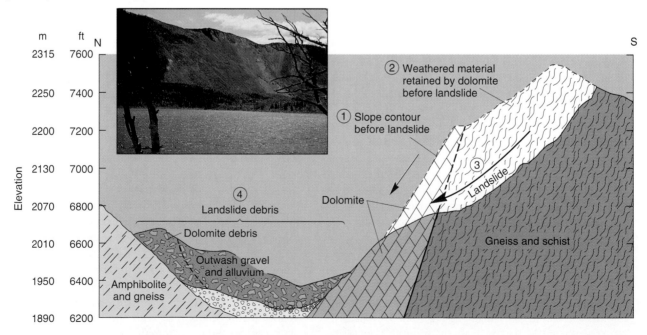

new temporary base level for the canyon. A channel was quickly excavated by the U.S. Army Corps of Engineers to prevent a disaster below the dam, for if Quake Lake overflowed the landslide dam, the water would quickly erode a channel and thereby release the entire contents of the new lake onto farmland downstream.

Classes of Mass Movements

For convenience, four basic classifications of mass movement are used: fall, slide, flow, and creep. Each involves the pull of gravity working on a mass until it falls, slides, flows, or creeps downward. The Madison River Canyon event was a type of *slide,* whereas the Nevado del Ruiz lahar mentioned earlier was a *flow.* Figure 10-17 summarizes the relationship between moisture content and rate of movement in producing the different classes of mass movement. (In contrast to these surface processes, submarine mass movements are disorganized, turbulent flows of mud accumulating as deep-sea depositional fans along the continental slope.)

Falls and Avalanches. A **rockfall** is simply a quantity of rock that falls through the air and hits a surface. During a rockfall individual pieces fall independently, and characteristically form a pile of irreg-

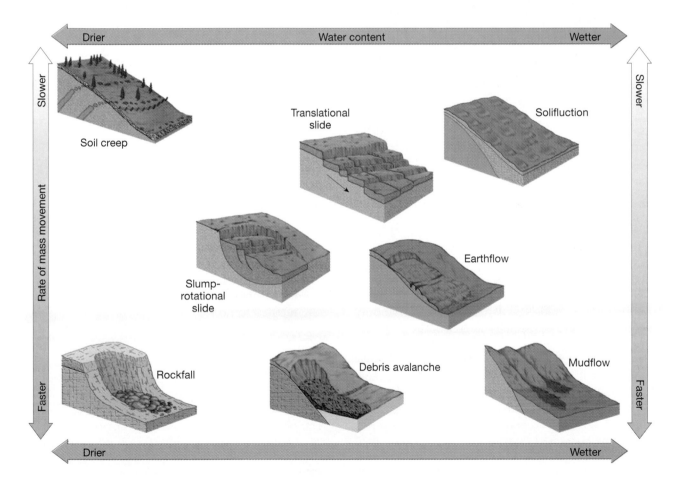

FIGURE 10-17
Principal types of mass movement and mass wasting events. Variations in water content and rates of movement produce a variety of forms.

FIGURE 10-18
Rockfall and talus cone at the
base of a steep slope. [Photo by
author.]

ular broken rocks called a *talus cone* at the base of a
steep slope (Figure 10-18).

A **debris avalanche** is a mass of falling and tumbling rock, debris, and soil. It is differentiated from a
debris slide or landslide by the tremendous velocity
achieved by the onrushing materials. These speeds
often result from ice and water that fluidize the debris. The extreme danger of a debris avalanche results
from these tremendous speeds and lack of warning.

In 1962 and again in 1970, debris avalanches roared
down the west face of Nevado Huascarán, the highest peak in the Peruvian Andes. The 1962 debris
avalanche contained an estimated 13 million m³ (460
million ft³) of material, burying the city of Ranrahirca
and eight other towns, killing 4000 people. The 1970
event was initiated by an earthquake. Upward of 100
million m³ (3.53 billion ft³) of debris buried the city
of Yungay, where 18,000 people perished. This
avalanche attained velocities of 300 kmph (185 mph),

which is especially incredible when you consider the
quantity of material involved and the fact that some
boulders weighed thousands of tons.

Another dramatic example of a debris avalanche,
one that led to no loss of life, occurred west of the
Saint Elias Range, north of Yakutat Bay in Alaska. Several dozen rockfalls, large debris avalanches, and
snow avalanches were triggered by a 7.1 Richter-scale
earthquake in 1979. Approximately 4.7 km² of the
Cascade Glacier was covered by the largest single
avalanche caused by the earthquake (Figure 10-19).
The photo reveals characteristic grooves, lobes, and
large rocks associated with these fluid avalanches.

Slide. A sudden rapid movement of a cohesive mass
of regolith and/or bedrock that is not saturated with
moisture is a **landslide**—a large amount of material
failing simultaneously. Surprise is what creates the danger, for the downward pull of gravity wins the strug-

334

"Debris Avalanche Destroys Colombian Town"

Debris avalanche dangers relate to the tremendous velocity of the rock and material moving downslope. Toez, Colombia, was buried by an avalanche of ice, rock, and mud on an afternoon in June 1994. The avalanche pressed air aside as it rushed downhill, producing hurricane-force winds.

The triggering mechanism of this killer event was a magnitude 6.4 Richter earthquake. The material traveled several kilometers down the slopes of Nevado del Huila, a 5785 m (18,975 ft) volcano, highest in South America. Hundreds of unsuspecting residents perished and many more were missing in yet another Andean debris avalanche.

gle for equilibrium in an instant of time. Slides occur in one of two basic forms—translational or rotational (Figure 10-17). *Translational slides* involve movement along a planar (flat) surface roughly parallel to the angle of the slope. The Madison Canyon landslide de-scribed earlier was a translational slide. Flow and creep patterns also are considered translational in nature.

Rotational slides occur when surface material moves along a concave surface. Frequently, underlying clay presents an impervious surface to percolating water.

FIGURE 10-19
A debris avalanche covers portions of the Cascade Glacier, west of the Saint Elias Range in Alaska. The one pictured here was the largest of several dozen triggered by a 1979 earthquake. [Photo by George Plafker.]

FIGURE 10-20
Anchorage, Alaska, landslide, illustrates mass movement patterns throughout a housing subdivision. [Courtesy of U.S. Geological Survey.]

As a result, water flows along the clay surface, undermining the overlying block. The surface may rotate as a single unit, or it may present a stepped appearance.

On March 27, 1964, Good Friday, the Pacific plate plunged a bit further beneath the North American plate near Anchorage, Alaska. More than 80 mass movement events were triggered by the resulting earthquake. One particular housing subdivision in Anchorage experienced a sequence of translational movements accompanied by rotational slumping. Figure 10-20 shows one view of the ground failure throughout the housing area.

Flow. When the moisture content of moving material is high, the suffix -flow is used, as in *earthflow* and *mudflow,* illustrated in Figure 10-17. Heavy rains frequently saturate barren mountain slopes and set them moving, as was the case east of Jackson Hole, Wyoming, in the spring of 1925. A slope above the Gros Ventre (pronounced "grow vaunt") River broke loose and slid downward as a unit. The slide occurred because sandstone formations rested on weak shale and siltstone, which became moistened and soft, offered little resistance to the overlying strata.

Because of melted snow and rain, the water content of the Gros Ventre landslide was high enough to classify it as an earthflow. About 37 million m³ (1.3 billion ft³) of wet soil and rock moved down one side of the canyon and surged 30 m (100 ft) up the other

side. The earthflow dammed the river and formed a lake. Two years later, the lake water broke through the temporary base-level dam, transporting a tremendous quantity of debris over the region downstream.

Creep. A persistent mass movement of surface soil is called **soil creep**. Individual soil particles are lifted and disturbed by the expansion of soil moisture as it freezes, by cycles of moistness and dryness, by daily temperature variations, or even by grazing livestock or digging animals. In the freeze-thaw cycle, particles are lifted at right angles to the slope by freezing soil moisture, as shown in Figure 10-21. However, when the ice melts, the particles are pulled straight downward by the force of gravity. As the process is repeated, the soil cover gradually creeps its way downslope. The overall wasting of a slope may cover a wide area and may cause fence posts, utility poles, and even trees to lean downslope. Various strategies are used to arrest the mass movement of slope material—grading the terrain, building terraces and retaining walls, planting ground cover—but the persistence of creep always wins.

Human-Induced Mass Movements

Every highway roadcut or surface-mining activity that exposes a slope can hasten mass wasting because the oversteepened surfaces are thrown into a new search

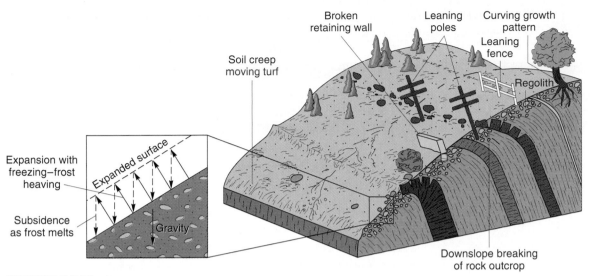

FIGURE 10-21
Soil creep and its effects.

for equilibrium. Imagine the disequilibrium in slope relationships created by the highway roadcut pictured in Figure 9-5b. Large open-pit strip mines—such as the Bingham Copper Mine west of Salt Lake City, the Berkeley Pit in Butte, Montana, and the extensive strip mining for coal in the East and West—are examples of human-induced mass movements, generally called **scarification**. At the Bingham Copper Mine, a mountain literally was removed. The disposal of *tailings* (mined ore of little value) and waste material is a significant problem with such large excavations because the tailing piles prove unstable and susceptible to further weathering, mass wasting, or wind dispersal. Wind dispersal is a particular problem with uranium tailings in the West due to their radioactivity.

Land subsidence and collapse in mined areas produce further mass movements of land. This is a major problem in portions of the Appalachians where buildings, utility poles, and streets, as well as drainage patterns, are affected.

SUMMARY—Weathering, Karst Landscapes, and Mass Movement

The exogenic system, powered by solar energy and gravity, tears down the landscape through processes of landmass **denudation** involving **weathering**, **mass movement**, **erosion**, *transportation*, and *deposition*. Agents of change include moving air, water, waves, and ice. **Geomorphology** is the science that analyzes and describes the origin, evolution, form, and spatial distribution of landforms.

Various models to characterize landmass denudation and **slope** reduction have been attempted. The principles of slope morphology (development and change) are guided by the **dynamic equilibrium model**. Slopes are shaped by the relationship between rate of weathering and breakup of slope materials and the rate of mass movement and erosion of those materials. A slope is thought of as stable if its strength is greater than these denudation processes; it is unstable if materials are weaker than these processes.

Both physical and chemical weathering derive material from parent bedrock that is essential for soil formation. **Physical weathering** refers to the actual breakup of rock into smaller pieces, whereas **chemical weathering** is the chemical altering of minerals in rock. The physical action of water when it freezes (expands) and thaws (contracts) is called **frost action**, a powerful agent in shaping the landscape. At the same time, the chemical action of **carbonation** (carbon dioxide dissolved in rainwater) attacks susceptible rock, such as limestone. **Karst topography** refers to distinctively pitted and weathered limestone landscapes. The formation of massive caverns is a result of these processes.

Mass movement of Earth's surface produces some dramatic incidents, including **rockfalls**, **landslides**, **mudflows**, and **soil creep**, all of which are mass movements of soil and debris and are important as-

pects of overall landmass denudation processes. In addition, human mining activity has created massive **scarification** of landscapes.

The next four chapters look at the work of rivers, wind, waves, and ice as geomorphic agents that use erosion, transport, and deposition processes to form unique landscapes.

KEY TERMS

base level	landslide
bedrock	mass movement
carbonation	oxidation
chemical weathering	parent material
debris avalanche	physical weathering
denudation	regolith
differential weathering	rockfall
dynamic equilibrium model	scarification
exfoliation dome	sediment
frost action	sheeting
geomorphic threshold	sinkholes
geomorphology	slopes
hydration	soil creep
hydrolysis	solution
joints	spheroidal weathering
karst topography	weathering

REVIEW QUESTIONS

1. Define geomorphology and describe its relationship to physical geography.
2. How do base level and local base level control stream erosion of its channel?
3. Define landmass denudation. What processes are included in the concept?
4. How does the construction of a reservoir alter the relationship between a stream and base level?
5. When a hillslope system reaches a geomorphic threshold, what occurs?
6. Relative to slopes, what is meant by an "angle of equilibrium"? Can you apply this to the photograph in Figure 10-4?
7. Describe the processes of physical and chemical weathering on an open expanse of bedrock. How does regolith develop? How is sediment derived?
8. What is the interplay between the resistance of rock structures and weathering variabilities?
9. Why is freezing water such an effective physical weathering agent?

10. What are the weathering processes that produce a granite dome? Describe the sequence of events that takes place.
11. Describe the development of limestone topography. What is the name applied to such landscapes? From what region was this name derived?
12. Differentiate among sinkholes, dolines, uvalas, and cockpit karst. Within which form is the radio telescope at Arecibo, Puerto Rico?
13. Name and describe the type of mudflow associated with volcanic eruptions.
14. Define the role of slopes in mass movements—angle of repose, driving force, resisting force, and so on.
15. What are the classes of mass movement? Describe each briefly and differentiate among these classes.
16. Describe the difference between a landslide and what happened on the slopes of Nevado Huascarán.
17. What is scarification, and why is it considered a type of mass movement? Give an example of scarification.

Drai Sap Falls, near Ban Me Thuot, Central Highlands, Vietnam. [*Photo by Wolfgang Kaehler*]

11

RIVERS AND RELATED LANDFORMS

FLUVIAL PROCESSES AND
  LANDSCAPES
  The Drainage Basin System
  Streamflow Characteristics
  Stream Gradient
  Stream Deposition
FLOODS AND RIVER MANAGEMENT
  Streamflow Measurement
SUMMARY
FYI REPORT 11-1 FLOODPLAIN MANAGEMENT
  STRATEGIES

Earth's rivers and waterways form vast arterial networks that both shape and drain the continents, transporting the by-products of weathering, mass movement, and erosion. To call them Earth's lifeblood is no exaggeration, inasmuch as rivers redistribute mineral nutrients important for soil formation and plant growth and serve society in many ways. Not only do rivers provide us with essential water supplies, but they also receive, dilute, and transport wastes, provide critical cooling water for industry, and form one of the world's most important transportation networks. Rivers have been of fundamental importance throughout human history. This chapter discusses the dynamics of river systems and their landforms.

At any one moment approximately 1250 km³ (300 mi³) of water flows through Earth's waterways. Even though this volume represents only 0.003% of all freshwater, the work performed by this energetic flow makes it the dominant agent of landmass denudation. Figure 6-14 portrays North America's surface runoff, which along with groundwater, is our basic water supply. Of the world's rivers, those with the greatest flow volume per unit of time, or *discharge,* are the Amazon and the Paraná of South America, the Zaire (Congo) of Africa, the Ganges of India, and the Chàng Chiang (Yangtze) of Asia (see Table 6-1). In North America, the greatest discharges are from the Missouri-Ohio-Mississippi, Saint Lawrence, and Mackenzie river systems.

Fluvial Processes and Landscapes

Stream-related processes are termed **fluvial** (from the Latin *fluvius,* meaning "river"). Geographers seek to describe recognizable stream patterns and the fluvial processes that created them. Fluvial systems, like all natural systems, have characteristic processes and produce predictable landforms. Yet, a stream system can behave with randomness, unpredictability, and disorder.

Insolation and gravity are the driving forces of fluvial systems because they power the hydrologic cycle. Individual streams vary greatly, depending on the climate in which they operate, the variety of surface composition and topography over which they flow, the nature of vegetation and plant cover, and the length of time they have been functioning in a specific setting.

Wind, water, and ice dislodge, dissolve, or remove surface material in the process called **erosion**. Thus, streams produce *fluvial erosion,* which supplies weathered sediment for **transport** to new locations, where it is laid down in a process known as **deposition**. A stream is a mixture of water and solids—carried in solution, suspension, and by mechanical transport. **Alluvium** is the general term for the clay, silt, and sand transported by running water.

The work of streams modifies the landscape in many ways. Landforms are produced by the erosive action of flowing water and the deposition of stream-transported materials. Rivers create floodplains as they overflow during episodic floods. Let's begin our study by examining a basic fluvial unit—the drainage basin.

The Drainage Basin System

Every stream has a **drainage basin**. A drainage basin is the spatial geomorphic unit occupied by a river system. It is defined by ridges that form *drainage divides;* that is, the ridges are the dividing lines that control into which basin precipitation drains. Drainage divides define a **watershed**, the catchment area of the drainage basin. The United States and Canada are divided by several **continental divides**; these are extensive mountain and highland regions that separate vast drainage basins, sending flows either to the Pacific, or to the Gulf of Mexico and the Atlantic, or to Hudson Bay and the Arctic Ocean (Figure 11-1). The principal drainage basins in the United States and Canada are mapped in Figure 11-2. These drainage basins provide a spatial framework for water studies and water planning.

A major drainage basin system, such as the one created by the Mississippi-Missouri-Ohio river system, is made up of many smaller drainage basins, which in turn comprise even smaller basins, each divided by specific watersheds. Each drainage basin gathers and delivers its precipitation and sediment to a larger

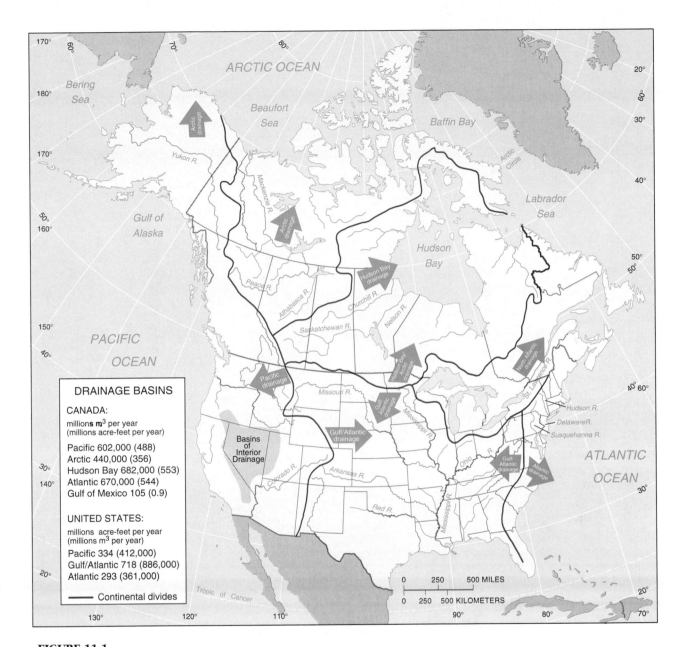

FIGURE 11-1
Continental divides (purple lines) separate the major drainage basins that empty
into the Pacific, Atlantic, Gulf of Mexico, and to the north through Canada into
Hudson Bay and the Arctic Ocean. [After U.S. Geological Survey and *The National
Atlas of Canada*, Energy, Mines, and Resources Canada.]

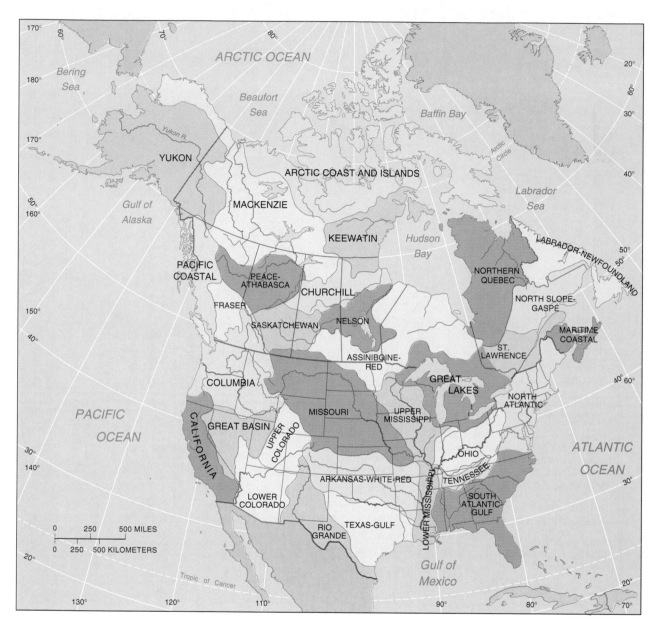

FIGURE 11-2
Major drainage basins in the United States and Canada. [Data from Environment
Canada, *Currents of Change–Inquiry on Federal Water Policy–Final Report 1985*,
and from U.S. Water Resources Council, *The Nation's Water Resources.*]

basin, concentrating the volume in the main stream. For example, rainfall in north-central Pennsylvania generates the streams that flow into the Allegheny River. The Allegheny River then joins with the Monongahela River at Pittsburgh to form the Ohio

River. The Ohio flows southwest and connects with the Mississippi River at Cairo, Illinois, and eventually flows on past New Orleans to the Gulf of Mexico. Each contributing tributary adds its discharge and sediment load to the larger river. In our example, sed-

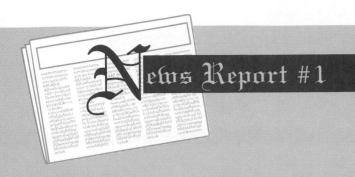
"Climate Change and a River Basin"

The Delaware River basin encompasses 33,060 km² (12,890 mi²) and includes parts of five states in the stream's length, which is 595 km (370 mi) from headwaters to the mouth. The entire basin lies within a humid, temperate climate and receives an average annual precipitation of 120 cm (47.2 in.). The river provides water for an estimated 20 million people in the region. Major conduits export water from the Delaware River. Several reservoirs in the drainage basin enhance water availability and control of the drainage basin system.

In 1988, the USGS launched a study of the Delaware River basin to research the potential impact of future climate change on water resources. The specific concern is global warming. The study includes projected changes in streamflow, irrigation demand, reduction in soil moisture storage, possible saltwater intrusion into the basin near the ocean, and problems associated with sea-level rise. A decrease in precipitation caused by changing climate patterns would cause significant disruption of the water supply to the surrounding metropolitan region.

iment weathered and eroded in north-central Pennsylvania is transported thousands of kilometers and accumulates as the Mississippi delta on the floor of the Gulf of Mexico.

Drainage basins are open systems whose inputs include precipitation, the minerals and rocks of the regional geology, and changes of energy with both the uplift and subsidence provided by tectonic activities. System outputs of water and sediment leave through the mouth of the river. Change that occurs in any portion of a drainage basin can affect the entire system, because the stream adjusts to carry the appropriate load of sediment relative to discharge (its volume of flow) and velocity. A stream drainage system exhibits a constant struggle toward an equilibrium among interacting variables of discharge, transported sediment load, channel shape, and channel slope.

Drainage Patterns. Initially, water moves downslope in a thin film as **sheet flow**, or overland flow. This surface runoff concentrates in *rills,* or small-scale indentations, which may develop further into *gullies* and *stream courses.* The resultant **drainage pattern** is an arrangement of channels which is determined by slope, differing rock resistance, climatic and hydrologic variability, relief of the land, and structural controls imposed by the landscape.

The *Landsat* image in Figure 11-3 is of the Ohio River area near the junction of the West Virginia, Ohio, and Kentucky borders. The high-density drainage pattern and intricate dissection of the land occur because of the region's generally level sandstone, siltstone, and shale strata, which are easily eroded, given the humid mesothermal climate. Fluvial action, along with other denudation processes, is responsible for this dissected topography.

The seven most common drainage patterns encountered in nature are represented in Figure 11-4. The most familiar pattern is *dendritic* (a); this tree-like pattern is similar to that of many natural systems, such as capillaries in the human circulatory system or the vein patterns in leaves. Energy expended by this drainage system is efficient because the overall length of the branches is minimized.

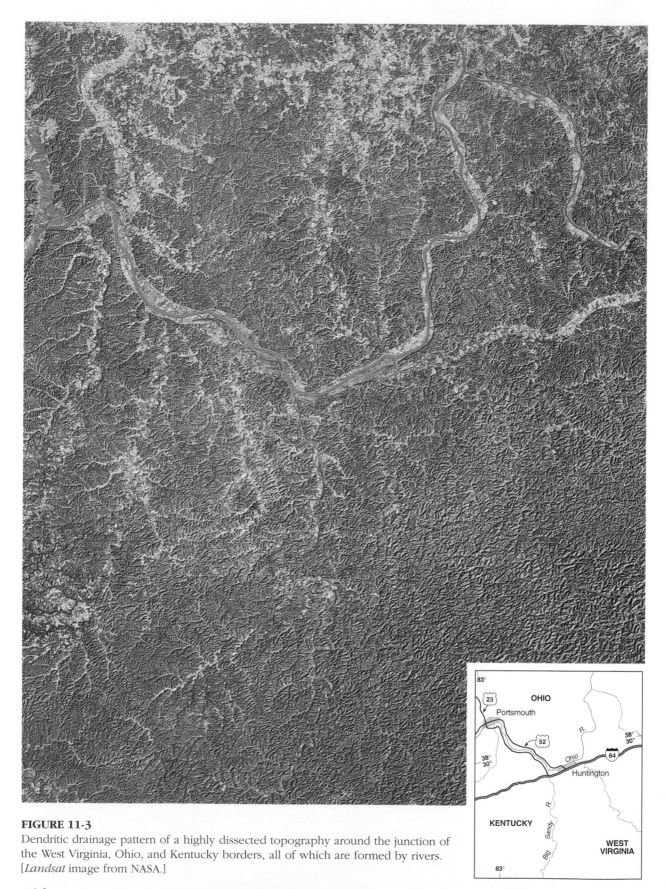

FIGURE 11-3
Dendritic drainage pattern of a highly dissected topography around the junction of
the West Virginia, Ohio, and Kentucky borders, all of which are formed by rivers.
[*Landsat* image from NASA.]

The *trellis* drainage pattern (c) is characteristic of dipping or folded topography, which exists in nearly parallel mountains of the Ridge and Valley Province of the East, where drainage patterns are influenced by rock structures of variable resistance and folded strata. Refer back to Figure 9-12, which presents a satellite image of this region that shows this distinctive drainage pattern. The principal streams are directed by the parallel folded structures, whereas smaller dendritic streams are at work on nearby slopes, joining the main streams at right angles.

The sketch in Figure 11-4c suggests that the headward-eroding part of one stream could break through a drainage divide and *capture* the headwaters of another stream in the next valley, and indeed this does happen. The sharp bends in two of the streams in the illustration are called *elbows of capture* and are evidence that one stream has breached a drainage divide. This type of capture, or *stream piracy*, can occur in other drainage patterns.

The remaining drainage patterns in Figure 11-4 are caused by other specific structural conditions. *Parallel* drainage (b) is associated with steep slopes and some relief. A *rectangular* pattern (d) is formed by a faulted and jointed landscape, directing stream courses in patterns of right-angle turns. A *radial* drainage pattern (e) results from streams flowing off a central peak or dome, such as occurs on a volcanic mountain. *Annular* patterns (f) are produced by structural domes, with concentric patterns of rock strata guiding stream courses. In areas having disrupted surface patterns, such as the glaciated shield regions of Canada and northern Europe, a deranged pattern (g) is in evidence, with no clear geometry in the drainage and no true stream valley pattern.

Streamflow Characteristics

A mass of water positioned above base level in a stream has potential energy. As the water flows downstream under the influence of gravity, this energy becomes kinetic energy. The rate of this potential energy conversion depends on the steepness of the stream channel.

Stream channels vary in *width* and *depth*. The streams that flow in them vary in *velocity* and in the sediment load they carry. All of these factors may increase with increasing discharge. Discharge is calculated by multiplying the velocity of the stream by

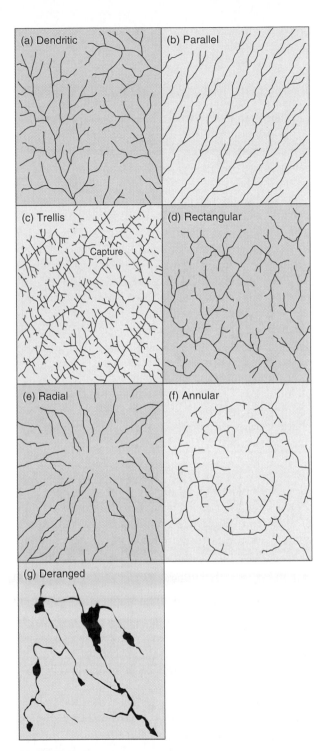

FIGURE 11-4
The seven most common drainage patterns. [After A. D. Howard, "Drainage Analysis in Geological Interpretation: A Summation," *Bulletin of American Association of Petroleum Geologists* 51, 1967, p. 2248. Adapted by permission.]

its width and depth for a specific cross section of the channel as stated in the simple expression:

$$Q = wdv$$

where Q = discharge; w = channel width; d = channel depth; and v = stream velocity. As Q increases, some combination of channel width, depth, and stream velocity increases. Discharge is expressed either in cubic meters per second (m³/s) or cubic feet per second (cfs).

Given the interplay of channel width, depth, and stream velocity, the cross-section of a stream varies over time, especially during heavy floods. Figure 11-5 shows changes in the San Juan River channel in Utah that occurred during a flood. The increase in discharge increases the velocity and therefore the carrying capacity of the river as the flood progresses. As a result, the river's ability to scour materials from its bed is enhanced. Such scouring represents a powerful clearing action, especially in the excavation of alluvium from the streambed by a stream in flood.

You can see in the figure that the San Juan River's channel was deepest on October 14, when floodwaters were highest (blue outline). Then, as the discharge returned to normal, the kinetic energy of the river was reduced, and the bed again filled as sediments were redeposited. You can see this in the progression of the red, green, and purple outlines. The process depicted in Figure 11-5 moved a depth of about 3 m (10 ft) of sediment from this cross section of the stream channel. Such adjustments in a stream channel occur as the stream system continuously works toward an equilibrium to balance discharge, velocity, and sediment load.

Stream Erosion. The erosional work of a stream carves and shapes the landscape through which it flows. Several types of erosional processes are operative. **Hydraulic action** is the work of *turbulence* in the water. Running water causes hydraulic squeeze-and-release action to loosen and lift rocks. As this debris moves along, it mechanically erodes the streambed further, through a process of **abrasion**, with rock particles grinding and carving the streambed.

The upstream tributaries in a drainage basin usually have small and irregular discharges, with most of the stream energy expended in turbulent eddies. As a result, hydraulic action in these upstream sec-

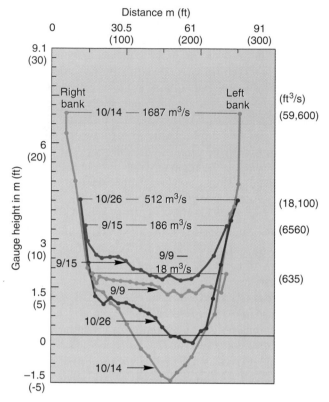

FIGURE 11-5
Stream channel cross sections showing the progress of a 1941 flood on the San Juan River near Bluff, Utah. [After Luna Leopold and Thomas Maddock, Jr., 1953, "The Hydraulic Geometry of Stream Channels and Some Physiographic Implications," *U.S. Geological Survey Professional Paper 252*, Washington, DC: Government Printing Office, p. 32.]

tions is at maximum, even though the coarse-textured load of such a stream is small. The downstream portions of a river, however, move much larger volumes of water past a given point and carry larger suspended loads of sediment. Thus, both volume and velocity are important determinants of the amount of energy expended in the erosion and transportation of sediment (Figure 11-6).

Stream Transport. You may have watched a river or creek after a heavy rainfall, the water colored brown by the heavy sediment load being transported. The amount of material available to a stream is dependent upon topographic relief, the nature of rock

(a)

(b)

FIGURE 11-6
The relation between volume and velocity in (a) a turbulent mountain stream, Alanje Rio, Costa Rica and (b) the downstream portion of the Mississippi River near Natchez, Mississippi. [(a) Photo by Stephen J. Krasemann/DRK Photos; (b) photo by author.]

and materials through which the stream flows, climate, vegetation, and the types of processes at work in a drainage basin. *Competence,* which is a stream's ability to move particles of specific size, is a function of stream velocity. The total possible load that a stream can transport is its *capacity.* Eroded materials are transported by four processes: solution, suspension, saltation, and traction (Figure 11-7).

Solution refers to the **dissolved load** of a stream, especially the chemical solution derived from minerals such as limestone or from soluble salts. The main

contributor of material in solution is chemical weathering. Sometimes the undesirable salt content that hinders human use of rivers comes from dissolved rock formations or from springs in the stream channel.

The **suspended load** consists of fine-grained, clastic particles (bits and pieces of rock) physically held aloft in the stream, with the finest particles not deposited until the stream velocity slows to near zero. Turbulence in the water, with random upward motions, is an important mechanical factor in holding a load of sediment in suspension.

FIGURE 11-7
Fluvial transportation of eroded materials through saltation, traction, suspension, and solution.

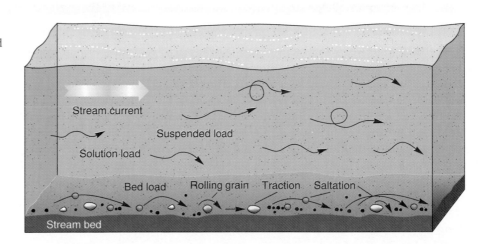

FIGURE 11-8
Braided stream pattern in Chitina River, Wrangell–Saint Elias National Park, Alaska. This stream reflects excessive sediment load associated with glacial meltwaters. [Photo by Tom Bean.]

The **bed load** refers to those coarser materials that are dragged along the bed of the stream by **traction** or are rolled and bounced along by **saltation** (from the Latin *saltim,* which means "by leaps or jumps"). With increased kinetic energy, parts of the bed load are rafted up and become suspended load. This was demonstrated in the flood-induced channel deepening of the San Juan River shown in Figure 11-5. Saltation is also a process in the transportation of materials by wind (see Chapter 12).

The first Spanish explorers to visit the Grand Canyon reported in their journals that they were kept awake at night by the thundering sound of boulders tumbling along the Colorado River bed. Such sounds today are substantially lessened because of the reduced velocity of the Colorado resulting from the many dams and control facilities that now trap sediments and reduce bed load capacity.

If the load in a stream exceeds its capacity, sediments accumulate as **aggradation** (the opposite of degradation) as the stream channel builds up through deposition. With excess sediment, a stream becomes a maze of interconnected channels that form a **braided stream** pattern. Braiding often occurs when reduced discharge affects a stream's transportation ability such as under seasonal conditions, or when a landslide occurs upstream, or where weak banks of sand or gravel exist. Locally, braiding also may result from a new sediment load, which frequently is associated with glacial meltwaters, such as in the Chitina River in Alaska, pictured in Figure 11-8.

Channel Patterns. Flow characteristics of a stream are best seen in a cross-section view. The greatest velocities in a stream are near the surface at the center, corresponding with the deepest part of the stream channel (Figure 11-9). Velocities decrease closer to the sides and bottom of the channel because of the frictional drag on the water flow. In a curving stream, the maximum velocity line migrates from side to side along the channel, deflected by the curves.

Where slopes are gradual, stream channels assume a sinuous (snakelike) form weaving across the landscape. This action produces a **meandering stream**, from the Greek *maiandros,* after the ancient Maiandros River in Asia Minor (the present-day Menderes River in Turkey) that had a meandering channel pattern. The outer portion of each meandering curve is subject to the greatest erosive action and can be the site of a steep bank called a **cut bank** (an undercut bank) (Figure 11-9). On the other hand, the inner portion of a meander receives sediment fill, forming a deposit called a **point bar**.

As meanders develop, the scour-and-fill features gradually work their way downstream. As a result, the landscape near a meandering river bears meander scars of residual deposits from the previous river channels. Former point-bar deposits leave low-lying ridges, creating a bar-and-swale relief (ridges and

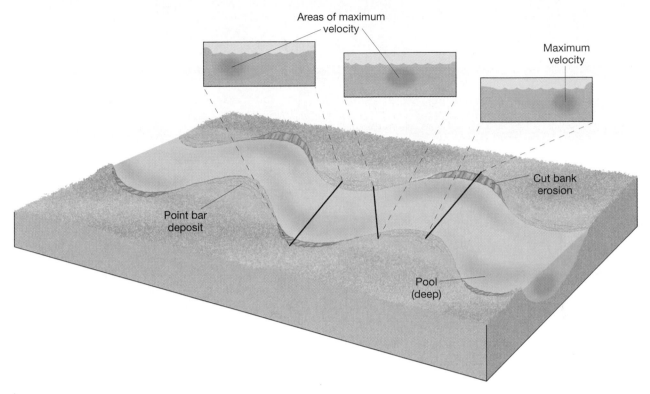

FIGURE 11-9
Aerial view and cross sections of a meandering stream, showing the location of
maximum flow velocity, point bar deposits, and areas of cut bank erosion.

slight depressions). The photograph of the Itkillik
River in Alaska, Figure 11-10a, shows both meanders
and meander scars.

Figure 11-10b shows (in four numbered steps)
how a stream meander can form a cutoff. The stream
erodes its outside bank as the curve migrates downstream (1), the neck of land created by the looping
meander (2) eventually erodes through and forms a
cutoff (3). When the former meander becomes isolated from the rest of the river, the resulting **oxbow
lake** (4) may gradually fill with silt or may again become part of the river when it floods. The Mississippi River is many miles shorter today than it was in
the 1830s because of artificial cutoffs that were
dredged across these necks of land to improve navigation and safety.

Streams often are used as natural political boundaries, but it is easy to see how disagreements might
arise when boundaries are based on river channels

that shift around. For example, the Ohio, Missouri,
and Mississippi rivers can shift their positions and,
therefore, the boundaries based upon them quite
rapidly during times of flood. Carter Lake, Iowa, provides us with a case in point (Figure 11-10c). The
Nebraska-Iowa border was originally placed mid-
channel in the Missouri River. In 1877, the meander
loop that curved around the town of Carter Lake was
cut off by the river, leaving the town "captured" by
Nebraska. The new oxbow lake was called Carter
Lake and still constitutes the state line.

Boundaries should always be fixed by surveys independent of river locations, because border dis-
putes result when river channels change course.
Such surveys have been completed along the Rio
Grande near El Paso, Texas, and along the Colorado
River, between Arizona and California, permanently
establishing political boundaries separate from
changing river locations.

(a)

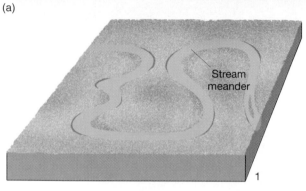

Stream meander

1

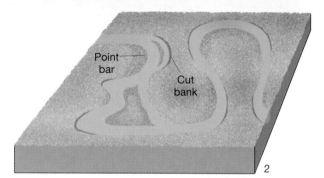

Point bar

Cut bank

2

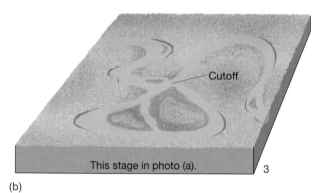

Cutoff

This stage in photo (a).

3

Oxbow lake

4

(b)

FIGURE 11-10

The evolution of meanders into an oxbow lake. (a) Itkillik River in Alaska; (b) development of a river meander and oxbow lake simplified in four stages. (c) Carter Lake, Iowa, sits within the curve of a former meander that was cut off by the Missouri River. The city and oxbow lake remain part of Iowa even though they are stranded within Nebraska. [(a) U.S. Geological Survey photo.]

352

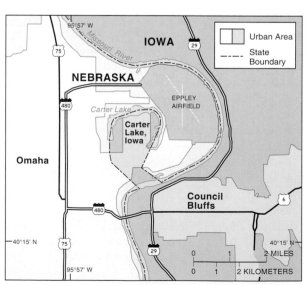

(c)

Idealized cross section of the longitudinal profile of a stream showing its gradient. Upstream segments have a steeper gradient; downstream the gradient is more gentle.

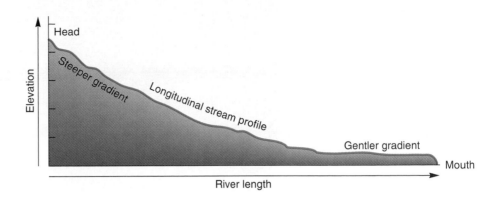

Stream Gradient

Every stream has a degree of inclination or **gradient**, which is the decline in elevation from its headwaters to its mouth. A stream's gradient generally forms a concave-shaped slope (Figure 11-11). Characteristically, the *longitudinal profile* of a stream (a side view) features a steeper slope upstream and a more gradual slope downstream. This curve assumes its shape for complex reasons related to the stream's ability to do just enough work to accomplish the transport of the load it receives.

A **graded stream** condition occurs when the load carried by the stream and the landscape through which it flows become mutually adjusted (balanced), forming a state of *dynamic equilibrium* among erosion, transported load, deposition, and the stream's capacity. Dynamic equilibrium infers that the stream and landscape work together to maintain this balance.

Both high-gradient and low-gradient streams can achieve a graded condition. The difference in gradient results from variation in each stream's discharge and the nature of the transported load. A stream's profile tells geographers about characteristics of slope, discharge, and load:

> A graded stream is one in which, over a period of years, *slope* is delicately adjusted to provide, with available *discharge* and with prevailing *channel characteristics,* just the velocity required for transportation of the load supplied from the drainage basin.*

---

* J. H. Mackin, "Concept of the Graded River," *Geological Society of America Bulletin,* 59 (1948): 463.

Attainment of a graded condition does *not* mean that the stream is at its lowest gradient, but rather that it represents a balance among erosion, transportation, and deposition over time along a specific portion of the stream.

One problem with applying the graded stream concept is that an individual stream can have both graded and ungraded portions. It may have graded sections without having an overall graded slope. In fact, variations and interruptions are the rule rather than the exception.

Stream gradient is affected by tectonic uplift of the landscape, or a change in base level. Such changes stimulate renewed erosional activity. If tectonic forces slowly lift the landscape the stream gradient will increase. Imagine this occurring to the landscape in Figure 10-2. The region and stream flowing through the landscape become *rejuvenated.* With rejuvenation, river meanders actively return to downcutting and become **entrenched meanders** in the landscape (Figure 11-12).

Nickpoints. When the longitudinal profile of a stream shows an abrupt change in gradient, such as at a waterfall or an area of rapids, the point of interruption is termed a **nickpoint** (also spelled knickpoint). At a nickpoint, the conversion of potential energy to kinetic energy is concentrated and in turn works to eliminate the nickpoint. Figure 11-13 shows a stream with two such interruptions. Nickpoints can result from a stream flowing across a zone of hard, resistant rock, or from various tectonic uplift episodes, such as might occur along a fault line. Temporary blockage in a channel, caused by a land-

FIGURE 11-12
Rejuvenated (uplifted) Colorado Plateau landscape is cut into by entrenched meanders of the San Juan River near Mexican Hat, Utah. [Aerial photo by Betty Crowell.]

slide or a logjam, also could be considered a nickpoint; when the logjam breaks, the stream quickly readjusts its channel to its former grade.

One of the more interesting and beautiful gradient breaks is a waterfall. At its edge, a stream becomes free-falling, moving at high velocity and causing increased abrasion on the channel below. This action generally undercuts the waterfall, and eventually the rock ledge at the lip of the fall collapses, causing the waterfall to shift a bit farther upstream (Figure 11-13). Thus, nickpoints migrate upstream. The height of the waterfall is reduced gradually as debris accumulates at its base.

At Niagara Falls on the Ontario-New York border, glaciers advanced and receded over the region, exposing resistant rock strata underlain by less resistant shales. As the less resistant material continues to weather away, the overlying rock strata collapse, allowing the falls to erode farther upstream toward Lake Erie (Figure 11-14). As this example demonstrates, a nickpoint should be thought of as a relatively temporary and mobile feature on the landscape.

Stream Deposition

Deposition is the next logical event after weathering, mass movement, erosion, and transportation. In *deposition,* a stream deposits alluvium, or unconsolidated sediments, thereby creating specific depositional landforms, such as floodplains, terraces, and deltas.

Floodplains. The flat low-lying area along a stream channel that is subjected to recurrent flooding is a **floodplain**. It is formed when the river overflows its channel during times of high flow. Thus, when floods occur, the floodplain is inundated. When the water recedes, alluvial deposits generally mask the under-

FIGURE 11-13
Longitudinal stream profile showing nickpoints produced by resistant rock strata. Potential energy is converted into kinetic energy and concentrated at the nickpoint, accelerating erosion, which eventually eliminates the feature.

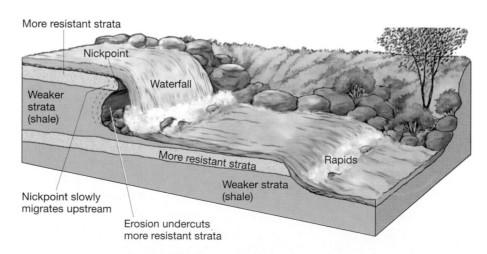

More resistant strata

Nickpoint

Waterfall

Weaker strata (shale)

More resistant strata

Rapids

Weaker strata (shale)

Nickpoint slowly migrates upstream

Erosion undercuts more resistant strata

lying rock. Figure 11-15 illustrates a characteristic floodplain, with the present river channel embedded in the plain's alluvial deposits.

On either bank of most streams, **natural levees** develop as by-products of flooding. When flood waters arrive, the river overflows its banks, loses velocity as it spreads out, and drops a portion of its sediment load to form the levees. Larger sand-sized particles drop out first, forming the principal component of the levees, with finer silts and clays deposited farther from the river. Next time you have an opportunity to see a river and its floodplain, look

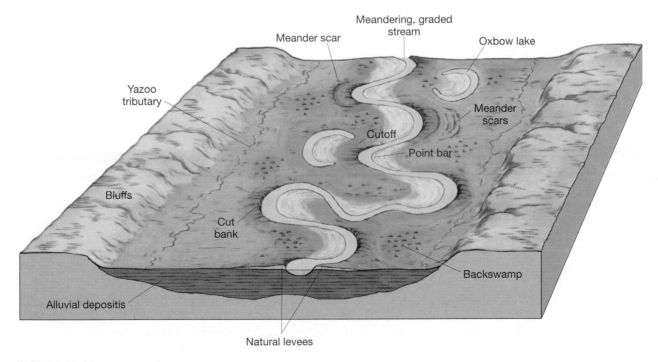

FIGURE 11-15
Typical floodplain landscape and related landscape features.

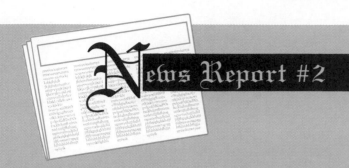

"Niagara Falls Closed for Inspection"

A waterfall is a break in stream gradient where energy is concentrated in the free-falling water. The ledge over which the waterfall cascades is undercut by this process. Niagara Falls is a fine example of this action, where these natural processes are laboring to eliminate the beautiful tourist attraction, reducing this portion of the river to a mere a series of rapids. The falls already have retreated more than 11 km (6.8 mi) from the steep face of the Niagara escarpment (cliff) during the past 12,000 years.

Millions of tourists visit these dramatic falls along the United States–Canadian border. Using control facilities upstream, flows over the American Falls at Niagara have been reduced at times to a trickle so that the area beneath the falls could be examined for possible reinforcement to save the waterfall. Although such inspections are infrequent, it is quite a shock to visitors to arrive and see the falls turned off!

---

for the levees. Successive floods increase the height of the levees and may even raise the overall elevation of the channel itself so that it is *perched* above the surrounding floodplain.

Notice on Figure 11-15 an area labeled *backswamp* and a stream called a *yazoo tributary*. The natural levees and elevated channel of the river prevent this tributary from joining the main channel, so it flows parallel to the river and through the backswamp area. (The name comes from the Yazoo River in the southern part of the Mississippi floodplain.)

People build cities on floodplains because they are nearly level and next to the water, despite the threat of flooding. People often are encouraged by government assurances of artificial protection from floods or of disaster assistance if floods occur. Government assistance may be provided in building artificial levees on top of natural levees. Artificial levees do increase the capacity in the channel, but they also lead to even greater floods when they are topped or when they fail. The catastrophic floods along the Mississippi River and its tributaries in 1993 illustrate the risk of floodplain settlement. Estimates of total damage from these floods exceed $30 billion (Figure 11-16).

Perhaps the best use of some floodplains is for agriculture, because inundation generally delivers nutrients to the land with each new alluvial deposit. A significant example is the Nile River in Egypt, where annual flooding enriches the soil. However, floodplains that are covered with coarse sediment—sand and gravel—are less suitable for agriculture. Are there river floodplains where you live? If so, how would you assess present land-use patterns, people's hazard perception, and local planning and zoning?

Stream Terraces. As explained earlier, several factors may rejuvenate stream energy and stream-landscape relationships so that a stream again scours downward with increased erosion. The resulting entrenchment of the river into its own floodplain produces **alluvial terraces** on either side of the valley, which look like topographic steps above the river. Alluvial terraces generally appear paired at similar elevations on either side of the valley (Figure 11-17). If more than one set of paired terraces is present, the valley probably has undergone more than one episode of rejuvenation.

If the terraces on either side of the valley do not match in elevation, then entrenchment actions must

CHAPTER 11 RIVERS AND RELATED LANDFORMS

FIGURE 11-16
1993 Midwest floods devastate neighborhoods in an Iowa city. [Photo by © Allan Tannenbaum, Sygma.]

have been continuous as the river meandered from side-to-side, with each meander cutting a terrace slightly lower in elevation. Thus, alluvial terraces represent an original depositional feature, a floodplain, which is subsequently eroded by a stream that has experienced changes in stream load and capacity.

River Deltas. The mouth of a river is where it reaches its base level. The river's forward velocity rapidly decelerates as it enters a larger body of water, with the reduced velocity causing its transported load to exceed its capacity. Coarse sediments like sand and gravel drop out first, with finer clays car-

ried to the extreme end of the deposit. This depositional plain formed at the mouth of a river is called a **delta**, named after the Greek letter *delta:* Δ, the triangular shape of which was perceived by Herodotus in ancient times to be similar to the shape of the Nile River delta.

Each flood stage deposits a new layer of alluvium over the surface of the delta so that it grows outward. At the same time river channels divide into smaller channels known as *distributaries,* which appear as a reverse of the dendritic drainage pattern discussed earlier. The Ganges River delta features an intricate braided pattern of distributaries. Alluvium

FIGURE 11-17
Alluvial terraces are formed as a stream cuts into alluvial fill. [After W. M. Davis, *Geographical Essays*, New York: Dover, 1964 (1909), p. 515.]

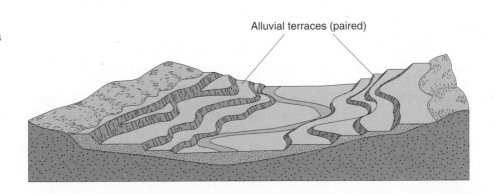

Alluvial terraces (paired)

"The 1993 Midwest Floods"

The Mississippi and Missouri Rivers are no strangers to flooding. The severity of each flood event is enhanced as more people settle on the vulnerable floodplains. The widespread floods of 1993 in the upper Mississippi and lower Missouri River Basins exceeded peak discharge records at nearly 100 gaging stations, making it one of the greatest floods in U.S. history.

In Spring 1993, a series of low-pressure systems (with their counterclockwise winds) stalled to the west, and high-pressure (with its clockwise winds) dominated the Eastern Seaboard, combining to produce a region of sustained convergence, instability, and thundershowers. Precipitation was one and one-half to two times normal for most cities in the flooded area. By late June, many reservoirs were filled and soils were saturated by record rainfall over the region.

As 10,000 km (6200 mi) of levees were overtopped, more than 1000 levees were breached, flooding many cities and towns. The result was a Presidential Disaster Declaration for 487 counties in Illinois, Iowa, Kansas, Minnesota, Missouri, the Dakotas, Nebraska, and Wisconsin. Some cities had prepared through improved levee construction and hazard zoning of susceptible floodplains. Many others had postponed local taxes for such action and had done little to prevent flood damage. This event was a painful reminder of the power of nature in our lives and the need for improved hazard perception.

carried from deforested slopes upstream provides excess sediment that forms the many deltaic islands (Figure 11-18). The Nile River forms an *arcuate* (arc-shaped) *delta* (Figure 11-19). Also arcuate in form are the Danube River delta in Romania as it enters the Black Sea and the Ganges and Indus river deltas. In another distinct form, the Tiber (Tevere) River in Italy has an *estuarian delta,* or one that is in the process of filling an **estuary**, which is the seaward mouth of a river where the river's freshwater encounters seawater.

The Mississippi River has produced a *bird-foot delta,* a long channel with many distributaries and sediments carried beyond the tip of the delta into the Gulf of Mexico. Over the past 120 million years the Mississippi has deposited sediments downstream all the way from southern Minnesota and Illinois. During the past 5000 years, the river has formed seven distinct deltaic complexes along the Louisiana coast. The seventh and current subdelta has been building for at least 500 years. Each lobe reflects distinct course changes in the Mississippi River, probably where the river broke through its natural levees. In 1966, Kolb and Lopik, two engineering geologists, prepared a map of this recent deltaic history, which shows the relatively smaller size and difference in configuration of the present delta (Figure 11-20a).

The Mississippi River delta is therefore dynamic over time. *Landsat* images taken in 1973 and 1989 demonstrate the changes that occurred over a 16-year span (Figure 11-20c and d). The main channel persists because of much effort and expense directed at maintaining the artificial levee system.

To further complicate this situation, the tremendous weight of the sediments in the Mississippi River is creating isostatic adjustments in Earth's crust. The entire region of the delta is subsiding, thereby placing ever-increasing stress on natural and artificial lev-

FIGURE 11-18
The complex distributary pattern in the "many mouths" of the Ganges River delta in Bangladesh and extreme eastern India is visible from orbit. [Space Shuttle photo courtesy of the National Aeronautics and Space Administration.]

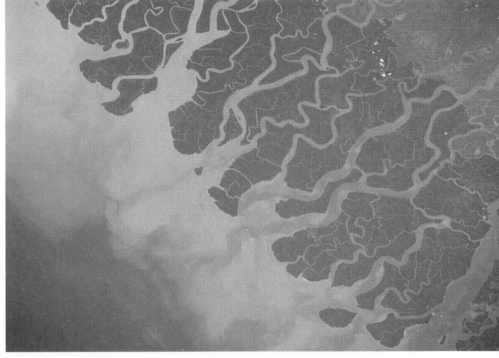

FIGURE 11-19
The arcuate Nile River delta. Intensive agricultural activity is noted in false-color (red) in the delta and along the Nile River floodplain. [Image courtesy of NASA.]

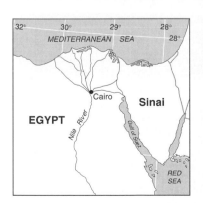

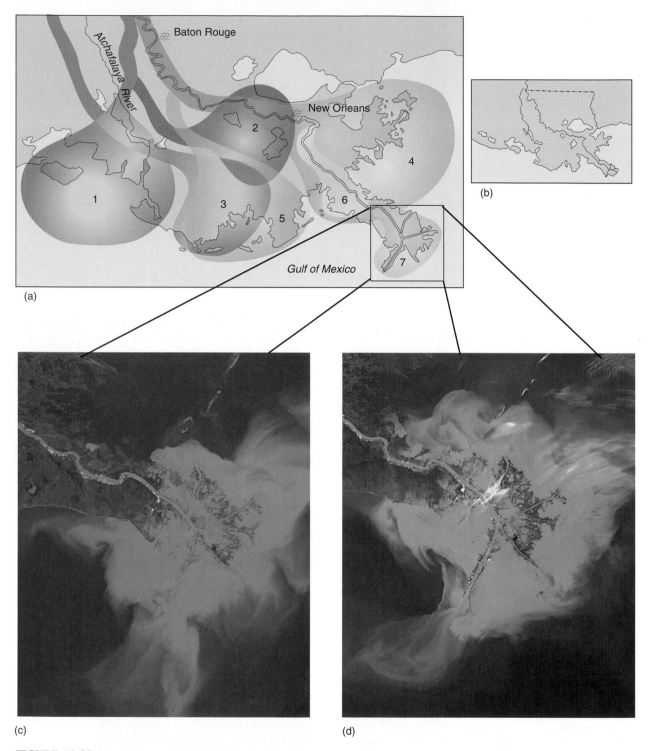

FIGURE 11-20
The Mississippi River delta. (a) Evolution of the present delta. (b) Location map and potential "capture point" (arrow) of Atchafalaya River. The bird-foot delta of the Mississippi River exhibits change over time, as shown in these two *Landsat* images from 1973 and 1989 (c and d). The continuous supply of sediments, focused by controlling levees, extends ever farther into the Gulf of Mexico, although subsidence of the delta and rising sea level have reduced the overall surface area. [(a) Adapted from Charles R. Kolb and Jack R. Van Lopik, "Depositional Environments of the Mississippi River Deltaic Plain," in Martha Lou Shirley, ed., *Deltas In Their Geologic Framework*, Houston: Houston Geological Society, 1966, p. 22. Adapted by permission. *Landsat* images courtesy of EROS Data Center.]

ees and other structures along the lower Mississippi. Severe problems are a certainty for existing and planned settlements unless further intervention or relocation efforts take place. Past protection and reclamation efforts by the U.S. Army Corps of Engineers apparently have only worsened the flood peril, as demonstrated by the 1993 floods in the Midwest.

An additional problem for the lower Mississippi Valley is the possibility, in a worst-case flood, that the river could break from its existing channel and seek a new route to the Gulf of Mexico. An obvious alternative is the Atchafalaya River, now blocked off from the Mississippi at 320 km (200 mi) from its mouth. It would be less than one-half the distance to the Gulf (see arrow in Figure 11-20b, inset). The principal causes of such an event would be more than just flood water and would include sediment deposition. Occurrence of a major flood is only a matter of time and residents should prepare for the river to change channel.

The Amazon River, which exceeds 175,000 m³/s (6.2 million cfs) discharge and carries sediments far into the Atlantic, lacks a true delta. Its mouth, 160 km (100 mi) wide, has formed an underwater deltaic plain deposited on a sloping continental shelf. As a result, the Amazon's mouth is braided into a broad maze of islands and channels (see Space Shuttle photo in Figure 6-15). Other rivers lack deltaic formations if they lack significant sediment or discharge into strong erosive currents. The Columbia River of the U.S. Northwest lacks a delta because of offshore currents.

Floods and River Management

Throughout history, civilizations have settled floodplains and deltas, especially since the agricultural revolution of 10,000 years ago when the fertility of floodplain soils was discovered. Early villages generally were built away from the area of flooding, or on stream terraces, because the floodplain was dedicated exclusively to farming. However, as commerce grew, sites near rivers became important for transportation. Port and dock facilities were built, as were river bridges. Also, because water is a basic industrial raw material used for cooling and for diluting and removing wastes, waterside industrial sites be-

came desirable. However, all these human activities on vulnerable flood-prone lands require planning to avoid disaster.

Catastrophic floods continue to be a threat especially in poor nations. In Bangladesh, intense monsoonal rains and tropical cyclones in 1988 and 1991 created devastating floods over the country's vast alluvial plain (130,000 km² or 50,000 mi²). One of the most densely populated countries on Earth, Bangladesh was more than *three-fourths* covered by floodwaters. Excessive forest harvesting in the upstream portions of the Ganges-Brahmaputra River watersheds increased runoff and added to the severity of the flooding. Over time the increased load carried by the river was deposited in the Bay of Bengal, creating new islands (see Figure 11-18). These islands, barely above sea level, became sites of new settlements, farming villages and many lost lives. When the recent floodwaters finally did recede, the lack of freshwater—coupled with crop failures, disease, and pestilence—led to famine and the death of tens of thousands of people. About 30 million people were left homeless and many of the alluvial islands had disappeared.

Floods and floodplains are rated statistically for the expected intervals between floods. A *10-year flood* is expected to occur once every 10 years (i.e., it has a 10% probability of occurring in any one year). Such a frequency labels a floodplain as one of moderate threat. A 50-year or 100-year flood is of greater and perhaps catastrophic consequence, but it is also less likely to occur in a given year. These statistical estimates are probabilities that events will occur randomly during any single year of the specified period. Of course, two decades might pass without a 10-year flood, or 10-year flood volumes could occur three years in a row. The record-breaking Mississippi River Valley floods in 1993 easily exceeded a 1000-year flood probability of occurrence. FYI Report 11-1 presents a closer look at floodplain management.

Streamflow Measurement

A **flood** is a high-water level that overflows the natural (or artificial) levees along any portion of a stream. Understanding flood patterns of a drainage

Floodplain Management Strategies

Detailed measurements of streamflows and floods have been kept rigorously in the United States only for about 100 years, in particular since the 1940s. At any selected location along any given stream, the *probable maximum flood* (PMF) is a hypothetical flood of such a magnitude that there is virtually no possibility it will be exceeded. Because floods are produced by the collection and concentration of rainfall, hydrologists speak of a corollary, the *probable maximum precipitation* (PMP) for a given drainage basin, which is an amount of rainfall so great that it will never be exceeded.

These parameters are used by hydrologic engineers to establish a *design flood* against which to take protective measures. For urban areas near creeks, planning maps often include survey lines for a 50-year or a 100-year floodplain; such maps have been completed for most U.S. urban areas. The design flood usually is used to enforce planning restrictions and special insurance requirements. Unfortunately, the scenario all too often goes like this: (1) minimal zoning precautions are not carefully supervised; (2) a flooding disaster occurs; (3) the public is outraged at being caught off guard; (4) businesses and homeowners are surprisingly resistant to stricter laws and enforcement; and (5) eventually another flood refreshes the memory and promotes more knee-jerk planning. As strange as it seems, *there is little indication that our risk perception improves as the risk increases.*

Strategies

A strategy in some larger river systems is to develop artificial floodplains. This is done by constructing *bypass channels* to accept seasonal or occasional floods. When not flooded, the bypass channel can serve as farmland, often benefiting from the occasional soil-replenishing inundations. When the river reaches flood stage, large gates called *weirs* are opened, allowing the water to enter the bypass channel. This alternate route relieves the main channel of the burden of carrying the entire discharge.

The benefit of any levee, bypass, or other project intended to prevent flood destruction is measured in *avoided damage* and is used to justify the cost of the protection facility. Thus, ever-increasing damage leads to the justification of ever-increasing flood control structures. All such strategies are subjected to cost-benefit analysis, but bias is a serious drawback because such an analysis usually is prepared by an agency or bureau with a vested interest in building more flood-control projects.

Dams and reservoirs are common streamflow control structures within a watershed. For conservation purposes, a *dam* holds back seasonal peak flows for distribution during low-water periods. In this way streamflows are regulated to assure year-round water supplies. Dams also are constructed for flood control, to hold back excess flows for later release at more moderate discharge levels. Adding *hydroelectric power* production to these functions of conservation and flood control can define a modern multipurpose reclamation project.

The function of a *reservoir* is to provide flexible storage capacity within a watershed to regulate river flows, especially in a region with variable precipitation. Figure 1 shows one reservoir during drought conditions and during a time of wetter weather six years later.

Reservoir Considerations

Unfortunately, reservoir construction also involves negative consequences. The area upstream from a dam becomes permanently drowned. In mountainous regions, this may mean loss of white water rapids and recreational sections of a river. In agricultural areas the ironic end result may be that *a hectare of farmland is inundated upstream in order to preserve a hectare of farmland downstream*. Furthermore, dams built in warm and arid climates lose substantial water to evaporation, compared to the free-flowing streams they replace. Reservoirs in the southwestern United States can lose 3–4 meters

(a)

(b)

FIGURE 1
Comparative photographs of the
New Hogan reservoir, central
California, during (a) dry and (b)
wet weather conditions. [Photos
by author.]

(10–13 feet) of water a year. Also, sedimentation can reduce the effective capacity of a reservoir and can shorten a dam's life span.

Large multi-purpose projects invariably produce political conflict over territorial rights and questions of public trust versus private right to the environment. Vast scenes of environmental disruption for the sake of economic gain no longer appear popular with the public.

The James Bay Project in central and northern Québec is a case in point. Launched over 20 years ago by Hydro-Québec and only one-third complete at this time, the project might eventually include 215 dams, 25 power stations, and 20 river diversions. Many unexpected environmental problems have arisen because *no*

environmental impact studies were completed at the outset. The early stages remain a huge experiment with fragile ecosystems. The public learned well into the planning phase of corporate interests seeking inexpensive public power supplies and that much of the power was for export to the United States, all at public expense to Canadians. The second major phase of the James Bay Project, the Great Whale project, may never be completed because of the success of conservation programs begun by utilities in the northeastern United States and court challenges to assess impacts first before further construction. In addition, the state of New York in 1992 withdrew its offer to buy 1000 megawatts of power from the Hydro-Québec project.

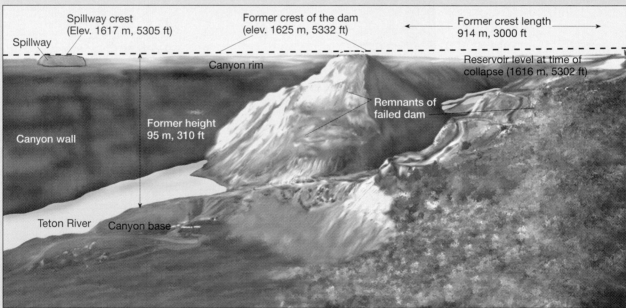

FIGURE 2
The failed remains of Teton Dam in Idaho. [Photo by author.]

Geologic Assessment. One final concern is the need for great care in geologic assessment of the dam site, for dam failures occur more often than many realize. For example, in 1972 two dams failed, one near Rapid City, South Dakota, and another at Buffalo Creek, West Virginia, killing 237 and 118 people, respectively. The General Accounting Office estimates that about 1900 unsafe dams exist near urban areas.

The Teton Dam, near Rexburg, Idaho, collapsed June 5, 1976, releasing more than 303 billion liters (80 billion gallons) of water, destroying 41,000 hectares (100,000 acres) of farmland, killing 16,000 head of livestock, and causing more than $1 billion in property damage (Figure 2). Congressional testimony at the time disclosed: "The principal human cause of failure of the Teton Dam was very poor site selection." After engineering surveys disclosed specific geologic problems at the chosen site, construction of Teton Dam continued anyway. The dam survived less than one month after filling began!

As suggested in an article titled "Settlement Control Beats Flood Control,"* there are other ways to protect populations than with enormous, expensive, sometimes environmentally disruptive projects. Strictly zoning the floodplain is one approach, but flat, easily developed floodplains near pleasant rivers are perceived as desirable for housing, and thus may weaken political resolve. A zoning strategy would set aside the floodplain for farming or passive recreation, such as a riverine park, golf course, or plant and wildlife sanctuary, or for other uses that are not hurt by natural floods. This study concludes that "urban and industrial losses would be largely obviated [avoided] by set-back levees and zoning and thus cancel the biggest share of the assessed benefits which justify big dams."

---

*Walter Kollmorgen, *Economic Geography* 29, no. 3 (July 1953): 215.

---

basin is as complex as understanding the weather, for floods and weather are equally variable, and both include a level of unpredictability. However, to develop the best possible flood management, the behavior of each large watershed and stream is measured and analyzed. Unfortunately, such data often are not available for small basins or for the changing landscapes of urban areas.

The key is to measure *streamflow*—the height and discharge of a stream (Figure 11-21). A *staff gauge,* a pole placed in a stream bank and marked with water heights, is used to measure stream level. With a fully measured cross section, stream level can be used to determine discharge (discharge is equal to width times depth times velocity). A *stilling well* is sited on the stream bank and a gauge is mounted in it to measure stream level. A movable current meter can be used to sample velocity at various locations. Approximately 11,000 stream gaging stations are used in the United States (an average of over 200 per state). Of these, 7000 have continuous recorders for stage and discharge, operated by the U.S. Geological Survey. Many of these stations automatically send telemetry data to satellites, from which information is retransmitted to regional centers. Environment Canada's Water Survey of Canada maintains more than 3000 gaging stations.

If we study streamflow measurements, we can understand channel characteristics as conditions vary. A graph of stream discharge over a time period for a specific place is called a **hydrograph**. The hydrograph in Figure 11-22 shows the relation between precipitation input and stream discharge. During dry periods, at low-water stages, the flow is described as *base flow* and is largely maintained by input from local groundwater. When rainfall occurs in some portion of the watershed, the runoff collects and is concentrated in streams and tributaries. The amount, location, and duration of the rainfall episode determine the *peak flow*. Also important is the nature of the surface in a watershed; for example, a hydrograph for a specific portion of a stream changes after a forest fire or urbanization of the watershed.

Human activities have enormous impact on water flow in a basin. The effects of urbanization are quite dramatic, both increasing and hastening peak flow as you can see by studying Figure 11-22. In fact, urban areas produce runoff patterns quite similar to

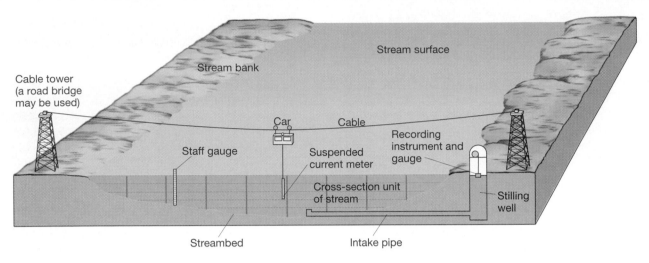

FIGURE 11-21

A typical streamflow measurement installation: staff gauge, stilling well with recording instrument, and suspended current meter.

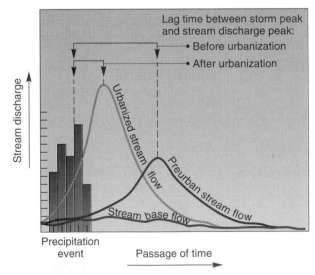

FIGURE 11-22

Effect of urbanization on a typical stream hydrograph. Normal base flow is indicated with a dark blue line. The purple line indicates discharge after a storm, prior to urbanization. Following urbanization, stream discharge dramatically increases and peaks earlier, as shown by the light blue line.

those of deserts. The sealed surfaces of the city drastically reduce infiltration and soil moisture recharge, behaving much like the hard, nearly barren surfaces of the desert.

SUMMARY—Rivers and Related Landforms

River systems, **fluvial** processes and landscapes, **floodplains**, and river control strategies are important topics in physical geography as populations inhabit risky areas and as demands for limited water resources increase. The basic fluvial system is a **drainage basin**, an open system. Denudation by wind, water, and ice that dislodges, dissolves, or removes surface material is called **erosion**. Thus, streams produce *fluvial erosion,* which supplies weathered and wasted sediments for **transport** to new locations, where they are laid down in a process known as **deposition**. A stream is a mixture of water and solids—carried in solution, suspension, and by mechanical transport. **Alluvium** is the general term for the clay, silt, and sand transported by running water.

Overland flow gathers and concentrates into the main stream channel in a manner consistent with topographic relief, the nature of rock and materials through which the stream flows, climate, vegetation, and the types of processes at work in a drainage basin.

Stream transport of materials occurs as **dissolved load** in solution, as **suspended load** held aloft in the stream, and as **bed load** dragged along the stream bed by **traction**, or rolled and bounced along by **saltation**.

All streams have a **gradient** and work to establish a graded profile over distance, with interruptions triggering adjustments. Slope is the critical factor in a **graded stream's** maintenance of an equilibrium condition between water and solid materials. Stream form and operation result from complex interactions of slope, discharge, load, and channel characteristics, all variable within different climates and with different rock types. These functional considerations are embodied in the *dynamic equilibrium* approach to understanding the fluvial landscape.

Various landforms are associated with the action of flowing water: **terraces**, **levees**, **deltas**, and **floodplains**. Floodplains have been an important site of human activity throughout history. Rich soils, bathed in fresh nutrients by floodwaters, attract agricultural activity and urbanization. Floodplains are settled, despite our knowledge of historical devastation by **floods**, raising issues of human hazard perception. Collective efforts by government agencies undertake to reduce flood probabilities. Society is still learning how to live in a sustainable way with Earth's dynamic river systems.

KEY TERMS

abrasion	floodplain
aggradation	fluvial
alluvial terrace	graded stream
alluvium	gradient
bed load	hydraulic action
braided stream	hydrograph
continental divide	meandering stream
cut bank	natural levee
delta	nickpoint
deposition	oxbow lake
dissolved load	point bar
drainage basin	saltation
drainage pattern	sheet flow
entrenched meander	suspended load
erosion	traction
estuary	transport
flood	watershed

REVIEW QUESTIONS

1. Define the term *fluvial*. What is a fluvial process?
2. What is the basic spatial geomorphic unit of an individual river system? How is this unit defined? Give an example of one described in the text.
3. Follow the river systems from Pennsylvania to the Gulf of Mexico in Figure 11-1, and analyze the pattern of tributaries, and describe the channel.
4. Describe common drainage basin patterns.
5. What was the impact of flood discharge on the channel of the San Juan River near Bluff, Utah? Why did these changes take place?
6. How does stream discharge do its erosive work? What are the processes at work on the channel?
7. Differentiate between stream competence and stream capacity.
8. How does stream transport of sediments occur? What processes are at work?
9. Describe the flow characteristics of a meandering stream. What is the nature of the flow in the channel, the erosional and depositional features, and the typical landforms created?
10. Explain these statements: (a) all streams have a gradient but not all streams are graded, and (b) graded streams may have ungraded segments.
11. How is Niagara Falls an example of a nickpoint? Without human intervention, what do you think would eventually take place at Niagara Falls?
12. Describe the evolution of a floodplain. How are natural levees, oxbow lakes, backswamps, and yazoo tributaries formed?
13. What is a river delta? What are the various deltaic forms? Give some examples.
14. Has the Mississippi delta been stable for long? Explain.
15. Describe the Ganges River delta. What factors upstream explain its form and pattern? Assess the consequences of settlement on this delta.
16. What do you see as the major consideration regarding floodplain management? How would you describe the general attitude of society toward natural hazards and disasters?
17. Specifically, what is a flood? How are such flows measured and tracked?
18. Differentiate between a hydrograph from a natural terrain and one from an urbanized area.

View over Canyonlands National Park to the distant Manti-La Sal Mountains. [*Photo by author.*]

12

WIND PROCESSES AND DESERT LANDSCAPES

THE WORK OF WIND
    Eolian Erosion
    Eolian Transportation
    Eolian Depositional Landforms
    Loess Deposits
OVERVIEW OF DESERT LANDSCAPES
    Desert Climates
    Desert Fluvial Processes
    Desert Landscapes
SUMMARY
FYI REPORT 12-1 THE COLORADO RIVER: A SYSTEM
    OUT OF BALANCE

Wind is an agent of erosion, transportation, and deposition. Its effectiveness has been the subject of much debate; in fact, wind at times was thought to produce major landforms. Presently, wind is regarded as a relatively minor weathering and erosion agent, but it is significant enough to deserve our attention. Wind processes modify and move material accumulations in deserts and along coastlines. The wind contributes to soil formation in distant places, bringing fine material from regions where glaciers deposited it. Elsewhere, fallow fields (those not planted) give up their soil resource to destructive wind erosion. In this chapter we examine the work of wind, associated processes, and resulting landforms.

Water remains the major erosional force in the desert, but an overall lack of moisture and stabilizing vegetation allows wind to create extensive sand seas and dunes of infinite variety. The polar regions are deserts as well, with unique features related to their cold, dry environment—aspects covered in Chapter 14.

Arid landscapes display unique landforms and life forms: ". . . instead of finding chaos and disorder the observer never fails to be amazed at a simplicity of form, an exactitude of repetition and a geometric order. . . "* in the desert.

The Work of Wind

The work of the wind—erosion, transportation, and deposition—is called **eolian** (also spelled *aeolian;* for Aeolus, ruler of the winds in Greek mythology). Much eolian research was accomplished by a British major, Ralph Bagnold, who was stationed in Egypt in 1925. An engineering officer who spent much of his time in the deserts west of the Nile, Bagnold measured, sketched, and developed hypotheses about the wind and desert forms. His often-cited work, *The Physics of Blown Sand and Desert Dunes,* was published in 1941 following the completion in London of wind-tunnel simulations of windy desert conditions.

*From Ralph A. Bagnold, *The Physics of Blown Sand and Desert Dunes,* Methuen, 1941.

The actual ability of wind to move materials is small compared with that of other transporting agents such as water and ice, because air is so much less dense than these other media. Yet over time, wind accomplishes enormous work. Bagnold studied the ability of wind to transport sand over the surface of a dune. Wind of 50 kmph (30 mph) can move approximately one-half ton of sand per day over a one-meter-wide section of dune!

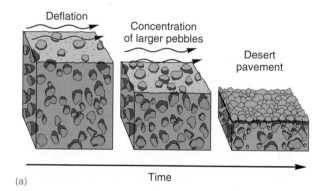

(a)

(b)

FIGURE 12-1
Desert pavement. (a) Desert pavement is formed from larger rocks and fragments left after deflation and sheetwash; (b) a typical desert pavement. [Photo by author.]

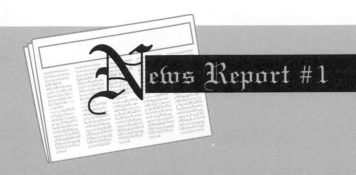

"War Tears up the Pavement"

A serious environmental impact of the 1991 Persian Gulf War was the disruption of desert pavement. Thousands of square kilometers of stable desert pavement were shattered by bombardment with many thousands of tons of explosives and disrupted by the movement of heavy equipment. The resulting loosened sand and silt is now available for deflation, and thus threatens cities and farms with increased dust and sand accumulations. Given the firepower available in modern technological warfare, environmental assessment would be a wise addition to strategic planning.

Eolian Erosion

Two principal wind-erosion processes are **deflation**, the removal and lifting of individual loose particles, and **abrasion**, the grinding of rock surfaces with a "sandblasting" action by particles captured in the air. Deflation and abrasion produce a variety of distinctive landforms and landscapes.

Deflation. Deflation literally blows away loose or noncohesive sediment. **Desert pavement** is formed from pebble and gravel concentration left behind after wind deflation and water washes away fine materials, and concentrates and cements remaining rock pieces. Resembling a cobblestone street, desert pavement protects underlying sediment from further deflation (Figure 12-1). Desert pavements are so common that many provincial names have been used for them—for example, *gibber plain* in Australia, *gobi* in China, and in Africa, *lag gravels* or *serir* (or *reg* desert if some fine particles remain).

Heavy recreational activity damages fragile desert landscapes. Over 14 million off-road vehicles (ORVs) are in use in the United States. Such vehicles crush plants and animals, disrupt desert pavement, promote deflation, and create ruts that easily concentrate sheetwash to form gullies. Measures to restrict their use to specific areas, preserving the remaining desert, are controversial.

Wherever wind encounters loose sediment, deflation also may form basins. Called **blowout depressions**, these range from small indentations of less than a meter up to areas hundreds of meters wide and many meters deep. Chemical weathering, although operating slowly in the desert, is important in the formation of a blowout, for it removes the cementing materials that give particles their cohesiveness. Large depressions occurring in the Sahara are at least partially formed by deflation. The enormous Qattara Depression just inland from the Mediterranean Sea in the Western Desert of Egypt, which covers 18,000 km² (6950 mi²), is now about 130 m (427 ft) below sea level at its lowest point.

Abrasion. Sandblasting is commonly used to clean stone surfaces on buildings or to remove unwanted markings from streets. Abrasion by wind-blown particles is nature's version of sandblasting and is especially effective at polishing exposed rocks when the abrading particles are hard and angular. Variables that affect the rate of abrasion include the hardness of vulnerable surface rocks and the wind velocity and constancy. Abrasive action is restricted to the area immediately above the ground, usually no more

"Yardangs from Mars"

Wind deflation and abrasion streamline rock structures and leave behind distinctive, elongated yardangs. On Earth some yardangs are large enough to be detected on satellite imagery. The Ica Valley of southern Peru has yardangs reaching 100 m (330 ft) in height and several kilometers in length, and yardangs in the Lut Desert of Iran attain 150 m (490 ft) height.

Spacecraft in orbit around Mars use remote sensors to study the Martian surface. Images have disclosed curious features on Mars that suggest yardangs, or wind-sculpted formations. This provides clues as to the nature of the Martian surface: a windy place of frequent dust storms, and enough wind-blown sediment to sandblast yardangs.

---

than a meter or two in height because sand grains are lifted only a short distance.

Rocks exposed to eolian abrasion appear pitted, grooved, or polished and usually are aerodynamically shaped in a specific direction, according to the flow of airborne particles. Rocks that bear such evidence of eolian erosion are called **ventifacts**.

On a larger scale, deflation and abrasion are capable of streamlining rock structures that are aligned parallel to the most effective wind direction, leaving behind distinctive, elongated ridges called *yardangs*. These can range from meters to kilometers in length and up to many meters in height. Abrasion is concentrated on the windward end of each yardang, with deflation operating on the leeward portions. The Sphinx in Egypt perhaps partially formed as a yardang, suggesting a head and body to the ancients. Some scientists think this shape led them to complete the bulk of the sculpture artificially with masonry.

Eolian Transportation

As mentioned in Chapter 4, atmospheric circulation is capable of transporting fine material such as volcanic debris worldwide within days. The distance that wind is capable of transporting particles varies greatly with particle size. Wind exerts a drag or frictional pull on surface materials. Only the finest dust particles travel significant distances, and so the finer material suspended in a *dust storm* is lifted much higher than the coarser particles of a *sand storm,* which may be lifted only about 2 m (6.5 ft). People living in areas of frequent dust storms are faced with very fine particles infiltrating their homes and businesses through even the smallest cracks. (Figure 2-23 illustrates such a dust storm in Nevada and blowing alkali dust in the Andes.) People living in desert regions, where frequent sand storms occur, contend with the sandblasting of painted surfaces and etched window glass.

Deflation and wind transport of soil produced a catastrophe called the American Dust Bowl of the 1930s. Overgrazing and intensive agricultural left soil vulnerable. The transported dust darkened the skies of midwestern cities and the lives of millions.

Through processes of weathering, erosion, and transportation, mineral grains are removed from parent rock and redistributed elsewhere. In Figure 10-5b, you can see the relationship between the composition and color of the sandstone in the background and the derived sandy surface in the foreground. Wind action is not significant in the weathering process that frees individual grains of sand from the parent rock, but it is active in relocating the weathered grains.

The term *saltation* was used in Chapter 11 to describe movement of particles by water. The term also describes the wind transport of grains along the ground, grains usually larger than 0.2 mm (0.008 in.). About 80% of wind transport of particles is accomplished by this skipping and bouncing action (Figure 12-2). Compared with fluvial transport, in which saltation is accomplished by hydraulic lift, eolian saltation is executed by aerodynamic lift, elastic bounce, and impact (compare Figures 12-2 and 11-7).

Saltating particles crash into other particles, knocking them both loose and forward. This causes **surface creep**, which slides and rolls particles too large for saltation and affects about 20% of the material being transported. Once in motion, particles continue to be transported by lower wind velocities. In a desert or along a beach, you can hear the myriad saltating grains of sand produce a slight hissing sound, almost like steam escaping, as they bounce along and collide with surface particles.

Sand erosion and transport from a beach are slowed by conservation measures such as the introduction of stabilizing native plants, the use of fences, and the restriction of pedestrian traffic to walkways (Figure 12-3).

Eolian Depositional Landforms

The smallest features shaped by individual saltating grains are *ripples* (Figure 12-4). Ripples form in crests and troughs positioned transversely (at a right angle) to the direction of the wind. Their formation is influenced by the length of time particles are airborne. Eolian ripples are different from fluvial ripples because the impact of saltating grains is very slight in water.

A common assumption is that most deserts are covered by sand. Instead, desert pavements predominate across most subtropical arid landscapes; only about 10% of desert areas are covered with sand. Sand grains generally are deposited as transient ridges or hills called dunes. A **dune** is a wind-sculpted accumulation of sand. An extensive area of dunes, such as that found in North Africa, is characteristic of an **erg desert**, which means **sand sea**. The Grand Erg Oriental in the central Sahara exceeds 1200 m (4000 ft) in thickness and covers 192,000 km² (75,000 mi²). This sand sea has been active for over 1.3 million years and has average dune heights of 120 m (400 ft). Similar sand seas are active in Saudi Arabia, the Ar Rub' al Khālī Erg (Figure 12-5) is an example.

FIGURE 12-2
Eolian saltation and surface creep are forms of sediment transportation.

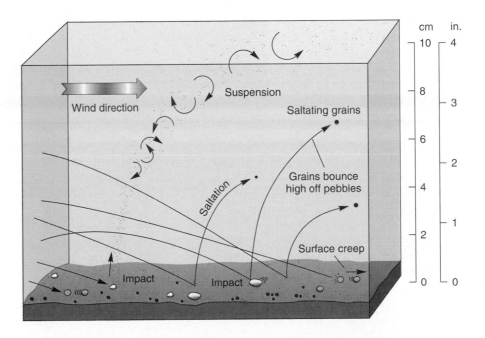

FIGURE 12-3
Preventing further erosion and
transport of coastal dunes
through stabilizing native plants,
fences, and walkways. [Photo
by author.]

FIGURE 12-4
Myriad sand ripple patterns later
may become lithified into fixed
patterns in rock. The area in the
photo is approximately 1 m wide.
[Photo by author.]

Dune Movement and Form. Dune fields, whether
in arid regions or along coastlines, tend to migrate in
the direction of effective, sand-transporting winds. In
this regard, stronger seasonal winds or winds from a
passing storm may prove more effective than average prevailing winds. When saltating sand grains encounter small patches of sand their kinetic energy
(motion) is dissipated, and they accumulate. As
height increases above 30 cm (12 in.) a **slipface** and
characteristic dune features form.

Study the dune in Figure 12-6 and you can see
that winds characteristically create a gently sloping
windward side (stoss side), and a more steeply
sloped *slipface* on the *leeward side*. A dune usually
is asymmetrical in one or more directions. The angle
of a slipface is the angle at which loose material is
stable—its *angle of repose*. Thus, the constant flow
of new material makes a slipface a type of
avalanche slope. As sand moves over the crest of the
dune to the brink, it builds up and avalanches as the
slipface continually adjusts, seeking its angle of repose (usually 30–34°). In this way, a dune migrates
downwind with the effective wind, as suggested by
the successive dune profiles in Figure 12-6.

Dunes that move actively are called *freedunes* and
reflect most dynamically the interaction between

FIGURE 12-5

The Grand Ar Rub' al Khālī Erg that dominates southern Saudi Arabia. Effective southwesterly winds shape the pattern and direction of the transverse and barchanoid dunes. [SPOT image by CNES, Reston, Virginia. Used by permission.]

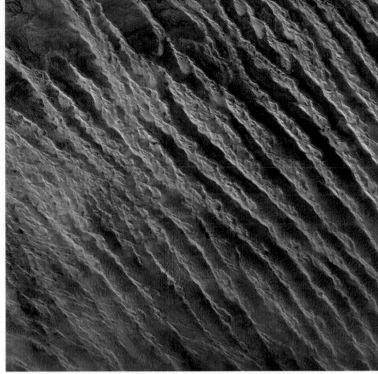

fluid atmospheric winds and moving sand. However, because the sand is moving close to the ground, it may encounter an obstruction such as stabilizing vegetation or a rock outcrop, resulting in a *tied dune,* or one that is fixed in place.

> . . . I see hills and hollows of sand like rising and falling waves. Now at midmorning, they appear paper white. At dawn they were fog gray. This evening they will be eggshell brown.*

*Janice E. Bowers, *Seasons of the Wind*. Flagstaff, AZ: Northland Press, 1985, p. 1.

The ever-changing form of these eolian deposits is part of their beauty, but their many wind-shaped styles have made classification difficult. We can simplify dune forms into three classes—*crescentic, linear,* and *star* dunes (Figure 12-7).

Crescentic dunes are crescent-shaped ridges of sand that form in response to a fairly unidirectional wind pattern. The crescentic group is most common, with related forms including *barchan* dunes (limited sand), *transverse* dunes (abundant sand), *parabolic* dunes (vegetation controlled), and *barchanoid* ridges (rows of coalesced barchans).

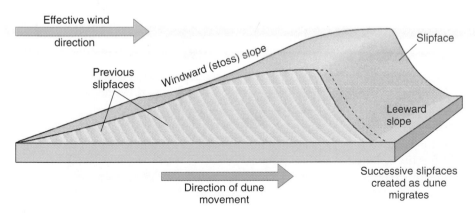

FIGURE 12-6

Dune cross section, showing the pattern of successive slipfaces as the dune migrates in the direction of the effective wind.

Class	Type	Description
Crescentic	Barchan	Crescent-shaped dune with horns pointed downwind. Winds are constant with little directional variability. Limited sand available. Only one slipface. Can be scattered over bare rock or desert pavement or commonly in dune fields.
	Transverse	Asymmetrical ridge, transverse to wind direction (right angle). Only one slipface. Results from relatively ineffective wind and abundant sand supply.
	Parabolic	Role of anchoring vegetation important. Open end faces upwind with U-shaped "blow-out" and arms anchored by vegetation. Multiple slipfaces, partially stabilized.
	Barchanoid Ridge	A wavy, asymmetrical dune ridge aligned transverse to effective winds. Formed from coalesced barchans; looks like connected crescents in rows with open areas between them.

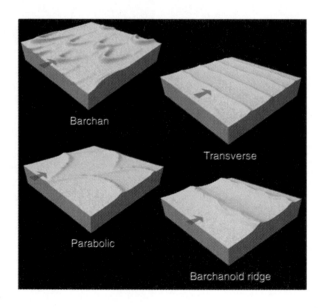

FIGURE 12-7

Major classes of dune forms. Arrows show wind direction. [Adapted from Edwin D. McKee, *A Study of Global Sand Seas*, U.S. Geological Survey Professional Paper 1052. Washington, DC: Government Printing Office, 1979.]

Linear dunes generally form long parallel ridges separated by sheets of sand or bare ground. Linear dunes characteristically are much longer than they are wide; some exceed 100 km (60 mi) in length. Winds producing these dunes are principally bidirectional, so the slipface alternates from side to side. Related forms include *longitudinal dunes* and the *seif* (Arabic for "sword") which is a sharper, narrower linear dune with a more sinuous crest.

Star dunes are the mountainous giants of the sandy desert. They form in response to complicated, changing wind patterns and have multiple slipfaces. They are pinwheel-shaped, with several radiating arms rising and joining to form a common central peak. The best examples are in the Sahara, where they approach 200 m (650 ft) in height (Figure 12-8).

These same dune-forming principles and terms (e.g., dune, barchan, slipface) apply to snow-covered

Class	Type	Description
Linear	Longitudinal	Long, slightly sinuous. ridge-shaped dune, aligned parallel with the wind direction; two slipfaces. Can be 100 m high and 100 km long. The "draas" at the extreme is up to 400 m high. Results from strong effective winds varying in one direction.
	Seif	After Arabic word for "sword"; a more sinuous crest and shorter than longitudinal dunes. Rounded toward upwind direction and pointed downwind.
Star Dune	Star	The giant of dunes. Pyramidal or star-shaped with 3 or more sinuous radiating arms extending outward from a central peak. Slipfaces in multiple directions. Results from effective winds shifting in all directions. Tends to form isolated mounds in high effective winds and connected sinuous arms in low effective winds.
Other	Dome	Circular or elliptical mound with no slipface. Can be modified into barchanoid forms.
	Reversing	Asymmetrical ridge form intermediate between star dune and transverse dune. Wind variability can alter shape between forms.

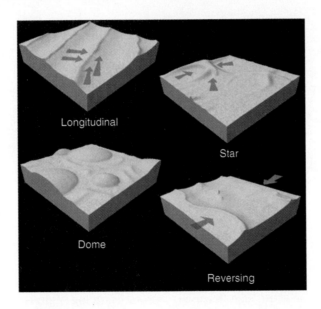

landscapes. *Snow dunes* form as wind deposits snow in drifts. In semiarid farming areas, fences, and tall stubble left in the fields capture drifting snow which contributes significantly to soil moisture upon melting.

Loess Deposits

Approximately 15,000 years ago, in several episodes, Pleistocene glaciers retreated in many parts of the world, leaving behind large glacial outwash deposits of fine-grained clays and silts (<0.06 mm or 0.0023 in.). These materials were blown great distances by the wind and redeposited in unstratified, homogeneous deposits, named **loess** by peasants working along the Rhine River Valley in Germany. No specific landforms were created; instead, loess covered existing landforms with a thick blanket of material that assumed the general topography of the existing landscape. Because of

FIGURE 12-8
Star dune in the Namib Desert of Namibia in southwestern Africa. [Photo by Comstock, Inc.]

its own binding strength, loess weathers and erodes into steep bluffs, or vertical faces. At Xian (Shaanxi) in China, a loess wall has been excavated for dwelling space (Figure 12-9a). When a bank is cut into a loess deposit, it generally will stand vertically, although it can fail if saturated (Figure 12-9b).

Figure 12-10 shows the worldwide distribution of loess deposits. Significant accumulations of loess throughout the Mississippi and Missouri valleys form continuous deposits 15–30 m (50–100 ft) thick. Loess deposits also occur in eastern Washington State and Idaho. This silt explains the fertility of the soils in these regions, for loess deposits are well drained, deep, and have excellent moisture retention. Loess deposits also cover much of Ukraine, central Europe, China, the Pampas-Patagonia regions of Argentina, and lowland New Zealand. These soils derived from loess are some of Earth's "bread basket" farming regions.

In Europe and North America, loess is thought to be derived mainly from glacial sources. The vast deposits of loess in China, covering more than 300,000 km² (115,800 mi²), are thought to be derived from wind-blown desert sediment rather than glacial sources. Accumulations in the Loess Plateau of China exceed 300 m (1000 ft) thickness, forming some complex weathered badlands and some good agricultural land. These wind-blown deposits are interwoven with much of Chinese history. New research also has discovered that plumes of wind-blown dust from African deserts have moved across the Atlantic Ocean to enrich soils of the Amazon rain forest of South America, the southeastern United States, and the Caribbean islands.

(a)

(b)

FIGURE 12-9

Loess deposits. (a) Loess formation in Xian (Shaanxi), China, has strong enough structure to be excavated for dwelling rooms. (b) A loess bluff along the Arikaree River in extreme northwestern Cheyenne County, Kansas. [(a) Photo by Betty Crowell, (b) photo by Steve Mulligan.]

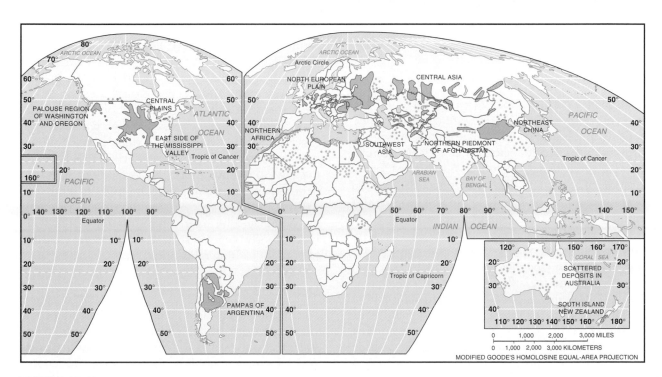

FIGURE 12-10

Worldwide loess deposits. Dots represent small scattered loess formations. [After R. E. Snead, *Atlas of World Physical Features*, p. 138, copyright © 1972 by John Wiley & Sons. Adapted by permission of John Wiley & Sons, Inc.]

"The Dust Bowl"

Deflation and wind transport of loess soils produced a catastrophe in the American Great Plains in the 1930s–the Dust Bowl. Over a century of overgrazing and intensive agriculture left soil susceptible to drought and eolian processes. The deflation of many centimeters of soil occurred in southern Nebraska, Kansas, Oklahoma, Texas, and eastern Colorado.

Fine sediments were lifted by winds to form severe dust storms. The transported dust darkened the skies of midwestern cities and drifted over farmland. Streetlights were left on throughout the day in Kansas City, St. Louis, and other midwestern cities and towns. Such episodes can devastate economies, cause tremendous loss of topsoil, and even bury farmsteads. Southeastern Australia experienced severe dust storms in 1993 that included consequences similar to those of the American Dust Bowl.

Overview of Desert Landscapes

Desert Climates

Dry climates occupy about 26% of Earth's land surface and, if all semiarid climates are considered, perhaps as much as 35% of all land, constituting the largest single climatic region on Earth (see Figure 7-4 and the climate map included with this text for the location of these *BW arid deserts* and *BS semiarid steppe* climate regions, and Figure 12-11 for the distribution of these desert environments).

The spatial distribution of these dry lands is related to subtropical high-pressure cells between 15° and 35° N and S (see Figures 4-13 and 4-15), to rain shadows on the lee side of mountain ranges (see Figure 5-21), and to areas at great distance from moisture-bearing air masses, such as central Asia. Figure 12-11 portrays this distribution according to the modified Köppen climate classification used in this text. These areas possess unique landscapes created by the interaction of intermittent precipitation events, weathering processes, and wind. Rugged, hard-edged desert landscapes of cliffs and scarps contrast sharply with the vegetation-covered, rounded and smoothed slopes characteristic of humid regions.

Desert environments experience high sensible heat conditions and intense ground heating. Such areas receive a high input of insolation through generally clear skies and experience high radiative heat losses at night. A typical desert water balance shows high potential evapotranspiration (POTET) demand, low precipitation supply, and prolonged seasonal deficits (see, for example, Figure 6-9b for Phoenix, Arizona). Fluvial processes in the desert generally are dominated by intermittent running water, with hard, poorly vegetated desert pavement yielding high runoff during rainstorms.

Desert Fluvial Processes

Precipitation events in a desert may be rare indeed, a year or two apart, but when they do occur, a dry streambed fills with a torrent called a **flash flood**. Such channels may fill in a few minutes and surge briefly during and after a storm. Depending on the region, such a dry streambed is known as a **wash**, an *arroyo* (Spanish), or a *wadi* (Arabic). A desert highway that crosses a wash usually is posted to warn dri-

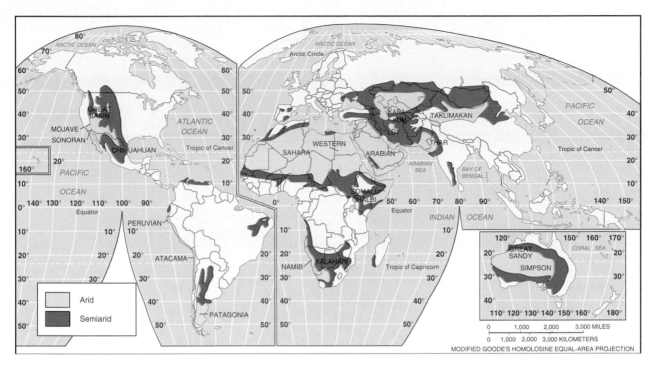

FIGURE 12-11
Worldwide distribution of arid lands (*BW arid desert climates*) and semiarid lands
(*BS semiarid steppe climates*) based on the Köppen climatic classification system.

vers not to proceed if rain is in the vicinity, for a flash flood can suddenly sweep away anything in its path.

When washes fill with surging flash flood waters, a unique set of ecological relationships quickly develops. Crashing rocks and boulders break open seeds that respond to the timely moisture and germinate. Other plants and animals also spring into brief life cycles as the water irrigates their limited habitats.

At times of intense rainfall, remarkable scenes fill the desert. Figure 12-12 shows two photographs taken just one month apart in a large sand dune field in Death Valley, California. The rainfall event that occurred produced 2.57 cm (1.01 in.) of precipitation in one day, in a place that receives only 4.6 cm (1.83 in.) in an average year. The runoff flowed for hours and then collected in low spots on the hard, underlying clay surfaces. The water was quickly consumed by the high evaporation demand so that in just a month these short-lived watercourses were dry and covered with accumulations of alluvial materials.

As runoff water evaporates, salt crusts may be left behind on the desert floor. This intermittently wet-and-dry low area in a region of closed drainage is called a **playa**; the site of an *ephemeral lake* when water is present. Accompanying our earlier discussion of evaporites, Figure 8-12 shows such a playa in Death Valley, covered with salt precipitate just one month after the record rainfall event mentioned here.

Permanent lakes and continuously flowing rivers are uncommon features in the desert, although the Nile River and the Colorado River are notable exceptions. Both these rivers are *exotic streams* with their headwaters in a wetter region. The Colorado River and its problem of overuse is described in detail in FYI Report 12-1.

Alluvial Fans. In arid climates, a prominent landform is the **alluvial fan**, which occurs at the mouth of a canyon where it exits into a valley. The fan is produced by flowing water that loses velocity as it leaves the constricted channel of the canyon and therefore drops layer upon layer of sediment along the base of the mountain block. Water then flows over the surface of the fan and produces a braided

(a)   (b)

FIGURE 12-12
The Stovepipe Wells dune field in Death Valley, California. (a) The day following a
2.57 cm (1.01 in.) rainfall and (b) the same location one month later. [Photos by author.]

drainage pattern, shifting from channel to channel with each precipitation event (Figure 12-13). A continuous apron, or **bajada** (Spanish for "slope") may form if individual alluvial fans coalesce into one sloping surface (see Figure 12-17). Fan formation of any sort is reduced in humid climates because perennial streams constantly carry away much of the alluvium.

An interesting aspect of an alluvial fan is the natural sorting of materials by size. Near the mouth of the canyon at the apex of the fan, coarser materials are deposited, grading slowly to pebbles and finer gravels with distance out from the mouth. Then sands and silts are deposited, with the finest clays and salts carried in suspension and solution all the way to the valley floor. Minerals dissolved in solution accumulate there as precipitates left by evaporation, and form evaporite deposits.

Well-developed alluvial fans also can be a major source of groundwater. Some cities—San Bernardino, California, for example—are built on alluvial fans and extract their municipal water supplies from them. However, because water resources in an alluvial fan are recharged from surface supplies and these fans are in arid regions, groundwater mining and overdraft beyond recharge rates are common. In other parts of the world, such water-bearing alluvial fans are known as *qanat* (Iran), *karex* (Pakistan), and *foggara* (western Sahara).

Desert Landscapes

Contrary to popular belief, deserts are not wastelands, for they abound in specially adapted plants and animals. Moreover, the limited vegetation, intermittent rainfall, intense insolation, and distant vistas produce starkly beautiful landscapes. And, all deserts are not the same: for example, North American deserts have more vegetation cover than do the generally barren Saharan expanses.

Mountainous deserts are found in interior Asia, from Iran to Pakistan, and in China and Mongolia. In South America, lying between the ocean and the Andes, is the rugged Atacama Desert. Deserts also occur as topographic plains, such as the Great Sandy and Simpson deserts of Australia, the Arabian Desert, the Kalahari Desert, and portions of the extensive Taklimakan Desert, which covers some 270,000 km² (105,000 mi²) in the central Tarim Basin of China.

The shimmering heat waves and related mirage effects in the desert are products of light refraction through layers of air that have developed a temperature gradient near the hot ground. The desert's enchantment is captured in the book *Desert Solitaire:*

> Around noon the heat waves begin flowing upward
> from the expanses of sand and bare rock. They
> shimmer like transparent, filmy veils between my

384

FIGURE 12-13

Alluvial fan in a desert landscape (right); topographic map of Cedar Creek alluvial fan (below). Topographic map is the Ennis Montana Quadrangle, 15-minute series, scale 1:62,500, contour interval = 40 ft; latitude/longitude coordinates for mouth of canyon are 45°22' N, 111° 35' W. [Photo by author; U.S. Geological Survey map.]

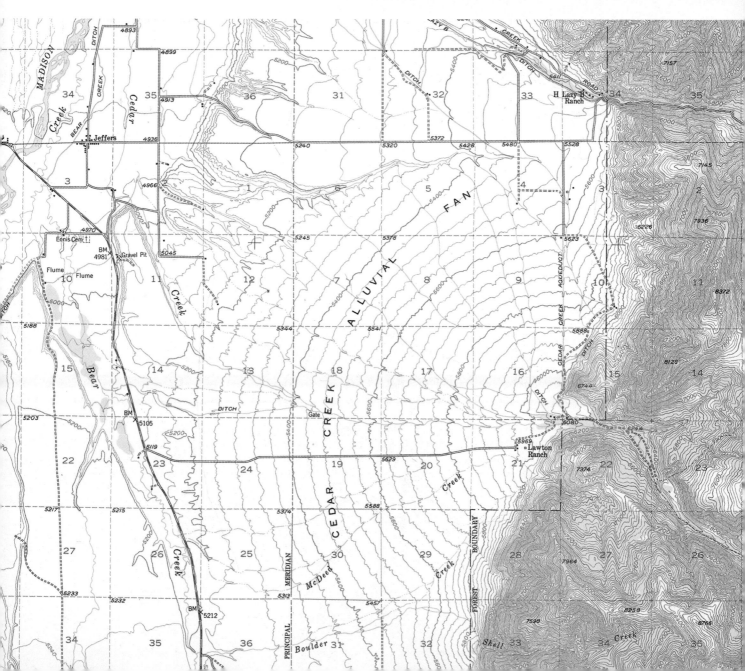

The Colorado River: A System Out of Balance

An exotic stream is one with headwaters in a humid region but flows mostly through arid lands for the rest of its journey to the sea. Exotic streams have fewer incoming tributaries and a smaller discharge downstream as compared to other rivers. The East African mountains and the Colorado Rockies represent such source areas for the Nile River and the Colorado River, respectively. The Colorado begins on the high slopes of Mount Richthofen (3962 m or 13,000 ft) in Rocky Mountain National Park and flows almost 2317 km (1440 mi) to where a trickle of water disappears in the sand, kilometers short of its former mouth in the Gulf of California (Figure 1).

Orographic precipitation totaling 102 cm (40 in.) per year (mostly snow) falls in the Rockies, feeding the Colorado headwaters. But at Yuma, Arizona, near the river's end, annual precipitation is a scant 8.9 cm (3.5 in.), an extremely small amount when compared to the annual potential evapotranspiration demand in the Yuma region of 140 cm (55 in.).

From its source region, the Colorado River quickly leaves the humid Rockies and spills out into the arid desert of western Colorado and eastern Utah. At Grand Junction, Colorado, on the Utah border, annual precipitation is only 20 cm (8 in.). After carving its way through the intricate labyrinth of canyonlands in Utah, the river enters Lake Powell, 945 m (3100 ft) lower in elevation than the river's source area upstream in the Rockies. The Colorado then flows through the Grand Canyon chasm, formed by its own erosive power. The Canyon's mystical beauty is evident to all who visit its rim and to those who venture within its depths.

West of the Grand Canyon, the river turns south, tracing its final 645 km (400 mi) as the Arizona-California border. Along this stretch sits Hoover Dam, just east of Las Vegas; Davis Dam, built to control the releases from Hoover; Parker Dam for the water needs of Los Angeles; three more dams for irrigation water (Palo Verde, Imperial, and Laguna); and finally Morelos Dam at the Mexican border. Mexico owns the end of the river and whatever water is left. Overall, the drainage basin encompasses 641,025 km² (247,500 mi²) of mountain, basin and range, plateau, canyon, and desert, in parts of seven states and two countries.

A Brief History

John Wesley Powell (1834–1902), the first person of record to successfully navigate the Colorado River through the Grand Canyon, was the first director of the U.S. Bureau of Ethnology and later director of the U.S. Geological Survey (1881–1892). Powell perceived that the challenge of the West was too great for individual efforts and believed that solutions to problems such as water availability could be met only through private cooperative efforts. His 1878 study (reprinted 1962), *Report of the Lands of the Arid Region of the United States,* is a conservation landmark.

Today, Powell probably would be skeptical of the intervention by government agencies in building large-scale reclamation projects. An anecdote in Wallace Stegner's *Beyond the Hundredth Meridian* relates that at an 1893 international irrigation conference held in Los Angeles, Powell observed development-minded delegates bragging that the entire West could be conquered and reclaimed from nature. Powell spoke against that sentiment: "I tell you, gentlemen, you are piling up a heritage of conflict and litigation over water rights, for there is not sufficient water to supply the land." He was booed from the hall. But history has shown Powell to be correct.

The Colorado River Compact was signed by six of the seven basin states in 1923. (The seventh, Arizona, signed in 1944, the same year as the Mexican Water Treaty.) With this compact the Colorado River basin was divided into an upper basin and a lower basin, arbitrarily separated for administrative purposes at Lees Ferry (Figure 1). Congress adopted the Boulder Canyon Act in 1928, authorizing Hoover Dam as the first major reclamation project on the river. Also authorized was the All-American Canal into the Imperial Valley, which required an additional dam.

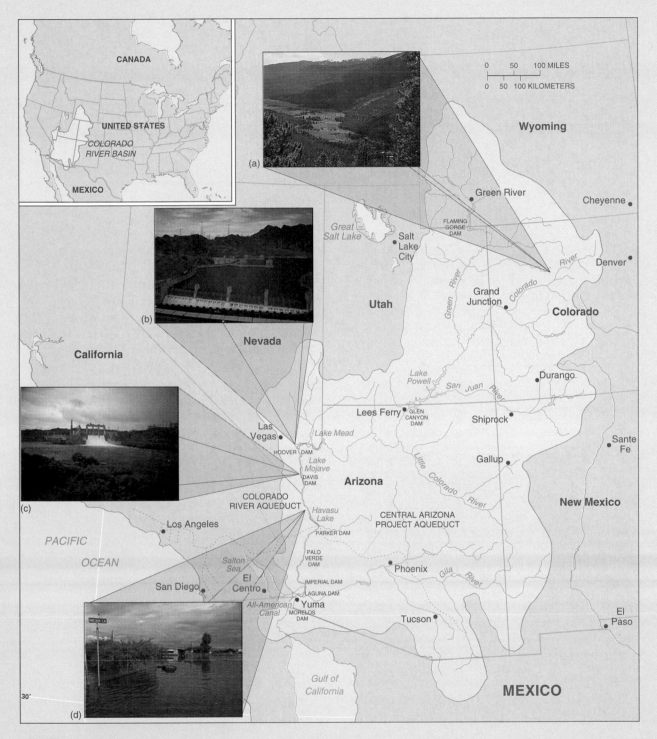

FIGURE 1

The Colorado River basin showing division of the upper and lower basins near Lees Ferry in northern Arizona. Photos portray (a) headwaters of the Colorado River near Mount Richthofen in the Colorado Rockies; (b) Hoover Dam spillways in rare operation, (c) Davis Dam in full release, and (d) the flooded Topock Estates subdivision near Needles, California, in 1983. [(a) Photo by Randall Christopherson; other photos by author.]

Table 1
Estimated Colorado River Budget

Water Demand	Quantity (maf)*
In-basin consumptive uses (75% agricultural)**	11.5
Central Arizona Project (rising to 2.8 maf)	1.0
Mexican allotment (1944 Treaty)	1.5
Evaporation from reservoirs	1.5
Bank storage at Lake Powell	0.5
Phreatophytic losses (water-demanding plants)	0.5
Budgeted total demand	16.5
(1930–1980 average flow of the river	**13.0)**

*1 million acre-feet = 325,872 gallons; 1.24 million liters.
**7.0 maf in lower basin; includes 5.5 pumped out-of-basin, 5.1 of which goes to California

Shortly after Hoover Dam was finished and downstream enterprises were thus offered flood protection, the other projects were quickly completed. Los Angeles then began its project to bring Colorado River water 390 km (240 mi) from still another dam and reservoir on the river to their city. There are now eight major dams on the river and many irrigation works. The latest effort to redistribute Colorado River water is the Central Arizona Project, which carries water to the Phoenix area.

Flow Characteristics

Exotic streamflows are highly variable, and the Colorado is no exception. The discharge measured at Lees Ferry in 1917 totaled 24 million acre-feet (maf), whereas in 1934 it reached only 5.03 maf. In 1977, the discharge dropped to 5.02 maf but it rose to an all-time high of 24.5 maf in 1984. In addition to this variability, approximately 70% of the year-to-year discharge occurs between April and July; the other 30% is spread over the balance of the year.

The average flows between 1906 and 1930 were almost 18 maf a year; averages dropped to 13 maf during the last 50 years. As a planning basis for the Colorado River Compact, the government used average river discharges from 1914 up to the treaty signing in 1923, an exceptionally high average of 18.8 maf. That amount was perceived as more than enough for the upper and lower basins each to receive 7.5 maf and, later, enough for Mexico to receive 1.5 maf in the 1944 Mexican Water Treaty.

We might question whether proper long-range planning should rely on the providence of high variability. Tree-ring analyses of past climates have disclosed that the only other time Colorado discharges were at the 1914–1923 level was between A.D. 1606 and 1625! However, government thinking about Colorado River flows seems never to have accepted this reality, for the de-

pendable flows of the river have been consistently overestimated. This predicament is shown in an estimated budget for the river (Table 1): clearly the situation is out of balance, for there is not enough discharge to meet budgeted demands.

Water Losses at Glen Canyon Dam

Glen Canyon Dam was completed and began water impoundment in 1963, 27 km (17 mi) north of the basin division point at Lees Ferry. The deep, fluted, inner gorges of Glen, Navajo, Labyrinth, Cathedral, and myriad other canyons slowly were flooded by advancing water. The Bureau of Reclamation stated that the dam's purpose was to regulate flows between the basins, although many thought that Lake Mead, behind Hoover Dam, could have served this administrative-regulatory function without the risk of additional water losses.

The porous Navajo sandstone that forms the bulk of the container for the Lake Powell reservoir at Glen Canyon Dam absorbs an estimated 4% of the river's overall annual discharge as *bank storage,* water that is not retrievable in any practical sense. The higher the lake level, the greater the loss into the sandstone. In addition, Lake Powell is an open body of water in an arid desert, where hot, dry winds accelerate evaporative losses. Another 4% of the river's overall discharge is lost annually in this manner from Lake Powell alone.

A third aspect of water loss involves the now-permanent sand bars and banks that have stabilized along the course of the regulated river. Water-demanding plants known as *phreatophytes* are now established and extracting an additional 4% of the flow of the river. Combined, these annual losses total 1.5 maf, or 12% of the long-term average flow—all attributable to Glen Canyon Dam and Lake Powell. Given the chronic deficits in the overall Colorado River budget, we can now seriously question the positive impact of this single project.

Flood Control in 1983

Intense precipitation and heavy snowpack in the Rockies, attributable to the 1982–1983 El Niño (see FYI Report 7-1), led to record-high discharge rates on the Colorado, testing the controllability of one of the most regulated rivers in the world. Federal reservoir managers were not prepared for the high discharge, having set aside Lake Mead's primary purpose (flood control) in favor of competing water and power interests. What ensued was a human-caused flood.

The only time the spillways at Hoover Dam had ever operated was more than 40 years earlier, when the reservoir capacity was artificially raised for a test; now they were opened to release the floodwaters. Davis Dam, which regulates releases from Hoover Dam, was within 30 cm (1 ft) of overflow by June 22, 1983, a real problem for a structure made partially of earth fill. In addition, Glen Canyon Dam was over capacity and at risk; it actually was damaged by the volume of discharge tearing through its spillways. The decision to increase releases to save these and other upstream facilities doomed towns and homeowners along the river, especially in subdivisions near Needles, California.

We might wonder what John Wesley Powell would think if he were alive today to witness such errant attempts to control the mighty and variable Colorado. He foretold such a " . . . heritage of conflict and litigation."

---

sanctuary in the shade and all the sun-dazzled world beyond. Objects and forms viewed through this tremulous flow appear somewhat displaced or distorted. . . . the great Balanced Rock floats a few inches above its pedestal, supported by a layer of superheated air. The buttes, pinnacles, and fins in the windows area bend and undulate beyond the middle ground like a painted backdrop stirred by a draft of air.*

The buttes, pinnacles, and mesas of arid landscapes are resistant horizontal rock strata that have eroded differentially. Removal of the less-resistant sandstone strata produces unusual desert sculptures—arches, pedestals, and delicately balanced rocks (Figure 12-14). Specifically, the upper layers of sandstone along the top of an arch or butte are more resistant to weathering and protect the sandstone rock beneath.

Desert landscapes are places where stark erosional remnants stand above the surrounding terrain as knobs, hills, or "island mountains." Such a bare, exposed rock, called an *inselberg,* or island mountain, is exemplified by Uluru (Ayers) Rock in Australia, pictured at the beginning of the book in Figure 1-1.

---

*Edward Abbey, *Desert Solitaire*. New York: McGraw-Hill Book Co., 1968, p. 154. Copyright © 1968 by Edward Abbey.

In a desert area, weak surface material may weather to a complex, rugged low topography, called a **badland**, probably so-named because it offered little economic value and was difficult to traverse in nineteenth-century wagons. The Badlands region of the Dakotas is of this form, as are portions of Death Valley, California, as illustrated in Figure 15-20.

Sand dunes that existed in some ancient deserts have lithified, forming sandstone structures that bear the imprint of cross-stratification. When such a dune was accumulating, sand cascaded down its slipface, and distinct bedding planes (layers) were established that remained after the dune lithified (Figure 12-15). Ripple marks, animal tracks, and fossils also are found preserved in these sandstones, which originally were eolian-deposited sand dunes.

Basin and Range Province. A *province* is a large region that shares several geologic or physiographic traits. The **Basin and Range Province** of the western United States consists of alternating basins and mountain ranges that lie in the rain shadow of mountains to the west. Thus, the province has a dry climate, few permanent streams, and *interior drainage patterns*—drainage basins that lack any outlet to the ocean (see Figures 11-1 and 11-2). The Basin and Range

FIGURE 12-14
Balanced Rock in Arches National Park, Utah, where
writer/naturalist Edward Abbey (quoted in text) worked
as a ranger years before it became a park. [Photo by
author.]

Province—almost 800,000 km² (300,000 mi²)—was a
major barrier to early settlers in their migration westward (Figure 12-16). The desert climate and north-
south trending mountain ranges were harsh challenges.

As the North American plate moved westward, it
overrode former oceanic crust and hot spots at such
a rapid pace that slabs of subducted material literally were run over. This stretched the crust, creating
a landscape fractured by many faults. The present
landscape consists of nearly parallel sequences of
horsts (upward-faulted blocks which are the
"ranges") and *grabens* (downward-faulted blocks
which are the "basins" or valleys). Figure 12-17 illustrates this pattern.

John McPhee captured the feel of this desert
province in his book *Basin and Range:*

> Supreme over all is silence. Discounting the cry of
> the occasional bird, the wailing of a pack of
> coyotes, silence—a great spatial silence—is pure in
> the Basin and Range. It is a soundless immensity
> with mountains in it. You stand . . . and look up at a
> high mountain front, and turn your head and look
> fifty miles down the valley, and there is utter
> silence.*

Basin-and-range relief is abrupt, and rock structures are angular and rugged. As the ranges erode,
transported materials accumulate to great depths in
the basins, gradually producing extensive desert
plains. The basin's elevation averages 1220 to 1525
m (4000–5000 ft) above sea level, with mountain
crests rising higher by some 915 to 1525 m
(3000–5000 ft). Death Valley is an extreme example;
its lowest basin has an elevation of –86 m (–282 ft).
However, to the west of the valley, the Panamint
Range rises to 3368 m (11,050 ft) at Telescope Peak—
over 3 vertical kilometers (2 mi) of desert relief!

Figure 12-17 illustrates typical basin-and-range terrain. Note the **bolson**, a slope-and-basin area between the crests of two adjacent ridges in a dry
region of interior drainage. The inset photo illustrates
these features in Death Valley, California. The figure
also identifies a *playa* (central salt pan), a *bajada*
(coalesced alluvial fans), and a *pediment,* the area
created as a mountain front retreats through weathering and erosional action. A pediment is an area of
bedrock that is layered with a thin veneer, or coat-
ing, of alluvium. It is an erosional surface, as opposed to the depositional surface of the bajada.

SUMMARY—Wind Processes and Desert Landscapes

Winds are produced by the movement of the atmosphere in response to pressure differences. Wind
is an agent of erosion, transportation, and deposition. **Eolian** processes erode soil from fields, move
sand accumulations along coastlines and in deserts,
and similarly move snow. In arid and semiarid cli-

*John McPhee, *Basin and Range*. New York: Farrar, Straus,
Giroux, 1981, p. 46.

FIGURE 12-15
Cross bedding in sandstone
formed from ancient patterns in
sand. [Photo by author.]

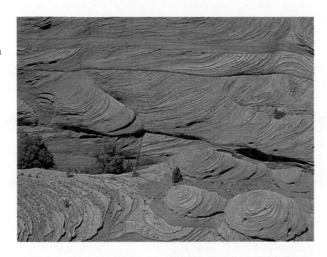

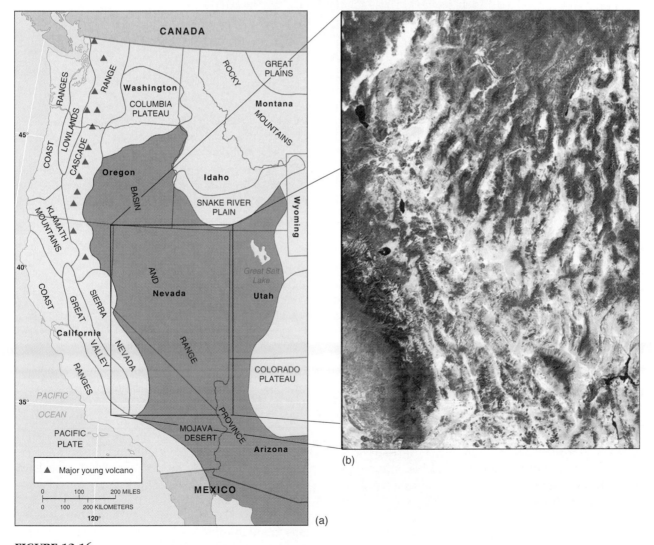

(a)

(b)

FIGURE 12-16
Basin and Range Province in the western United States. (a) Map and (b) *Landsat*
image of the area. Recent scientific discoveries demonstrate that this province
extends south through northern and central Mexico. [(b) Image from NASA.]

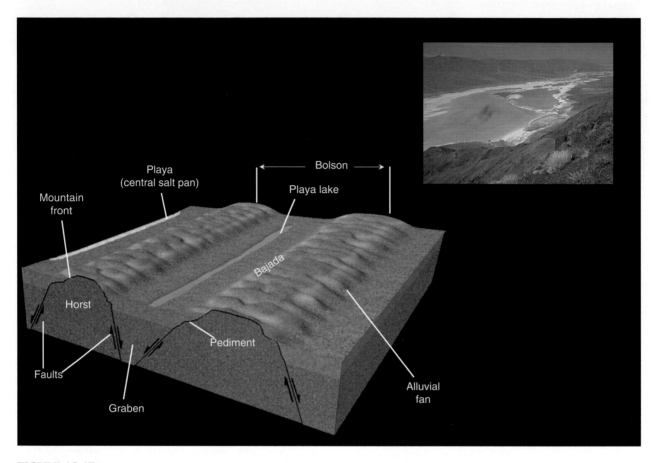

FIGURE 12-17
A bolson in the mountainous desert landscape of the Basin and Range Province.
Parallel normal faults produce a series of horsts (ranges) and grabens (basins). The
inset photograph is in Death Valley National Park, California, and shows the
central playa, parallel mountain ranges, and bajadas along the base of the range.
[Photo by author.]

mates and along some coastlines, available sand accumulates in **dunes**, which appear in a variety of
forms.

Two principal wind-erosion processes are **deflation**, the removal and lifting of individual loose particles, and **abrasion**, the grinding of rock surfaces
with a "sandblasting" action by particles captured in
the air. Most deserts in the world are covered by
desert pavement, which resembles a cobblestone
street. Eolian-transported materials contribute to soil
formation in distant places. Wind-blown **loess** deposits occur worldwide and provide a basis for soil
development. These fine-grained clays and silts are

moved by the wind many kilometers, where they are
redeposited in unstratified, homogeneous deposits
that cover the existing landscape.

Dry climates occupy about 26% of Earth's land
surface and, if all semiarid climates are considered,
perhaps as much as 35% of all land, constituting the
largest single climatic region on Earth. Desert landscapes are regions of special plant and animal adaptations and unique rock structures that appear stark
and angular, bared of vegetation and other cover-
ings found in the more humid regions of the world.
Wind and water operate together in dry regions. Although water events are infrequent, running water

is still the major erosional agent in deserts, sometimes producing dramatic **flash floods**. In arid climates, with their intermittent water flow, a notable fluvial landform is the **alluvial fan**, that occurs at the mouth of a canyon where it exits into a valley.

Rapid population growth and urbanization in the arid southwestern United States, as well as the Middle East, are producing much stress on available water resources. The Colorado River, an exotic stream, is budgeted beyond its average annual discharge, an issue of growing concern in the West. Fragile desert environments also are susceptible to disruption, as evidenced in the Persian Gulf region following the 1991 war.

KEY TERMS

abrasion	eolian
alluvial fan	erg desert
badland	flash flood
bajada	loess
Basin and Range Province	playa
blowout depression	sand sea
bolson	slipface
deflation	surface creep
desert pavement	ventifact
dune	wash

REVIEW QUESTIONS

1. Explain the term eolian and its application in this chapter. How would you characterize the ability of the wind to move material?
2. Describe the erosional processes associated with moving air. Compare them to the erosional processes involving moving water.
3. Explain deflation and the evolutionary sequence that produces desert pavement, ventifacts, and yardangs.
4. Differentiate between a dust storm and a sand storm.
5. What is the difference between eolian saltation and fluvial saltation?
6. What is the difference between an erg and a reg desert? Which type is a sand sea? Are all deserts covered by sand? Explain.
7. What are the three classes of dune forms? Describe the basic types of dunes within each class. What do you think is the major shaping force for sand dunes?
8. Which form of dune is the mountain giant of the desert? What are the characteristic wind patterns that produce such dunes?
9. How are loess materials generated? What form do they assume when deposited? Name a few examples of significant loess deposits.
10. Characterize desert energy and water balance regimes. Where are the world's deserts and how do their locations relate to these regimes?
11. How would you describe the water budget of the Colorado River? What was the basis for agreements regarding distribution of the river? Why has thinking about river flows been so optimistic?
12. Describe a desert bolson from crest to crest. Draw a simple sketch with the components of the landscape labeled.

Coastal cliffs, Nullarbor National Park, South Australia. [*Photo by M. P. Kahl/DRK photos.*]

13

COASTAL PROCESSES AND LANDFORMS

COASTAL SYSTEM COMPONENTS
  The Coastal Environment and Sea Level
COASTAL SYSTEM ACTIONS
  Tides
  Waves
  Sea Level Changes
COASTAL SYSTEM OUTPUTS
  Erosional Coastal Processes and Landforms
  Depositional Coastal Processes and Landforms
  Emergent and Submergent Coastlines
  Organic Processes: Coral Formations
  Salt Marshes and Mangrove Swamps
HUMAN IMPACT ON COASTAL ENVIRONMENTS
SUMMARY
FYI REPORT 13-1 AN ENVIRONMENTAL APPROACH TO SHORELINE PLANNING

The interaction of vast oceanic, atmospheric, and lithospheric systems is dramatic along a shoreline. At times, the ocean attacks the coast in a stormy rage of erosive power; at other times, the moist sea breeze, salty mist, and repetitive motion of the water are gentle and calming. Few have captured the confrontation between land and sea as well as Rachel Carson:

> The edge of the sea is a strange and beautiful place. All through the long history of Earth it has been an area of unrest where waves have broken heavily against the land, where the tides have pressed forward over the continents, receded, and then returned. For no two successive days is the shoreline precisely the same. Not only do the tides advance and retreat in their eternal rhythms, but the level of the sea itself is never at rest. It rises or falls as the glaciers melt or grow, as the floors of the deep ocean basins shift under its increasing load of sediments, or as the Earth's crust along the continental margins warps up or down in adjustment to strain and tension. Today a little more land may belong to the sea, tomorrow a little less. Always the edge of the sea remains an elusive and indefinable boundary.

Commerce and access to sea routes, fishing, and tourism prompt many people to settle near the ocean. A population survey by the United Nations estimates that about two-thirds of Earth's present population lives in or very near coastal regions. Therefore, an understanding of coastal processes and landforms is important to humanity. And because these processes along coastlines often produce dramatic change, they are essential to consider in planning and development.

You will find in this chapter coastal processes organized in a system of specific inputs (components and driving forces), actions (movements and processes), and outputs (results and consequences). We conclude with a look at the considerable human impact on coastal environments.

*From "The Marginal World," in *The Edge of the Sea* by Rachel Carson. Copyright © 1955 by Rachel Carson, © renewed 1983 by Roger Christie. Boston: Houghton Mifflin, p. 11.

Coastal System Components

Earth's surface features, like mountains and crustal plates, were formed over millions of years. However, most of Earth's coastlines are relatively new, existing in their present state as the setting for continuous change. The land, ocean, atmosphere, Sun, and Moon interact to produce tides, currents, waves, erosional features, and depositional features along the continental margins.

Inputs to the coastal environment include many of the elements already discussed in this text. *Solar energy* remains the driving force of both the atmosphere and the hydrosphere. *Prevailing winds* and *active weather patterns* are produced by conversion of insolation to heat energy and mechanical energy. In turn, *ocean currents and waves* are generated principally by atmospheric winds. *Climatic regimes* are influential in coastal geomorphic processes. The *nature of coastal rock* is important in determining rates of erosion and sediment production.

All of these inputs occur within the ever-present influence of *gravity's pull,* not only from Earth but also from the Moon and the Sun. Gravity provides the potential energy of position for materials in motion and produces the tides. A dynamic equilibrium among all these components produces coastline features of infinite variety.

The Coastal Environment and Sea Level

The coastal environment is called the **littoral zone**. (*Littoral* comes from the Latin word for shore.) The littoral zone spans some land as well as water. Landward, it extends to the highest water line that occurs on shore during a storm. Seaward, it extends to the point at which storm waves can no longer move sediments on the seafloor (usually at depths of approximately 60 m or 200 ft). The specific contact line between the sea and the land is the *shoreline,* and adjacent land is considered the *coast.* Figure 13-1 illustrates the littoral zone and includes specific components that are discussed later in the chapter.

Because the level of the ocean varies, the littoral zone naturally shifts position from time to time. A rise

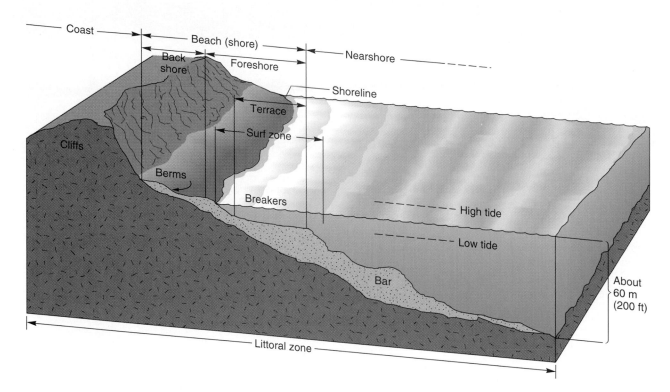

FIGURE 13-1
The littoral zone includes the coast, beach, and nearshore environments.

in sea level causes *submergence* of land, whereas a drop in sea level produces coastline *emergence*. Submergent and emergent coastlines each have distinctive features which are discussed later in the chapter. Changes in the littoral zone also are initiated by uplift and subsidence of the land itself.

Sea level is a relative term and reflects changes in both water quantity and land elevations. Because sea level is not a straightforward measurement, there exists no international system to determine exact sea level over time. NOAA is establishing a Global Absolute Sea Level Monitoring System. The goal is to accurately monitor global changes in absolute sea level.

Mean sea level is a value based on average tidal levels recorded hourly at a given site over a period of at least 19 years, which is one full lunar tidal cycle. Mean sea level varies because of ocean currents and waves, tides, air temperature and pressure differences, ocean temperature variations, slight variations in Earth's gravity, and changes in oceanic volume.

At present, the overall mean sea level for the United States is calculated at approximately 40 locations along the coastal margins of the continent. These sites are being upgraded with new equipment in the Next Generation Water Level Measurement System, specifically along the United States and Canadian Atlantic coasts, Bermuda, and the Hawaiian Islands. These measurements are augmented by new remote sensing technology, including the *TOPEX/Poseidon* satellite launched in August 1992. Of great assistance are the *NAVSTAR* satellites that make up the new Global Positioning System (GPS), which allow correlation from a network of ground- and ocean-based measurements.

Coastal System Actions

The complex fluctuation of tides, the movement of solar energy-driven winds, waves, and ocean currents, and the occasional impact of storms all are

"Sea Level Varies Along the U.S. Coastline"

The mean sea level of the U.S. Gulf Coast is about 25 cm (10 in.) higher than that of the east coast of Florida, which has the lowest mean sea level in North America. Sea level rises as one moves northward up the east coast, measuring about 38 cm (15 in.) higher in Maine than in Florida. Along the U.S. west coast, mean sea level in San Diego is about 58 cm (23 in.) higher than it is in Florida, rising to about 86 cm (34 in.) higher in Oregon than in Florida. Overall, the Pacific coast of North America has a mean sea level that averages about 66 cm (26 in.) higher than the average sea level along the Atlantic coast. Differences in ocean currents, air pressure and wind patterns, water density, and water temperature affect the height of mean sea level.

powerful actions within the coastal system. These forces shape landforms from gentle beaches to steep cliffs, and sustain delicate ecosystems.

Tides

Earth's relation to the Sun and the Moon and the reasons for the seasons are discussed in Chapter 2. These astronomical relationships also produce the pattern of **tides**, the complex daily oscillations in sea level that are experienced to varying degrees around the world. Tides also are influenced by the size, depth, and topography of ocean basins, by latitude, and by shoreline shape.

Tidal action is a distinctive energy agent for geomorphic change. As tides *flood* (rise) and *ebb* (fall), the daily migration of the shoreline landward and seaward causes significant changes that affect sediment erosion and transportation. Figure 13-2 illustrates the relationship among the Moon, the Sun, and Earth and the generation of variable tidal bulges on opposite sides of the planet.

Tides are produced by the gravitational pull exerted on Earth and its oceans by both the Sun and the Moon. The Sun's influence is only about half that of the Moon's because of the Sun's greater distance from Earth, although it still is a significant force.

The gravitational pull of both bodies on Earth actually pulls Earth's atmosphere, oceans, and lithosphere on the side that is facing the Sun or Moon at the moment. This raises a "tidal bulge" in the ocean. For complex reasons, gravitational force and inertia also creates a bulge on the opposite side of Earth. These opposing bulges are always there in the alignments of Earth, Sun, and Moon as shown in the figure.

Every 24 hours and 50 minutes, any given point on Earth rotates through these two bulges. Thus, every day, most coastal locations experience two high (rising) tides known as *flood tides,* and two low (falling) tides known as *ebb tides.* The difference between consecutive high and low tides is the *tidal range.*

When Earth, Sun, and the Moon are aligned as shown in (a) and (b), their combined gravitational pull creates a greater *spring tide* (meaning to "spring forth," not the season). When positioned as in (c) and (d), their weaker pulls create a lesser *neap tide.*

Tides also are influenced by other factors, causing a great variety of tidal ranges. For example, some locations may experience almost no difference between high and low tides. The highest tides occur

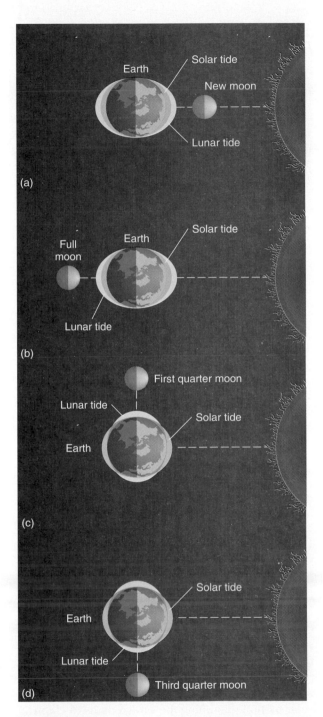

FIGURE 13-2
Sun, Moon, and Earth relationships combine to produce spring tides (a), (b) and neap tides (c), (d). (The tides depicted are not illustrated to scale.)

where open water is forced into partially enclosed gulfs or bays. The Bay of Fundy in Nova Scotia records the greatest tidal range on Earth, a difference of 16 m (52.5 ft) (Figure 13-3a and b). Tidal prediction is especially important to ships because the entrance to many ports is limited by shallow water and thus high tide is required for passage. At other times, tall-masted ships need a low tide to clear overhead bridges. Tides also exist in large lakes, but the tidal range is small, making tides difficult to distinguish from changes caused by wind. Lake Superior, for instance, has a tidal variation of only about 5 cm (2 in.).

Tidal Power. The fact that sea level changes daily with the tides suggests an opportunity: could these predictable flows be harnessed to produce electricity? The answer is yes, given the right conditions. The bay or estuary under consideration must have a narrow entrance suitable for the construction of a dam with gates and locks, and it must experience a tidal range of flood and ebb tides large enough to turn turbines, at least a 5 m (16 ft) range.

About 30 locations in the world are suited for tidal power generation, although at present only three of them are actually producing electricity. Two are outside North America—an experimental 1-megawatt station in Russia at Kislaya-Guba Bay, on the White Sea, since 1969, and a facility in the Rance River estuary on the Brittany coast of France, since 1967. The tides in the Rance estuary fluctuate up to 13 m (43 ft), and power production has been almost continuous there, providing an electrical generating capacity of a moderate 240 megawatts (about 20% of the capacity of Hoover Dam).

The third site is on the Bay of Fundy. According to the Canadian government, the present cost of tidal power at ideal sites is economically competitive with that of fossil fuels, although certain environmental concerns must be addressed. At one of several favorable sites on the Bay of Fundy, the Annapolis Tidal Generating Station was built in 1984 as a test. Nova Scotia Power Incorporated operates this 20-megawatt plant (Figure 13-3c). Further expansion depends on the growth in electrical demand, how competitive fossil fuels remain, the continuing problems with nuclear power, and environmental demands.

(a)

(c)

(b)

FIGURE 13-3

Tidal range is great in some bays and estuaries, such as Halls Harbor near the Bay of Fundy at flood tide (a) and ebb tide (b). The flow of water between high and low tide is ideal for turning turbines and generating electricity, as is done at the Annapolis Tidal Generating Station (c), near the Bay of Fundy in Nova Scotia, in operation since 1984. [(a) and (b) Photos by Michael Melford, (c) photo courtesy of Nova Scotia Power Incorporated.]

Waves

Wind friction on the surface of the ocean generates undulations of water called **waves**. They travel in *wave trains,* or groups of waves. On a small scale, a moving boat creates a wake, which consists of waves; at a larger scale, storms around the world generate large groups of wave trains. A stormy area at sea is called a *generating region* for these waves, which radiate outward in all directions. As a result, the ocean is crisscrossed with intricate patterns of waves traveling in all directions. The waves seen along a coast may be the product of a storm center thousands of kilometers away.

Regular patterns of smooth, rounded waves are called **swells**—these are the mature undulations of the open ocean. As waves leave the generating region, wave energy continues to run in these swells, which can range from small ripples to very large flat-crested waves. A deep-water wave leaving a generating region tends to extend its wavelength many

meters (see Figure 13-4 noting crest, trough, wavelength, and wave height).

Water within a wave in the open ocean is not really migrating but is transferring energy from molecule to molecule through the water in simple cyclic undulations, or *waves of transition* (Figure 13-4). Individual water particles move forward only slightly, forming a vertically circular pattern. The diameter of the paths formed by the orbiting water particles decreases with depth. As a deep-ocean wave approaches the shoreline and enters shallower water (10–20 m or 30–65 ft), the drag of the bottom shortens the wavelength and slows the speed of the circular motion, flattening the orbital paths to elliptical shapes. As the peak of each wave rises, a point is reached when its height exceeds its vertical stability and the wave falls into a characteristic **breaker**, crashing onto the beach (Figure 13-4b).

In a breaker the orbital motion of transition gives way to form *waves of translation,* in which energy

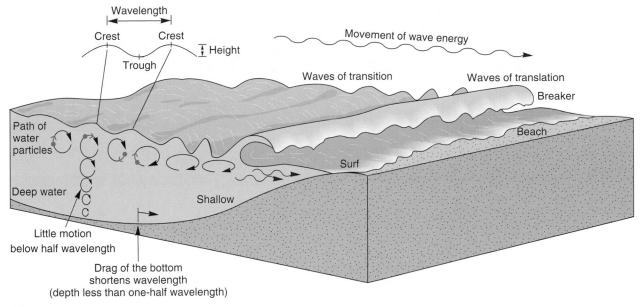

(a)

FIGURE 13-4

(a) Wave structure and the orbiting tracks of water particles change from circular motions and swells in deep *waves of transition* to more elliptical orbits in shallow *waves of translation*. (b) Cascades of waves and breakers attack the shore. [Photo by Bobbé Christopherson.]

(b)

and water both move forward toward shore as water cascades down from the wave crest. In conjunction with the energy of arriving waves, the slope of the shore determines wave style: plunging breakers indicate a steep bottom profile, whereas spilling breakers indicate a gentle, shallow bottom profile.

Wave Refraction. Generally, wave action results in coastal straightening. As waves approach an irregular coast, they bend around **headlands**, which are protruding landforms generally composed of resistant rocks (Figure 13-5). The submarine topography refracts (bends) approaching waves. This produces patterns that focus energy around headlands, and dissipate energy in coves and the submerged coastal valleys between them. Thus, headlands represent a specific focus of wave attack along a coastline. Waves tend to disperse their energy in *coves* and *bays* on either side of headlands. This **wave refraction** (wave bending) along a coastline redistributes wave energy so that different sections of the coastline are subjected to variations in erosion potential.

As waves approach a coast, they usually arrive at some angle other than parallel to it (Figure 13-6). As the waves enter shallow water, they are refracted and generate a current parallel to the coast, zigzagging in the prevalent direction of the incoming waves. This **longshore current**, or *littoral current,* depends on

401

"Killer Waves Strike Beachcombers"

Signs along portions of the California, Oregon, Washington, and British Columbia coastline warn beachcombers to watch for "killer waves." Such warnings make it a good idea to check conditions before you venture along the shore.

As various wave trains move along in the open sea, they interact by *interference*. These interfering waves sometimes align so that the wave crests and troughs from one wave train are in phase with those of another. When this *in-phase* condition occurs, the height of the waves is increased, sometimes dramatically. The resulting "killer waves" or "sleeper waves" can sweep in unannounced and overtake unsuspecting victims.

On the other hand, when wave trains are *out of phase* with each other, they cancel each other and their heights are reduced. Maybe you have noticed the changing pattern of the surf beat caused by such interference.

wind direction and wave direction. A longshore current is generated only in the surf zone and works in combination with wave action to transport large amounts of sand, gravel, sediment, and debris along the shore.

Particles on the beach also are moved along as **beach drift**, shifting back and forth between water and land with each *swash* and *backwash* of surf. Individual sediment grains trace arched paths along the beach (Figure 13-6). You have perhaps stood on a beach and heard the sound of myriad grains and seawater, especially in the backwash of surf. These dislodged materials are available for transport and eventual deposition in coves and inlets and can represent a significant volume.

Tsunami, or Seismic Sea Wave. A particular type of wave that greatly influences coastlines is the **tsunami**. *Tsunami* is Japanese for "harbor wave," named for its devastating effect in harbors. Statistically, this sea wave occurs an average of every 3.5 years in the Pacific basin. Specifically in Hawaii, the U.S. Army Corps of Engineers has reported 41 damaging occurrences during the past 145 years. Tsunami often are in-

correctly referred to as "tidal waves," but they have no relation to the tides. They are formed by sudden and sharp motions in the seafloor, caused by earthquakes, submarine landslides, or eruptions of undersea volcanoes. Thus they are referred to as *seismic sea waves*.

A large undersea disturbance usually generates a solitary wave of great wavelength. These waves generally exceed 100 km (60 mi) in wavelength but only a meter or so in height. They travel at great speeds in deep-ocean water—velocities of 600–800 kmph (375–500 mph) are not uncommon—but often pass unnoticed on the open sea because their great length makes the slow rise and fall of water hard to observe.

However, as a tsunami approaches a coast, the shallow water forces the wavelength to shorten. As a result, the wave height may increase up to 15 m (50 ft) or more. Such a wave has the potential for coastal devastation, resulting in property damage and death. For example, in September 1992 the citizens of Casares, Nicaragua, were surprised by a 12 m (39 ft) tsunami that took 270 lives. An earthquake along an offshore subduction zone triggered this killer wave. In the Alaskan earthquake of 1964, a tsunami was generated by undersea landslides and

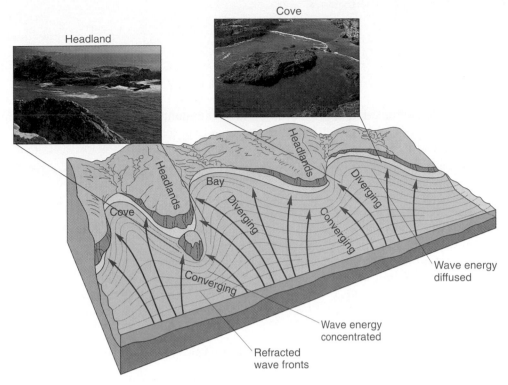

FIGURE 13-5
The process of coastal straightening brought about by wave refraction. Wave
energy is concentrated as it converges on headlands and is diffused as it diverges
in coves and bays. [Inset photos by author.]

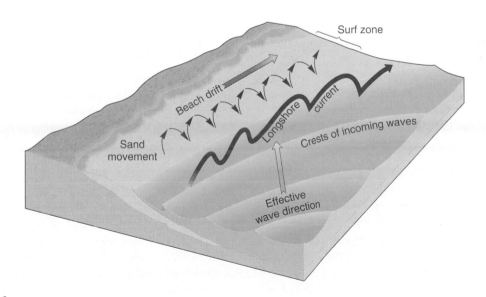

FIGURE 13-6
Longshore currents are produced as waves approach the surf zone and shallower
water. Littoral and beach drift result as substantial volumes of material are moved
along the shore.

radiated across the Pacific Ocean. It reached Crescent City, California, in 4 hours 3 minutes; Hilo, Hawaii, in 5 hours 24 minutes; Hokkaido, Japan, in 6 hours 44 minutes; and La Punta, Peru, in 15 hours 34 minutes. Curious onlookers died in both California and Oregon as a result of the tsunami generated by the distant Alaskan earthquake.

Because tsunami travel so quickly and are undetectable in the open ocean, accurate forecasts are difficult and occurrences often are unexpected. A warning system now is in operation for nations surrounding the Pacific, where the majority of tsunami occur. Warnings always should be heeded, despite the many false alarms, for the causes lie beneath the ocean and are difficult to monitor in any consistent manner.

Sea Level Changes

Over the long term, sea level fluctuations expose a great range of coastal landforms to tidal and wave processes. Sea level changes are initiated either by tectonic forces that raise or lower coastlines, or by glacio-eustatic processes that affect the quantity of water stored as ice. Because of complex interactions, it is difficult to isolate individual factors. However, as average global temperatures cycle through cold or warm climatic spells, the quantity of ice locked up in Antarctica, Greenland, and alpine glaciers can increase or decrease accordingly. Chapter 7 presents the climatic implications of a greenhouse-effect warming pattern on high-latitude and polar environments, which are critical areas because they contain enormous deposits of ice.

If Antarctica and Greenland ever were ice-free (ice sheets completely melted), sea level would rise at least 65 m (215 ft). Indeed, sea level has risen throughout the most recent geologic epoch, the Holocene (the last 10,000 years), and generally ever since the end of the Pleistocene Ice Age, which began about 1.65 million years ago. At the peak of the last Pleistocene glaciation about 18,000 years B.P. (before the present), sea level was about 130 m (430 ft) lower than it is today. Just 100 years ago sea level was 38 cm (15 in.) lower along the coast of southern Florida. Venice, Italy, has experienced a rise of 25 cm (10 in.) since 1890. Elsewhere, increases of 10–20 cm (4–8 in.) this century are common. This rate of rise of 20–40 cm (8–16 in.) per

100 years is about *six times faster* than any observed from historical records or recent geologic time.

A 1 m (3.2 ft) rise could inundate 15% of Egypt's arable land, 17% of Bangladesh, many island nations and communities, and 20,000 km^2 (7800 mi^2) of land along U. S. shores, at a staggering loss of $650 billion in North America alone. However, great uncertainty exists in these forecasts. The Intergovernmental Panel on Climate Change (IPCC, discussed in Chapter 7) places the estimated range of expected sea-level rise at between 30 and 110 cm (1–4 ft). Despite any uncertainty in these forecasts, planning should start now along coastlines worldwide because preventive strategies are cheaper than the increasing costs of possible destruction. Insurance underwriters have begun the process by refusing coverage for shoreline properties vulnerable to rising sea level.

Coastal System Outputs

Coastlines are active portions of continents, with energy and sediment being continuously delivered to a narrow environment. The action of tides, currents, wind, waves, and changing sea level produces a variety of erosional and depositional landforms. We look first at erosional coastlines like the West Coast, then at depositional coastlines as found along the East and Gulf Coasts.

Erosional Coastal Processes and Landforms

The active margins of the Pacific Ocean along the North and South American continents are characteristic coastlines affected by erosional landform processes. Erosional coastlines tend to be rugged, of high relief, and tectonically active, as expected from their association with the leading edge of drifting lithospheric plates (see plate tectonics discussion in Chapter 8). Figure 13-7 presents features commonly observed along an erosional coast.

Sea cliffs are formed by the undercutting action of the sea. As indentations are produced at water level, such a cliff becomes notched, leading to subsequent collapse and retreat of the cliff. Other erosional forms evolve along cliff-dominated coastlines, including *sea caves, sea arches,* and *sea stacks.* As erosion continues, arches may collapse, leaving isolated stacks in the water. The coasts of southern England

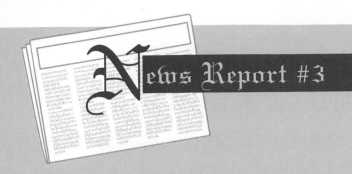

"Warmer Water Raises Sea Level"

As much as 25% of the present sea level rise is caused by increasing ocean temperatures. This is because of the thermal expansion of warmer water (water is densest at 4°C or 39°F, and expands at higher temperatures).

Measurements by the Scripps Institute of Oceanography demonstrate a 0.8C° (1.44F°) warming since 1950 off southern California. Similar ocean warming trends are being reported worldwide. In addition, sea level height is enhanced by the melting of glacial ice as global temperatures increase.

Given these warming trends and the predicted climatic change, sea level could rise as much as 6 m (20 ft) during the next century. However, it is generally believed that the rise will be less than 2 m (6.5 ft). Nonetheless, even this amount represents a potentially devastating change for many coastal locations, because a rise of only 0.3 m (1 ft) would cause shorelines worldwide to move inland an average of 30 m (100 ft)!

and Oregon are prime examples of such erosional landscapes (Figure 13-7).

Wave action can cut a horizontal bench in the tidal zone, extending from a sea cliff out into the sea. Such a structure is called a **wave-cut platform**, or *wave-cut terrace*. If the relationship between the land and sea level has changed over time, multiple platforms or terraces may rise like stairsteps back from the coast. These marine terraces are remarkable indicators of an emerging coastline, with some terraces more than 365 m (1200 ft) above sea level. A tectonically active region, such as the California coast, has many examples of multiple wave-cut platforms (Figure 13-8).

Depositional Coastal Processes and Landforms

Depositional coasts generally are located along land of gentle relief, where sediments are available from many sources. Such is the case with the Atlantic and Gulf coastal plains of the United States, which lie along the relatively passive, trailing edge of the North American lithospheric plate. These depositional coasts also are influenced by erosional processes and inundation, particularly during storm activity.

Characteristic wave- and current-deposited landforms are illustrated in Figure 13-9. A **barrier spit** consists of material deposited in a long ridge extending out from a coast attached at one end; it partially crosses and blocks the mouth of a bay. Classic examples include Sandy Hook, New Jersey (south of New York City), and Cape Cod, Massachusetts. The photo in Figure 13-9 shows a typical barrier spit forming part way across the mouth of Prion Bay in Southwestern National Park in Tasmania. A spit becomes a **bay barrier**, sometimes referred to as a *baymouth bar,* if it completely cuts off the bay from the ocean and forms an inland **lagoon**. Spits and barriers are made up of materials that have been eroded and transported by *littoral drift* (beach and longshore drift combined). For much sediment to accumulate, offshore currents must be weak. Tidal flats and salt marshes are characteristic low-relief features wherever tidal influence is greater than wave action.

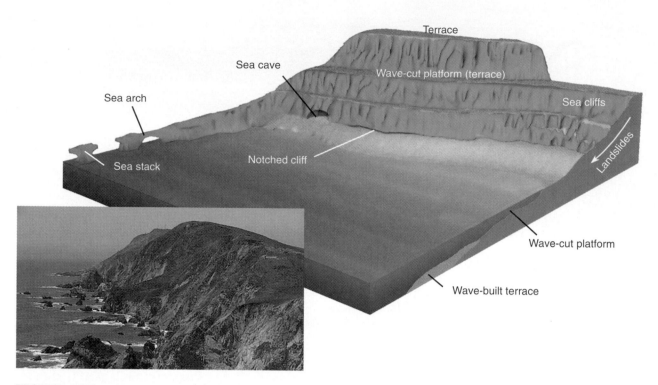

FIGURE 13-7

Characteristic coastal erosional landforms. The photo shows the central California coast, a typical emergent coastline. [Photo by Ray Atkeson.]

A *tombolo* occurs when sediment deposits connect the shoreline with an offshore island or sea stack (Figures 13-9 and 13-10). It forms when sediments accumulate on an underwater wave-built terrace.

Beaches. Of all the features associated with a depositional coastline, beaches probably are the most familiar. Beaches vary in type and permanence, especially along coastlines dominated by wave action. Technically, a **beach** is that place along a coast where sediment is in motion, deposited by waves and currents. Material from the land temporarily resides there while it is in active transit along the shore. You may have experienced a beach at some time, along a sea coast, a lake shore, or even a stream. Perhaps you have even built your own "landforms" in the sand, only to see them washed away by the waves.

The beach zone ranges, on average, from 5 m (16 ft) above high tide to 10 m (33 ft) below low tide (Figure 13-1). The specific definition varies greatly along individual shorelines. Worldwide, beaches are dominated by sands of quartz (SiO_2) because it resists weathering, and therefore remains after other minerals are removed. In volcanic areas, beaches are derived from wave-processed lava. These black-sand beaches are found in Hawaii and Iceland, for example.

Many beaches, such as those in southern France and western Italy, are composed of pebbles and cobbles—a type of "shingle beach." Some shores have no beaches at all; scrambling across boulders and rocks may be the only way to move along the coast. The coast of Maine and portions of the Atlantic provinces of Canada are classic examples, composed of resistant granite rock that is scenically rugged but with few beaches.

A beach acts to stabilize a shoreline by absorbing wave energy, as is evident by the amount of material that is in almost constant motion (see "sand movement" in Figure 13-6). Some beaches are stable. Others cycle seasonally; they accumulate during the summer only to be moved offshore by winter storm waves, forming a submerged bar, and are replaced the following summer. Protected areas along a coastline tend to accumulate sand.

Figure 13-1 shows the location of beach elements—the backshore, berm, and foreshore (which includes the surf zone). Beach profiles fall into two broad categories, according to prevailing conditions. During quiet,

FIGURE 13-8
Marine terraces formed as wave-cut platforms along the Pacific coast. [Photo by Steve Lambros.]

FIGURE 13-9
Characteristic coastal depositional landforms. The photo depicts a typical barrier spit developing across the mouth of the New River, forming the New River Lagoon and Prion Bay at southwestern National Park in Tasmania. [Photo by Reg Morrison/ Auscape International.]

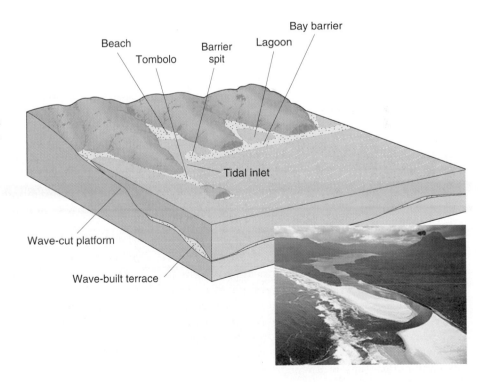

FIGURE 13-10
A tombolo at Point Sur along the central California coast where sediment deposits connect the shore with an island. [Photo by author.]

low-energy periods, both backshore and foreshore areas experience net accumulations of sediment and are well defined. However, a storm brings increased wave action and high-energy conditions to the beach, usually reducing its profile to a sloping foreshore only. Sediments from the backshore area do not disappear but may be moved by wave action to the nearshore (submerged) area, only to be returned by a later low-energy surf. Therefore, beaches go through constant change, especially in this era of rising sea levels.

Changes in coastal sediment transport can disrupt human activities—beaches are lost, harbors closed, and coastal highways and beach houses can be inundated with sediment. Consequently, various strategies are employed to interrupt longshore currents and beach drift. The goal is either to halt sand accumulation or to force accumulation in a desired way through construction of engineered structures—"hard" shoreline protection.

Figure 13-11 illustrates common approaches: a *jetty* to block material from harbor entrances; a *groin* to slow drift action along the coast, and a *breakwater* to create a zone of still water near the coastline. However, interrupting the coastal drift, that naturally replenishes beaches, may lead to unwanted changes in sediment distribution downcurrent. Careful planning and impact assessment should be part of any strategy for preserving or altering a beach.

Beach nourishment refers to the artificial placement of sand along a beach. Through such efforts, a beach that normally experiences a net loss of sediment will instead show a net gain. In contrast to hard structures this hauling of sand to replenish a beach is considered "soft" shoreline protection. (See News Report #4.)

Barrier Forms. The image in Figure 13-12 illustrates the many features of barrier chains. These chains are long, narrow, depositional features, generally of sand, that form offshore roughly parallel to the coast. Common forms are **barrier beaches**, or the broader, more extensive landform, **barrier islands**. Tidal variation in the area usually is moderate to low, with adequate sediment supplies coming from nearby coastal plains. On the landward side of a barrier formation are tidal flats, marshes, swamps, lagoons, coastal dunes, and beaches. Barrier beaches appear to adjust to sea level and may naturally shift position from time to time in response to wave action and longshore currents. Breaks in a barrier, visible in the image form inlets, connecting a bay with the ocean. Their name "barrier" is appropriate, for they take the brunt of storm energy and actually act as protection for the mainland.

Barrier beaches and islands are quite common worldwide, lying offshore of nearly 10% of Earth's

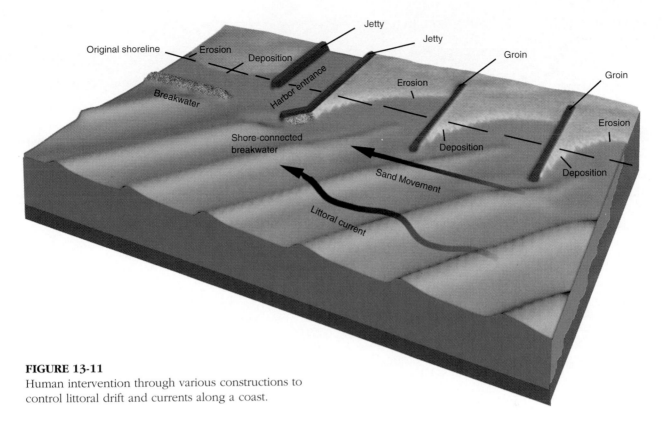

FIGURE 13-11
Human intervention through various constructions to
control littoral drift and currents along a coast.

coastlines. Examples are found offshore of Africa,
India (east coast), Sri Lanka, Australia, the north slope
of Alaska, and the shores of the Baltic and Mediterranean Seas. The most extensive chain of barrier islands is along the Atlantic and Gulf Coast states,
extending some 5000 km (3100 mi) from Long Island
to Texas and Mexico. A particularly interesting portion of this chain, shown in Figure 13-12, is the Outer
Banks off North Carolina's coast. It includes Cape
Hatteras, across Pamlico Sound from the mainland.
The area presently is designated as one of three national seashore reserves supervised by the National
Park Service. Figure 13-12 displays key depositional
forms: spit, island, beach, lagoon, sound, and inlet.

Barrier islands are an unwise choice for homesites
or commercial building because they shift and are
directly exposed to severe storms. Construction activities alone cause damage to the islands.

Emergent and Submergent Coastlines

Coastlines can be classified in several ways. Most significant is the relation between rising or falling sea
level and whether the continental margins are uplifting or subsiding. The following discussion simplifies
coastlines to two general types—emergent and sub-

mergent. Please note that coastal landforms may exhibit traits of both emergence and submergence.

Emergent Coastlines. Emergent coastlines are indicated by sea cliffs and wave-cut platforms, that appear raised above water level (Figure 13-7). The U.S.
West Coast is an example. Such coastlines develop
either because sea level lowers or the land rises.
Emergent coastlines are common during glaciations,
when more of Earth's water is tied up as ice, thus
lowering worldwide sea level. Emergent coasts also
may be of tectonic origin, especially common
around the Pacific Rim; old depositional features
thus may become elevated high above sea level.
Other emergent coastlines may be isostatically responding (rising) to the melting and retreat of great
expanses of ice by uplifting submerged coastal valleys. Areas of northern Canada and Scandinavia are
still emerging in this way at a centimeter (0.4 inch)
or so a year, which is a rather rapid rate, geologi-
cally speaking.

Submergent Coastlines. Submergent coasts are
drowned by the sea, either because of sea-level rise or
land subsidence. Characteristically, submergent coastlines form *embayments,* or the inundation of lowlands

"Engineers Nourish A Beach"

The city of Miami, Florida, and surrounding Dade County have spent almost $70 million since the 1970s in a continuing effort to rebuild their beaches. Sand is transported to the replenishment area. To maintain a 200 m-wide beach (660 ft), net sand loss per year is determined and a schedule for replenishment is set. In Miami Beach, an eight-year replenishment cycle is maintained. During Hurricane Andrew in 1992 the replenished Miami Beach is thought to have prevented millions of dollars in shoreline structural damage.

Unforeseen environmental impact may accompany the addition of new and foreign sand types to a beach not matched to the ecological relationships of the site. If the new sands do not match the existing varieties, disruption of coastal marine life is possible. The U.S. Army Corps of Engineers, which operates the Miami replenishment program, is running out of "borrowing areas" for sand that matches the natural sand of the beach. A proposal to haul a different type of sand from the Bahamas is being studied as to possible environmental consequences.

---

near the coast by the sea, generally near the mouth of a river. Examples of this type of coastline are those along the U.S. East and Gulf Coasts. Evidence relates such coastlines to interglacial periods and times of higher temperatures, with corresponding lower volumes of ice and rising sea levels. A Spanish term, *ria* (for "river"), is applied to this type of coast, which often is penetrated by a river entering the sea and therefore forming an *estuary,* where seawater and freshwater mix. Familiar estuaries include the flooded lower reaches of the Hudson River around New York City, and Delaware Bay and Chesapeake Bay to the south, which also contain tributary estuaries.

If sea level continues to rise over the next century, many estuaries will expand inland, and new ones could form, submerging even more coastline. Despite isostatic rising, the increase in sea level since the last ice age has left many glaciated coasts drowned with glacially carved valleys called *fjords.* Their great depth, to hundreds of meters, precludes any development of depositional features. Fjords exist in South Island of New Zealand, southern Chile, British Columbia, Alaska, Greenland, and Scandinavia.

410

Organic Processes: Coral Formations

Not all coastlines form due to purely physical processes. Some form as the result of organic processes, such as coral growth. A **coral** is a simple marine animal with a small, cylindrical, saclike body (polyp); it is related to other marine invertebrates, such as anemones and jellyfish. Corals secrete calcium carbonate ($CaCO_3$) from the lower part of their bodies, forming a hard external skeleton. Corals function in a *symbiotic* (mutually helpful) relationship with algae—that is, they live in close association with the algae, and each depends on the other for survival. Algae photosynthesize some of the food for the coral, and in turn the corals provide some nutrients and shelter for the algae.

Figure 13-13 shows the distribution of currently living coral formations. Corals thrive in warm tropical oceans, so the difference in ocean temperature between the western coasts and eastern coasts of continents is critical to their distribution. Western coastal waters tend to be cooler, thereby discouraging coral activity, whereas eastern coastal currents are warmer

FIGURE 13-12
Landsat image of barrier island chain along the North Carolina coast. Hurricane Emily swept past Cape Hatteras in August 1993, causing damage and beach erosion. [Image from NASA.]

FIGURE 13-13
Worldwide distribution of living coral formations. Yellow areas also include prolific reef growth and atoll formation. [After J. L. Davies, *Geographical Variation In Coastal Development*. Essex, England: Longman House, 1973. Adapted by permission.]

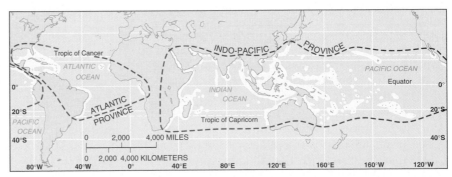

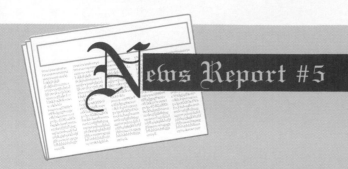

"Hurricane Hugo Devastates the Beachfront"

The hazard represented by settlement of barrier islands was made graphically clear when Hurricane Hugo assaulted South Carolina in 1989. This storm made a destructive pass over Puerto Rico, the Virgin Islands, the northeastern Caribbean, and then turned northwestward. The storm attacked the Grand Strand barrier islands off the northern half of South Carolina's coastline, hitting Charleston and the South Strand portion of the islands hardest. The storm made landfall as the worst hurricane to strike there in 35 years.

Beachfront houses, barrier-island developments, and millions of tons of sand were swept away; up to 95% of the single-family homes in Garden City were destroyed. The southern portion of Pawleys Island was torn away, and one of every four homes was destroyed. Some homes that had escaped the last major storm, Hurricane Hazel in 1954, were lost to Hugo's peak winds of 209 kmph (130 mph). In comparable dollars, Hugo caused nearly $4 billion in damage, compared to Hazel's $1 billion. The increased damage partially resulted from expanded construction during the intervening years between the two storms. Due to increased development and real estate appreciation, each future storm can be expected to cause ever-increasing capital losses to these barrier and beachfront landscapes.

and thus enhance coral growth. Living colonial corals range in distribution from about 30° N to 30° S and occupy a very specific ecological zone: 10–55 m (30–180 ft) depth, 27–40‰ (parts per thousand) salinity, and 18 to 29°C (64–85°F) water temperature. Corals require clear, sediment-free water and consequently do not locate near the mouths of sediment-charged freshwater tributaries (note the lack of these formations along the U.S. Gulf Coast).

Coral Reefs. Some corals are colonial and their skeletons accumulate in enormous structures. Through many generations, live corals near the ocean's surface build on the foundation of older coral skeletons, which in turn may rest upon a volcanic seamount or some other submarine feature built up from the ocean floor. In the process *coral reefs* are formed.

The principal shapes of reefs are fringing reefs, barrier reefs, and atolls. In 1842 Charles Darwin pre-sented a hypothesis for the evolution of reef formation: as reefs develop around a volcanic island and the island itself gradually subsides, an equilibrium is maintained between the subsidence of the island and the upward growth of coral. This generally accepted idea is portrayed in Figure 13-14 with specific examples of each reef stage: *fringing reefs* (platforms of surrounding coral rock), *barrier reefs* (forming enclosed lagoons), and *atolls* (circular, ring-shaped).

Salt Marshes and Mangrove Swamps

Some coastal areas have great biological productivity stemming from trapped organic matter and sediments. Such a rich coastal marsh environment can greatly outproduce a wheat field in raw vegetation per acre. These wetland ecosystems are quite fragile and are threatened by human development.

Coastal wetlands are of two general types—salt marshes and mangrove swamps. In the Northern Hemisphere, **salt marshes** tend to form north of the 30th parallel, whereas **mangrove swamps** form equatorward of that point. This is dictated by the occurrence of freezing conditions, which control the survival of mangrove seedlings. Roughly the same latitudinal limits apply in the Southern Hemisphere.

Salt marshes usually form in estuaries and behind barrier beaches and spits. An accumulation of mud produces a site for the growth of *halophytic* (salt-tolerant) plants. Plant growth then traps additional alluvial sediments and adds to the salt marsh area. Because salt marshes are in the intertidal zone (between the farthest reaches of high and low tides), sinuous, branching channels are produced as tidal waters flood into and ebb from the marsh (Figure 13-15).

Sediment accumulation on tropical coastlines provides the site for mangrove trees, shrubs, and other vegetation. The prop roots of the mangrove are con-

stantly finding new anchorages (Figure 13-16b). They are visible above the water line but reach below the water surface, providing a habitat for a multitude of specialized lifeforms. Mangrove swamps often secure and fix enough material to form islands.

Mangrove losses are estimated by the World Resources Institute and the U.N. Environment Programme at between 40% (e.g., Cameroon and Indonesia) to nearly 80% (e.g., Bangladesh and Philippines) since pre-agricultural times. Deliberate removal was a common practice by many governments because of a falsely conceived fear of disease or pestilence in these swamplands.

Human Impact on Coastal Environments

The development of modern societies is closely linked to estuaries, wetlands, barrier beaches, and coastlines. Estuaries are important sites of human settlement be-

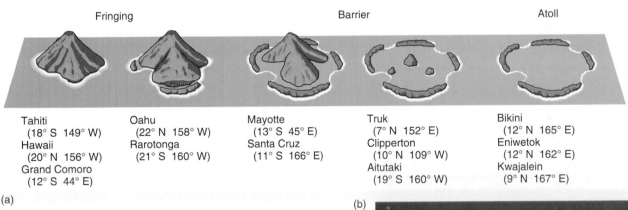

| Fringing | | Barrier | | Atoll |

Tahiti	Oahu	Mayotte	Truk	Bikini
(18° S 149° W)	(22° N 158° W)	(13° S 45° E)	(7° N 152° E)	(12° N 165° E)
Hawaii	Rarotonga	Santa Cruz	Clipperton	Eniwetok
(20° N 156° W)	(21° S 160° W)	(11° S 166° E)	(10° N 109° W)	(12° N 162° E)
Grand Comoro			Aitutaki	Kwajalein
(12° S 44° E)			(19° S 160° W)	(9° N 167° E)

(a)

(b)

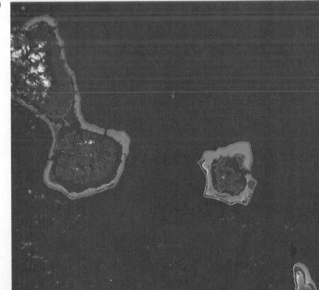

FIGURE 13-14

(a) Common coral formations in a sequence of reef growth formed around a subsiding volcanic island: fringing reefs, barrier reefs, and an atoll. (b) Space Shuttle photograph of the atolls in Bora Bora, Society Islands, 16° 30' S, 151° 45' W. [(a) After D. R. Stoddart, *The Geographical Magazine* LXIII (1971): 610. (b) Space Shuttle photo from NASA.]

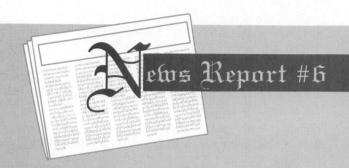

News Report #6

"Worldwide Coral Bleaching Worsens"

Corals are host to colorful algae that live within and upon the coral. Normally colorful corals have turned stark white as nutrient-supplying algae are expelled by the host coral. Scientists at the University of Puerto Rico and NOAA are tracking an unprecedented bleaching and dying-off of corals worldwide. The Caribbean, Australia, Japan, Indonesia, Kenya, Florida, Texas, and Hawaii are experiencing this phenomenon.

Exactly why the coral ejects its living partner is unknown. Possibilities include local pollution, disease, sedimentation, and changes in salinity. Another possible cause is the 1 to 2C° (1.8–3.6F°) warming of sea-surface temperatures, as stimulated by greenhouse warming of the atmosphere. During the 1982–1983 ENSO ("El Niño"), areas of the Pacific Ocean were warmer than normal and widespread coral bleaching occurred. Coral bleaching worldwide is continuing as average ocean temperatures climb higher. Further bleaching through the decade will effect most of Earth's coral-dominated reefs and may be an indicator of serious and enduring environmental trauma. If present trends continue most of the living corals on Earth could perish around A.D. 2000!

cause they provide natural harbors, a food source, and convenient sewage and waste disposal. Society depends upon daily tidal flushing of estuaries to dilute the pollution created by waste disposal. Thus, the estuarine and coastal environment is vulnerable to abuse and destruction if development is not carefully planned. In the United States and Canada about 50% of estuarine zones now have been altered or destroyed, and by the end of this decade most barrier islands will have been developed and occupied. A useful comparison is the fact that only 28 of 280 barrier islands in the United States were developed to any degree before World War II.

Barrier islands and coastal beaches do migrate over time. Houses that were more than a mile from the sea in the Hamptons along the southeastern shore of Long Island, New York, are now within only 30 m (100 ft) of it. The time remaining for these homes can be quickly shortened by a single hurricane or by a sequence of storms with the intensity of a series that occurred in 1962 or the severe storm that struck the U.S. and Canadian Atlantic coast in December 1992.

Despite our understanding of beach and barrier-island migration, the effect of storms, and warnings from scientists and government agencies, coastal development proceeds. Society behaves as though beaches and barrier islands are stable, fixed features, or as though they can be engineered to be permanent. Experience has shown that severe erosion generally cannot be prevented, as we have seen with the cliffs of southern California, the shores of the Great Lakes, the New Jersey shore, the Gulf coast, and the devastation of coastal South Carolina by Hurricane Hugo. Shoreline planning is the topic of FYI Report 13-1.

SUMMARY—Coastal Processes and Landforms

Rachel Carson called the edge of the sea "A strange and beautiful place." The coastal environment is called the **littoral zone** and exists where the tide-driven, wave-driven sea confronts the land. Indeed, physical processes in operation along coastlines pro-

FIGURE 13-15
Salt marsh, a productive ecosystem commonly occurring poleward of 30° latitude
in both hemispheres. This is Bolinas Lagoon on the Pacific coast in central
California. [Photo by author.]

(a)

FIGURE 13-16
Mangroves tend to grow equatorward of 30° latitude.
(a) Mangroves along the East Alligator River (12° S
latitude) in Kakadu National Park, Northern Territory,
Australia. (b) Mangroves retain sediments and can
form anchors for island formations such as Aldabra
Island (9° S), Seychelles. [(a) Photo by Belinda
Wright/DRK Photo, (b) photo by Wolfgang Kaehler.]

(b)

An Environmental Approach to Shoreline Planning

Coastlines are places of wonderful opportunity. They also are zones of specific constraints. Poor understanding of this resource and a lack of environmental analysis often go hand in hand. Ecologist and landscape architect Ian McHarg, in *Design with Nature,* discusses the New Jersey shore. He shows how proper understanding of a coastal environment could have avoided problems from major storms and coastal development. Much of what is presented here applies to coastal areas elsewhere. Figure 1 illustrates the New Jersey shore from ocean to back bay. Let's walk across this fragile landscape and discover how it should be treated under ideal conditions.

Beaches and Dunes—Where to Build?

Sand beaches are the primary natural defense against the ocean; they act as bulwarks against the pounding of a stormy sea. But they are susceptible to pollution and require environmental protection to control nearshore dumping of dangerous materials. The shoreline tolerates recreation, but not construction because of its shifting, changing nature during storms, daily tidal fluctuations, and the potential effects of rising sea level. In recent years, high water levels in the Great Lakes and along the southern California coast attest to the vulnerability of shorelines to erosion and inundation. Wave erosion at-

FIGURE 1

Coastal environment—a planning perspective from ocean to bay. [After *Design with Nature* by Ian McHarg. Copyright © 1969 by Ian L. McHarg. Adapted by permission of Ian McHarg.]

Ocean	Beach	Primary dune	Trough	Secondary dune	Backdune	Bayshore	Bay
Tolerant	Tolerant	Intolerant	Relatively tolerant	Intolerant	Tolerant	Intolerant	Tolerant
Intensive recreation	Intensive recreation	No passage, breaching, or building	Limited recreation	No passage, breaching, or building	Most suitable for development	No filling	Intensive recreation
Subject to pollution controls	No building		Limited structures				
	Intolerant of construction						

416

tacks cliff formations and undermines, bit by bit, the foundations of houses and structures built too close to the edge (Figure 2).

The primary dune along a coast also is intolerant of heavy use, even the passage of people trekking to the beach; it is fragile, easily disturbed, and vulnerable to erosion. Delicate plants struggle to hold the sand in place. Primary dunes are like human-made dikes in the Netherlands; they are the primary defense against the sea. Carefully controlled access points to the beach should be enforced (see Figure 12-3).

The trough behind the primary dune is relatively tolerant, with limited recreation and building potential. The plants that fix themselves to the surface send roots down to fresh groundwater reserves. Thus, if construction should inhibit the surface recharge of that supply, the natural protective ground cover could fail and destabilize the environment; or subsequent saltwater intrusion might contaminate well water for human use. Clearly, groundwater resources and the location of recharge aquifers must be considered in planning.

Behind the trough is the secondary dune, a second

line of defense against the sea. It, too, is tolerant of some use yet is vulnerable to destruction through improper use.

The backdune is more suitable for development than any zone between it and the sea. Further inland are the bayshore and the bay, where ideally there should be no dredging and filling and only limited dumping of treated wastes and toxics. In reality, the opposite of such careful assessment and planning prevails.

A Scientific View and a Political Reality

Prior to an intensive government study completed in 1962, no analysis of coastal hazards had been done outside academic circles. Thus, what is common knowledge to geographers, botanists, biologists, and ecologists in the classroom and laboratory still has not filtered through to the general planning and political processes. As a result, on the New Jersey shore and along much of the Atlantic and Gulf coasts, improper development of the fragile coastal zone led to extensive destruction during storms in 1962, 1992, and numerous hurricanes of this century.

When the storms of 1962 hit, scientists were studying the area. They forecast a strong probability that the Cape May portion of the New Jersey shore could be 90 m (300 ft) farther inland by the year 2010, less than two decades from now. There is a 15% chance that the shore could be even twice as far inland (180 m or 600 ft) by that time. Similar estimates persist along the entire East Coast, making it clear that society must synthesize *ecology* and *economics* if these coastal environments are to be sustained.

South Carolina enacted their Beach Management Act of 1988 (modified 1990) to apply some of McHarg's principles. Vulnerable coastal areas are protected from new construction or rebuilding. Now guided by law are structure size, replacement limits for damaged structures, and placement of structures on lots, although liberal interpretation is allowed. The act sets standards for different coastal forms: beach and dunes, eroding shorelines, or a hypothetical baseline along a coast that lacks dunes or has unstabilized inlets. In the first two years more than 70 lawsuits protested the act as an invalid seizure of private property without compensation. Implementation has been difficult, political pressure intense, and results mixed. Similar measures in other states have faced the same difficult path.

The key to protective environmental planning and zoning is the allocation of responsibility and cost in the event of a disaster. An ideal system places a hazard tax on land based on assessed risk and restricts the government's responsibility to fund reconstruction or an individual's right to reconstruct on frequently damaged sites. Comprehensive mapping of erosion-hazard areas would help avoid ever-increasing costs from recurring disasters. This type of planning should include coastlines, floodplains, earthquake and volcanic hazard areas, sites with toxic contamination, and important scenic and scientific areas. These changes require amending the National Flood Insurance Act.

Ian McHarg summarizes the situation well:

May it be that these simple ecological lessons will become known and incorporated into ordinance [law] so that people can continue to enjoy the special delights of life by the sea.*

---

*From *Design with Nature* by Ian McHarg, p. 17. Copyright © 1969 by Ian L. McHarg. Published by Doubleday, a division of Bantam Doubleday Dell Publishing Group, Inc.

---

duce relatively rapid change, much faster than the rate of change inland. Physical geographers are interested in the spatial aspects of these processes, the landscapes created, and the implications of each unique coastline to planning and possible development.

Coastal processes operate as open systems with *inputs* of solar energy through the fluid agents of air and water and sea level, *actions* of **tides** and **mean sea-level** variation and **wave** motion, and *outputs* of erosional, depositional, and organic landforms and physical features. These features include terraces, embayments, **lagoons**, cliffs, and beaches. A **beach** is that place along a coast where sediment is in motion, deposited by waves and currents and attached to the margin of the continent along its entire length. Throughout geologic time, coastal margins have been subjected to numerous changes in sea level, with each change altering wave and tidal processes.

Headlands are protruding landforms along a coast generally composed of resistant rocks, and they are a specific focus for waves attacking a coastline. As waves enter shallow water, they are refracted and generate a current parallel to the coast called a **longshore current**. Reefs, **barrier islands**, and coastlines are produced by organic formations built by corals, salt marshes, and mangroves. Presently, sea level is rising and sea-surface temperatures are increasing. The varied interaction of ocean and continent maintains a fragile balance. It can be hazardous to humans, and at the same time damaged by human occupation. The coastal zone, like the river floodplain, requires careful assessment of natural factors for responsible use. Ecological planning can enhance long-term economic goals and our enjoyment of Earth's wonderful coastal environments.

KEY TERMS

barrier beach

barrier island

barrier spit

bay barrier

beach

beach drift

breaker

coral

headland

lagoon

littoral zone

longshore current

mangrove swamp

mean sea level

salt marsh

swells

tides

tsunami

wave-cut platform

wave refraction

wave

REVIEW QUESTIONS

1. What are the key terms used to describe the coastal environment?
2. Define mean sea level. How is this value determined? Is it constant or variable around the world? Explain.
3. Is tidal power being used anywhere to generate electricity? Explain briefly how such a plant would utilize the tides to produce electricity. Are there any sites in North America?
4. What is a wave? How are waves generated, and how do they travel across the ocean? Discuss the process of wave formation and transmission.
5. Describe the refraction process that occurs when waves reach an irregular coastline.
6. What is meant by an erosional coast? What are the expected features of such a coast?
7. What are some of the depositional features encountered along a coastline?
8. How do people attempt to modify littoral and beach drift? What are the positive and negative impacts of these actions?
9. Describe a beach—its form, composition, function, and evolution.
10. What success has Miami had with beach replenishment? Is it a practical strategy?
11. Based on the information in the text and any other sources at your disposal, do you think barrier islands and beaches should be used for development? If so, under what conditions? If not, why not?
12. After the Grand Strand off South Carolina was destroyed by Hurricane Hazel in 1954, settlements were rebuilt, only to be hit by Hurricane Hugo 35 years later, in 1989. Why do these recurring events happen to human populations? Compare the impact of the two storms.
13. Describe the western and eastern coasts of North America as emergent or submergent coastlines.
14. How are corals able to construct reefs and islands?
15. Why are the coastal wetlands poleward of 30° N and S latitude different from those that are equatorward? Describe the differences.
16. What type of environmental analysis is needed for rational development and growth in a region like the New Jersey shore? Evaluate South Carolina's approach to coastal hazards and protection.

Icefall and waterfall at terminus of Bow Glacier, Banff National Park, Alberta. [Photo by Martin G. Miller.]

14

GLACIAL AND PERIGLACIAL LANDSCAPES

RESERVOIRS OF ICE
  Types of Glaciers
GLACIAL PROCESSES
  Formation of Glacial Ice
  Glacial Mass Balance   Glacial Movement
GLACIAL LANDFORMS
  Erosional Landforms Created by Alpine
    Glaciation
  Depositional Landforms Created by Alpine
    Glaciation
  Erosional and Depositional Features of
    Continental Glaciation
PERIGLACIAL LANDSCAPES
  Permafrost   Frozen Ground Phenomena
  Humans and Periglacial Landscapes
THE PLEISTOCENE ICE AGE EPOCH
  Pluvial Periods and Paleolakes
ARCTIC AND ANTARCTIC REGIONS
  The Antarctic Ice Sheet
SUMMARY
FYI REPORT 14-1 DECIPHERING PAST CLIMATES:
  PALEOCLIMATOLOGY

Alarge measure of the freshwater on Earth is frozen, with the bulk of that ice sitting restlessly in just two places—Greenland and Antarctica. The remaining ice covers various mountains and fills some alpine valleys. More than 29 million km³ (7 million mi³) of water is tied up as ice, or about 77% of all freshwater. These deposits of ice provide an extensive frozen record of Earth's climatic history over the past several million years and perhaps some clues to its climatic future.

This chapter focuses on Earth's extensive ice deposits—their formation, movement, and the ways in which they function to produce various erosional and depositional landforms. Glaciers are transient landforms themselves and leave in their wake a variety of landscape features. The fate of glaciers is intricately tied to changes in global temperature that ultimately concerns us all.

Approximately 20% of Earth's land area is subject to freezing conditions and frost action characteristic of periglacial regions (areas near glaciers). We examine in this chapter the cold, near-glacial world of permafrost and periglacial processes. Finally, we see reminders of the last ice age over many parts of the globe. An FYI Report details methods used to decipher past climates—the science of *paleoclimatology*.

Reservoir of Ice

A **glacier** is a large mass of ice, resting on land or floating shelflike in the sea adjacent to land. Glaciers are not frozen lakes or groundwater ice but form by the continual accumulation of snow that recrystallizes into an ice mass under its own weight. They move under the pressure of their own great mass and the pull of gravity. Today, about 11% of Earth's land area is dominated by these slowly flowing streams of ice. During colder episodes in the past, as much as 30% of continental land was covered by glacial ice. Through these "ice ages," below-freezing temperatures prevailed at lower latitudes, allowing snow to accumulate year after year.

A *snowline* is the lowest elevation where snow remains year-round—specifically, the lowest line

where winter snow accumulation persists throughout the summer. Glaciers form in such areas of permanent snow, both at high latitudes and high elevations. Glaciers even form on some high mountains along the equator, such as in the Andes Mountains of South America or on Mount Kilimanjaro in Tanzania. In equatorial mountains, the snowline is around 5000 m (16,400 ft); on midlatitude mountains, such as the European Alps, snowlines average 2700 m (8850 ft); and in southern Greenland snowlines are down to 600 m (1970 ft).

Types of Glaciers

Glaciers are as varied as the landscape itself. They fall within two general groups, based on their form, size, and flow characteristics: alpine glaciers and continental glaciers.

Alpine Glaciers. With few exceptions, a glacier in a mountain range is called an **alpine glacier**, or *mountain glacier*. The name comes from the Alps of central Europe, where such glaciers abound. Alpine glaciers form in several subtypes. One prominent type is a **valley glacier**, an ice mass confined within a valley that originally was formed by stream action. Such glaciers range in length from only 100 m (325 ft) to over 100 km (60 mi). The *snowfield* that continually feeds the glacier with new snow is at a higher elevation. In Figure 14-1a, at least a dozen valley glaciers are identifiable in the *Landsat* image of the Alaska Range and several are named. Figure 14-1b is a high-altitude photograph of the Eldridge and Ruth glaciers, filling valleys as they flow from source areas near Mount McKinley.

As a valley glacier flows slowly downhill, the mountains, canyons, and river valleys beneath its mass are profoundly altered by its erosive passage. Some of the debris created by the glacier's excavation is transported in and on the ice, visible as dark streaks of material being transported for deposition elsewhere.

Most alpine glaciers originate in a mountain snowfield that is confined in a bowl-shaped recess. This scooped-out erosional landform at the head of a val-

FIGURE 14-1
Glaciers in south-central Alaska. (a) Valley glaciers in the Alaska Range of Denali National Park; (b) oblique infrared image of Eldridge and Ruth glaciers, with Mount McKinley at upper left, taken at 18,300 m (60,000 ft). [(a) *Landsat* image from NASA; (b) Alaska High Altitude Aerial Photography from EROS Data Center.]

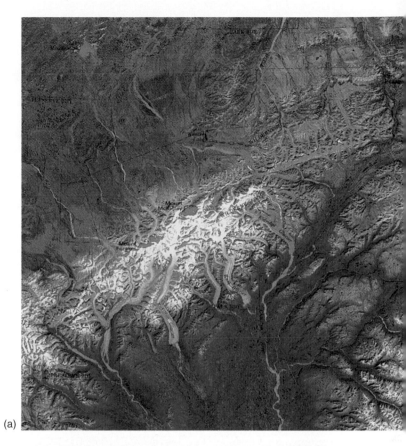

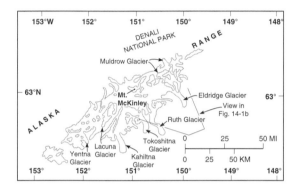

(a)

(b)

ley is called a **cirque**. A glacier that forms in a cirque is called a *cirque glacier*. Several cirque glaciers may jointly feed a valley glacier. Numerous cirques and cirque glaciers appear in the high-altitude photo in Figure 14-1b. See if you can identify these glaciers feeding into tributary valley glaciers.

Wherever several valley glaciers pour out of their confining valleys and coalesce at the base of a mountain range, a *piedmont glacier* is formed and spreads freely over the nearby lowlands. A *tidal glacier,* such as the Columbia Glacier on Prince William Sound in Alaska, ends in the sea, *calving* (breaking off) to form floating ice called **icebergs**. Icebergs usually form wherever glaciers meet the ocean.

Continental Glaciers. On a larger scale than individual alpine glaciers, a continuous mass of ice is known as a **continental glacier** and in its most extensive form is called an **ice sheet**. Most glacial ice exists in the snow-covered ice sheets that blanket 80% of Greenland and 90% of Antarctica. Antarctica alone has 91% of all the glacial ice on the planet.

These two ice sheets represent such an enormous mass that large portions of each landmass have been isostatically depressed (under the weight of the ice) more than 2000 m (6500 ft) below sea level. Each ice sheet is more than 3000 m (9800 ft) deep, burying all but the highest peaks. An ice sheet is the most extensive form of continental glacier.

Two additional types of continuous ice cover associated with mountain locations are designated as *ice caps* and *ice fields*. An **ice cap** is roughly circular and covers an area of less than 50,000 km² (19,300 mi²). An ice cap completely buries the underlying landscape. The volcanic island of Iceland features several ice caps (Figure 14-2a).

An **ice field** extends in a characteristic elongated pattern in a mountainous region with ridges and peaks visible above the buried terrain. Figure 14-2b shows the southern Patagonian ice field in the Andes, one of Earth's largest, stretching 360 km (224 mi) in length between 46° and 51° S latitude and up to 90 km (56 mi) in width. An ice field is not extensive enough to form the characteristic dome of an ice cap.

Glacial Processes

A glacier is composed of dense ice that is formed from snow and water through a process of compaction, recrystallization, and growth. A glacier is an open system, with *inputs* of snow and moisture and *outputs* of melting ice and evaporation. Please refer to Figure 14-3 for an overview of how glaciers operate. These inputs and outputs produce a *mass balance*, readily visible in the glacier's physical stature. A glacier's *budget* consists of net gains or losses of glacial ice, which determine whether the glacier expands or retreats. A glacier is a dynamic body, moving relentlessly downslope, greatly modifying the landscape through which it flows. Let's look at glacial ice formation, mass balance, movement, and erosion before we discuss specific landforms produced by these processes.

Formation of Glacial Ice

The essential input to a glacier is precipitation that accumulates in a *snowfield,* a glacier's accumulation zone (Figure 14-3a and b). Snowfields usually are at the highest elevation of an ice sheet, ice cap, or head of a valley glacier, usually in a cirque. Highland snow accumulation sometimes is the product of orographic processes, as discussed in Chapter 5. Avalanches from surrounding mountain slopes can add to the snowfield. As the snow accumulation deepens in sedimentary-like layers, the increasing thickness results in increased weight and pressure on underlying portions. Rain and summer snowmelt then contribute water, which stimulates further melting, and that meltwater seeps down into the snowfield and refreezes.

Snow that survives the summer and into the following winter begins a slow transformation into glacial ice. Air spaces among ice crystals are pressed out as snow packs to a greater density. The ice crystals recrystallize and consolidate under pressure. In a transition step to glacial ice, snow becomes **firn**, which has a compact, granular texture.

As this process continues, many years pass before denser glacial ice is produced. Formation of **glacial ice** is analogous to formation of metamorphic rock:

FIGURE 14-2
Ice cap and ice field. (a) The Vatnajökull ice cap in southeastern Iceland (*jökull* means ice cap in Danish). (b) The southern Patagonian ice field of Argentina. [(a) *Landsat* image from NASA; (b) photo by Cosmonauts G. M. Greshko and Yu V. Romanenko, *Salyut 6.*]

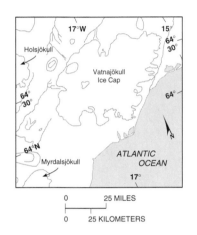

(a)

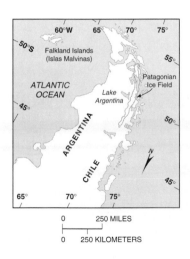

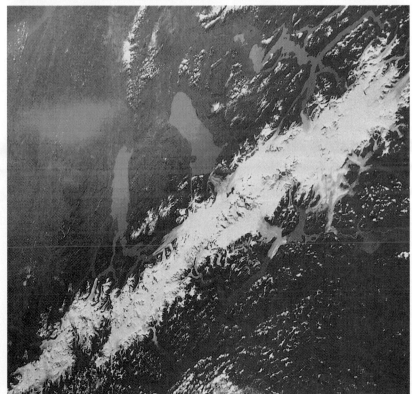

(b)

FIGURE 14-3

(a) Annual mass balance of a glacial system, showing the relationship between accumulation and ablation and the location of the equilibrium line. (b) Cross section of a typical retreating alpine glacier. (c) John Hopkins Glacier, Glacier Bay National Park, Alaska, demonstrates many of the features in the illustration. [(c) Photo by Frank S. Balthis.]

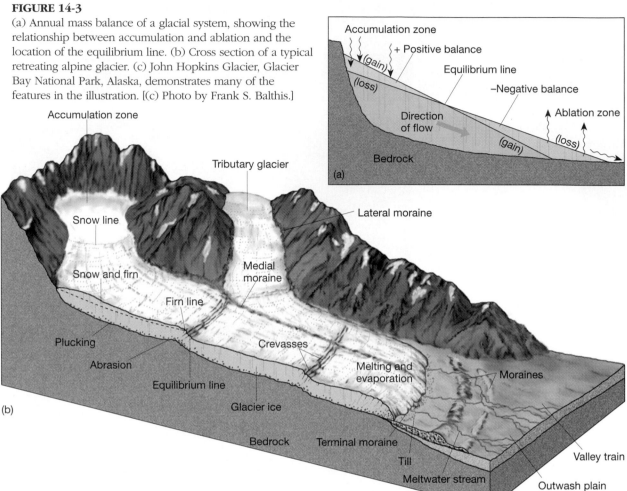

sediments (snow and firn) are pressured and recrystallized into a dense metamorphic rock (glacial ice). In Antarctica, glacial ice formation may take 1000 years because of the dryness of the climate (minimal snow input), whereas in wet climates this time is reduced to just a few years because of the volume of new snow constantly being added to the system.

Glacial Mass Balance

A glacier is fed by snowfall and other moisture sources and is wasted (reduced) by losses from its upper and lower surfaces and along its margins. A snowline (fresh snow-covered area) called a **firn line** is visible across the surface of a glacier, indicating where the winter snow and ice accumulation survived the summer melting season. Toward a glacier's lower end, losses of mass occur because of surface melting, internal and basal melting, sublimation (recall from Chapter 5, "Heat Properties," that this is the direct evaporation of ice), wind removal by *deflation*, and the calving of ice blocks. The combined effect of these losses is called **ablation**.

The zone where accumulation gain balances ablation loss is the **equilibrium line** (Figure 14-3a). This area of a glacier generally coincides with the firn line (Figure 14-3b). Glaciers achieve a positive net balance of mass—grow larger—during colder periods with adequate precipitation, and the equilibrium line shifts down-glacier. In warmer times, the equilibrium line migrates up-glacier and the glacier retreats—grows smaller—due to its negative net balance. Internally, gravity continues to move a glacier forward even though its lower terminus might be in retreat due to ablation.

Highways of ice that flow from accumulation areas high in the mountains are marked by trails of transported debris called *moraines* (Figure 14-3b and c). A *lateral moraine* accumulates along the sides. A *medial moraine* forms down the middle when two glaciers merge and their lateral moraines combine.

Glacial Movement

We generally think of ice as those small, brittle cubes from the freezer, but glacial ice is quite different. In fact, glacial ice behaves in a plastic manner, for it distorts and flows in its lower portions in response to weight and pressure from above and the degree of slope below. In contrast the glacier's upper portions are quite brittle. Rates of flow range from almost nothing to a kilometer or two per year on a steep slope. The rate of accumulation of snow in the formation area is critical to the speed.

The movement of a glacier is *not* like a block of rock sliding downhill. The greatest movement within a valley glacier occurs *internally,* below the rigid surface layer, which fractures as the underlying zone moves plastically forward (Figure 14-4a). At the same time, the base creeps and slides along, varying its speed with temperature and the presence of any lubricating water beneath the ice. This *basal slip* usually is much less rapid than the internal plastic flow of the glacier, so the upper portion of the glacier flows ahead of the lower portion.

In addition, unevenness in the landscape beneath the ice may vary the pressure, melting some of the basal ice by compression at one moment, only to have it refreeze later; this process is an important factor in glacial down-slope movement. This melting/refreezing incorporates rock debris into the glacier. Consequently, the basal ice layer, which can extend tens of meters above the base of the glacier, has a much greater debris content than in the ice above.

A flowing glacier can develop vertical cracks known as **crevasses** (Figure 14-4b). These result from friction with valley walls, or tension from stretching as the glacier passes over convex slopes, or compression as the glacier passes over concave slopes. Traversing a glacier, whether an alpine glacier or an ice sheet, is dangerous because a thin veneer of snow sometimes masks the presence of a crevasse. One type of crevasse occasionally associated with a cirque glacier is known as a *bergschrund,* or *headwall crevasse;* it forms when snow and firn compact and pull away from ice that remains frozen to rock at the head of the glacier.

Tributary valley glaciers merge to form a *compound* valley glacier. The flowing movement of a compound valley glacier is different from that of a river with tributaries. Tributary glaciers flow into a compound glacier and *merge alongside* one another by extending and thinning, rather than blending as do rivers. You can see this because each tributary

"South Cascade Glacier Loses Mass"

The net mass balance of the South Cascade Glacier in Washington State, demonstrated significant losses between 1955 and 1992. In just one year (September 1991 to October 1992) the terminus of the glacier retreated 38 m (125 ft) and resulted in other major changes to the surface and sides of the glacier. This represents more than a 2% loss in the glacier's mass in just one year.

The reasons for these losses are complex. Possible causes are increasing average air temperature and decreasing precipitation. In comparing the trend of this glacier's mass balance to others in the world, apparent temperature changes are causing widespread reductions in middle-and lower-elevation glacial ice. The wastage (losses of ice) of alpine glaciers worldwide is thought by scientists potentially to contribute over 25% to sea level rise.

---

maintains its own patterns of transported debris (dark streaks). This is visible on the John Hopkins Glacier (Figure 14-3c).

Glacier Surges. Although glaciers flow plastically and predictably most of the time, some will lurch forward with little or no warning in a **glacial surge**. This is not quite as abrupt as it sounds; in glacial terms, a surge can be tens of meters per day. The Jakobshavn Glacier in Greenland, for example, is known to move between 7 and 12 km (4.3 and 7.5 mi) a year.

In the spring of 1986 Hubbard Glacier and its tributary Valerie Glacier surged across the mouth of Russell Fjord in Alaska, cutting it off from contact with Yakutat Bay. This area in southeastern Alaska is fed by annual snowfall in the Saint Elias mountain range that averages more than 8.5 m (28 ft) a year, so the surge event had been predicted. But the rapidity of the surge was surprising. The fjord was dubbed Russell Lake for the time it remained isolated (Figure 14-5). The glacier's movement exceeded 34 m (112 ft) per day during the peak surge, an enormous increase over its normal rate of 15 cm (6 in.) per day.

The exact cause of such a glacial surge is being studied. Some surge events result from a buildup of water pressure under the glacier, sometimes enough to actually float the glacier slightly. As a surge begins, icequakes are detectable, and ice faults are visible.

Glacial Erosion. Glacial erosion is similar to a large excavation project, with the glacier hauling debris from one site to another for deposition. The passing glacier mechanically plucks rock material and carries it away. Rock pieces actually freeze to the basal layers of the glacier in this *glacial plucking* process and, once embedded, allow the glacier to scour and sandpaper the landscape as it moves, a process called **abrasion**. Abrasion and gouging produce a smooth surface on exposed rock, which shines with *glacial polish* when the glacier retreats. Larger rocks in the glacier act much like chisels, working the underlying surface to produce glacial striations (scratches) in the flow direction (Figure 14-6).

FIGURE 14-4
(a) Cross section of a glacier, showing the nature of its forward motion at the surface and along its basal layer. (b) Surface crevasses and cracks are evidence of a glacier's forward motion toward the top and upper-right portion of the photo (near the Don Sheldon Amphitheater, Denali National Park, Alaska). [(b) Photo by Michael Collier.]

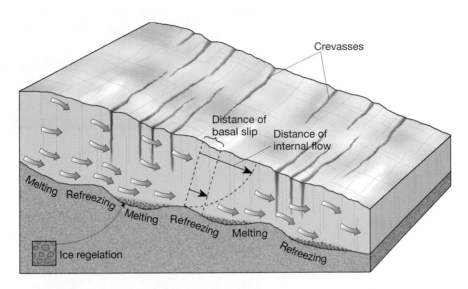

Glacial Landforms

Glacial erosion and deposition produce unique landforms. You might expect all glaciers to create the same landforms, but this is not so. Alpine and continental glaciers each generate their own characteristic landscapes. We look first at erosional landforms created by alpine glaciers, then at their depositional landforms. Finally, we examine the results of continental glaciation.

Erosional Landforms Created by Alpine Glaciation

Geomorphologist William Morris Davis characterized the stages of a valley glacier in a set of drawings

FIGURE 14-5
The Hubbard Glacier surged across Russell Fjord in Alaska, effectively damming the fjord and temporarily creating Russell Lake (lower right). [*Landsat* IR image from EROS Data Center.]

FIGURE 14-6
Glacial polish and striations are examples of glacial erosion. [Photo courtesy of U.S. Geological Survey.]

published in 1906 and redrawn here in Figure 14-7. Illustration (a) shows a typical river valley with characteristic **V**-shape and stream-cut tributary valleys that exist before glaciation. Illustration (b) shows that same landscape during a later period of active glaciation. Glacial erosion and transport are actively removing much of the regolith (weathered bedrock) and the soils that covered the stream valley landscape. Illustration (c) shows the same landscape at a later time when climates have warmed and ice has retreated. The glaciated valleys now are **U**-shaped, greatly changed from their previous stream-cut form. You can see the oversteepened sides and the straightened course of the valley. The physical weathering associated with the freeze-thaw cycle has loosened much rock along the steep cliffs, and the rock has fallen to form *talus cones* along the valley sides during the postglacial period.

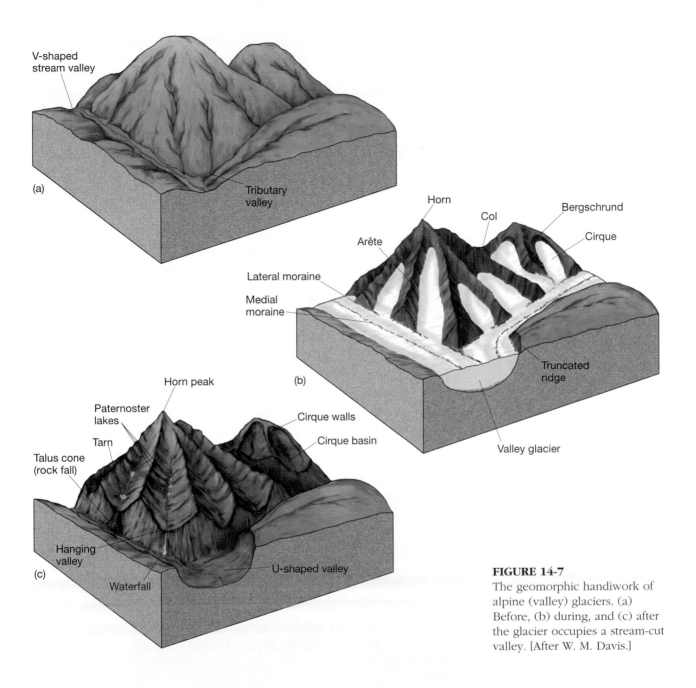

FIGURE 14-7
The geomorphic handiwork of alpine (valley) glaciers. (a) Before, (b) during, and (c) after the glacier occupies a stream-cut valley. [After W. M. Davis.]

The cirques where the valley glacier originated are clearly visible in (c). As the cirque walls wore away, an **arête** formed, a sharp ridge that divides two cirque basins (b). Arêtes ("knife-edge" in French) become the sawtooth and serrated ridges in glaciated mountains. Two eroding cirques may reduce an arête to form a pass or saddlelike depression, called a **col**. A pyramidal peak called a **horn** results when several cirque glaciers gouge an individual mountain summit from all sides. The most famous example is the Matterhorn in the Swiss Alps, but many others occur.

A small mountain lake, especially one that collects in a cirque basin behind risers of rock material, is called a *tarn* (c). Small, circular, stair-stepped lakes in a series are called *paternoster lakes* because they look like a string of rosary (religious) beads forming in individual rock basins aligned down the course of a glaciated valley. Such lakes may have been formed by the differing resistance of rock to glacial processes or by the damming effect of glacial deposits.

Figure 14-7c also shows tributary glacier valleys left stranded high above the glaciated valley floor, following the removal of previous slopes and stream courses by the glacier. These *hanging valleys* are the sites of spectacular waterfalls as streams plunge down the steep cliffs (Figure 14-8).

Where a glacial trough intersects the ocean, the glacier can erode the landscape below sea level. As the glacier retreats, the trough floods and forms a deep **fjord** in which the sea extends inland, filling the lower reaches of the steep-sided valley. The fjord may be flooded further by a rising sea level or by changes in the elevation of the coastal region. All along the glaciated coast of Alaska, glaciers now are in retreat, thus opening many new fjords that previously were blocked by ice. Coastlines with notable fjords include those of Norway, Greenland, Chile, the South Island of New Zealand, Alaska, and British Columbia.

Depositional Landforms Created by Alpine Glaciation

You have just seen how glaciers excavate tremendous amounts of material and create fascinating landforms in the process. Glaciers also produce another set of landforms when they melt and deposit their eroded and transported debris cargo, which is carried toward the *terminus* (end) of the glacier (Figure 14-3b). Where the glacier melts, this debris accumulates to mark the former margins of the glacier—the end and sides. **Glacial drift** is the general term for *all* glacial deposits, both unsorted and sorted. Direct ice deposits are unstratified and *unsorted* and are called **till**. Sediments deposited by glacial meltwater are *sorted* and are termed **stratified drift**.

Moraine is the name for specific landforms produced by the deposition of these sediments. Several types of moraines are exhibited in Figure 14-9. A **lateral moraine** forms along each side of a glacier. If two glaciers with lateral moraines join, a **medial moraine** may form as lateral moraines combine as in the figure. (Can you identify these elements on the active glacier in Figure 14-3c?)

Eroded debris that is dropped at the glacier's farthest extent is called a *terminal moraine*. However,

FIGURE 14-8
Bridalveil Creek hanging valley and Bridalveil Falls, part of the scenic beauty of California's Yosemite National Park and its glaciated stream valleys. [Photo by author.]

FIGURE 14-9
Medial, lateral, terminal (end) and ground moraine deposits are evident in this photo. These were produced by the Hole-in-the-Wall Glacier, Wrangell-Saint Elias National Park, Alaska. Also, note the kettle pond in the foreground. [Photo by Tom Bean.]

there also may be *end moraines,* formed wherever a glacier pauses after reaching a new equilibrium. Both of these forms are clearly visible in Figure 14-9. Lakes may form behind terminal and end moraines following a glacier's retreat, with the moraine acting as a dam. If a glacier is in retreat, individual deposits are called *recessional moraines.* A deposition of till generally spread across a surface is called a *ground moraine,* or *till plain,* and may hide the former landscape. Such plains are found in portions of the Midwest.

All of these types of till are unsorted and unstratified debris. In contrast, streams of glacial meltwater can deposit sorted and stratified glacial drift beyond a terminal moraine. This type of meltwater-deposited material downvalley from a glacier is called a *valley train deposit.* Peyto Glacier in Alberta produces such a valley train that continues into Peyto Lake (Figure 14-10). Distributary stream channels appear braided

across its surface. The picture also shows the milky meltwater associated with glaciers, laden with finely ground "rock flour". Meltwater is produced by glaciers at all times, not just when they are retreating.

Erosional and Depositional Features of Continental Glaciation

The extent of the most recent continental glaciation in North America and Europe, 18,000 years ago, is portrayed in Figure 14-20. When these huge sheets of ice advanced and retreated, they produced many of the erosional and depositional features characteristic of alpine glaciation. However, because continental glaciers form under different circumstances across broad, open landscapes the intricately carved alpine features, lateral moraines, and medial moraines all are lacking in continental glaciation.

"Glacial Erratic Marks Famous Grave"

The grave of Louis Agassiz (1807-1873), an early glacial theorist and the author of two books on glaciers, is marked by a polished and striated boulder. Retreating glaciers leave behind large rocks (sometimes house-sized), boulders, and cobbles, that are "foreign"—different in composition and origin from the ground on which they were deposited. These *glacial erratics*, lying in strange locations with no obvious means of transport, were an early clue to Agassiz and others that blankets of ice once had covered the land. Dr. Agassiz was an early proponent of a continental glaciation hypothesis. It is fitting that compelling evidence of Earth's "ice age" marks his resting place.

Figure 14-11 illustrates some of the most common erosional and depositional features associated with the retreat of a continental glacier. A **till plain** forms behind end moraines, featuring *unstratified* coarse till, low, rolling relief, and deranged drainage patterns (see Chapter 11, Figure 11-4g). Beyond the morainal deposits, **outwash plains** of *stratified drift* feature stream channels that are meltwater-fed, braided, and overloaded with debris deposited across the landscape.

Figure 14-11a shows a sinuously curving, narrow ridge of coarse sand and gravel called an **esker**. It forms along the channel of a meltwater stream that flows beneath a glacier, in an ice tunnel, or between ice walls beneath the glacier. As a glacier retreats, the steep-sided esker is left behind in a pattern roughly parallel to the path of the glacier. The ridge may not be continuous and in places may even appear branched, following the course set by the subglacial watercourse. Commercially valuable deposits of sand and gravel are quarried from some eskers.

Sometimes an isolated block of ice, perhaps more than a kilometer across, persists in a ground moraine, an outwash plain, or valley floor after a glacier retreats. Perhaps 20 to 30 years are required for it to melt. In the interim, sediment continues to accumulate around the melting ice block. When the block finally melts, it leaves behind a steep-sided hole which frequently fills with water. This is called a **kettle** (Figure 14-11b). Thoreau's Walden Pond (quotation in Chapter 5), is such a kettle.

Another feature of outwash plains is a **kame**, a small hill, knob, or mound of poorly sorted sand and gravel that is deposited directly by water, by ice in crevasses, or in ice-caused indentations in the surface (Figure 14-11c). Kames also can be found in deltaic forms and in terraces along valley walls.

Glacial action also forms streamlined hills, one erosional, called a roche moutonnée, and the other depositional, called a drumlin. A *roche moutonnée* ("sheep rock" in French) is an asymmetrical hill of exposed bedrock. Its gently sloping upstream side (*stoss* side) has been polished smooth by glacial action, whereas its downstream side (*lee* side) is abrupt and steep where rock pieces were plucked by the glacier (Figure 14-12).

A **drumlin** is deposited till that has been streamlined in the direction of continental ice movement, blunt end upstream and tapered end downstream. Multiple drumlins (called *swarms*) occur across the landscape in portions of New York and Wisconsin, among other areas. Sometimes their shape is that of

FIGURE 14-10
Peyto Glacier in Alberta, Canada. Note valley train, braided stream, and glacial meltwater. [Photo by author.]

FIGURE 14-11
Common depositional landforms produced by glaciers. (a) An esker through farmland near Campbellsport, Wisconsin; (b) a kettle surrounded by forest in the Chippewa moraine, Chippewa County, Wisconsin; (c) a kame covered by a woodlot near Campbellsport, Wisconsin. [(a) Photos by Tom Bean/DRK Photos, (b) and (c) by Tom Bean.]

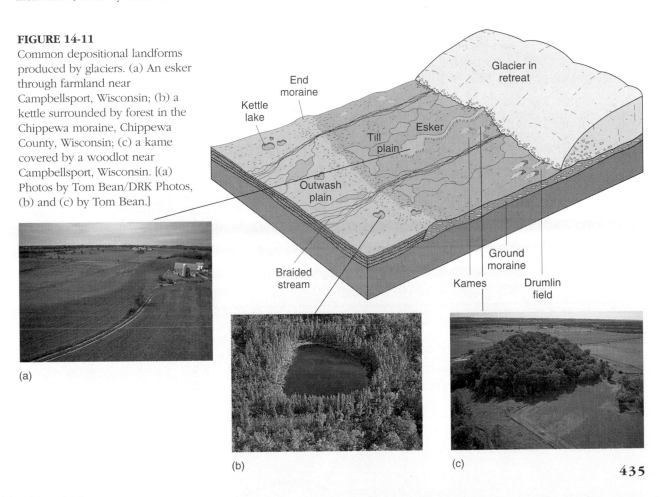

435

FIGURE 14-12
Roche moutonnée, as exemplified by Lambert Dome in the Tuolumne Meadows area of Yosemite National Park, California. [Photo by author.]

an elongated teaspoon bowl, lying face down. They attain lengths of 100–5000 m (330 ft to 3.1 mi) and heights up to 200 m (650 ft). Figure 14-13 shows a portion of a topographic map for the area south of Williamson, New York, which experienced conti- nental glaciation. In studying the map, can you iden- tify the numerous drumlins? In what direction do you think the continental glaciers moved across this re- gion? (Look for more gentle, tapered slopes in the downstream direction as the glacier retreated.)

FIGURE 14-13
Topographic map south of Williamson, New York, featuring numerous drumlins (7.5-minute series quadrangle map, originally produced at a 1:24,000 scale, 10 ft contour interval). [Courtesy of U.S. Geological Survey.]

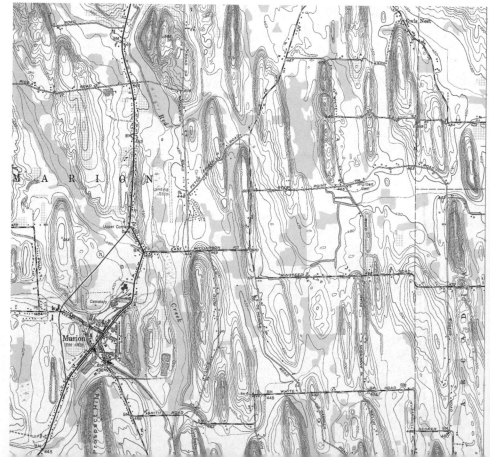

Along the margins of permanent ice, cold-climate processes dominate. Let us now look at these cold-dominated landscapes.

Periglacial Landscapes

The term **periglacial** was coined by scientist W. V. Lozinski in 1909 to describe cold-climate processes, landforms, and topographic features along the margins of glaciers, past and present. These periglacial regions occupy over 20% of Earth's land surface (Figure 14-14). The areas are either near permanent ice or at high elevation, and have ground that is seasonally snow free. Under these conditions, a unique set of periglacial processes operate, including permafrost, frost action, and ground ice.

Climatologically, these regions are in *Dfc, Dfd subarctic,* and *E polar* climates (especially *ET tundra* climate). Such climates occur either at high latitude (tundra and boreal forest environments) or high elevation in lower-latitude mountains (alpine environments). These periglacial regions are dominated by processes that relate to physical weathering, and mass movement (Chapter 10), climate (Chapter 7), and soil (Chapter 15).

Permafrost

When soil or rock temperatures remain below 0°C (32°F) for at least two years, a condition of **permafrost** develops. An area that has permafrost but is not covered by glaciers is considered periglacial. Note that this criterion is based solely on temperature and not on how much water is present. Other

FIGURE 14-14

Distribution of permafrost in the Northern Hemisphere. Alpine permafrost is noted except for small alpine occurrences in Hawaii, Mexico, Europe, and Japan that are too small to show on the map. Sub-sea permafrost occurs in the ground beneath the Arctic Ocean along the margins of the continents noted. [Adapted from Troy L. Péwé, "Alpine Permafrost in the Contiguous United States: A Review," *Arctic and Alpine Research* 15, no. 2 (May 1983): 146. © Regents of the University of Colorado. Used by permission.]

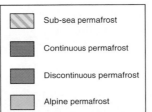

▨	Sub-sea permafrost
■	Continuous permafrost
■	Discontinuous permafrost
▨	Alpine permafrost

than high latitude and low temperatures, two other factors contribute to permafrost: the presence of fossil permafrost from previous ice-age conditions and the insulating effect of snow cover or vegetation that inhibits heat loss.

Permafrost regions are divided into two general categories, continuous and discontinuous, that merge along a general transition zone. *Continuous permafrost* describes the region of the most severe cold and is perennial, (purple areas in Figure 14-14). Continuous permafrost affects all surfaces except those beneath deep lakes or rivers and may exceed 1000 m (over 3000 ft) in depth, averaging approximately 400 m (1300 ft).

Discontinuous permafrost occurs in unconnected patches that gradually merge poleward toward the continuous permafrost zone. Equatorward, along lower-latitude margins, permafrost becomes scattered or sporadic until it gradually disappears (dark blue area on the map). As much as 50% of Canada and 80% of Alaska are affected by permafrost of either type. In addition to these two types, zones of high-altitude *alpine permafrost* extend to lower latitudes, as shown on the map.

Frozen Ground Phenomena

In regions of permafrost, subsurface water that is frozen is termed **ground ice**. The moisture content of areas with ground ice may vary from nearly none in regions of drier permafrost to almost 100% in saturated soils. Frozen water in the soil initiates *frost action*, the expansion of water volume as it freezes.

Frost Action Processes. The 9% expansion of water as it freezes produces strong mechanical forces that fracture rock and disrupt soil at and below the surface. If sufficient water freezes, the saturated soil and rocks are subjected to *frost-heaving* (vertical movement) and *frost-thrusting* (horizontal motions). Boulders and slabs of rock are thrust to the surface. Layers of soil (soil horizons) may appear disrupted as if stirred or churned. Frost action also produces a contraction in soil and rock volume, opening up cracks for ice wedges to form.

An **ice wedge** develops when water enters a crack in the permafrost and freezes. Thermal contraction in ice-rich soil forms a tapered crack—wider at the top, narrowing toward the bottom. Repeated seasonal freezing and melting progressively expands the wedge, which may widen from a few millimeters to 5–6 m and up to 30 m (100 ft) in depth (Figure 14-15). Widening may be small each year, but after many years the wedge can become significant.

The expansion and contraction of frost action results in the transport of stones and boulders. As the water-ice volume changes and the ice wedge deepens, coarser particles are moved toward the surface. An area with a system of ground ice and frost action develops sorted and unsorted accumulations of rock at the surface that take the shape of polygons called **patterned ground** (Figure 14-16).

Hillslope Processes. Soil drainage is poor in areas of permafrost and ground ice. When the surface thaws seasonally, the upper layer of soil and regolith is saturated with soil moisture and the whole layer will flow if the landscape is sloped. Such soil flows are in general called **solifluction**; in the presence of ground ice, the more specific term *gelifluction* is applied.

The cumulative effect of this landflow can be an overall flattening of a rolling landscape, with sagging surfaces and scalloped and lobed patterns in the downslope soil movements. Periglacial mass movement processes are related to slope dynamics and processes discussed in Chapter 10.

Humans and Periglacial Landscapes

Human populations in areas that experience permafrost and frozen ground phenomena face various difficulties. Because thawed ground above the permafrost zone frequently shifts, highways and rail lines become warped, twisted, and fail, and utility poles are disrupted. In addition, any building placed directly on frozen ground literally will melt itself into the defrosting soil, creating subsidence in structures (Figure 14-17).

Construction in periglacial regions dictates placing structures above the ground to allow air circulation beneath. This airflow allows the ground to cycle through its normal annual temperature pattern. Utilities such as water and sewer lines must be built above ground in "utilidors" to protect them from freezing and thawing ground (Figure 14-18). Like-

FIGURE 14-15
An ice wedge and ground ice in northern Canada. [Photo by Hugh M. French.]

FIGURE 14-16
Patterned ground. Aerial view of polygonal nets formed in Alaska, a fairly widespread frozen-ground phenomenon. [Photo courtesy of U.S. Geological Survey.]

FIGURE 14-17
Building failure due to the melting of permafrost south of Fairbanks, Alaska. [Adapted from U.S. Geological Survey. Photo courtesy of O. J. Ferrians, Jr., illustration based on U.S. Geological Survey pamphlet "Permafrost" by Louis L. Ray.]

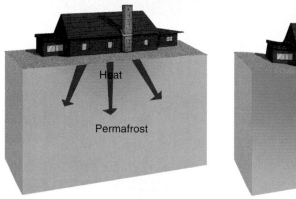

Heat

Permafrost

FIGURE 14-18
Proper construction in periglacial environments requires placement of buildings above ground, and water and sewage lines in elevated "utilidors," Inuvik, Northwest Territories. [Photo by Joyce Lundberg.]

wise, the Alaskan oil pipeline was constructed above ground on racks to avoid melting the frozen ground, causing shifting that could rupture the line.

The Pleistocene Ice Age Epoch

The most recent episode of cold climatic conditions began about 1.65 million years ago, launching the Pleistocene epoch. At the height of the Pleistocene, ice sheets and glaciers covered 30% of Earth's land area, amounting to more than 45 million km² (17.4 million mi²). Periglacial regions during the last ice age covered about twice their present areal extent. The Pleistocene is thought to have been one of the more prolonged cold periods in Earth's history. It featured not just one glacial advance and retreat, but at least 18 expansions of ice over Europe and North America, each obliterating and confusing the evidence from the one before.

The term **ice age** is applied to any such extended period of cold (not a single cold spell). It is a period of generally cold climate that includes one or more *glacials,* interrupted by brief warm spells known as *interglacials.* Each glacial and interglacial is given a name that is usually based on the location where evidence of the episode is prominent—for example, "Wisconsinan glacial."

Currently, glaciologists acknowledge the Illinoian glacial and Wisconsinan glacial, with the Sangamon interglacial between them during the past 300,000 years (Figure 14-19). The glacial/interglacial stages on the chart are numbered back to stage 23 at approximately 900,000 years ago.

The continental ice sheets over Canada, the United States, Europe, and Asia are illustrated on the polar map projection in Figure 14-20. In North America, the Ohio and Missouri river systems mark the south-

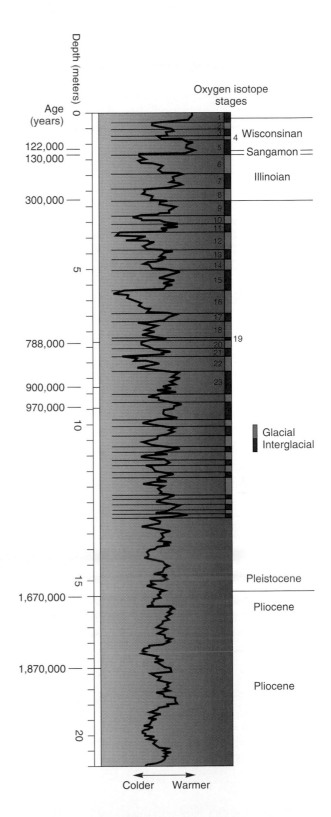

FIGURE 14-19
Pleistocene temperatures, determined by oxygen-isotope fluctuations in fossils found in deep-sea cores. Twenty-three stages cover 900,000 years, with names assigned for the past 300,000 years. [After N. J. Shackelton and N. D. Opdyke, *Oxygen-Isotope and Paleomagnetic Stratigraphy of Pacific Core V28-239, Late Pliocene to Latest Pleistocene,* Geological Society of America Memoir 145. Copyright © 1976 by the Geological Society of America. Adapted by permission.]

ern terminus of continuous ice. The edge of an ice sheet is not even; instead it expands and retreats in ice lobes. Such lobes were positioned where the Great Lakes are today and contributed to the formation of these lakes when the ice retreated from their gouged, isostatically depressed basins.

As these glaciers retreated, they exposed a drastically altered landscape that includes: the rocky soils of New England, the polished and scarred surfaces of Canada's Atlantic Provinces, the sharp crests of the Sawtooth Range and Tetons of Idaho and Wyoming, the scenery of the Canadian Rockies and the Sierra Nevada, and the Great Lakes of the United States and Canada, and much more. Study of these glaciated landscapes is important, for we can better understand paleoclimatology (past climates) and discover the mechanisms that produce ice ages and climatic change. FYI Report 14-1 elaborates on this.

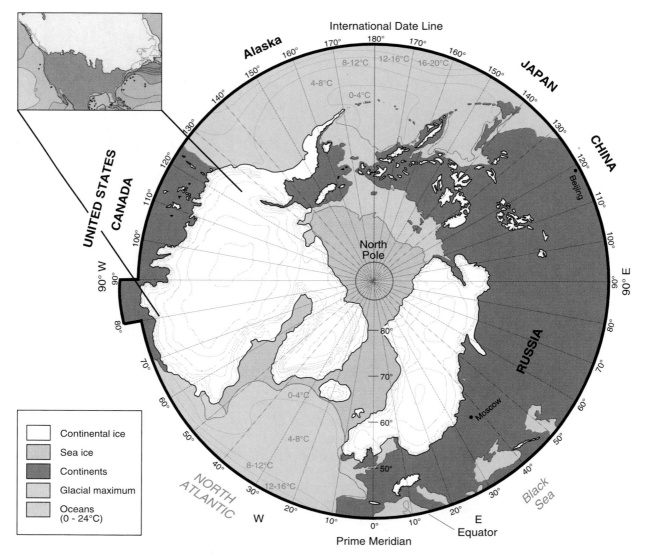

FIGURE 14-20
Extent of Pleistocene glaciation in the Northern Hemisphere 18,000 years ago viewed from a polar perspective. [From Andrew McIntyre, *CLIMAP Project*, Lamont-Dougherty Earth Observatory. Copyright © 1981 by the Geological Society of America.]

Deciphering Past Climates: Paleoclimatology

The Vikings were favored by a medieval warming episode as they sailed out into the less frozen North Atlantic to settle Iceland and Greenland between A.D. 900 and 1200, and inadvertently ventured onto the North American continent as well. However, between the years 1250 and 1740, a "little ice age" took place, burying many key mountain passes of Europe under deep ice, and lowering snowlines about 200 m (650 ft). Evidence that Earth's climate has fluctuated in and out of warm and cold ages now is being traced back through ice cores, layered deposits of silts and clays, changing paleomagnetism recorded in the rocks, the geologic sequences locked in rocks, and radioactive dating methods. And one interesting fact is emerging from these studies: humans (*Homo erectus* and *Homo sapiens* of the last 1.9 million years) have never experienced Earth's "normal" climate, that is, the climate most characteristic of Earth's 4.6-billion-year span.

Apparently, Earth's climates were quite moderate until the last 1.2 billion years, when temperature patterns with cycles of 200–300 million years became more pronounced. The most recent cold episode was the Pleistocene epoch, which began in earnest 1.65 million years ago and through which we may still be progressing. The Holocene epoch began 10,000 years ago and may represent either an end to the Pleistocene, or merely an interglacial period.

The mechanisms that bring on an ice age are the subject of much research and debate. Because past occurrences of low temperatures appear to have followed a pattern, researchers have looked for causes that also are cyclic in nature. They have identified a complicated mix of interacting variables that appear to influence long-term climatic trends.

Galactic and Earth-Sun Relationships

As our Solar System revolves around the distant center of the Milky Way, it crosses the plane of the galaxy approximately every 32 million years. At that time Earth's plane of the ecliptic aligns parallel to the galaxy's plane, and we pass through regions in space of increased interstellar dust and gas, which may have some climatic effect.

Other possible astronomical factors were proposed by Milutin Milankovitch (1879–1954), a Yugoslavian astronomer who studied Earth-Sun orbital relationships. Milankovitch wondered whether the development of an ice age was related to seasonal astronomical factors—Earth's revolution, rotation, and tilt—extended over a longer time span. Earth's elliptical orbit about the Sun is not constant. The shape of the ellipse varies by more than 17.7 million km (11 million mi) during a 100,000-year cycle, stretching out to an extreme ellipse (Figure 1a). In addition, Earth's axis "wobbles" through a 26,000-year cycle, in a movement much like that of a spinning top winding down. Earth's slowly wobbling axis inscribes an inverted cone in space, a movement called *precession* (Figure 1b). This precession changes the orientation of hemispheres and landmasses to the Sun.

Finally, Earth's present axial tilt of 23.5° varies from 22° to 24° during a 40,000-year period (Figure 1c). Milankovitch calculated, without the aid of today's computers, that the interaction of these Earth-Sun relationships creates a 96,000-year climatic cycle. His model assumes that changes in astronomical relationships affect the amounts of insolation received.

Milankovitch died in 1954, his ideas still not accepted by a skeptical scientific community. Now, in the era of modern computers, remote sensing satellites, and worldwide efforts to decipher past climates, Milankovitch's valuable work has stimulated much research to explain climatic cycles and has experienced a measure of confirmation.

Geophysical Factors

Major glaciations also can be associated with the migration of landmasses to higher latitudes. Chapters 8 and 9 explain that the shape and orientation of landmasses and ocean basins have changed greatly during Earth's history. Continental plates have drifted from equatorial locations to polar regions and vice versa, thus producing cold and warm episodes on the land. Gondwana (the southern half of Pangaea) experienced extensive glaciation that left its mark on the rocks of parts of Africa, South America, India,

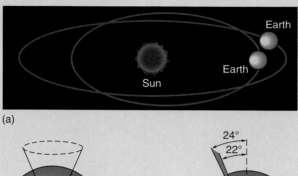

(a)

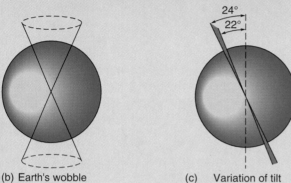

(b) Earth's wobble   (c)   Variation of tilt

FIGURE 1

Astronomical factors that possibly affect broad climatic cycles. (a) Earth's elliptical orbit varies widely during a 100,000-year cycle, stretching out to an extreme ellipse. (b) Earth's 26,000-year axial wobble. (c) Variation in Earth's axial tilt every 40,000 years.

Antarctica, and Australia. Landforms in the Sahara, for example, bear the markings of even earlier glacial activity, partly explained by the fact that portions of Africa were centered near the South Pole during the Ordovician period, 465 million years ago (see Figure 8-16a).

Geographical-Geological Factors

Geographical-geological factors in glaciation include the terrestrial causes of atmospheric variables, increases in relief associated with mountain building, sea level changes, and alterations of oceanic circulation patterns. One such terrestrial source, a volcanic eruption, might produce lower summer temperatures for a year or two, leading to a buildup of long-term snow cover at high latitudes. These high-albedo snow surfaces then would reflect more insolation away from Earth, to further enhance cooling. The climatic effects from the volcanic eruptions in Japan (Mount Unzen) and the Philippines (Mount Pinatubo) in the summer of 1991 are being closely studied.

Furthermore, episodes of mountain building over the past billion years have forced mountain summits above the snowline, where snow remains after the summer melt. Mountain chains can affect downwind weather patterns and jet stream circulation, which in turn guides weather

systems. In addition, oceanic circulation patterns have changed. For example, the Isthmus of Panama formed about 3 million years ago and effectively separated the circulation of the Atlantic and Pacific oceans. Such changes in circulation and ocean basin configuration affect air mass formation and temperature patterns.

Also linked to a terrestrial source is the fluctuation of atmospheric greenhouse gases, triggering higher or lower temperatures. An ice core taken at Vostok, the Russian research station near the geographic center of Antarctica, contained trapped air samples from approximately 160,000 years ago and showed carbon dioxide levels varying from more than 290 ppm to a low of nearly 180 ppm (Figure 2). Higher levels of CO_2 generally are thought to correlate with each interglacial. During the 1990s, CO_2 reached 360 ppm, higher than at any time in the past 160,000 years, principally due to anthropogenic (human) sources. A more extensive ice core drilled in Greenland (GISP 2), over 3000 m (9000 ft) deep and spanning 250,000 years, was completed in 1993 and promises to unlock more climatic clues.

Thus, Earth's climate is unfolding as a multicyclic system controlled by an interacting set of cooling and warming processes, all founded on galactic alignments, Earth-Sun relationships, geophysical, and geographical-geological factors. Within this complex of change, scientists are attempting to decipher past climatic rhythms and present and future climatic trends.

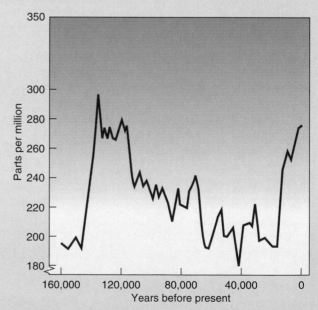

FIGURE 2

Ancient CO_2 levels revealed in air bubbles trapped in Antarctic ice were measured from the Vostok ice core. [Adapted by permission from *Nature* 329, p. 408. Copyright © 1988 Macmillan Magazines Ltd.]

Pluvial Periods and Paleolakes

Figure 14-21 portrays the West dotted with large lakes 15,000–25,000 years ago. Except for the Great Salt Lake in Utah (a remnant of the former Lake Bonneville noted on the map) and a few smaller lakes, these lakes are gone today. Only dry basins and lake sediments remain. These ancient lakes are called **paleolakes**. The term **pluvial** describes this period of wetter conditions, principally during the Pleistocene epoch. During pluvial periods, lake levels increased in arid regions. The drier periods between pluvials are called *interpluvials*. Interpluvials are marked with **lacustrine deposits**, lake sediments that form terraces along former shorelines.

Earlier researchers attempted to correlate pluvial and glacial ages, given their coincidence during the Pleistocene. However, few sites actually demonstrate such a simple relationship. For example, in the western United States, the estimated volume of melted ice is only a small portion of the water volume of pluvial lakes of the region. Also, these lakes tend to predate glacial times and correlate instead with periods of wetter climate, or periods thought to have had lower evaporation rates.

In North America, the two largest late Pleistocene paleolakes were in the Basin and Range Province of the West. Figure 14-21 portrays these—Lake Bonneville and Lake Lahontan—and other paleolakes at their highest levels. Pluvial lakes also occurred in Mexico, South America, Africa, Asia, and Australia.

The Great Salt Lake, near Salt Lake City, Utah, and the Bonneville Salt Flats in western Utah are remnants of Lake Bonneville, which at its greatest extent covered more than 50,000 km² (19,500 mi²) and reached depths of 300 m (1000 ft), spilling over into the Snake River drainage to the north.

Arctic and Antartic Regions

Climatologists use environmental criteria to define the Arctic and the Antarctic regions (Figure 14-22). *The Arctic region* is defined by the 10°C (50°F) isotherm for July (Northern Hemisphere summer) because it coincides with the visible treeline, which is the boundary between the northern forests and tundra climates. The Arctic Ocean is covered by floating sea ice (frozen seawater) and glacier ice (frozen freshwater) in icebergs that formed at the edge of the surrounding land. This ice pack thins in the summer months and sometimes breaks up. For the first time two ice breaker ships successfully broke through the ice pack to the North Pole during the 1994 summer season.

The Antarctic region is defined by the *antarctic convergence*, a narrow zone that extends around the continent as a boundary between colder antarctic water and warmer water at lower latitudes. This boundary follows roughly the 10°C (50°F) isotherm for February (Southern Hemisphere summer) and is located near 60° S latitude. The Antarctic region that is covered with sea ice represents an area greater than

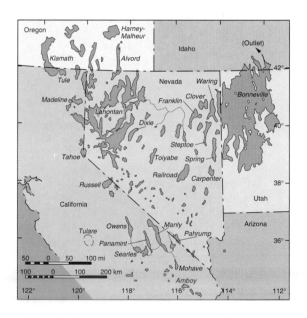

FIGURE 14-21
Paleolakes of the western United States at their greatest extent 15,000 to 25,000 years ago, during a recent pluvial period. Lake Lahontan and Lake Bonneville were the largest. The Great Salt Lake in Utah is a remnant of Lake Bonneville; Pyramid Lake in western Nevada and Honey Lake in California are remnants of ancient Lake Lahontan; Mono Lake in California is what's left of pluvial Lake Russell. With few exceptions, most paleolakes are dry basins today. [After R. F. Flint, *Glacial and Pleistocene Geology.* Copyright © 1957 by John Wiley & Sons. Adapted by permission of John Wiley & Sons, Inc.]

"The Great Salt Lake Floods"

Great Salt Lake in Utah is experiencing a dynamic change during the 1980s and 1990s. The lake level began rising in 1981, following periods of increased rainfall and snowpack meltwaters from the adjoining Wasatch Mountains. Shifts in storm tracks favored increased precipitation during the 1982–1986 period. Evaporation is the lake's only output from the closed Salt Lake basin, so the lake is filling.

By the mid-1980s, the lake achieved its highest level in historic times, flooding an interstate highway, transcontinental railroad tracks, resorts, and thousands of acres of grazing land. Fortunately, the summers following the highest lake level also had high evaporation rates and precipitation dropped to 50% below average. Raising the beds of the railroad tracks and the interstate highway will prevent their loss if the lake level again rises. This ancient paleolake vividly demonstrates how climate change can effect levels in water bodies—especially those that lack natural drainage.

North America, Greenland, and Western Europe combined!

The Antarctic Ice Sheet

Antarctica is a continental-sized landmass and therefore is much colder overall than the Arctic, which is an ocean. In simplest terms, Antarctica can be thought of as a continent covered by a single enormous glacier, although it contains distinct regions such as the East Antarctic and West Antarctic ice sheets, which respond differently to slight climatic variations.

The edges of the ice sheet that enter bays along the coast form extensive *ice shelves,* with sharp ice cliffs rising up to 30 m (100 ft) above the sea. Large tabular islands of ice are formed when sections of the shelves break off and move out to sea (Figure 14-23). These ice islands can be very large; one in the late 1980s exceeded the size of Rhode Island.

Individual glaciers operate around the periphery of the ice sheet in Antarctica. The Byrd Glacier moves through the Transantarctic Mountains, draining an area of more than 1,000,000 km² (386,000 mi²) as it flows 180 km (112 mi) and drops 5000 m (3.1 mi) in elevation. This glacier exceeds 22 km (13.7 mi) in width at its narrowest point and, where its thickness is 1000 m (3300 ft), it maintains a flow rate of approximately 840 m (2750 ft) per year, as fast as some surging glaciers.

SUMMARY—Glacial and Periglacial Landscapes

More than 77% of the freshwater on Earth is frozen. Ice presently covers about 11% of Earth's surface, and periglacial features occupy another 20% of ice-free, cold-dominated landscapes. **Glaciers** form by the accumulation and recrystallization of snow into dense glacial ice that flows plastically. They move under the pressure of their own mass and the pull of gravity. Large **ice sheets** cover Greenland and Antarctica, and less-extensive **ice caps** and **ice fields** dominate mountain landscapes. **Alpine glaciers** occupy valleys at high elevations and high

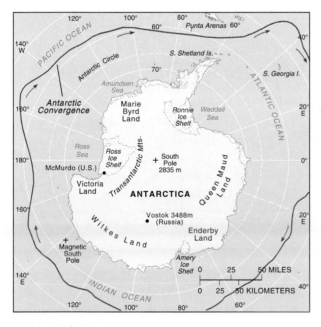

FIGURE 14-22
The Arctic and Antarctic regions. Note the 10°C (50°F) isotherm for mid-summer in the Arctic.

latitudes. A continuous mass of ice sitting on land and adjoining a coastline is known as a **continental glacier**.

A glacier is an open system with inputs and outputs that can be analyzed through observation of the expansion and retreat of the glacier itself. A glacier is fed by snowfall and loses ice from its upper and lower surfaces and along its margins through **ablation**. Snow that survives the summer and into the following winter undergoes a slow transformation into glacial ice. Formation of **glacial ice** is analogous to the formation of metamorphic rock: sediments (snow and **firn**) are pressured and recrystallized into a dense metamorphic rock (glacial ice).

Extensive **valley glaciers** have reshaped mountains worldwide, carving **V**-shaped stream valleys into **U**-shaped glaciated valleys. The passage of ice over a landscape excavates and creates profound changes, producing many distinctive erosional and depositional landforms. **Glacial drift** is the general term for all glacial deposits, both unsorted and sorted. Direct ice

FIGURE 14-23
Icebergs break off from the Filchner Ice Shelf (lower left) into the Weddell Sea, Antarctica (78° S, 40°W). [SPOT Image courtesy of CNES. All rights reserved. Used by permission.]

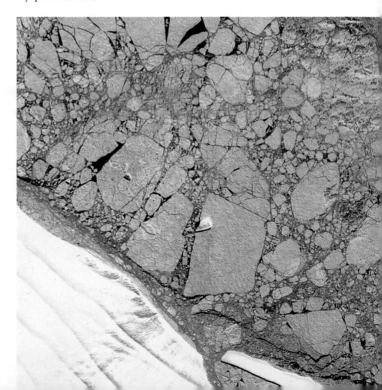

deposits appear unstratified and unsorted and are called **till**. Sediments deposited by glacial meltwater are sorted and termed **stratified drift**.

Modern active glaciers and the present physical landscape as molded by past glaciers provide evidence for the study of past climates—*paleoclimatology*. During the Pleistocene epoch, extensive continental ice covered portions of the United States, Canada, Europe, and Russia in an **ice age**. Glaciers worldwide are being watched more closely than ever because of their direct link to changing trends in world climates and the potential increase in mean sea level that can be brought on by the melting of even small amounts of stored ice.

KEY TERMS

ablation
abrasion
arête
cirque
col
crevasse
drumlin
equilibrium line
esker
firn
firn line
fjord
glacial drift
glacial ice
glacial surge
glacier
    alpine glacier
      continental glacier
      valley glacier
ground ice
horn
ice age

ice cap
ice field
ice sheet
ice wedge
icebergs
kame
kettle
lacustrine deposit
moraine
    lateral moraine
      medial moraine
outwash plain
paleolakes
patterned ground
periglacial
permafrost
pluvial
solifluction
stratified drift
till
till plain

REVIEW QUESTIONS

1. What is a glacier? What is implied about existing climate patterns in a glacial region?
2. Differentiate between an alpine glacier and a continental glacier.
3. Name the three types of continental glaciers. What is the basis for dividing continental glaciers into types? Which type is Antarctica?
4. Trace the evolution of glacial ice from fresh fallen snow.
5. What is meant by glacial mass balance? What are the basic inputs and outputs underlying that balance?
6. What is meant by a glacial surge? What do scientists think produces surging episodes?
7. Describe the evolution of a **V**-shaped stream valley to a **U**-shaped glaciated valley. What kinds of features are visible after a glacier has retreated?
8. Differentiate between two forms of glacial drift—till and outwash.
9. What is a morainal deposit? What specific forms of moraines are created by alpine and continental glaciers?
10. What are some common depositional features encountered in a till plain?
11. In terms of climatic types, describe the areas on Earth where periglacial landscapes occur. Include both higher latitude and higher altitude climate types.
12. What is the difference between permafrost and ground ice?
13. Relate some of the specific problems humans encounter in developing periglacial landscapes.
14. What is paleoclimatology? Describe Earth's past climatic patterns. Are we experiencing a normal climate pattern in this era or have scientists noticed any significant trends?
15. Define an ice age. When was the most recent one on Earth? How is our present age classified? Explain "glacial" and "interglacial" in your answer.
16. Give an overview of what science has learned about the various causes of ice ages by listing and explaining at least four possible climate change factors.

PART 4

BIOGEOGRAPHY

CHAPTER 15
The Geography of Soils

CHAPTER 16
Ecosystems and Biomes

CHAPTER 17
Earth, Humans, and the New Millennium

Hoh Rain Forest, Olympic National Park. [Photo by Jack W. Dykinga.]

Earth is the home of the only known biosphere in the Solar System—a unique, complex, and interactive system of abiotic and biotic components working together to sustain a tremendous diversity of life. Energy enters the system through conversion of solar energy by means of photosynthesis in the leaves of plants. Soil is the essential link among the lithosphere, plants, and the rest of Earth's physical systems. Thus, soil helps sustain life. Life is organized into a feeding hierarchy from producers to consumers, ending with decomposers. Taken together, the soils, plants, animals, and all nonliving components produce aquatic and terrestrial ecosystems, known as biomes. Today we face crucial issues—preservation of the diversity of life in the biosphere and the survival of the biosphere itself—important applied topics considered in Part 4.

Farm plots and vegetable fields near Dalat, in the central highlands of Vietnam. [*Photo by Wolfgang Kaebler.*]

15

THE GEOGRAPHY OF SOILS

SOIL CHARACTERISTICS
    Soil Profiles
    Soil Horizons
    Soil Properties
    Soil Chemistry
    Soil Formation Factors and Management

SOIL CLASSIFICATION
    A Brief History of Soil Classification
    Diagnostic Soil Horizons
    The Eleven Soil Orders of Soil Taxonomy

SUMMARY

Earth's landscape generally is covered with soil. **Soil** is a dynamic natural material composed of fine particles in which plants grow, and which contains both mineral and organic matter. If you have ever planted a garden, tended a house plant, or been concerned about famine, this chapter will interest you. A knowledge of soil is at the heart of agriculture and food production.

Physical and chemical weathering of rocks in the upper lithosphere provides the raw mineral ingredients for soil formation. These are called *parent materials,* and their composition, texture, and chemical nature are important. Air, water, nutrients, and solid materials all cycle through soil and provide a critical sustaining structure for plants, animals, and human life.

Soil fertility is the ability of soil to sustain plants. Soil has fertility when it contains organic substances and clay minerals that absorb water and certain elements needed by plants. Billions of dollars are expended to create fertile soil conditions, yet the future of Earth's most fertile soils is threatened because soil erosion is increasing worldwide. Some 35% of farmlands are losing soil faster than it can form—a loss exceeding 22.75 billion metric tons (25 billion tons) per year. Increased crop production from artificial fertilizers and new crop hybrids partially mask this effect, but such compensations are nearing the end of their capacity. Soil depletion and loss are at record levels from Iowa to China, Peru to Ethiopia, the Middle East to the Americas. The impact on society is potentially disastrous as population and food demands increase.

Soil science is interdisciplinary, involving physics, chemistry, biology, mineralogy, hydrology, taxonomy, climatology, and cartography. Physical geographers are interested in the spatial patterns formed by soil types and the physical factors that interact to produce them. As an integrative science, physical geography is well suited for this task.

Soil science must deal with a substance whose characteristics vary from kilometer to kilometer, and even centimeter to centimeter. Thus, people interested in soil need information specific to their area. In many locales, an *agricultural extension service* can provide this information and perform a detailed soil analysis. Soil surveys and local soil maps are available for most counties in the United States and the Canadian provinces.

Soil Characteristics

Classifying soils is similar to classifying climates. Both involve complex interacting variables. Simple theoretical models are the basis for soil classifications. The following section presents general assumptions about soil structure, soil properties, and soil formation processes, forming a basis for soil classification.

Soil Profiles

Just as a book cannot be judged by its cover, so soils cannot be evaluated at the surface only, for this can produce a misleading picture. Instead, a soil profile should extend from the surface to the deepest extent of plant roots, or to where regolith or bedrock is encountered. Such a profile, a **pedon**, is conceptually a hexagonal column from 1 to 10 m² in surface area (Figure 15-1).

A pedon is the basic sampling unit used in soil surveys. Many pedons together in one area make up a *polypedon,* which has distinctive characteristics differentiating it from surrounding polypedons. A **polypedon** is an essential soil individual, constituting an identifiable *series* of soils in an area. A polypedon is the soil unit used in preparing local soil maps.

Soil Horizons

Each layer exposed in a pedon is a **soil horizon**. A horizon is roughly parallel to the pedon's surface and has characteristics distinctly different from horizons directly above or below. The boundary between horizons usually is visible in the field, using the properties of color, texture, structure, consistence (meaning soil consistency), porosity, the presence or absence of certain minerals, moisture, and chemical processes. Soil horizons are the building blocks of soil classification. However, because they can vary in an almost endless number of ways, a simple

FIGURE 15-1
A soil pedon (sampling unit) is derived from a polypedon (mapping unit).

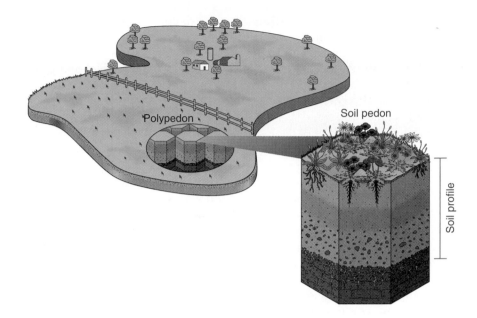

model is useful. Figure 15-2 presents an ideal pedon and *soil profile*, with letters assigned to each soil horizon for identification.

At the top of the soil profile is the O (organic) horizon, named for its organic composition, derived from plant and animal litter that was deposited on the surface and transformed into humus. **Humus** is a mixture of decomposed organic materials and is usually dark in color. The O horizon is 20–30% or more organic matter, important because of its water-absorbing ability and its nutrients.

At the bottom of the soil profile is the R (rock) horizon, consisting of either unconsolidated (loose) material or consolidated bedrock. The A, E, B, and C horizons mark important mineral strata between O and R. These middle layers are composed of sand, silt, clay, and other weathered by-products.

In the A horizon, the presence of humus and clay particles is particularly important, for they provide essential chemical links between soil nutrients and plants. The A horizon usually is darker and richer in organic content than lower horizons. It grades into the E horizon, made up of coarse sand, silt, and resistant minerals. From the E horizon, clays and oxides of aluminum and iron are leached (removed by water) and are carried to lower horizons with water as it percolates through the soil. This process of water rinsing upper horizons and removing fine particles

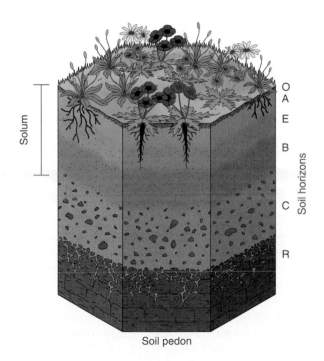

FIGURE 15-2
Typical O, A, E, B, C, and R soil horizons within a developed soil pedon.

and minerals is termed **eluviation**; thus the designation E for this horizon. The greater the precipitation in an area, the higher the rate of eluviation.

In contrast to the A and E horizons, B horizons accumulate clays, aluminum, iron. B horizons are dominated by **illuviation**, a depositional process. They may exhibit reddish or yellowish hues because of the illuviated presence of mineral and organic oxides. In the humid tropics, these layers often develop to some depth.

The combination of the A and E horizons with their *eluviation* removals and the B horizon with its *illuviation* accumulations is designated the **solum**, considered the true definable soil of the pedon. The A, E, and B horizons experience the most active soil processes.

The C horizon is weathered bedrock or weathered parent material, excluding the bedrock itself. The C horizon is not much affected by soil operations in the solum and lies outside the biological influences in the shallower horizons. Plant roots and soil microorganisms are rare in the C horizon. In dry climates, calcium carbonates commonly form the cementing material of these hardened layers.

Soil Properties

Observing a real soil profile will help you identify color, texture, structure, and other soil properties. A good place to see a soil profile is at a construction site or excavation, perhaps on your campus, or at a roadcut.

Soil Color. Color is important, for it sometimes suggests composition and chemical makeup. If you look at exposed soil, color may be the most obvious trait. Among the many possible hues are the red and yellow soils of the southeastern United States, which are high in ferric (iron) oxides; the black-prairie, richly organic soils of portions of the U.S. grain-growing regions and Ukraine; or the white-to-pale soils attributable to the presence of aluminum oxides and silicates. Color can be deceptive too, for soils of high humus content are often dark, yet clays of warm-temperate and tropical regions with less than 3% organic content are some of the world's blackest soils.

Soil Texture. Soil texture, perhaps a soil's most permanent attribute, refers to the size and organization of particles in the soil. Individual mineral particles are called *soil separates*. Particles range from the finest clays, to silts, to coarser sands, and to larger pebbles and gravels.

Figure 15-3 is a soil triangle, showing the relation of sand, silt, and clay concentrations distributed along each side. Each corner of the triangle represents a soil consisting solely of the particle size noted (rarely are true soils composed of a single separate). Every soil on Earth is defined somewhere in this triangle. The figure includes the common designation **loam**, which is a balanced mixture of sand, silt, and clay beneficial to plant growth. A sandy loam with clay content below 30% (lower left) usually is considered ideal by farmers because of its water-holding characteristics and cultivation ease. Soil texture is important in determining water retention and water transmission traits.

Soil Structure. Soil structure refers to the *arrangement* of soil particles. Structure can partially modify the effects of soil texture. The term *ped* describes an individual unit of soil particles; it is a tiny natural lump or cluster of particles held together. The shape of these peds determines which of the structural types the soil exhibits (Figure 15-4). Peds separate from each other along zones of weakness, creating voids (pores) that are important for moisture storage and drainage. More rounded peds have more pore space and greater permeability. They are therefore more productive for plant growth than are coarse, blocky, prismatic, or platy peds, despite comparable fertility.

Soil Consistence. In soil science, the term *consistence* is used to describe the consistency of soil particles. It refers to cohesion in soil, a product of texture and structure. Consistence reflects a soil's resistance to breaking and manipulation under varying moisture conditions. *Wet soils* are sticky when held between the thumb and forefinger, ranging from a little adherence to either finger, to sticking to both fingers, to stretching when the fingers are moved apart. Plasticity, the quality of being molded, is roughly measured by rolling a piece of soil between your fingers and thumb to see whether it rolls into a thin strand.

Moist soil is soil that is filled to about half of field capacity (the usable water capacity of soil), and its

FIGURE 15-3
Soil texture triangle. As an example, points 1, 2, and 3 designate samples taken at three different horizons in the *Miami silt loam* in Indiana. A sample taken near the surface in the A horizon is recorded at point 1; in the B horizon at point 2; and in the C horizon at point 3. Note that silt dominates the surface, clay the B horizon, and sand the C horizon. [After USDA Soil Conservation Service, *Soil Taxonomy*, Agricultural Handbook No. 436, Washington, DC: U.S. Government Printing Office, 1975.]

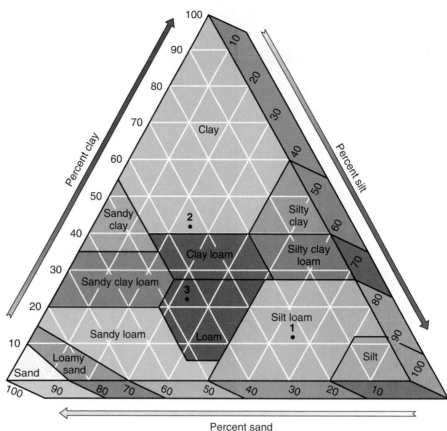

Percent clay

Percent silt

Percent sand

Clay
Sandy clay
Silty clay
Clay loam
Silty clay loam
Sandy clay loam
Sandy loam
Loam
Silt loam
Silt
Loamy sand
Sand

consistence grades from loose (noncoherent), to *friable* (easily pulverized), to firm (not crushable between thumb and forefinger). A *dry soil* is typically brittle and rigid, with consistence ranging from loose, to soft, to hard, to extremely hard.

Soil Porosity. Soil porosity, permeability, and moisture storage are discussed in Chapter 6. Pores in the soil horizon control the flow of water, its intake and drainage, and air ventilation. Important porosity factors are pore *size, continuity* (whether the pores are interconnected), shape (whether pores are spherical, irregular, or tubular), *orientation* (whether pore spaces are vertical, horizontal, or random), and *pore location* (whether within or between soil peds).

Porosity is improved by the biotic actions of plant roots, animal activity such as the tunneling of gophers or worms, and human intervention through soil manipulation (plowing, adding humus or sand, or plant-ing soil-building crops). Much of a farmer's soil preparation work is done to improve soil porosity.

Soil Moisture. Reviewing Figures 6-6 and 6-7 will help you understand this section. Plants operate most efficiently when the soil is at *field capacity*, that is, maximum water availability for plant use. Field capacity is determined by soil type. The effective rooting depth of a plant determines the amount of soil moisture to which the plant's roots are exposed. If soil moisture is drawn below field capacity, plants use available water inefficiently until the plant's wilt-ing point is reached, beyond which plants cannot extract the water they need.

Soil Chemistry

The soil atmosphere is mostly nitrogen, oxygen, and carbon dioxide. Nitrogen concentrations in the soil are about the same as in the atmosphere. Oxygen is

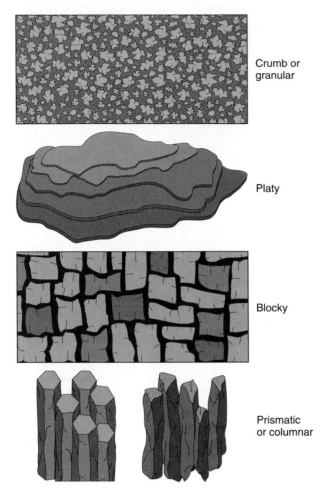

Crumb or granular

Platy

Blocky

Prismatic or columnar

FIGURE 15-4
Types of soil structure.

less and carbon dioxide is greater in soil than the atmosphere because of ongoing respiration processes.

Water present in soil pores is called *soil solution.* It is the medium in the soil for chemical reactions. This solution is critical to plants as their source of nutrients and is the foundation of *soil fertility.* Carbon dioxide and various organic materials combine with water to produce carbonic and organic acids, respectively. These acids are then active in soil processes, as are dissolved alkalies and salts.

To understand how the soil solution behaves, let's go through a quick chemistry review. An *ion* is an atom, or group of atoms, that carries an electrical charge. An ion has either a positive charge or a neg-

ative charge. For example, when NaCl (sodium chloride) is dissolved in solution, it separates into two ions: Na^+, a *cation* (positively charged ion), and Cl^-, an *anion* (negatively charged ion). Some ions in soil carry single charges, whereas others carry double or even triple charges (e.g., sulfate, SO_4^{2-}; aluminum, $Al3^+$).

Retaining Ions for Plants. Ions in soil are retained by **soil colloids**. These tiny particles of clay and organic material carry a negative electrical charge and consequently attract any positively charged ions in the soil (Figure 15-5). The positive ions, many metallic, are critical to plant growth. If it were not for the soil colloids, the positive ions would be leached away by the soil solution and thus would be unavailable to plant roots.

Individual clay colloids are thin and platelike, with parallel surfaces that are negatively charged. Cations attach to the surfaces of the colloids by **adsorption** (not *ab*sorption). The metallic cations are adsorbed onto the soil colloids. Colloids can exchange cations between their surfaces and the soil solution, because of their **cation-exchange capacity (CEC)**. A high CEC means good soil fertility (unless there is a complicating factor, such as the soil being too acid).

Soil Acidity and Alkalinity. A soil solution may contain significant hydrogen ions (H^+), cations, which stimulate acid formation. The result is an *acid soil.* On the other hand, a soil high in base cations (calcium, magnesium, potassium, sodium) is a *basic* or *alkaline soil.* Acidity and alkalinity is expressed on the pH scale (Figure 15-6).

Pure water is nearly *neutral,* with a pH of 7.0. Readings below 7.0 represent increasing acidity. Readings above 7.0 represent increasing alkalinity. Acidity usually is regarded as strong at 5.0 or lower on the pH scale, whereas 10.0 or above is considered strongly alkaline.

Today, the major contributor to soil acidity is acid precipitation (rain, snow, fog). Acid rain actually has been measured below pH 3.0—an incredibly low value for natural precipitation; as acid as lemon juice. Because most crops are sensitive to specific pH levels, acid soils below pH 6.0 require treatment to raise the pH. This is done by the addition of bases in the form of minerals that are rich in base cations, usually

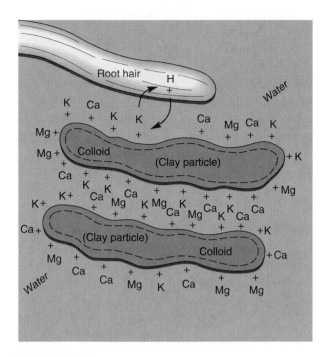

FIGURE 15-5
Typical soil colloids retaining mineral ions by adsorption (opposite charges attract).

lime (calcium carbonate). Increased acidity in the soil solution accelerates the chemical weathering of mineral nutrients and increases their depletion rates, as in soils in the region of the northern forests.

Soil Formation Factors and Management

There are three soil-forming factors: *dynamic* (climatic and biologic), *passive* (parent material, topography and relief, and time), and the *human* component.

These three factors work together as a system to form soils. The role of parent material in providing weathered minerals to form soils was discussed earlier in this chapter.

Natural Factors. Climate types correlate closely with soil types worldwide. The moisture, evaporation, and temperature regimes of climates determine the chemical reactions, organic activity, and eluviation rates of soils. Not only is the present climate important, but many soils exhibit the imprint of past climates, sometimes over thousands of years. Most notable is the effect of glaciations, as described and mapped in Chapter 14.

The organic content of soil is determined by vegetation and by animal and bacterial activity. The chemical makeup of the vegetation contributes to the acidity or alkalinity of the soil solution. For example, broadleaf trees tend to increase alkalinity, whereas needleleaf trees tend to produce higher acidity. Thus, when civilization moves into new areas and alters the natural vegetation by logging or plowing, the affected soils are likewise altered. In many cases they are permanently changed.

Topography also affects soil formation, mainly through slope and orientation. Slopes that are too steep cannot have full soil development because gravity and erosional processes remove materials. Slopes that are nearly level inhibit soil drainage. Relative to slope orientation, in the Northern Hemisphere, a south facing slope is warmer because it receives direct sunlight. This affects water-balance relationships. North-facing slopes are colder, causing slower snowmelt and lower evaporation rates, which result in more moisture for plants than is available on south-facing slopes.

FIGURE 15-6
Soil pH scale, measuring acidity (lower pH) and alkalinity (higher pH).

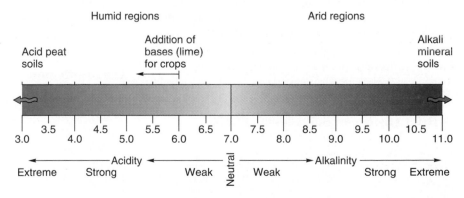

The Human Factor. Human intervention has a major impact on soils. Millennia ago, farmers in every culture on Earth learned to plow slopes "on the contour"—to plow around a slope at the same elevation, not vertically up and down the slope. Contour plowing reduces *sheet erosion* (which removes topsoil) and simultaneously improves water retention in the soil. However, the demands of today's large farm equipment have forced farmers to alter their practice of contouring because the equipment cannot make the tight turns that such proper plowing requires. Other effective soil-holding practices also are being eliminated to accommodate the larger equipment. Methods in decline include terracing (similar to contour plowing), planting tree rows as wind breaks, and planting shelter belts.

All of the identified factors in soil development (climate, biological activity, parent material, landforms and topography, and human activity) require *time* to operate. A few centimeters thickness of prime farmland soil may require *500 years* to mature. Yet these same soils are being lost at a few centimeters *per year* to sheet flow and gullying when the soil-holding vegetation is removed. Over the same period, exposed soils may be completely leached of needed cations, thereby losing their fertility. Figure 15-7 maps regions of soil losses and related concerns. Cooperative international action is needed to save Earth's productive soils.

Soil Classification

Classification of soils is complicated by the variety of interactions that create thousands of distinct soils—well over 15,000 soil types in the United States and Canada alone. Not surprisingly, a number of different classification systems are in use worldwide. The United States, Canada, the United Kingdom, Germany, Australia, Russia, and the United Nations Food and Agricultural Organization (FAO) each has its own soil classification system.

A Brief History of Soil Classification

The Russians contributed greatly to modern soil science because they were first to consider soil an independent natural body with a definite individual soil genesis. By contrast, American and European scientists were still considering soil as a geologic product, or simply a mixture of chemical compounds. The Russians determined that broad interrelationships exist among the physical environment, vegetation, and soils.

The first formal U.S. soil classification system is credited to Dr. Curtis F. Marbut. Published in the 1930s, Marbut's classification system recognized the importance of soils as products of dynamic natural processes. However, as the database of soil information grew, the system became inadequate. In 1951 the U.S. Soil Conservation Service began researching a new soil classification system. That process went through seven exhaustive reviews, culminating in what is called the Seventh Approximation in the 1960s. The current version was published in 1975: *Soil Taxonomy—A Basic System of Soil Classification for Making and Interpreting Soil Surveys,* called simply **Soil Taxonomy**. Much of the information in this chapter is derived from that keystone publication.

U.S. Soil Taxonomy.

The U.S. Soil Taxonomy system is based on soil properties actually seen in the field. Thus, it is open to addition, change, and modification as the sampling database grows. An important aspect of the system is that it recognizes the interaction of humans and soils and the changes that humans have introduced, both purposely and inadvertently.

The classification system divides soils into six categories, creating a hierarchical sorting system. Each *soil series* (the smallest, most detailed category numbering 13,000) ideally includes only one polypedon, but may include portions continuous with adjoining polypedons in the field. In sequence from smallest to largest categories, including the number of occurrences within each, the Soil Taxonomy recognizes: *soil families* (6000), *soil subgroups* (1200), *soil great groups* (230), *soil suborders* (47), and *soil orders* (11).

Diagnostic Soil Horizons

To identify specific soil types within the Soil Taxonomy, the U.S. Soil Conservation Service describes *diagnostic horizons* in a pedon. A diagnostic horizon reflects a distinctive physical property (color, texture, structure, consistence, porosity, moisture), or a dominant soil process (discussed with the soil types).

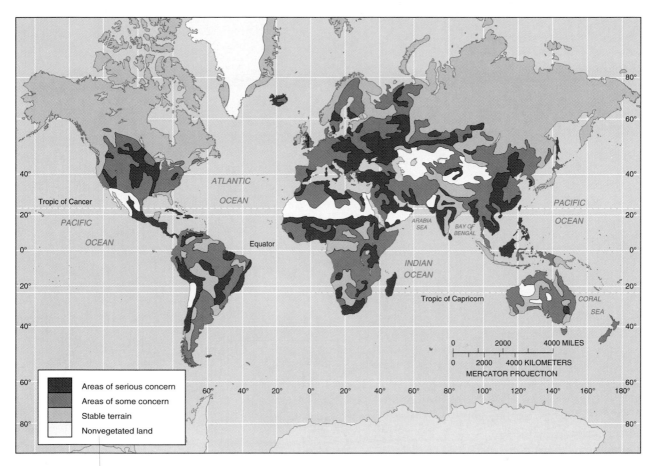

FIGURE 15-7
Approximately 1.2 billion hectares (3.0 billion acres) of Earth's soils suffer some degree of degradation through erosion caused by human misuse and abuse. [*A Global Assessment of Soil Degradation*, adapted from United Nations Environment Programme, International Soil Reference and Information Centre, "Map of Status of Human-induced Soil Degradation," Sheet 2, Nairobi, Kenya, 1990.]

In the solum, two diagnostic horizons may be identified:

- The **epipedon** (literally, over the soil) is the diagnostic horizon at the surface. It is visibly darkened by organic matter and sometimes is leached of minerals.
- The **subsurface diagnostic horizon** originates below the surface at varying depths.

The presence or absence of either diagnostic horizon usually distinguishes a soil for classification. Soil Conservation Service publications identify many different types within each diagnostic horizon using a myriad of terminology.

The Eleven Soil Orders of Soil Taxonomy

At the heart of the Soil Taxonomy are eleven general soil orders, which are described in Table 15-1 along with other information. Their worldwide distribution is shown in Figure 15-8. Please consult this table and map as you read the following descriptions. Because the taxonomy evaluates each soil

FIGURE 15-8
Worldwide distribution of the 11 soil orders. Volcanic ash related soils (Andisols) are finely distributed and are not completely represented on this map. [Adapted from U.S. Soil Conservation Service.]

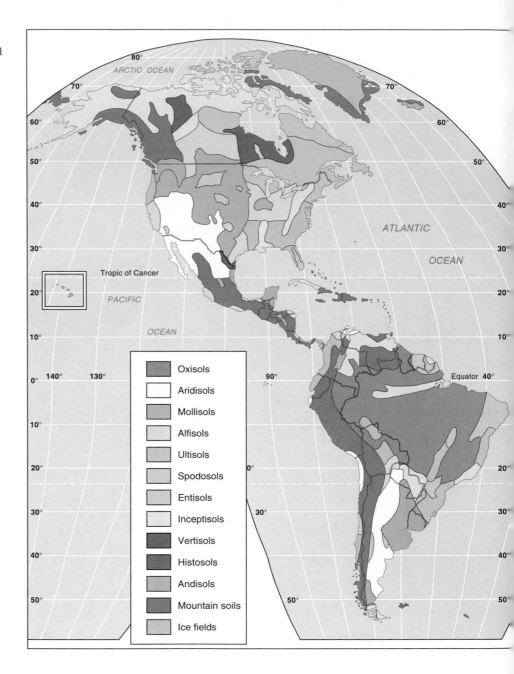

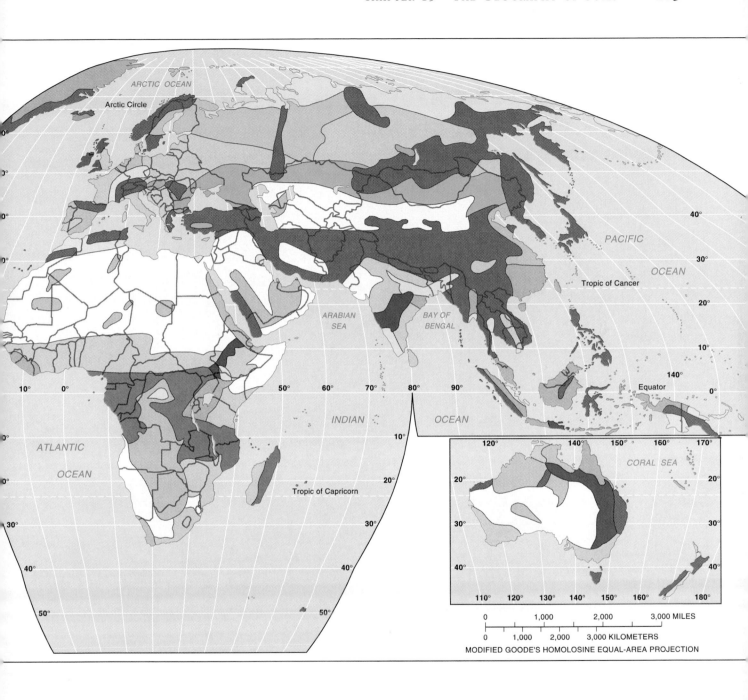

ARCTIC OCEAN

Arctic Circle

PACIFIC

OCEAN

40°

30°

Tropic of Cancer

20°

ARABIAN
SEA

BAY OF
BENGAL

10°

140°

Equator

0°

INDIAN

OCEAN

10°

ATLANTIC

OCEAN

Tropic of Capricorn

20°

30°

40°

50°

CORAL SEA

20°

30°

40°

0   1,000   2,000   3,000 MILES

0   1,000   2,000   3,000 KILOMETERS

MODIFIED GOODE'S HOMOLOSINE EQUAL-AREA PROJECTION

Table 15-1
Soil Taxonomy Soil Orders

Order	Pronunciation	Derviation of Term	Marbut Equivalent (Canadian)	General Location and Climate	World Land Area (%)	U.S. Land Area (%)	Description
Oxisols	ox	Fr. oxide, "oxide" Gr. oxide, "acid or sharp"	Latosois lateritic soils	Tropical soils, hot humid areas	9.2	0.02	Maximum weathering and eluviation, oxic horizon, continuous plinthite layer
Aridisols	arid	L. aridos, "dry"	Reddish desert, gray desert, sierozems	Desert soils, hot dry areas	19.2	11.5	Limited alteration of parent material, low climate activity, ochric epipedon, light color, subsurface illuviation of carbonates
Mollisols	mollify	L. mollis, "soft"	Chestnut, chernozem (Chernozemic)	Grassland soils subhumid, semiarid lands	9.0	24.6	Noticeably dark with organic material, base saturation high, mollic epipedon; friable surface with well structured horizons
Alfisols	alfalfa	Invented syllable	Gray-brown podzolic, degraded chernozem (Luvisol)	Moderately weathered forest soils, humid temperate forests	14.7	13.4	B horizon high in clays, moderate to high degree of base saturation, argillic or natric horizons, no pronounced color change of depth
Ultisols	ultimate	L. ultimus, "last"	Red-yellow podzolic, reddish-yellow lateritic	Highly-weathered forest soils, subtropical forests	8.5	12.9	Similar to Alfisols' B horizon high in clays, generally low amount of base saturation, strong weathering in argillic horizons, redder than Alfisols

Order	Mnemonic	Etymology	Equivalents	Occurrence	%	%	Characteristics
Spodosols	*odd*	Gr. *spodos* or L. *spodos*, "wood ash"	Podzols, brown podzolic (Podzol)	Northern conifer forest soils, cool humid forests	5.4	5.1	Illuvial B horizon of Fe/Al clays, humus; without structure, partially cemented; spodic horizon; highly leached, strongly acid; coarse texture of low bases
Entisols	*recent*	Invented syllable	Azonal soils, tundra	Recent soils profile undeveloped, all climates	12.5	7.9	Limited development; inherited properties from parent material; ochric epipedon; few specific properties; young soils lacking horizons
Inceptisols	*inception*	L. *inceptum*, "beginning"	Ando, subarctic brown forest lithisols, some humic gleys (Brunisol, Cryosol with permafrost, Gleysol wet)	Weakly developed soils, humid regions	15.8	18.2	Intermediate development; embryonic soils, cambic horizon, but few diagnostic features, further weathering possible
Vertisols	in*vert*	L. *verto*, "to turn"	Grumusols (1949) tropical black clays	Expandable clay soils; subtropics, tropics; sufficient dry period	2.1	1.0	Forms large cracks on drying, self-mixing action, contains >30% in swelling clays, ochric epipedon
Histosols	*histology*	Gr. *histos*, "tissue"	Peat, muck, bog (Organic)	Organic soils, wet places	0.8	0.5	Peat or bog, >20% organic matter muck with clay >40 cm thick, surface organic layers, no diagnostic horizons
Andisols	*and, andesite*	L. *Ando*, "volcanic ash"	—	Areas affected by frequent volcanic activity (formerly within Inceptisols and Entisols)	<1.0	<1.0	Volcanic parent materials, particularly ash and volcanic glass; weathering and mineral transformation important; high CEC and organic content, generally fertile

"Soil Losses Estimated"

The U.S. General Accounting Office recently estimated that from 1.2 to 2 million hectares (3–5 million acres) of prime farmland are lost *each year* in the U.S. through mismanagement or conversion to nonagricultural uses. A 1984 report to the Canadian Environmental Advisory Council estimated that the organic content of cultivated prairie soils has declined by 37 to 48%, compared to noncultivated native soils. In Ontario and Quebec, losses of organic content increased to 50% and even higher in the Atlantic provinces, which were naturally low in organic content at the outset of cultivation several centuries ago.

About half of all cropland is experiencing excessive soil erosion in the United States and Canada. This assessment is mirrored worldwide, where approximately 1.2 billion hectares (3 billion acres) of soil are suffering from human-induced erosion and loss to nonagricultural uses, depletion, and contamination.

order on its own characteristics, there is no priority to the classification. You will find the 11 orders loosely arranged here by latitude, beginning along the equator as is done in Chapter 7, "Earth's Climates," and Chapter 16, "Ecosystems and Biomes."

Oxisols. The persistent moisture, intense temperature, and uniform daylength of equatorial latitudes greatly affect soils. These generally old landscapes, are deeply developed and have greatly altered minerals (Figure 15-9). These soils are called **Oxisols** (tropical soils) because they have a distinctive horizon with a mixture of iron and aluminum oxides. Related vegetation is the luxuriant and diverse tropical and equatorial rain forest.

Heavy precipitation leaches soluble minerals and soil constituents (such as silica) from the A horizon. Typical Oxisols are reddish and yellowish from the iron and aluminum oxides left behind, with a weathered clay-like texture, sometimes in a granular soil structure that is easily broken apart. The high degree of eluviation removes basic cations and colloidal material to lower illuviated horizons. Thus, Oxisols are low in CEC (cation-exchange capacity) and fertility,

except in regions augmented by alluvial or volcanic materials. Figure 15-10 illustrates **laterization**, the leaching process that operates in well-drained soils in warm, humid tropical and subtropical climates.

The subsurface diagnostic horizon in an Oxisol is highly weathered, containing iron and aluminum oxides, at least 30 cm (12 in.) thick and within 2 m (6.5 ft) of the surface (Figure 15-9). If these horizons are subjected to repeated wetting and drying, a *hardpan* (hardened soil layer) develops, called a **plinthite** (from the Greek *plinthos,* meaning "brick"). This form of soil, also called a *laterite,* can be quarried in blocks and used as a building material (Figure 15-11).

Simple agricultural activities can be conducted in these soils with care. Early *slash-and-burn shifting cultivation* practices were adapted to these soil conditions and formed a unique style of crop rotation. The scenario went like this: people in the tropics cut down (slashed) and burned the rain forest in small tracts and then cultivated the land with stick and hoe. After several years the soil lost fertility, the people shifted cultivation to another tract and repeated the process. After many years of moving from tract to tract, the group returned to the first patch to begin

(a)

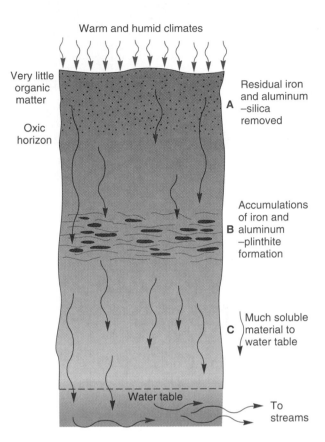

FIGURE 15-10
Laterization process characteristic of tropical and subtropical climatic regimes.

The labels in Figure 15-10 read:

Warm and humid climates

Very little organic matter

Oxic horizon

Residual iron and aluminum –silica removed — A

Accumulations of iron and aluminum –plinthite formation — B

Much soluble material to water table — C

Water table

To streams

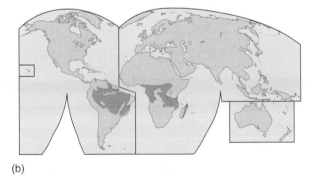

(b)

FIGURE 15-9
Oxisol. (a) Deeply weathered Oxisol profile in central Puerto Rico and (b) general map of worldwide distribution. [Photo from Marbut Collection, Soil Science Society of America, Inc.]

FIGURE 15-11
Plinthite quarrying in India.
[Photo by Henry D. Foth.]

the cycle again. This practice protected the limited fertility of the soils somewhat, allowing periods of recovery.

However, this orderly native pattern of land rotation was halted by the invasion of foreign plantation interests, development by local governments, vastly increased population pressures, and conversion of vast new tracts to pasturage. Permanent tracts of cleared land put tremendous pressure on the remaining forest and brought disastrous consequences. When Oxisols are disturbed, soil loss can exceed a thousand tons per square kilometer per year, not to mention the greatly increased rate of extinction of plant and animal species that accompanies such destruction. The regions dominated by the Oxisols and rain forests are rightfully the focus of much worldwide environmental attention.

Aridisols. The largest single soil order occurs in the world's dry regions. **Aridisols** (desert soils) occupy approximately 19% of Earth's land surface and some 12% of the land in the United States (Figure 15-8). A pale, light soil color near the surface is diagnostic (Figure 15-12a).

Not surprisingly for desert soils, the water balance in Aridisol regions has marked periods of soil mois-

ture deficit and generally inadequate soil moisture for plant growth. High potential evapotranspiration and low precipitation produce very shallow soil horizons. Usually there is no period greater than three months when the soils have adequate moisture. Aridisols also lack organic matter of any consequence. Although the soils are not leached often because of the low precipitation, they are leached easily when exposed to excessive water, for they lack a significant colloidal structure.

Salinization is a soil process that occurs in Aridisols. **Salinization** results from excessive potential evapotranspiration rates in the deserts and semiarid regions of the world. Salts dissolved in soil water are brought to surface horizons and deposited as surface water evaporates. These deposits will kill plants when the salts accumulate near the root zone.

Obviously, salinization complicates farming in Aridisols. The introduction of irrigation water may either waterlog poorly drained soils or lead to salinization. Nonetheless, vegetation does grow where soils are well drained and kept low in salt content. If large capital investments are made in water, drainage, and fertilizers, Aridisols possess much agricultural potential (Figure 15-13). In the Nile and Indus River valleys, for example, Aridisols are intensively farmed with a careful balance of these en-

FIGURE 15-12
Aridisol. (a) Profile from central Arizona and (b) general map of worldwide distribution. [Photo from Marbut Collection, Soil Science Society of America, Inc.]

(a)

(b)
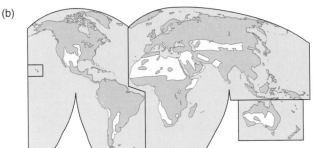

FIGURE 15-13
Agriculture in arid lands of the Coachella Valley of southeastern California. Providing drainage for excessive water application in such areas often is necessary to prevent salinization of the rooting zone. [Photo by author.]

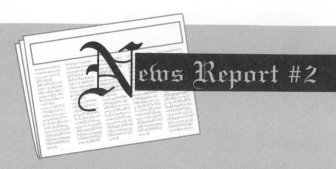

"Death of the Kesterson National Wildlife Refuge"

Drainage of agricultural wastewater, which contain salts, chemicals, and the element selenium, is a special problem in semiarid and arid lands because river discharge is inadequate to dilute and remove field runoff. In trace amounts selenium is necessary in human nutrition but in higher concentrations it is toxic. One solution to prevent unwanted accumulations and waterlogging is to place drains (perforated pipes) beneath the soil at depths of 3 or 4 meters (10–13 feet). Such drains collect percolating irrigation water from fields that have been purposely overwatered to keep salts and chemicals away from the rooting depth of the crops.

But drainage water must go somewhere, and for the San Joaquin Valley of central California this problem triggered a 15-year controversy. Potential drain outlets to the ocean, San Francisco Bay, or the central portion of the valley all failed to pass environmental impact assessments under Environmental Policy Act requirements. Nonetheless, by the late 1970s, about 80 miles of a drain were finished, even though no formal plan nor adequate funding had been completed for its outlet! In the absence of such a plan, large-scale irrigation continued, supplying the field drains with salty, selenium-laden runoff that made its way to the Kesterson National Wildlife Refuge in the northern portion of the San Joaquin Valley east of San Francisco, where the unfinished drain emptied.

It only took three years before the wildlife refuge was officially declared a contaminated toxic waste site. Aquatic life-forms (e.g., marsh plants, plankton, and insects) had taken in the selenium, which thus made its way into the diets of higher life-forms in the refuge. According to U.S. Fish and Wildlife Service scientists, the toxicity moved through the food chain and genetically damaged and killed wildlife. For example, birth defects and death were widely reported in all varieties of birds that nested at Kesterson; approximately 90% of the exposed birds perished or were injured. Because this wildlife refuge was a major migration flyway and stopover point for birds from throughout the hemisphere, this destruction of the refuge also violated several multinational protection treaties.

vironmental factors; although thousands of acres of once-productive land, not so carefully treated, now sit idle and salt-encrusted.

Mollisols. **Mollisols** (grassland soils) are some of Earth's most significant agricultural soils. The dominant diagnostic horizon is a dark, organic surface layer some 25 cm (10 in.) thick (Figure 15-14). Mollisols are soft, even when dry, with granular or crumbly peds, loosely arranged when dry. These humus-rich organic soils are high in basic cations (calcium, magnesium, and potassium) and have a high CEC (therefore, high fertility). In terms of water balance, these soils are intermediate between humid and arid soil moisture.

Soils of the steppes and prairies of the world belong to this soil group: the North American Great Plains, the Pampas of Argentina, and the region from Manchuria in China through to Europe. Agriculture ranges from large-scale commercial grain farming to grazing along the drier portions of the soil order.

(a)

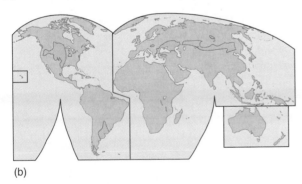

(b)

FIGURE 15-14
Mollisol. (a) Profile from central Iowa and (b) general map of worldwide distribution. [Photo from Marbut Collection, Soil Science Society of America, Inc.]

With fertilization or soil-building practices, high crop yields are common. The "fertile triangle" of Ukraine, Russia, and western portions of the Commonwealth of Independent States is of this soil type.

In North America, the Great Plains straddle the 98th meridian, which is coincident with the 51 cm (20 in.) isohyet of annual precipitation—wetter to the east and drier to the west. The Mollisols here mark the historic division between the short- and tall-grass prairies (Figure 15-15).

A soil process characteristic of some Mollisols, and adjoining areas of Aridisols, is calcification. **Calcification** is an illuviated accumulation of calcium carbonate or magnesium carbonate in the B and C horizons (Figure 15-16). When cemented or hardened, these deposits are called **caliche**, or *kunkur;* they occur in widespread soil formations in central and western Australia, the Kalahari region of interior Southern Africa, and the High Plains of the west-central United States, among other places.

Alfisols. Alfisols (moderately weathered forest soils) are the most widespread of the soil orders, extending from near the equator to high latitudes. Representative Alfisol areas include Boromo and Burkina Faso (interior western Africa); Fort Nelson, British Columbia; the states near the Great Lakes, and the valleys of central California. Most Alfisols have a pale, grayish brown-to-reddish epipedon and are considered moist versions of the Mollisol soil group. Moderate eluviation is present, as well as a subsurface horizon of illuviated white clays because of increased precipitation (Figure 15-17).

Alfisols have moderate-to-high reserves of basic cations and are fertile. However, productivity depends on moisture and temperature. Alfisols usually are supplemented by a moderate application of lime and fertilizers in areas of active agriculture. Some of the best farmland in the United States stretches from Illinois, Wisconsin, and Minnesota through Indiana, Michigan, and Ohio to Pennsylvania and New York. This land produces grains, hay, and dairy products. These naturally productive soils are farmed intensively for subtropical fruits, nuts, and special crops that can grow only in a few locales worldwide (e.g., California production of grapes, citrus, artichokes, almonds, figs).

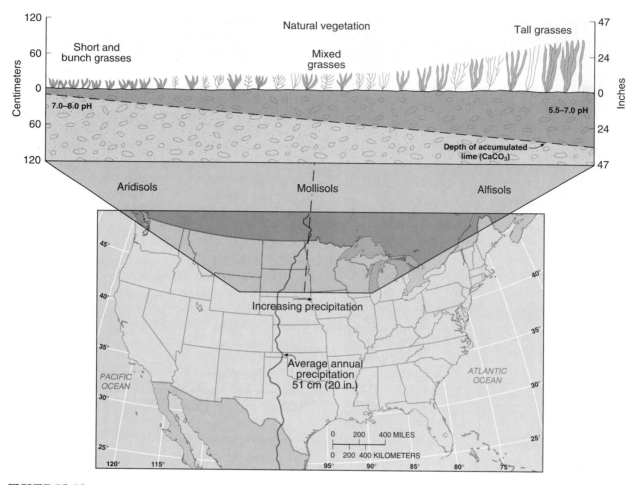

FIGURE 15-15

Aridisols (to the west), Mollisols, and Alfisols (to the east)—a soil continuum in the north-central United States. Graduated changes that occur in soil pH and the depth of available lime are shown. [Adapted with permission of Macmillan Publishing Company from *The Nature and Properties of Soils*, 10th ed., by Nyle C. Brady. Copyright © 1990 by Macmillan Publishing Company.]

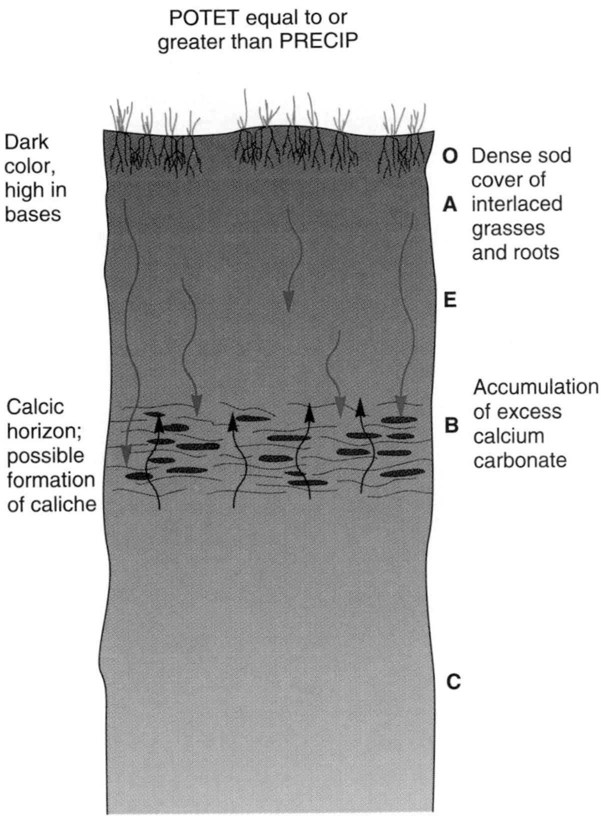

POTET equal to or
greater than PRECIP

Dark
color,
high in
bases

O Dense sod
A cover of
interlaced
grasses
and roots

E

Calcic
horizon;
possible
formation
of caliche

Accumulation
of excess
B calcium
carbonate

C

FIGURE 15-16
Calcification process in Aridisol/Mollisol soils in climatic
regimes that have a potential evapotranspiration equal to
or greater than precipitation.

Ultisols. Farther south in the United States are the
Ultisols, highly weathered forest soils. An Alfisol
might degenerate into an Ultisol, given time and exposure to increased weathering under moist conditions. These soils tend to be reddish because of
residual iron and aluminum oxides in the A horizon
(Figure 15-18).

The increased precipitation in Ultisol regions
means greater mineral alteration, more eluvial
leaching, and therefore a lower level of basic
cations, leading to infertility. Fertility is further reduced by certain agricultural practices and the effect of soil-damaging crops such as cotton and
tobacco, which deplete nitrogen and expose soil
to erosion. However, these soils respond well if

(a)

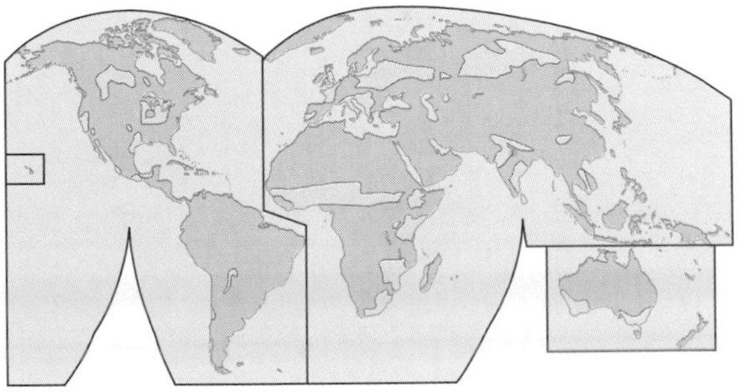

(b)

FIGURE 15-17
Alfisol. (a) Profile from central California and (b) general
map of worldwide distribution. [Photo from Marbut
Collection, Soil Science Society of America, Inc.]

(a)

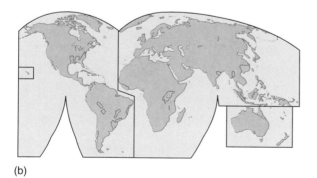

(b)

FIGURE 15-18
Ultisol. (a) Profile from the upper coastal plain of central
North Carolina and (b) general map of worldwide
distribution. [Photo from Marbut Collection, Soil Science
Society of America, Inc.]

subjected to good management—for example, crop
rotation that restores nitrogen and cultivation practices that prevent sheetwash and soil erosion.
Much needs to be done to achieve sustainable
management of these soils.

Spodosols. The **Spodosols** (northern coniferous forest soils) occur generally to the north and
east of the Alfisols. They are in cold and forested
moist regimes (*Dfb humid continental mild summer* climates) in northern North America and Eurasia, Denmark, the Netherlands, and southern
England. Because there are no comparable climates in the Southern Hemisphere, this soil type is
not identified there. Spodosols form from sandy
parent materials, shaded under evergreen forests
of spruce, fir, and pine. Spodosols with more
moderate properties form under mixed or deciduous forests (Figure 15-19a and b).

Spodosols lack humus and clay in the A horizons.
An eluviated white horizon, sandy and leached of
clays and irons, lies in the A horizon instead and
overlies a horizon of illuviated organic matter and
iron and aluminum oxides (Figure 15-19c). The surface horizon receives organic litter from base-poor,
acid-rich trees, which contribute to acid accumulations in the soil. The low pH (acidic) soil solution
effectively removes clays, iron, and aluminum, which
are passed to the upper diagnostic horizon. An
ashen-gray color is common in these subarctic forest
soils and is characteristic of a formation process
called **podzolization**.

When agriculture is attempted, the low base-cation
content of Spodosols requires the addition of nitrogen, phosphate, and potash (potassium carbonate),
and perhaps crop rotation as well. A soil *amendment*
such as limestone can significantly increase crop production by raising the pH of these acidic soils. For
example, the yields of several crops (corn, oats,
wheat, and hay) grown in specific Spodosols in New
York State were increased up to a third with the application of 1.8 metric tons (2 tons) of limestone per
0.4 hectare (1.0 acre) during each 6-year rotation.

Entisols. The **Entisols** (recent, undeveloped
soils) lack vertical development of horizons. The
presence of Entisols is not climate dependent, for
they occur in many climates worldwide. Entisols are

true soils that have not had sufficient time to generate the usual horizons.

Entisols generally are poor agricultural soils, although those formed from river silt deposits are quite fertile. The conditions that have inhibited complete development also have prevented adequate fertility—too much or too little water, poor structure, and insufficient accumulation of weathered nutrients. Active slopes, alluvium-filled floodplains, poorly drained tundra, tidal mud flats, dune sands and erg (sandy) deserts, and plains of glacial outwash all are characteristic regions that have these soils. Figure 15-20 shows an Entisol in a desert climate where shales formed the parent material.

Inceptisols. Inceptisols (weakly developed soils) are inherently infertile. They are weakly developed young soils, although more developed than

(a)

FIGURE 15-19
Spodosol. (a) Profile from northern New York and (b) general map of worldwide distribution; (c) podzolization process, typical in cool and moist climatic regimes. [Photo from Marbut Collection, Soil Science Society of America, Inc.]

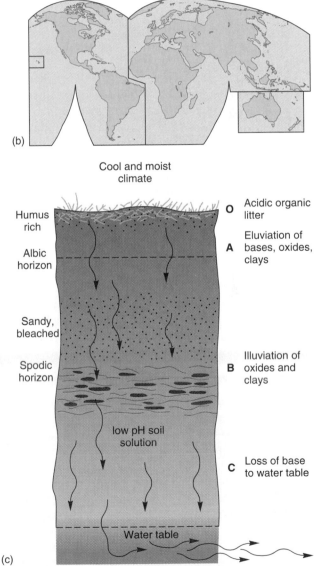

(b)

(c)

FIGURE 15-20
A characteristic Entisol forming
from a shale parent material in
the desert near Zabriskie Point,
Death Valley. [Photo by author.]

the Entisols. Inceptisols include a wide variety of different soils, all holding in common a lack of maturity with evidence of weathering just beginning. Inceptisols are associated with moist soil regimes and are regarded as eluvial because they demonstrate a loss of soil constituents throughout their profile but retain some weatherable minerals. This soil group has no distinct illuvial horizons.

Inceptisols include the soils of most of the arctic tundra; glacially derived till and outwash materials from New York down through the Appalachians; and alluvium on the Mekong and Ganges floodplains.

Andisols. Andisols (volcanic parent materials) occur in areas of volcanic activity. These soils formerly were classified as Inceptisols and Entisols, but in 1990 they were placed in this new order. Andisols are derived from volcanic ash and glass. Previous soil horizons frequently are found buried by ejecta from repeated volcanic eruptions. Andisols are unique in their mineral content and in their recharge by eruptions.

Weathering and mineral transformations are important in this soil order. Volcanic glass weathers readily into a clay colloid and oxides of aluminum and iron. Andisols feature a high CEC and high water-holding ability and develop moderate fertility. The fertile fields of Hawaii produce sugar cane and pineapple as important cash crops in Andisols (Figure 15-21). Andisol distribution is small in areal extent; however, such soils are locally important around the volcanic "ring of fire" in the Pacific Rim.

Vertisols. Vertisols (expandable clay soils) are heavy clay soils. They contain more than 30% *swelling clays* (which swell significantly when they absorb water). They are located in regions experiencing highly variable soil moisture balances through the seasons. These soils occur in areas of subhumid-to-semiarid moisture and moderate-to-high temperature. Vertisols frequently form under savanna and grassland vegetation in tropical and subtropical climates and are sometimes associated with a distinct dry season following a wet season. Although widespread, individual Vertisol units are limited in extent.

Vertisol clays are black when wet (but not because of organics, rather because of specific mineral content) and range to brown and dark gray. These deep clays swell when moistened and shrink when dried. In the process, vertical cracks form, which widen and deepen as the soil dries, producing cracks 2–3 cm (0.8–1.2 in.) wide and up to 40 cm (16 in.) deep. Loose material falls into these cracks, only to disappear when the soil again expands and the cracks close. After many such cycles, soil contents tend to invert or mix vertically, bringing lower horizons to the surface. (Figure 15-22).

Despite the fact that clay soils are plastic and heavy when wet, with little available soil moisture for plants, Vertisols are high in bases and nutrients and thus are some of the better farming soils wherever they occur. For example, they occur in a narrow zone along the coastal plain of Texas and in a section along the Deccan region of India. Vertisols often are planted with grain sorghums, corn, and cotton.

FIGURE 15-21

Fertile Andisols planted with sugar cane, Kauai, Hawaii. Hawaii is the largest producer of sugar cane in the United States. [Photo by Wolfgang Kaehler.]

FIGURE 15-22

Vertisol. (a) Profile in the Lajas Valley of Puerto Rico and (b) general map of worldwide distribution. [Photo from Marbut Collection, Soil Science Society of America, Inc.]

(a)

(b)
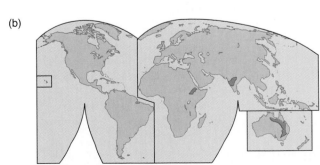

Histosols. Histosols (organic soils) are formed from accumulations of thick organic matter. In the midlatitudes, when conditions are right, beds of former lakes may turn into Histosols, with water gradually replaced by organic material to form a bog and layers of peat. (Lake succession and bog/marsh formation are discussed in Chapter 16.) Histosols also form in small, poorly drained depressions where conditions are ideal for significant deposits of *sphagnum peat* to form. This material can be cut, baled, and sold as a soil amendment. Dried peat has served for centuries as a low-grade fuel. The area southwest of Hudson Bay in Canada is typical of Histosol formation.

SUMMARY—The Geography of Soils

Soil is the portion of the land surface in which plants can grow. It is a dynamic natural body composed of fine materials and contains both mineral and organic matter. Soils range in depth from a few millimeters to many meters, and in some locales they may be completely absent. The basic soil mapping unit is the **polypedon**, made up of numerous sampling units called **pedons**. Soils are structured in vertical horizons that permit analysis and classification. The key properties for analysis are color, texture, structure, consistence, porosity, moisture, and chemistry.

Environmental factors that affect soil formation include parent materials, climate, vegetation, topography, and time. Human influence is having great impact on Earth's prime soils. A new survey completed by the United Nations Environment Programme identified significant areas of serious soil erosion and loss caused by human misuse and abuse.

The past century is marked by repeated efforts to develop soil classifications. Presently, the **Soil Taxonomy** classification system is used in the United States, built around an analysis of various diagnostic horizons and eleven soil orders.

Soil fertility (the ability to sustain plant productivity) is critical to society. Essential soils for agriculture and their fertility are threatened by mismanagement, destruction, and conversion to other uses. Much soil loss is preventable through the application of known technologies, improved agricultural practices, and sustainable government policies.

KEY TERMS

adsorption

calcification

caliche

cation-exchange capacity (CEC)

eluviation

epipedon

humus

illuviation

laterization

loam

pedon

plinthite

podzolization

polypedon

salinization

soil

soil colloid

soil fertility

soil horizon

Soil Taxonomy

    Alfisols

    Andisols

    Aridisols

    Entisols

    Histosols

    Inceptisols

    Mollisols

    Oxisols

    Spodosols

    Ultisols

    Vertisols

solum

subsurface diagnostic horizon

REVIEW QUESTIONS

1. Define polypedon and pedon, the basic units of soil analysis and classification.
2. Characterize the principal aspects of each soil horizon. Where does the main accumulation of organic material occur? The formation of humus? What is the eluviated layer? The illuviated layer? The solum, or true soil?
3. Define a soil separate. What are the various sizes of particles in soil? What is loam? Why is loam regarded so highly by agriculturalists?
4. What is a possible method for determining soil consistence?
5. What are soil colloids? How do they relate to cations and anions in the soil? Explain the cation-exchange capacity (CEC).
6. Briefly describe the contribution of the following factors and their effect on soil formation: parent material, climate, vegetation, landforms, time, and humans.
7. Summarize the brief history of soil classification described in this chapter. What led soil scientists to develop the new Soil Taxonomy classification system?
8. Define an epipedon and a subsurface diagnostic horizon.
9. Locate each soil order on the map as you give a general description of it.
10. How was slash-and-burn shifting cultivation, as practiced in the past, a form of crop and soil rotation?
11. Describe the salinization process in arid and semiarid soils. What soil horizons develop?
12. Which of the soil orders are associated with Earth's most productive agricultural areas?
13. Describe the podzolization process associated with northern coniferous forest soils. What characteristics are associated with the surface horizons? What strategies might enhance these soils?
14. What former Inceptisols were placed in a new soil order (1990)? Describe these soils: location, nature, and formation processes. Why do you think they were separated into their own order?

Flowering bromeliads in El Yunque Rain Forest, Puerto Rico. [*Photo by Tom Bean.*]

16

ECOSYSTEMS AND BIOMES

ECOSYSTEM COMPONENTS
   AND CYCLES
   Communities
   Plants: The Essential Biotic Component
   Abiotic Ecosystem Components
   Biotic Ecosystem Operations

STABLITY AND SUCCESSION
   Ecosystem Stability and Diversity
   Ecological Succession

EARTH'S ECOSYSTEMS
   Terrestrial Ecosystems

EARTH'S MAJOR TERRESTRIAL
   BIOMES
   Equatorial and Tropical Rain Forest
   Tropical Seasonal Forest and Scrub
   Tropical Savanna
   Midlatitude Broadleaf and Mixed Forest
   Northern Needleleaf Forest and
     Montane Forest
   Temperate Rain Forest
   Mediterranean Shrubland
   Midlatitude Grasslands
   Desert Biomes   Arctic and Alpine Tundra

SUMMARY
FYI REPORT 16-1 BIODIVERSITY AND
   BIOSPHERE RESERVES

Diversity is an impressive feature of the living Earth. The diversity of organisms is a response to the interaction of the atmosphere, hydrosphere, and lithosphere, which produces a variety of conditions within which the biosphere exists.

This sphere of life and organic activity extends from the ocean floor to about 8 km (5 mi) in the atmosphere. The biosphere includes myriad ecosystems from simple to complex, each operating within general spatial boundaries. An **ecosystem** is a self-regulating association of living plants and animals and their nonliving physical environment. In an ecosystem, a change in one component causes changes in others, as systems adjust to new conditions.

Earth's surface environment itself is the largest ecosystem within the natural boundary of the atmosphere. Natural ecosystems are open systems for both energy and matter, with almost all ecosystem boundaries functioning as transition zones rather than as sharp demarcations.

Ecology is the study of the relationships between organisms and their environment and among the various ecosystems in the biosphere. The word *ecology*, developed by German naturalist Ernst Haeckel in 1869, is derived from the Greek *oikos* (household, or place to live) and *logos* (study of). **Biogeography**, essentially a spatial ecology, is the study of the distribution of plants and animals, the diverse spatial patterns they create, and the physical and biological processes, past and present, that produce this distribution across Earth.

The degree to which modern society understands ecosystems will help determine our success as a species and the long-term survival of a habitable Earth:

> The time is ripe to step up and expand current efforts to understand the great interlocking systems of air, water, and minerals nourishing the Earth. . . . Moreover, without vigorous action toward that goal, nations will be seriously handicapped in trying to cope with proven and suspected threats to ecosystems and to human health and welfare resulting from alterations in the cycles of carbon, nitrogen, phosphorus, sulfur, and related

materials. . . . Society depends upon this life-support system of planet Earth.*

Humans are Earth's major biotic (living) agent, because we influence all ecosystems. In this chapter we explore ecosystems and concepts of stability and succession. We conclude with an examination of Earth's major terrestrial ecosystems, their appearance and structure, location, and their present status.

Ecosystem Components and Cycles

An ecosystem is a complex of many variables, all functioning independently yet in concert, with complicated flows of energy and matter (Figure 16-1). An ecosystem includes both biotic (living) and abiotic (nonliving) components. Nearly all depend upon an input of solar energy; the few limited ecosystems that exist in dark caves or on the ocean floor depend upon chemical reactions (chemosynthesis). Ecosystems are divided into subsystems, with the biotic portion composed of producers, consumers, and decomposers. The abiotic flows in an ecosystem include gaseous and sedimentary nutrient cycles. Figure 16-2 illustrates the essential elements of an ecosystem.

Communities

A convenient biotic subdivision within an ecosystem is a **community**, which is formed by interactions among populations of living animals and plants. An ecosystem is the interaction of many communities with the abiotic physical components of its environment. For example, in a forest ecosystem, a specific community may exist on the forest floor, whereas another community functions in the canopy of leaves high above. Similarly, within a lake ecosystem, the plants and animals that flourish in the bottom sediments form one community, whereas those

---

*Gilbert F. White and Mostafa K. Tolba, "Global Life Support Systems," *United Nations Environment Programme Information*, no. 47, Nairobi, Kenya: United Nations, 1979, p. 1.

FIGURE 16-1

The intricate web of a spider mirrors
the web of life.

"Life devours itself: everything that
eats is itself eaten; everything that
can be eaten is eaten; every
chemical that is made by life can be
broken down by life; all the
sunlight that can be used is used....
The web of life has so many
threads that a few can be broken
without making it all unravel, and if
this were not so, life could not have
survived the normal accidents of
weather and time, but still the
snapping of each thread makes the
whole web shudder, and weakens
it.... You can never do just one
thing: the effects of what you do in
the world will always spread out
like ripples in a pond."
[From Friends of the Earth and Amory
Lovins, The United Nations Stockholm
Conference: *Only One Earth*. London:
Earth Island Limited, 1972, p. 20.
Photo by author.]

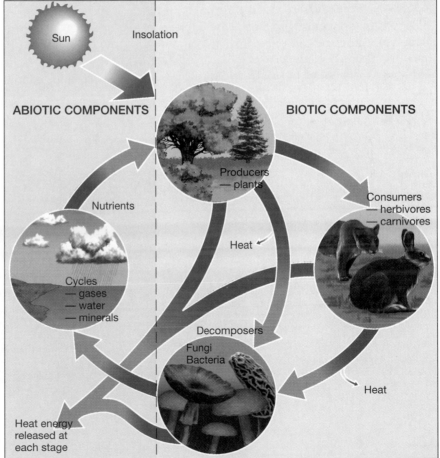

FIGURE 16-2
Abiotic and biotic components of
ecosystems. Solar energy is the
input that drives the biosphere.
Heat is the energy output from
this life system.

near the surface form another. A community is identified in several ways—by its physical appearance, the number of species and the abundance of each, the complex patterns of their interdependence, and the trophic (feeding) structure of the community.

Within a community, two concepts are important: habitat and niche. **Habitat** is the specific physical location of an organism, the type of environment in which it resides or is biologically adapted to live. In terms of physical and natural factors, most species have specific habitat requirements with definite limits and a specific regimen of sustaining nutrients.

Niche (French *nicher*, to nest) refers to the function, or occupation, of a life form within a given community. It is the way an organism obtains and sustains the physical, chemical, and biological factors it needs to survive. An individual species must satisfy several aspects in its niche. Among these are a *habitat* niche, a *trophic* (food) niche, and a *reproductive* niche. For example, the red-wing blackbird *(Agelaius phoeniceus)* occurs throughout the United States and most of Canada in *habitats* of meadow, pastureland, and marsh where they nest. Their trophic niche is weed seeds and cultivated seed crops throughout the year, adding insects to their diet during the nesting season.

Similar habitats produce comparable niches. In a stable community, no niche is left unfilled. The *competitive exclusion principle* states that no two species can occupy the same niche (food or space) successfully in a stable community. Thus, closely related species are separated spatially from one another. In other words, each species operates to reduce competition.

Some species have *symbiotic relationships,* an arrangement that mutually benefits and sustains each organism. For example, *lichen* (pronounced "liken") is made up of algae and fungi living together. The algae is the producer and food source, and the fungus provides structure and physical support. Their mutually beneficial relationship (mutualism) allows the two to occupy a niche in which neither could survive alone (Figure 16-3). The partnership of corals and algae discussed in Chapter 13 is another example of a symbiotic relationship.

By contrast, *parasitic relationships* eventually may kill the host, thus destroying the parasite's own niche and habitat. An example is mistletoe *(Phoradendron),* which lives on and may kill various kinds of trees. Some scientists are questioning whether our human society and the physical systems of Earth constitute a global-scale symbiotic relationship (sustainable) or a parasitic one (nonsustainable).

FIGURE 16-3
Lichen, an example of a symbiotic relationship. [Photo by Bobbé Christopherson.]

Plants: The Essential Biotic Component

Plants are the critical biotic link between life and solar energy. *Ultimately, the fate of the biosphere rests on the success of plants and their ability to capture sunlight.*

Beginnings. The first plants were simple single- or multiple-celled structures known as *cyanobacteria* (formerly known as blue-green algae). They began the harvest of sunlight and release of free oxygen about 3.3 billion years ago. Long, slow development ensued as these early aquatic forerunners of plants migrated from the subsurface to the water surface, increasingly protected by the evolving atmosphere.

The last billion years of Earth's history have seen an explosion of biotic diversity. Land plants and animals became common about 430 million years ago, according to fossilized remains. At present there are almost 250,000 species of plants.

Leaf Activity. Leaves are solar-powered chemical factories. Flows of carbon dioxide, water, light, and oxygen enter and exit the surface of each leaf (see Figure 1-4). They flow in and out through small pores called **stomata** (singular: stoma). Each stoma is surrounded by small guard cells that open and close the pore, depending on the plant's needs at the moment. Veins in each leaf connect to the stems and branches of the plant and thus to its main circulation system. The veins bring in water and nutrients and carry off the sugars produced by photosynthesis.

Water that moves through a plant exits the leaves through the stomata and evaporates from leaf surfaces, thereby assisting heat regulation within the plant. As water evaporates from the leaves, a pressure deficit is created that allows atmospheric pressure to push water up through the plant all the way from the roots in the same manner that a soda straw works. We can only imagine the complex operation of a 100 m (330 ft) tree!

Photosynthesis and Respiration. Under the influence of certain wavelengths of visible light, **photosynthesis** unites carbon dioxide and oxygen (derived from water in the plant). The process releases oxygen and produces energy-rich organic material. The name is descriptive: *photo-* refers to

sunlight, and *-synthesis* describes the reaction of materials within plant leaves.

The largest concentration of light-responsive, photosynthetic cells rests below the upper layers of the leaf. These are *chloroplasts*, and within each resides a green, light-sensitive pigment called **chlorophyll**. Within this pigment, light stimulates photochemistry. Competition for light is a dominant factor in the formation of plant communities. This competition is expressed in their height, orientation, and structure. Photosynthesis essentially follows this equation:

$$CO_2 \quad + \quad H_2O \quad \xrightarrow[\text{energy}]{\text{light}} \quad x(CH_2O) \quad + \quad O_2$$

(Carbon dioxide) (Water) (Carbo-hydrate–sugar) (Oxygen)

From the equation, you can see that photosynthesis removes carbon (in the form of CO_2) from Earth's atmosphere (Figure 16-4). The quantity is enormous: approximately 91 billion metric tons (100 billion tons) per year. Carbohydrates, the organic result of the photosynthetic process, are combinations of carbon, hydrogen, and oxygen. They form simple sugars, such as glucose ($C_6H_{12}O_6$). Glucose, in turn, is used by plants to build starches, which are more complex carbohydrates and the principal food storage substance in plants. *Primary productivity* refers to the rate at which energy is stored in such organic substances.

Plants not only store energy; they also must consume some of this energy by converting carbohydrates to derive energy for their other operations. Thus, **respiration** is essentially a reverse of the photosynthetic process:

$$x(CH_2O) \quad + \quad O_2 \longrightarrow CO_2 \quad + \quad H_2O \quad + \text{energy}$$

(Carbo-hydrate) (Oxygen) (Carbon dioxide) (Water) (Heat)

In respiration, plants oxidize carbohydrates, releasing carbon dioxide, water, and energy as heat (Figure 16-4). The overall growth of a plant depends on a surplus of carbohydrates beyond what is lost through plant respiration. Figure 16-4 presents a simple schematic of this process, which yields plant growth. The difference between photosynthetic production and respiration loss is called *net photosynthesis*.

FIGURE 16-4
The balance between photosynthesis and respiration determines plant growth.

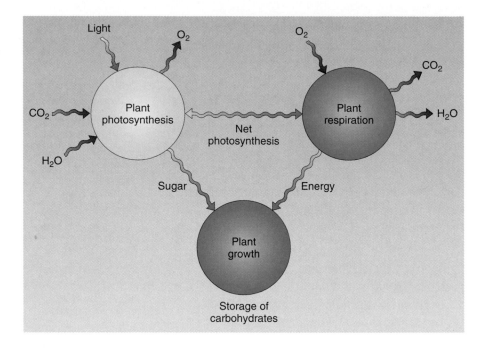

Net Primary Productivity The net photosynthesis for an entire plant community is its **net primary productivity**. This is the amount of stored chemical energy (biomass) that the community generates for the ecosystem. **Biomass** is the net dry weight of organic material; it is biomass that feeds the food chain.

Net primary productivity is mapped in terms of *fixed carbon per square meter per year*. ("Fixed" means chemically bound into plant tissues.) Study Figure 16-5 and you can see that on land, net primary production tends to be highest in the tropics and decreases toward higher latitudes. But precipitation also affects productivity, as evidenced on the map by the correlations of abundant precipitation with high productivity (tropics) and reduced precipitation with low productivity (subtropical deserts). Even though deserts receive high amounts of solar radiation, other controlling factors are more important, namely water availability and soil conditions.

In the oceans, productivity is limited by differing nutrient levels. Regions with nutrient-rich upwelling currents generally are the most productive (off western coastlines). The map shows that the tropical ocean and areas of subtropical high pressure are quite low in productivity.

In temperate and high latitudes, the rate at which carbon dioxide is fixed by vegetation varies seasonally. It increases in spring and summer as plants flourish with increasing solar input and, in some areas, more available (nonfrozen) water, and decreases in late fall and winter. Rates in the tropics are high throughout the year, and turnover in the photosynthesis-respiration cycle is faster, exceeding by many times the rates experienced in a desert environment or in the far northern limits of the tundra. A lush hectare (2.5 acres) of sugar cane in the tropics might fix 45 metric tons (50 tons) of carbon in a year, whereas desert plants in an equivalent area might achieve only 1% of this amount.

Table 16-1 lists various ecosystems, their net primary productivity per year, and an estimate of net total biomass worldwide. World net primary productivity is estimated at 170 billion metric tons of dry organic matter per year. Compare the various ecosystems, especially cultivated land, with most of the natural communities.

Abiotic Ecosystem Components

Critical in each ecosystem is the flow of energy and the cycling of nutrients and water. Nonliving abiotic components set the stage for ecosystem operations.

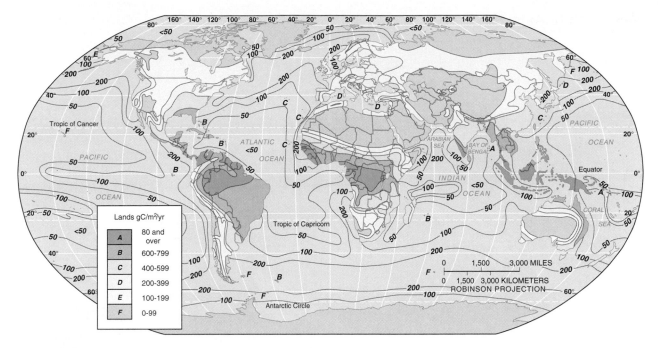

FIGURE 16-5
Worldwide net primary productivity in grams of carbon per square meter per year
(approximate values). [After D. E. Reichle, *Analysis of Temperate Forest Ecosystems.*
Heidelberg, Germany: Springer-Verlag, 1970. Adapted by permission.]

Table 16-1
Net Primary Production and Plant Biomass on Earth

Ecosystem	Area (10⁶ km²)*	Net Primary Productivity per Unit Area (g/m²/yr)** Normal Range	(mean)	World Net Primary Production (10⁹/t/yr)***
Tropical rain forest	17.0	1000–3500	2200	37.4
Temperate deciduous forest	7.0	600–2500	1200	8.4
Boreal forest	12.0	400–2000	800	9.6
Woodland and shrubland	8.5	250–1200	700	6.0
Savanna	15.0	200–2000	900	13.5
Temperate grassland	9.0	200–1500	600	5.4
Tundra and alpine region	8.0	10–400	140	1.1
Desert and semidesert scrub	18.0	10–250	90	1.6
Cultivated land	*14.0*	*100–3500*	*650*	*9.1*
Swamp and marsh	2.0	800–3500	2000	4.0
Lake and stream	2.0	100–1500	250	0.5
Open ocean	332.0	2–400	125	41.5
Algal beds and reefs	0.6	500–4000	2500	1.6
Estuaries	1.4	200–3500	1500	2.1

*1km² = 0.39 mi²
**1 g per m² = 8.9 lb per acre
***1 metric ton = 1.1023 tons
SOURCE: Helmut Lieth and R.H. Whittaker, editors, *The Primary Production of the Biosphere*. New York: Springer-Verlag, 1975. Reprinted by permission.

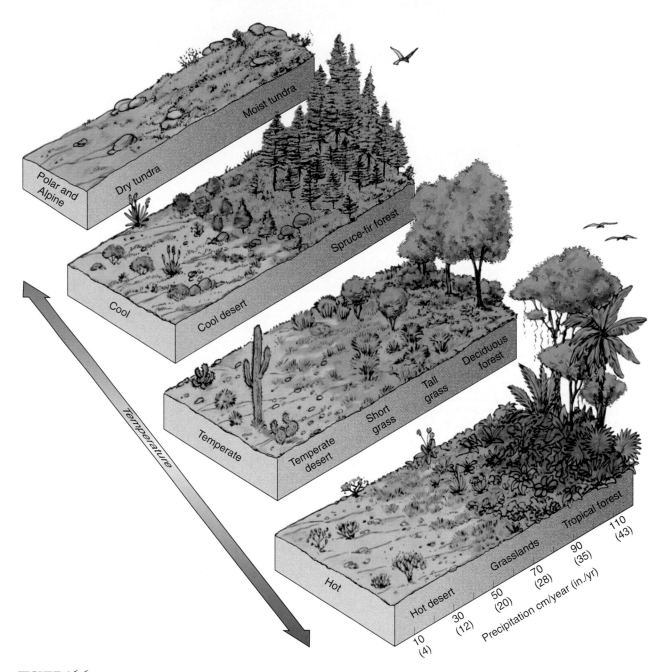

FIGURE 16-6
Abiotic climate control of ecosystem types: the generalized relationship among
rainfall, temperature, and vegetation. [Adapted from Daniel D. Chiras,
Environmental Science, 2d ed. Copyright © 1988, Benjamin/Cummings Publishing
Co., Menlo Park, CA. Used by permission.]

Light, Temperature, Water, and Climate. The pattern of solar energy receipt is crucial in both terrestrial and aquatic ecosystems. As explained, solar energy enters an ecosystem by way of photosynthesis, with heat dissipated from the system at many points. The duration of Sun exposure is the **photoperiod**. Along the equator, days are almost always 12 hours in length; however, with increasing distance from the equator, seasonal effects become pronounced, as discussed in Chapter 2. Plants have adapted their flowering and seed germination to seasonal changes in insolation.

Air and soil temperatures determine the rates at which chemical reactions proceed. Significant temperature factors are seasonal variation, duration and pattern of minimum and maximum temperatures, and average temperature (Chapter 3).

Operation of the hydrologic cycle and water availability depend on precipitation/evaporation rates and their seasonal distribution (see Chapters 5 and 6). Water quality is essential—its mineral content, salinity, and levels of pollution and toxicity. Also,

daily weather patterns over time create regional climates (see Chapter 7), which in turn affect the pattern of vegetation and ultimately influence soil development. All of these factors work together to establish the parameters (limits) for ecosystems that may develop in a given location.

Figure 16-6 illustrates the general relationship among temperature, precipitation, and vegetation. In general terms, can you identify the characteristic vegetation type and related temperature and moisture relationship that fits the area of your home or school?

Alexander von Humboldt (1769–1859), an explorer, geographer, and scientist, deduced that plants and animals recur in related groupings wherever similar conditions occur in the abiotic environment. After several years of study in the Andes Mountains of Peru, he described a distinct relationship between altitude and plant communities, his *life zone* concept. As he climbed the mountains, he noticed that the experience was similar to that of traveling away from the equator toward higher latitudes (Figure 16-7).

FIGURE 16-7

Progression of plant community life zones with increasing altitude or latitude.

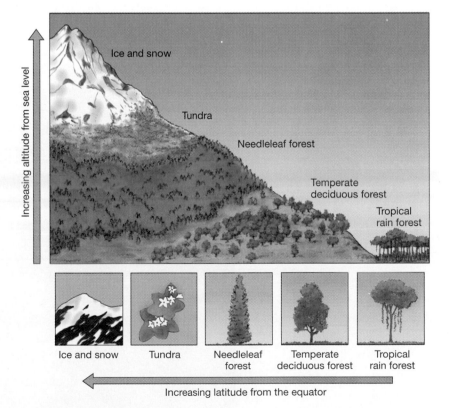

This zonation of plants with altitude is noticeable on any trip from lower valleys to higher elevations. Each **life zone** possesses its own temperature, precipitation, and insolation relationships and therefore its own biotic communities.

Beyond these general conditions, each ecosystem further produces its own *microclimate,* specific to individual sites. For example, in forests the insolation reaching the ground is reduced. A pine forest cuts light by 20–40%, whereas a birch-beech forest reduces it by as much as 50–75%. Forests also are about 5% more humid than nonforested landscapes, have warmer winters and cooler summers, and experience reduced winds. Such highly localized *microecosystems* are evident along a mountain trail, where changes in exposure and moisture can be easily seen (Figure 16-8).

Gaseous and Sedimentary Cycles. The most abundant natural elements in living matter are hydrogen (H), oxygen (O), and carbon (C). Together, these elements make up more than 99% of Earth's biomass; in fact, all life (organic molecules) contains hydrogen and carbon. In addition, nitrogen (N), cal-

cium (Ca), potassium (K), magnesium (Mg), sulfur (S), and phosphorus (P) are significant *nutrients,* elements necessary for the growth and development of a living organism.

Several key chemical cycles function in nature. Oxygen, carbon, and nitrogen each have *gaseous cycles,* part of which are in the atmosphere. Other elements have *sedimentary cycles* which principally involve the mineral and solid phases (major ones include phosphorus, calcium, and sulfur). Some elements combine gaseous and sedimentary cycles. These recycling processes are called **biogeochemical cycles**, because they involve chemical reactions in both living and nonliving systems.

Oxygen and carbon cycles we consider together because they are so closely intertwined through photosynthesis and respiration (Figure 16-9). The atmosphere is the principal reserve of *available* oxygen. Larger reserves of oxygen exist in Earth's crust, but they are unavailable, chemically bound to other elements.

As for carbon, the greatest pool of it is in the ocean—about 39,000 billion tons, or about 93% of Earth's total carbon. However, all of this carbon is

FIGURE 16-8
Fern and forest trail, indicative of a microecosystem. The microclimate at the forest floor is drastically changed when the forest is cleared. [Photo by author.]

FIGURE 16-9
The carbon and oxygen cycles, simplified.

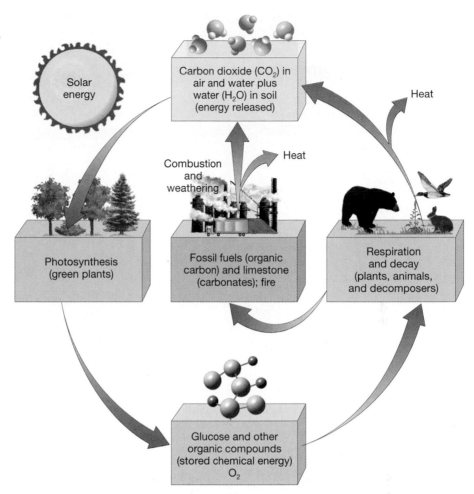

bound chemically in carbon dioxide, calcium carbonate, and other compounds. The ocean absorbs carbon dioxide through photosynthesis by small phytoplankton. Carbon is stored in certain carbonate minerals, such as limestone.

The atmosphere, which is the integrating link in the cycle, contains only about 700 billion tons of carbon (as carbon dioxide) at any moment. This is far less than in fossil fuels and oil shales (12,000 billion tons, as hydrocarbon molecules) or living and dead organic matter (2275 billion tons, as carbohydrate molecules). Carbon dioxide in the atmosphere is produced by the respiration of plants and animals, volcanic activity, and fossil fuel combustion by industry and transportation.

The *nitrogen cycle* involves the major constituent of the atmosphere, 78.084% of each breath we take. Nitrogen also is important in the makeup of organic molecules, especially proteins, and therefore is essential to living processes. A simplified view of the nitrogen cycle is portrayed in Figure 16-10. However, this vast atmospheric reservoir is inaccessible *directly* to most organisms. The key link to life is provided by *nitrogen-fixing bacteria,* which live principally in the soil and are associated with the roots of certain plants—for example, the *legumes* such as clover, alfalfa, soybeans, peas, beans, and peanuts. Bacteria colonies reside in nodules on the legume roots and chemically combine the nitrogen from the air in the form of nitrates (NO_3) and ammonia (NH_3). Plants use these chemically bound forms of nitrogen to produce their own organic matter. Anyone or anything feeding on the plants thus ingests the nitrogen. Finally, the nitrogen in the organic wastes of the consuming organisms is freed by *denitrifying bacteria,* which recycle it back to the atmosphere.

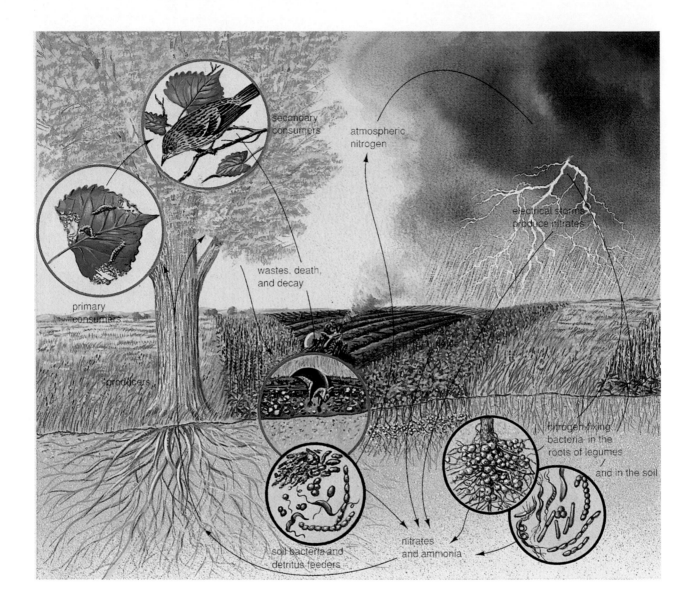

FIGURE 16-10

The nitrogen cycle. Atmospheric nitrogen gas is chemically fixed by bacteria to produce ammonia. Lightning and forest fires produce nitrates, and fossil fuel combustion forms nitrogen compounds that are washed from the atmosphere by precipitation. Plants absorb nitrogen compounds and produce organic material. The atmosphere is the essential reservoir of gaseous nitrogen. [Adapted from Gerald Audesirk and Teresa Audesirk, *Biology, Life On Earth*, 3rd ed., Figure 45-9, p. 997. Copyright © Macmillan Publishing Company. Used by permission.]

"Humans Dump Carbon into the Atmosphere"

In the last two decades, human activity has added to the atmospheric pool an amount of carbon equivalent to more than 25% of the total amount added since 1880. Emissions by society now have reached 6 billion tons each year and represent our greatest single atmospheric waste product. Furthermore, about 50% of the carbon dioxide emitted since the beginning of the industrial revolution, and not absorbed by oceans and organisms, is *still* in the atmosphere, enhancing Earth's natural greenhouse effect. The carbon dioxide dumped into the atmosphere by human activity constitutes a vast geochemical experiment, using the real-time atmosphere as a laboratory.

Limiting Factors. The term **limiting factor** identifies the one physical or chemical abiotic component that most inhibits biotic operations, through its lack or excess. A few examples include low temperatures at high elevations, the lack of water in a desert, the excess water in a bog, the amount of iron in ocean surface environments, the phosphorus content of soils in the eastern United States, or the general lack of active chlorophyll above 6100 m (20,000 ft).

Each organism possesses a range of tolerance for each limiting factor in its environment. This is illustrated vividly in Figure 16-11, which shows the geographic range for two tree species and two bird species. The coast redwood *(Sequoia sempervirens)* is limited to a narrow section of the Coast Ranges in California, covering barely 9500 km² (3667 mi²) and concentrated in areas that receive necessary summer advection fog. The red maple *(Acer rubrum),* on the other hand, thrives over a large area under varying conditions of moisture and temperature, thus demonstrating a broader tolerance to environmental variations.

Biotic Ecosystem Operations

The abiotic components of energy, atmosphere, water, weather, climate, and minerals support the biotic components of each ecosystem and their constituent soils, vegetation, and life forms.

Producers, Consumers, and Decomposers. Organisms that are capable of using carbon dioxide as their sole source of carbon are called *autotrophs,* or **producers**. They chemically fix carbon through photosynthesis. Organisms that depend on autotrophs for their carbon are called *heterotrophs,* or **consumers**.

Plants are the essential producers in an ecosystem—capturing light and generating heat energy and converting it to chemical energy, incorporating carbon, forming new plant tissue and biomass, and freeing oxygen, all as a part of photosynthesis.

From these producers, which manufacture their own food, energy flows through the system along a circuit called the **food chain**, reaching consumers and eventually decomposers. Ecosystems generally are structured in a **food web**, a complex network of interconnected food chains. In a food web, consumers participate in several different food chains. Organisms that share the same basic foods are said to be at the same *trophic* (feeding, nutrition) *level*.

As an example, look at the food web based on krill (Figure 16-12). Krill is a shrimplike crustacean about 7 cm (2.8 in.) long that is a major food for an interre-

493

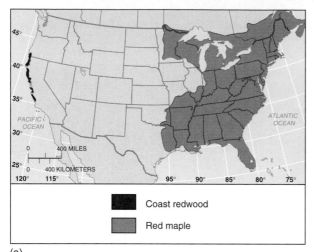

Coast redwood

Red maple

(a)

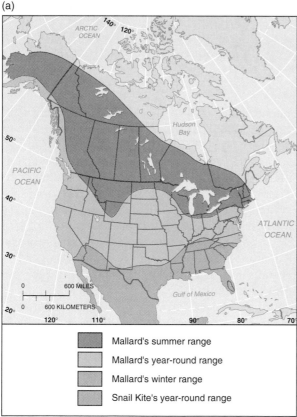

Mallard's summer range

Mallard's year-round range

Mallard's winter range

Snail Kite's year-round range

(b)

FIGURE 16-11
(a) Biotic distribution of coast redwood and red maple
are examples of limiting factors. (b) The effect of
limiting factors also is demonstrated by the mallard duck
and snail kite. The mallard is a generalist and feeds
widely. In contrast, the snail kite is limited to a single
type of snail for food.

lated group of organisms, including whales, fish, seabirds, seals, and squid in the Antarctic region. All of these organisms participate in numerous other food chains as well, some consuming and some being consumed. Because krill are a protein-rich, plentiful food, they increasingly are sought by commercial factory ships, such as those from Japan and Russia. The annual harvest currently approaches a million tons.

Primary consumers feed on producers. Because producers are always plants, the primary consumer is called a **herbivore**, or plant eater. A **carnivore** is a *secondary consumer* and primarily eats meat. A *tertiary consumer* eats primary and secondary consumers and is referred to as the "top carnivore" in the food chain, like the sperm whale in the krill food web. A consumer that feeds on both producers (plants) and consumers (meat) is called an **omnivore**—a role occupied by humans, among others.

Any assessment of world food resources depends on the level of consumer being targeted. Using humans as an example, many can be fed from the amount of wheat harvested from an acre of land. An acre of land (0.41 hectare) produces about 810 kg (1800 pounds) of grain. However, if herbivores eat that grain, only 82 kg (180 pounds) of biomass is produced, which can feed far fewer people (Figure 16-13). In terms of energy, *only about 10% of the calories in plant matter survive from the primary to the secondary trophic level.* When humans consume the meat, there is a further loss of biomass and added inefficiency. More energy is lost to the environment at each progressive step in the food chain. You can see that an omnivorous diet—ours—is quite expensive in terms of biomass and energy.

Food chain concepts are becoming politicized as world food issues grow more critical. Today, approximately half of the cultivated acreage in the United States and Canada is planted for animal consumption. This includes more than 80% of the annual corn and nonexported soybean harvest. In addition, some lands cleared of rain forests in Central and South America have been converted to pasture to produce beef for export to some restaurants, stores, and fast-food outlets in developed countries. Thus, life-style decisions and dietary patterns in North America and Europe are perpetuating certain food chains, not to mention the destruction of valuable resources, both here and overseas.

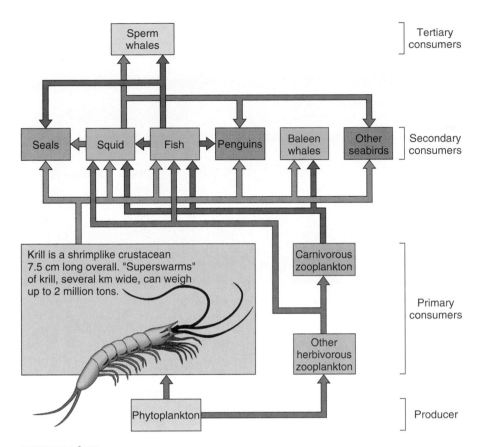

FIGURE 16-12

The food web of the krill in Antarctic waters, from phytoplankton *producers* (bottom) through various *consumers*. Phytoplankton begin this chain by using solar energy in photosynthesis. The krill, along with other organisms, feed on the phytoplankton. Krill in turn are fed upon by the next trophic level. [After *State of the Ark* by Lee Durrell. Copyright © 1986 by Gaia Books Ltd. Adapted by permission of Gaia Books Ltd.]

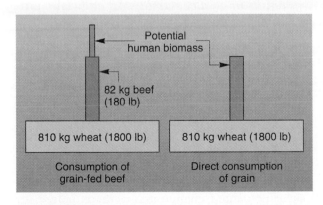

FIGURE 16-13

Biomass pyramids illustrating the difference between direct and indirect consumption of grain.

Decomposers are the final link in the chain. They are the organisms—bacteria, fungi, insects, worms, and others—that digest and recycle the organic debris and waste in the environment. Waste products, dead plants and animals, and other organic remains are their principal food source. Material is released by the decomposers and enters the food chain—and the cycle continues.

Ecological Relationships. A study of food chains and webs is a study of who eats what, and where they do the eating. Figure 16-14 demonstrates the summertime distribution of populations in two ecosystems, grassland and temperate forest. The

FIGURE 16-14
Ecological pyramids for 0.1
hectares (0.25 acres) of grassland
and forest in the summer.
Numbers of consumers are
shown. [Adaptation of Figure 3-
15a from *Fundamentals of
Ecology* by Eugene P. Odum,
copyright © 1971 by Saunders
College Publishing, a division of
Holt, Rinehart and Winston, Inc.
Adapted by permission of the
publisher.]

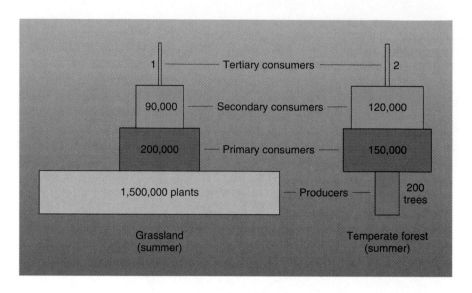

stepped population pyramid is characteristic of summer conditions in such ecosystems. You can see the
decreasing number of organisms at each successive
higher trophic level. The base of the temperate forest pyramid is narrow, however, because most of the
producers are large, highly productive trees and
shrubs, which are outnumbered by the consumers
they can support in the chain.

Stability and Succession

Far from being static, Earth's ecosystems are dynamic
and ever-changing. Over time, communities of plants
and animals have adapted to such variation, evolved,
and in turn shaped their environments. Each ecosystem operates in dynamic equilibrium, constantly adjusting to changing conditions to maintain stability.
The concept of *change* is key to the study of ecosystem stability.

Ecosystem Stability and Diversity

Any ecosystem moves toward maximum biomass
and stability to survive. However, the tendency for
birth and death rates to balance and the composition of species to remain stable, *inertial stability,*
does not necessarily foster the ability to recover from
change, *resilience.* Examples of stable communities
include a redwood forest, a pine forest at a high el-

evation, and a tropical rain forest near the equator.
Yet, cleared tracts recover slowly (if ever) and therefore have poor resilience. Figure 16-15 shows how
clear-cut tracts of former forest have altered drastically the microclimatic conditions, making regrowth
of the same species difficult. In contrast, a midlatitude grassland is low in stability; yet when burned,
its resilience is high because the community recovers rapidly.

Another aspect related to stability is **diversity**.
The more diverse the species population (both in
number of species and quantity of each species), the
more risk is spread over the entire community, because several food sources exist at each trophic
level. In other words, *greater diversity in an ecosystem results in greater stability*. An artificially produced *monoculture* community, such as a field of
wheat, is singularly vulnerable to failure due to
weather or attack from insects or plant disease. A
modern *agricultural ecosystem* not only is vulnerable
to failure due to its lack of ecological diversity, but
also to such practices as harvesting and removing
biomass from the land that interrupts the cycling of
materials into the soil. This net loss of nutrients must
be artificially replenished.

Humans simplify communities by eliminating diversity and in this way we place more ecosystems at
risk of unwanted change and perhaps failure. In
some regions, simply planting multiple crops brings
more stability to the ecosystem.

FIGURE 16-15
Natural and human disruption of stable communities. (a) Natural disruption of the
forest by the Mount Saint Helens volcanic eruption in 1980 and by logging in
numerous clear-cut tracts of national forest. About 10% of the old-growth forests
remain in the Northwest, as identified through orbital images and GIS analysis. (b)
A logged pine forest—an example of clear-cut timber harvesting disrupting a
stable community and producing drastic changes in microclimatic conditions. [(a)
Landsat image of a portion of the Gifford Pinchot National Forest and the Mount
Saint Helens National Volcanic Monument, April 29, 1992, centered at
approximately 46.5° N, 122° W, courtesy of Compton J. Tucker, NASA Goddard
Space Flight Center, Greenbelt, Maryland. (b) Photo by author.]

Climate Change. Certainly the distribution of plant
species is affected by changes in climate. Plant communities have survived wide climate swings in the
past. As examples, consider the beginning of the Tertiary period, 75 million years ago. Warm, humid conditions and tropical forests dominated all the way to
southern Canada, pines grew in the Arctic, and
deserts were few. Then, between 15 and 50 million
years ago, deserts began developing in the southwestern United States. Mountain-building processes

were creating higher elevations, causing rain-shadow aridity and affecting plant distribution. Recall too, that the movement of Earth's tectonic plates created climate changes that played an important role in the evolution and distribution of plants and animals (see Figure 8-16).

The key question is: as temperature patterns change, how fast can plants either adapt to the new conditions, or migrate through succession (location change) to remain within their specific habitats?

Adaptation is key to evolution. Through mutation and natural selection, species have either adapted, or failed to adapt, to changing environmental conditions over millions of years. The current pace of global change is occurring fast, at the rate of decades instead of millions of years. Thus, we see die-out along disadvantageous habitat margins with a succession to different species. The displaced species may colonize new regions made more advantageous by climate change. Agricultural lands producing wheat, corn, soybeans, and other commodities also will shift. Society will have to adapt to new cropping patterns. Rapid environmental change can lead to outright extinction of plants and animals that are unable to adapt or disperse.

A study completed by biologist Margaret Davis on North American forests suggests that trees will have to respond quickly if temperatures increase. Changes in the climate inhabited by certain species could shift 100–400 kilometers (60–250 mi) during the next 100 years. Davis prepared a map showing the possible impact of increasing temperatures on the distribution of beech trees (Figure 16-16).

> Changes in the geographical distributions of plants and animal species in response to future greenhouse warming threaten to reduce biotic diversity. . . . The risk posed by CO_2-induced warming depends on the distances that regions of suitable climate are displaced northward [in the Northern Hemisphere] and on the rate of displacement. . . . If the change occurs too rapidly for colonization of newly available regions, population sizes may fall to critical levels, and extinction will occur.*

*M. B. Davis and C. Zabinski, "Changes in the Geographical Range Resulting from Greenhouse Warming: Effects on Biodiversity in Forests," in R. L. Peters and T. E. Lovejoy, eds., *Global Warming and Biological Diversity*. New Haven: Yale University Press, 1992, p. 297.

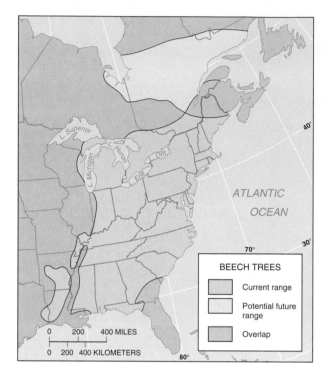

FIGURE 16-16
Present and predicted distribution of beech trees in North America, given estimated climate changes resulting from a doubling of CO_2 using the Goddard Fluid Dynamics Laboratory (GFDL) general circulation model. [After *Science* 243 (10 February 1989): 735. "How Fast Can Trees Migrate?" by Leslie Roberts, 1989. Copyright 1989 by the AAAS.]

Ecological Succession

Ecological succession occurs when older communities of plants and animals (usually simpler) are replaced by newer communities (usually more complex). Each successive community of species modifies the physical environment in a manner suitable for the establishment of a later community of species. Changes apparently move toward a more stable and mature condition, to an optimum for a specific environment. This end product in an area is traditionally called the *ecological climax,* with plants and animals forming a *climax community*—a stable, self-sustaining, and symbiotically functioning community with balanced birth, growth, and death.

However, given the complexity of natural ecosystems, real succession involves much more than a se-

FIGURE 16-17
Succession and recovery of pioneer species in the devastated area north of Mount Saint Helens, three years after the 1980 eruption. [Photo by author.]

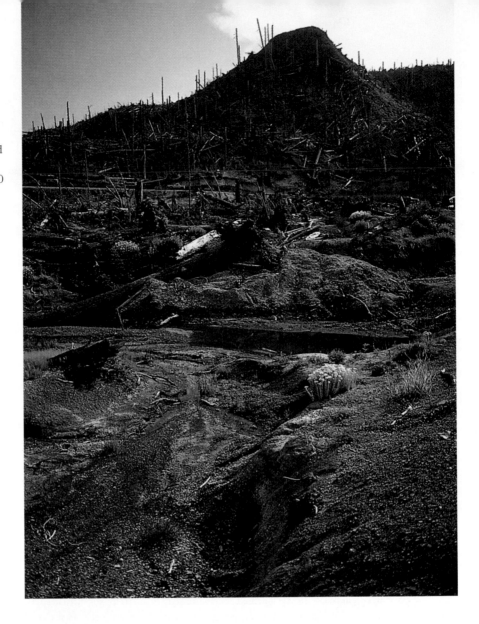

ries of predictable stages ending with a specific "monoclimax" community. Instead, there may be several final stages, or a "polyclimax" condition. Climax communities are properly thought of as being in *dynamic equilibrium,* and at times even may be out of phase with the immediate physical environment due to a lag time in adjustment.

Succession often requires an initiating disturbance, for instance strong winds, a volcanic eruption, or a practice such as prolonged overgrazing. When existing organisms are disturbed or removed, new communities can emerge. At such times of transition, the interrelationships among species produce elements of chance, and species having an adaptive edge will succeed in the competitive struggle for light, water, nu-

trients, space, time, reproduction, and survival. Thus, the succession of plant and animal communities is an intricate process with many interactive variables.

Terrestrial Succession. An area of bare rock and soil without any vestige of a former community can be a site for *primary succession*. The initial community is called a *pioneer community*. It may occur in sites such as a new surface created by mass movements of land, in areas exposed by a retreating glacier, on devastated lands such as those north of Mount Saint Helens in Washington State (Figure 16-17), on cooled lava flows, or on lands disturbed by surface mining, clear-cut logging, and land development. However, succession from a previously

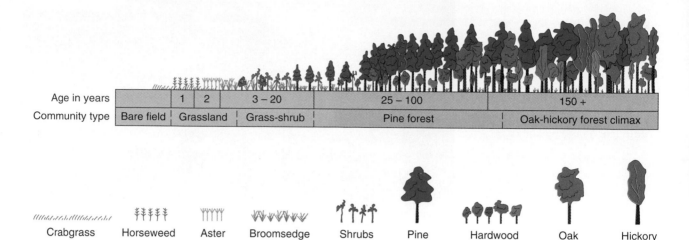

Age in years		1	2	3 – 20	25 – 100	150 +
Community type	Bare field	Grassland		Grass-shrub	Pine forest	Oak-hickory forest climax

Crabgrass Horseweed Aster Broomsedge Shrubs Pine Hardwood understory Oak Hickory

FIGURE 16-18

Typical secondary succession of principal plants in the southeastern United States. [Adaptation of Figure 9-4 from *Fundamentals of Ecology* by Eugene P. Odum, copyright © 1971 by Saunders College Publishing, a division of Holt, Rinehart and Winston, Inc. Adapted by permission of the publisher.]

functioning community is more common. An area whose natural community has been destroyed or disturbed, but still has the underlying soil intact, may experience *secondary succession* (Figure 16-18).

In terrestrial ecosystems, secondary succession begins with pioneer species and further soil development. As succession progresses, a different set of plants and animals with different niche requirements may adapt. Figure 16-18 illustrates such a sequence in the southeastern United States.

Secondary succession begins on an abandoned farm of formerly plowed fields (left side of illustration). Crabgrass and ragweed quickly take hold and do well in direct sunlight. These slowly give way to grasses and shrubs that invade and stabilize the soil, adding nutrients and organic matter. Pines eventually dominate the land from year 25 through the first century. The shade created by the pine forest produces conditions in which seed germination becomes more difficult. Shade-tolerant, slow-growing oak and hickory hardwoods readily take root beneath the pines. As these hardwoods grow they eventually shade the pine forest, which slowly dies back in reduced light conditions. A fairly stable climax forest of oak and hickory is in place between 150 and 200 years (right side of illustration). But despite the convenient categorization shown in the figure, you should think of succession and disruptions

in it as continuous, with succeeding communities overlapping in time and space.

Fire Ecology. Successional stages may be interrupted by fire. It is estimated that one-fourth of Earth's land area experiences fire each year. Over the past 50 years, **fire ecology** has been the subject of much scientific research and experimentation. Today, fire is recognized as a natural component of most ecosystems and not the enemy of nature it once was popularly considered. In fact, in many forests, undergrowth is purposely burned in controlled "cool fires" to remove fuel that could enable a catastrophic and destructive "hot fire." In contrast, when fire prevention strategies are rigidly followed, they can lead to abundant undergrowth accumulation, which allows total destruction of a forest by a major fire (Figure 16-19).

Fire ecology imitates nature by recognizing fire as a dynamic ingredient in community succession. The U.S. Department of Agriculture's Forest Service first recognized the principle of fire ecology in the early 1940s and formally implemented the practice in 1972. Their challenge became one of controlling fires to secure reproduction and to prevent accumulation of forest undergrowth. Controlled ground fires now are widely regarded as wise forest management practice and are used across the country.

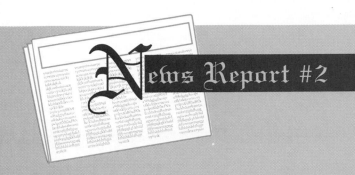

"Yellowstone Fire Assessment and Fire Ecology"

The concept of fire ecology was the center of heated debate after some 72,000 forest fires scorched the western United States in 1988. Especially significant were those fires that hit portions of the popular Yellowstone National Park. An outcry was heard from forestry and recreational interests. The demand was for the Forest Service and ecologists to admit they were wrong and to abandon fire-ecology principles. Critics called fire ecology practices the government's "let burn" policy. About 20% of the acreage in Yellowstone Park actually burned, and only about half of that area experienced the worst fire damage, far less than originally reported in the media.

In its final report on the fire, a government interagency task force concluded that, "an attempt to exclude fire from these lands leads to major unnatural changes in vegetation . . . as well as creating fuel accumulation that can lead to uncontrollable, sometimes very damaging wildfire." Thus, participating federal land managers and others reaffirmed their stand that fire ecology is a fundamentally sound concept.

Aquatic Succession. Lakes and ponds exhibit another form of ecological succession. A lake experiences successional stages as it fills with nutrients and sediment and as aquatic plants take root and grow, capturing more sediment and adding organic debris to the system (Figure 16-20). This gradual enrichment of water bodies is known as **eutrophication**. For example, in moist climates, a floating mat of vegetation grows outward from the shore to form a bog. Cattails and other marsh plants become established, and partially decomposed organic material accumulates in the basin, with additional vegetation bordering the remaining lake surface. A meadow may form as water is displaced by the peat bog; willow trees follow, and perhaps cottonwood trees; and eventually the lake may evolve into a forest community.

FIGURE 16-19
A view of the great fire of 1988 in Yellowstone National Park in northwestern Wyoming.
[Courtesy of the National Park Service.]

FIGURE 16-20
Lake bog succession.

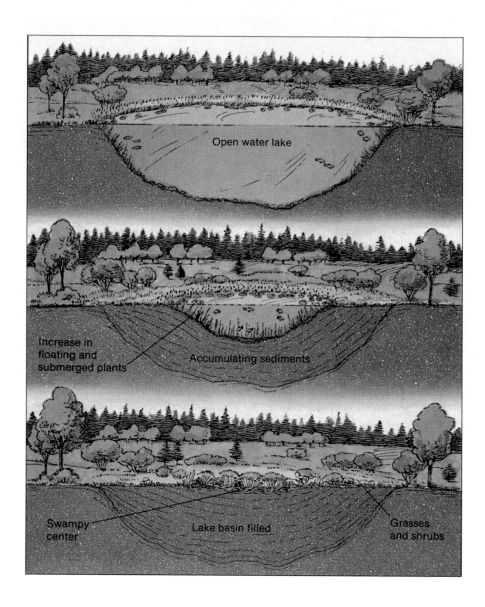

Open water lake

Increase in
floating and
submerged plants

Accumulating sediments

Swampy
center

Lake basin filled

Grasses
and shrubs

Now that we have examined ecosystem components and cycles, let's look at ecosystem patterns produced across Earth's surface.

Earth's Ecosystems

Earth's major ecosystems are conveniently divided into aquatic and terrestrial. To review, an *ecosystem* is a self-regulating association of living plants and animals and their nonliving environment.

Oceans, estuaries, and freshwater bodies are **aquatic ecosystems**. The *photic layer* is the upper zone of water that receives sufficient light for *phytoplankton* to survive. These are the one-celled, chlorophyll-based primary producers. Primary consumers that consume the phytoplankton include innumerable herbivores, some as small as 0.2 mm (0.008 in.). These form the basis of a productive and complex food chain. The health of aquatic ecosystems is critical to the biosphere. In contrast to these aquatic systems we focus next on terrestrial ecosystems.

"Large Marine Ecosystems: A Management Tool"

Poor understanding of aquatic ecosystems has led to the failure of some species, such as herring in the Georges Bank prime fishing area of the Atlantic, and an overall decline in fisheries worldwide. A *large marine ecosystem* (*LME*) is a planning/preservation concept. An LME is a distinctive oceanic region having unique organisms, floor topography, currents, areas of nutrient-rich upwelling circulations, or areas of significant preda-tion, including human. Examples of identified LMEs include the Gulf of Alaska, California Current, Gulf of Mexico, Northeast Continental Shelf, and the Baltic and Mediterranean seas. Thirty LMEs, each encompassing more than 200,000 km² (77,200 mi²), are presently defined worldwide.

The LME was conceived to encourage resource planners and managers to consider *complete ecosystems*, not just a targeted species. A goal is to im-prove management and to improve the sustainability of both biotic and abiotic aquatic resources. Also, LMEs are designated to insure long-term data collection for management and further research.

The largest of eleven protected areas (by government sanctions) in North America is the Monterey Bay National Marine Sanctuary, established in 1992 within the California Current LME. (The Florida Keys Marine Sanctuary is second largest.) Stretch-ing along 645 km (400 mi) of coast and covering 13,500 km² (5300 mi²), the sanctuary is home to 27 species of marine mammals, 94 species of shore-birds, 345 species of fish, and the largest sampling of invertebrates in any one place in the Pacific. This represents one of the most species-rich and diverse marine communities on Earth. In the United States these protected areas are administered by the Sanctuaries and Reserves Division of NOAA.

Terrestrial Ecosystems

A **terrestrial ecosystem** is a self-regulating association of plants and animals and their abiotic environment that is characterized by specific plant formations. In their growth, form, and distribution, plants reflect Earth's physical systems: its energy patterns; atmospheric composition; temperature and winds; air masses; water quantity, quality, and seasonal timing; soils; regional climates; geomorphic processes; and ecosystem dynamics.

A **biome** is a large, stable terrestrial ecosystem characterized by specific plant and animal communities. Each biome usually is named for its *dominant vegetation*. We can generalize Earth's wide-ranging plant species into six broad biomes: *forest, savanna, grassland, shrubland, desert, and tundra*. Because plant distributions respond to environmental conditions and reflect variations in climate and soil, the biome-related world climate map (inside the front cover) is a helpful reference for this chapter.

These general biomes are divided into more specific **formation classes**, vegetation units that refer to the dominant plants in a terrestrial ecosystem. Examples are, equatorial rain forest, northern needle-leaf forest, Mediterranean shrubland, and arctic tundra. Each formation class includes numerous plant communities, and each community includes innumerable plant habitats. Within those habitats, Earth's diversity is expressed in 250,000 plant species. Despite this intricate complexity, we can generalize Earth's numerous formation classes into 10 global terrestrial biome regions, as portrayed in Figure 16-22 and detailed in Table 16-2.

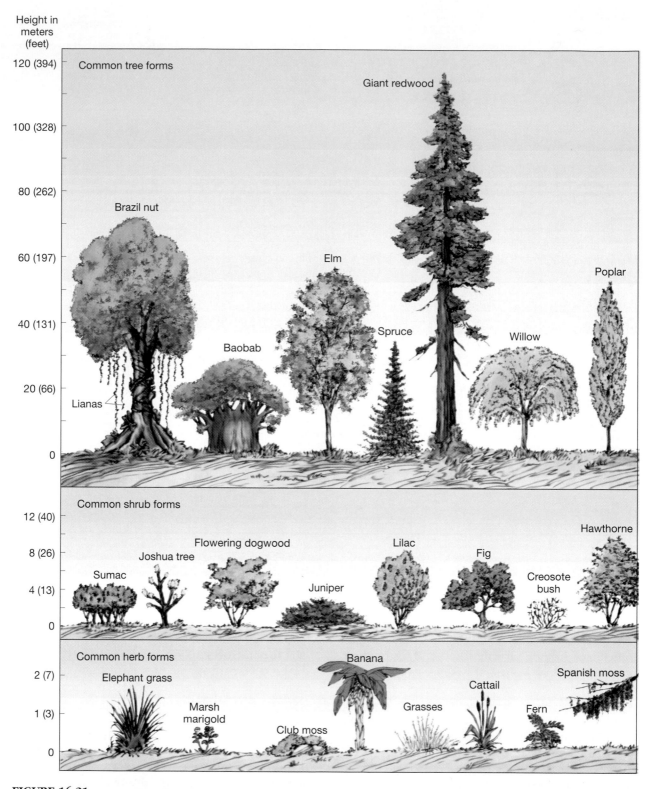

FIGURE 16-21
Examples of plant life-forms. [After William M. Marsh and J. Dozier, *Landscape: An Introduction to Physical Geography*, copyright © 1981, p. 273. Adapted by permission of John Wiley & Sons, Inc.]

A boundary transition zone between adjoining ecosystems is an **ecotone**. Because ecotones are defined by different physical factors, they vary in width. Climatic ecotones usually are more gradual than physical ecotones, whereas differences in soil or topography sometimes form abrupt boundaries. An ecotone between prairies and northern forests may occupy many kilometers of land. The ecotone is an area of tension as similar species of plants and animals compete for the resource base.

More specific characteristics are used for the structural classification of plants. *Life-form* designations are based on the outward physical properties of individual plants or the general form and structure of a vegetation cover. These physical life forms, portrayed in Figure 16-21, include *trees* (large woody main trunk, perennial, usually exceeding 3 m or 10 ft); *lianas* (woody climbers and vines); *shrubs* (smaller woody plants; branching stems at the ground); *herbs* (small plants without woody stems above ground); *bryophytes* (mosses, liverworts); *epiphytes* (plants growing above the ground on other plants, using them for support); and *thallophytes,* which lack true leaves, stems, or roots (bacteria, fungi, algae, lichens).

Earth's Major Terrestrial Biomes

Few natural communities of plants and animals remain; most biomes have been greatly altered by human intervention. Thus, the "natural vegetation" identified on many biome maps reflects ideal climax vegetation potential, given the physical factors in a region. Even though human practices have greatly altered these ideal forms, it is valuable to study the natural (undisturbed) biomes to better understand the natural environment and to assess the extent of human-caused alteration.

In addition, knowing the ideal in a region guides us to a closer approximation of natural vegetation in the plants we introduce. In the United States and Canada, we humans are perpetuating a type of transition community, somewhere between a grassland and a forest. We plant trees and lawns and then must invest energy, water, and capital to sustain such artificial modifications of nature.

The global distribution of Earth's ten major terrestrial biomes, based on vegetation *formation classes,* is portrayed in Figure 16-22. Table 16-2 describes each biome on the map and summarizes other pertinent information from throughout this text—a compilation from many chapters, for Earth's biomes are a synthesis of the environment and biosphere.

Equatorial and Tropical Rain Forest

Earth is girdled with a lush biome—the **equatorial and tropical rain forest**. In a climate of consistent daylength (12 hours), high insolation, and average annual temperatures around 25°C (77°F), plant and animal populations have responded with the most diverse body of life on the planet. The Amazon region is the largest tract of equatorial and tropical rain forest, also called the *selva*. In addition, rain forests cover equatorial regions of Africa, Indonesia, the margins of Madagascar and Southeast Asia, the Pacific coast of Ecuador and Colombia, and the east coast of Central America, with small discontinuous patches elsewhere. The cloud forests of western Venezuela are such tracts of rain forest at high elevation, perpetuated by high humidity and cloud cover. Undisturbed tracts of rain forest are rare.

Rain forests feature ecological niches distributed vertically rather than horizontally, because of the competition for light. The canopy is filled with a rich variety of plants and animals. Lianas (vines) stretch from tree to tree, entwining them with cords that can reach 20 cm (8 in.) in diameter. Epiphytes flourish there too: such plants as orchids, bromeliads, and ferns that live entirely above ground, supported physically but not nutritionally by the structures of other plants. Windless conditions on the forest floor make pollination difficult—insects, other animals, and self-pollination predominate.

The rain forest canopy forms three levels—see Figure 16-23. The upper level is not continuous but features tall trees whose high crowns rise above the middle canopy. The middle canopy is the most continuous, with its broad leaves blocking much of the light and creating a darkened forest floor. The lower level is composed of seedlings, ferns, bamboo, and the like, leaving the litter-strewn ground level in deep shade and fairly open.

FIGURE 16-22
The 10 major global terrestrial
biomes.

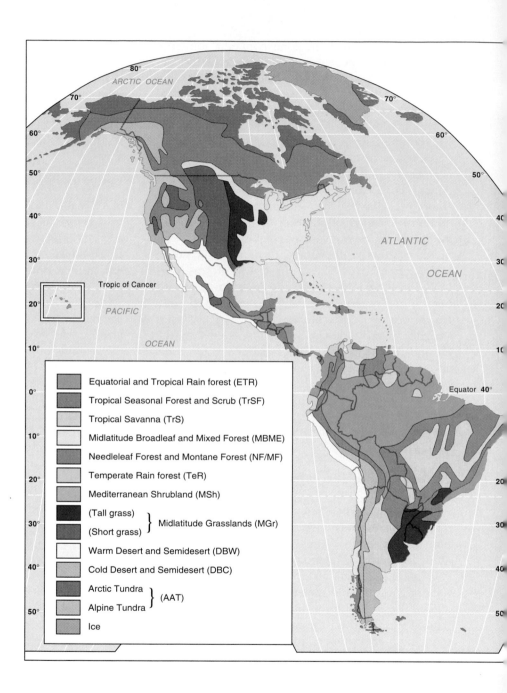

Equatorial and Tropical Rain forest (ETR)

Tropical Seasonal Forest and Scrub (TrSF)

Tropical Savanna (TrS)

Midlatitude Broadleaf and Mixed Forest (MBME)

Needleleaf Forest and Montane Forest (NF/MF)

Temperate Rain forest (TeR)

Mediterranean Shrubland (MSh)

(Tall grass) } Midlatitude Grasslands (MGr)
(Short grass)

Warm Desert and Semidesert (DBW)

Cold Desert and Semidesert (DBC)

Arctic Tundra } (AAT)
Alpine Tundra

Ice

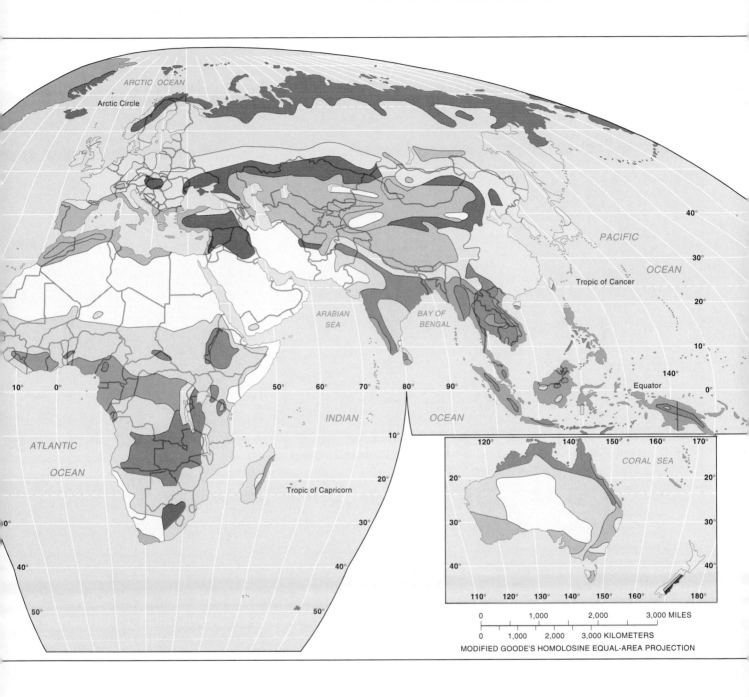

ARCTIC OCEAN

Arctic Circle

PACIFIC

OCEAN

40°

30°

Tropic of Cancer

20°

ARABIAN
SEA

BAY OF
BENGAL

10°

140°

10°   0°

Equator   0°

INDIAN   OCEAN

ATLANTIC

OCEAN   10°

20°

Tropic of Capricorn   CORAL SEA

20°

30°

Tropic of Capricorn   20°

30°

40°

40°

30°

50°

50°

40°

120°   140°   150°   160°   170°

110°   120°   130°   140°   150°   160°   180°

0   1,000   2,000   3,000 MILES

0   1,000   2,000   3,000 KILOMETERS

MODIFIED GOODE'S HOMOLOSINE EQUAL-AREA PROJECTION

Table 16-2

Major Terrestrial Biomes and Their Characteristics

Biomes and Ecosystems (map symbol)	Vegetation Characteristics	Soil Orders	Köppen Climate Designation	Annual Precipitation Range	Temperature Patterns	Water Balance
Equatorial and Tropical Rain Forest (ETR) Evergreen broadleaf forest Selva	Leaf canopy thick and continuous; broadleaf evergreen trees; vines (lianas), epiphytes, tree ferns, palms	Oxisols Ultisols (on well-drained uplands)	Af Am (limited dry season)	180–400 cm (>6 cm/mo)	Always warm (21–30°C; avg. 25°C)	Surpluses all year
Tropical Seasonal Forest and Scrub (TrSF) Tropical monsoon forest Tropical deciduous forest Scrub woodland and thorn forest	Transitional between rain forest and grasslands; broadleaf, some deciduous trees; open parkland to dense undergrowth acacias and other thorn trees in open growth	Oxisols Ultisols Vertisols (in India) Some Alfisols	Am Aw Borders BS	130–200 cm (<40 rainy days during 4 driest months)	Variable, always warm (>18°C)	Seasonal surpluses and deficits
Tropical Savanna (TrS) Tropical grassland Thorn tree scrub Thorn woodland	Transitional between seasonal forests, rain forests, and semiarid tropical steppes and desert; trees with flattened crowns, clumped grasses, and bush thickets; fire association	Alfisols (dry: Ustalfs) Ultisols Oxisols	Aw BS	90–150 cm, seasonal	No cold weather limitations	Tends toward deficits, therefore fire and drought susceptible
Midlatitude Broadleaf and Mixed Forest (MBME) Temperate broadleaf Midlatitude deciduous Temperate needleleaf	Mixed braodleaf and needleleaf trees; deciduous broadleaf, losing leaves in winter; southern and eastern evergreen pines demostrate fire association	Ultisols Some Alfisols Podzols (red yellows)	Cfa Cwa Dfa	75–150 cm	Temperate, with cold season	Seasonal pattern with summer maximum PRECIP and POTET; no irrigation needed
Northern Needleleaf Forest and Montane Forest (NF/MF) Taiga Boreal forest Other montane forests and highlands	Needleleaf conifers, mostly evergreen pine, spruce, fir; Russian larch, a deciduous needleleaf	Spodosols Histosols Inceptisols Gleysols Alfisols (Boralfs: cold) Podzols	Subarctic Dfb Dfc Dfd	35–100 cm	Short summer, cold winter	Low POTET, moderate PRECIP, moist soils, some waterlogged and frozen in winter; no deficits

508

Biome	Vegetation Characteristics	Soils	Köppen	Annual Precipitation	Temperature	Water Balance
Temperate Rain Forest (TeR) West coast forest Coast redwoods (U.S.)	Narrow margin of lush evergreen and deciduous trees on windward slopes; redwoods, tallest trees on Earth	Spodosols Inceptisols (mountainous environs) Podzols	Cfb Cfc	150–500 cm	Mild summer and mild winter for latitude	Large surpluses and runoff
Mediterranean Shrubland (MSh) Sclerophyllous shrubs Australian Eucalyptus forest	Short shrubs, drought adapted, trending to grassy woodlands; chaparral	Alfisols (Xeralfs) Mollisols (Xerolls) Luvisols	Csa Csb	25–65 cm	Hot, dry summer, cool winters	Summer deficits, winter surpluses
Midlatitude Grasslands (MGr) Temperate grassland Sclerophyllous shrub	Tall grass prairies and short grass steppes, highly modified by human activity; major areas of commercial grain farming; plains, pampas, and veld	Mollisols Aridisols Chernozemic	Cfa Dfa	25–75 cm	Temperate continental regimes	Soil moisture utilization and recharge balanced; irrigation nad dry farming in drier areas
Warm Desert and Semidesert (DBW) Subtropical desert and scrubland	Bare ground graduating into xerophytic plants including succulents, cacti, and dry shrubs	Aridisols Entisols (sand dunes)	BWh BWk	<2 cm	Average annual temperature, around 18°C, highest temperatures on Earth	Chronic deficits, irregular precipitation events, PRECIP <1/2 POTET
Cold Desert and Semidesert (DBC) Midlatitude desert, scrubland, and steppe	Cold desert vegetation includes short grass, and dry shrubs	Aridisols Entisols	BSh BSk	2–25 cm	Average annual temperature around 18°C	PRECIP >1/2 POTET
Arctic and Alpine Tundra (AAT)	Treeless; dwarf shrubs, stunted sedges, mosses, lichens, and short grasses; alpine, grass meadows	Inceptisols Histosols Entisols (permafrost) Organic Cryosols	ET Dwd	15–80 cm	Warmest month <10°C, only 2 or 3 months above freezing	Not applicable most of the year, poor drainage in summer
Ice			EF			

FIGURE 16-23

The three levels of a rain forest canopy. In his 1960 book *The Forest and the Sea*, Marston Bates compared rain forests and oceans and found remarkable similarities in these very dissimilar environments. The sea and the forest both have a vertical distribution of life, dependent on a competitive struggle for sunlight. Biomass in a rain forest is concentrated in the upper high-level and middle-level canopies, just as life in the sea is concentrated in the upper layer near the ocean's surface. The floor of the rain forest and the floor of the deep ocean also are roughly parallel in that both are dark or deeply shaded, and a place of fewer life forms.

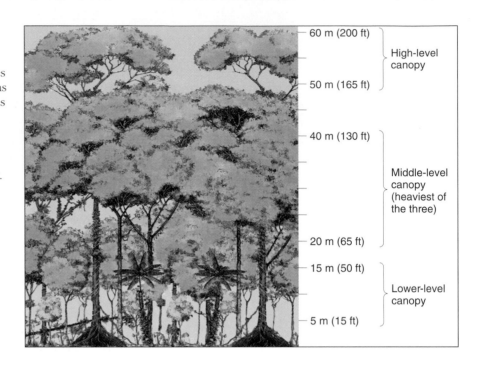

FIGURE 16-24

The rain forest. (a) Equatorial rain forest is thick along the Amazon River where light breaks through to the surface, producing a rich gallery of vegetation. (b) The rain forest floor in Corcovado, Costa Rica, with typical buttressed trees and lianas. [(a) Photo by Wolfgang Kaehler; (b) photo by Frank S. Balthis.]

(a)

(b)

A look at aerial photographs of a rain forest, or views along river banks covered by dense vegetation, aided by the false Hollywood-movie imagery of the jungle, makes it difficult to imagine the shadowy environment of the actual rain forest floor, which receives only about 1% of the sunlight arriving at the canopy (Figure 16-24a). The constant moisture, rotting fruit and moldy odors, strings of thin roots and vines dropping down from above, windless air, and echoing sounds of life in the trees, together create a unique environment.

The smooth, slender trunks of rain forest trees are covered with thin bark and buttressed by large wall-like flanks that grow out from the trees to brace the trunks (Figure 16-24b). These buttresses form angular open enclosures, a ready habitat for various animals. There are usually no branches for at least the lower two-thirds of the tree trunks.

The wood of many rain forest trees is extremely hard, heavy, and dense—in fact, some species will not even float. (Exceptions are balsa and a few others, which are very light.) Varieties of trees include mahogany, ebony, and rosewood. Logging is difficult because individual species are widely scattered; a species may occur only once or twice per square kilometer. Selective cutting is required for species-specific logging, whereas pulpwood production takes everything. Conversion of the forest to pasture usually is accomplished by setting destructive fires.

Rain forests represent approximately one-half of Earth's remaining forests, occupying about 7% of the total land area worldwide. This biome is stable in its natural state, resulting from the long-term residence of these continental plates near equatorial latitudes and their escape from glaciation. The soils, principally Oxisols, are essentially infertile, yet they support rich vegetation.

The animal and insect life of the rain forest is diverse, ranging from small decomposers (bacteria) working the surface to many animals living exclusively in the upper stories of the trees. These tree dwellers are referred to as *arboreal,* from the Latin for tree, and include sloths, monkeys, lemurs, parrots, and snakes. Beautiful birds of many colors, tree frogs, lizards, bats, and a rich insect life that includes over 500 species of butterflies are found in rain forests. Surface animals include pigs (bushpigs and the giant for-

est hog in Africa, wild boar, bearded pig in Asia, and peccary in South America), species of small antelopes (bovids), and mammal predators (tiger in Asia, jaguar in South America, and leopard in Africa and Asia). The present human assault on Earth's rain forests has put this diverse fauna and the varied flora at risk.

Deforestation of the Tropics. More than one-half of Earth's original rain forest is gone, cleared for pasture and cattle grazing, timber, fuel, and farming. An area slightly smaller than Wisconsin is lost each year (168,900 km², 65,200 mi²) and about a third more is disrupted by selective cutting of canopy trees. When astronauts in orbit look down on the rain forests at night, they see thousands of human-set fires. During the day the lower atmosphere in these regions appears choked with the smoke (Figure 16-25a). These fires are used to clear land for agriculture, which is intended to feed the domestic population as well as to produce cash exports of beef, rubber, coffee, and other commodities.

However, poor soil fertility means that the lands are quickly exhausted under intensive farming and are then generally abandoned in favor of newly burned and cleared lands, unless the lands are artificially maintained. Unfortunately, the dominant trees require from 100 to 250 years to establish themselves after major disturbances. Following clearing, the massive growth of low bushes intertwined with vines and ferns means that a new forest may be slow to follow.

Figure 16-25b is a satellite image of a portion of western Brazil recorded in 1987, the peak year for losses in the Brazilian rain forest. This image gives a sense of the level of rain forest destruction in progress. Scientists with Goddard Space Flight Center and the Brazilian government recently completed a digitized database to facilitate analysis of these rain forest losses. FYI Report 16-1 looks more closely at efforts to curb increasing rates of species extinction, much of which is directly attributable to the loss of rain forests.

Tropical Seasonal Forest and Scrub

A varied biome on the margins of the rain forest is the **tropical seasonal forest and scrub**, which occupies regions of low and erratic rainfall. The shift-

FIGURE 16-25
(a) The Shuttle *Discovery* captured this photograph October 1, 1988 showing thousands of fires producing smoke (gray areas) that hide the Amazon rain forest from view. Individual smoke plumes rise from the rain forest fires; a single fire can exceed the area burned in the 1988 Yellowstone fire. (b) A 1987 *Landsat 5* image of a portion of the State of Rondônia in western Brazil. Highway BR364, the main artery in the region, passes through the towns of Jaru and Ji-Paraná near the center of the image at about 11°S, 62°E. The branching pattern of feeder roads and cleared forest (linear patterns) appears in a light blue-grey color. [(a) Space Shuttle photo from NASA; (b) courtesy of GEOPIC, Earth Satellite Corporation.]

(a)

(b)

Biodiversity and Biosphere Reserves

When the first European settlers landed in the Hawaiian Islands in the late 1700s, 43 species of birds were counted. Today, 15 of those species are extinct, and 19 more are threatened or endangered, with only one-fifth of the original species relatively healthy. Humans have great impact on animals and plants.

As more is learned about Earth's ecosystems and their related communities, more is known of their value to civilization and our interdependence with them. Natural ecosystems are a major source of new foods, new chemicals, new medicines, and specialty woods, and of course they are indicators of a healthy, functioning biosphere.

International efforts are underway to study and preserve specific segments of the biosphere: Rain Forest Action Network, Natural World Heritage Sites, Wetlands of International Importance, the Nature Conservancy, and Biosphere Reserves are just a few of the active organizations. The motivation to set aside natural sanctuaries is directly related to concern over the increase in the rate of species extinctions. We are facing a loss of genetic diversity that may be unparalleled in Earth's history, even compared to the major extinctions in the geologic record.

Species Threatened

Scientists have classified only 1.4 million species of plants and animals of an estimated 10–30 million overall; the latter figure represents an increase in what scientists once thought to be the diversity of life on Earth. And those yet-to-be-discovered species represent a potential future resource for society. However, in 1982, the *Global 2000 Report,* prepared by the Council of Environmental Quality under a commission originally established by President Carter, stated that between 15 and 20% of all plant and animal species on Earth would be extinct by the year 2000. That percentage equals 500,000 to 2 million species, many of which have not yet even been identified. This rate of species extinction averages one every 30 minutes! The effects of pollution,

loss of wild habitats, excessive grazing, poaching, and collecting are at the root of such devastation. Approximately 60% of the extinctions are attributable to the clearing and loss of rain forests alone.

Both black rhinos and white rhinos in Africa exemplify species in jeopardy. Rhinos once grazed over much of the savanna grasslands and woodlands. Today, they survive only in protected districts and are threatened even there (Figure 1). There are less than two dozen white rhinos in existence now in sub-Saharan ranges; the number of black rhinos is approximately 8000, but that is a little over half the number in 1980. These large land mammals are nearing extinction and will survive only as a dwindling zoo population. The limited genetic pool that remains complicates further reproduction.

Biosphere Reserves

To slow this type of extinction, formal natural reserves are a possible strategy. Setting up such a *biosphere reserve* involves principles of *island biogeography*. Island communities are special places for study because of their spatial isolation and the relatively small number of species present. They resemble natural experiments because the impact of individual factors, such as civilization, can be more easily assessed on islands than over larger continental areas. According to the equilibrium theory of island biogeography, developed by biogeographers R. H. MacArthur and E. O. Wilson, the number of species should increase with the size of the island, decrease with increasing distance from the nearest continent, and remain about the same over time, even though composition may vary. Clearly, human intervention throws natural balances out of equilibrium.

Studies of islands also can assist in the study of mainland ecosystems, for in many ways a park or biosphere reserve, surrounded by modified areas and artificial boundaries, is like an island. Indeed, a biosphere reserve is conceived as an ecological island in the midst of change. The intent is to establish a core in which genetic

FIGURE 1

The rhinoceros in Africa. (a) Failing population and receding distribution of both the black and white rhinos. (b) Black rhino and young in Tanzania, escorted by oxpecker birds. [(a) Illustration after *State of the Ark* by Lee Durrell. Copyright © 1986 by Gaia Books Ltd. Adapted by permission of Gaia Books Ltd. (b) Photo by Stephen J. Krasemann/DRK Photo.]

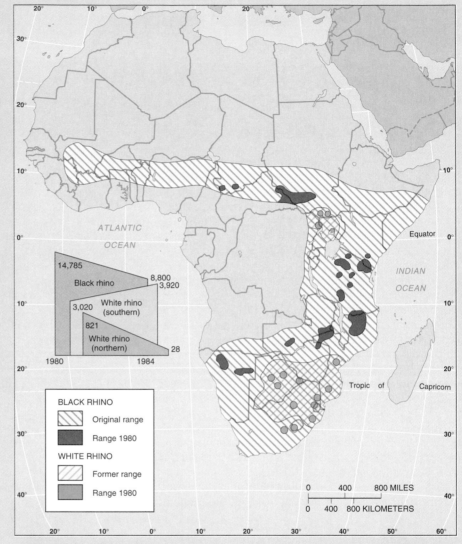

material is protected from outside disturbances, surrounded by a buffer zone that is, in turn, surrounded by a transition zone and experimental research areas.

It is predicted that new, undisturbed reserves will not be possible after A.D. 2010, because pristine areas will be gone. Coordinated by the Man and the Biosphere (MAB) Programme of UNESCO, nearly 300 such biosphere reserves covering some 12 million hectares (30 million acres) are now operated voluntarily in 76 countries. Not all protected areas are ideal bioregional entities. Some are simply imposed on existing park space and some remain in the planning stage although they are officially designated.

The ultimate goal, about half achieved, is to establish at least one reserve in each of the 194 distinctive *biogeographical communities* presently identified. The United Nations list of national parks and protected areas is compiled by UNESCO and the World Conservation Monitoring Centre. Presently, 6930 areas covering 657 million hectares (1.62 billion acres) are designated in some protective form, representing about 4.8% of national land area on the planet. Even though these are not all set aside to the degree of biosphere reserves, they do demonstrate progress toward preservation of our planet's plant and animal heritage.

> The preservation of species diversity is a problem that must today be confronted by one species, *Homo sapiens.* . . . the diversity of species is worth preserving because it represents a wealth of knowledge that cannot be replaced. Moreover, today's extinctions are unlike those in previous eras, in which long periods of recovery could follow extinctions. The present situation is an inexorably irreversible one in which human overpopulation will destroy most species unless we plan for protection immediately. Accepting that the goal is worthwhile requires that more energy be devoted to planning and priorities and less to emotionalism and indignation [on all sides].*

---

* Daniel E. Koshland, Jr., "Preserving Biodiversity," editorial, *Science* 253, no. 5021 (August 16, 1991): 717.

ing intertropical convergence zone (ITCZ) brings precipitation with the seasonally shifting high Sun and dryness with the low Sun, producing a seasonal pattern that features moisture deficits, some leaf loss, and dry-season flowering. The term *semideciduous* applies to some of the broadleaf trees that lose their leaves during the dry season. Areas of this biome have fewer than 40 rainy days during their four consecutive driest months, yet heavy monsoon downpours characterize their summers (see Figure 4-22). The Köppen climates *Am tropical monsoon* and *Aw tropical savanna* apply to these transitional communities between rain forests and tropical grasslands.

Portraying such a varied biome is difficult. In many areas the natural biome is disturbed by humans, so that the savanna grassland adjoins the rain forest directly. The biome does include a gradation from wetter to drier areas: monsoonal forests, to open woodlands and scrub woodland, to thorn forests, to drought-resistant scrub species (Figure 16-26). The monsoonal forests average 15 m (50 ft) high with no continuous canopy of leaves, graduating into open orchardlike parkland with grassy openings or into areas choked by dense undergrowth. In more open tracts, a common tree is the acacia, with its flat-topped appearance and usually thorny stems.

Local names are given to these communities: the *caatinga* of the Bahia State of northeast Brazil, the *chaco* area of Paraguay and northern Argentina, the *brigalow scrub* of Australia, and the *dornveld* of southern Africa. The map in Figure 16-22 shows areas of this biome in Africa, extending from eastern Angola through Zambia to Tanzania; in southeast Asia and portions of India, from interior Myanmar (Burma) through northeastern Thailand; and in parts of Indonesia. The trees throughout most of this biome are not good for lumber but may be valuable for fine cabinetry, especially teak wood. In addition, some of the plants with dry season adaptations produce usable waxes and gums, such as carnauba and palm-hard waxes.

Tropical Savanna

Large expanses of grassland interrupted by trees and shrubs aptly describes the **tropical savanna** (Figure 16-27). This is a transitional biome between the tropical forests and semiarid tropical steppes and

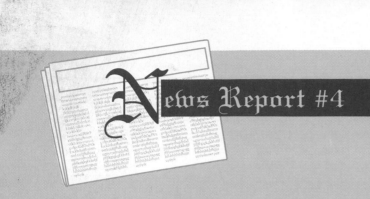

"Annual Losses to the Rain Forest Estimated"

The United Nations Food and Agricultural Organization (FAO) estimates that every year approximately 16.9 million hectares (41.7 million acres) of rain forest are destroyed, and more than 5 million hectares (12.3 million acres) are selectively logged. This total—averaged for 1980–1991 in the 76 countries that contain 97% of all rain forest—represents a 0.9% loss of equatorial and tropical rain forest worldwide each year (up from the previous average of 0.6%).

If this destruction continues unabated, these forests will be completely removed by about A.D. 2050! By continent, forest losses are estimated at more than 50% in Africa, over 40% in Asia, and 40% in Central and South America. Brazil, Colombia, Mexico, and Indonesia lead the list of lesser-developed countries that are removing their forests at record rates, although these statistics are disputed by the countries in question.

deserts. The savanna biome also includes treeless tracts of grasslands. The trees of the savanna woodlands are characteristically flat topped.

Savannas covered more than 40% of Earth's land surface before human intervention but were especially modified by human-caused fire. Fires occur annually throughout the biome. The timing of these fires is important. Early in the dry season they are beneficial and increase tree cover; if late in the season they are very hot and kill trees and seeds. Savanna trees are adapted to resist the "cooler" fires. Elephant grasses averaging 5 m (16 ft) high and forests once penetrated much farther into the dry regions, for they are known to survive there when pro-

FIGURE 16-26
Open thorn forest and savanna in the Samburu Reserve, Kenya. [Photo by Stephen J. Krasemann/DRK Photo.]

FIGURE 16-27
Savanna landscape of the Serengeti Plains, with wildebeest, zebras, and thorn forest. [Photo by Stephen Cunha.]

tected. Savanna grasslands are much richer in humus than the wetter tropics and are better drained, thereby providing a better base for agriculture and grazing. Sorghums, wheat, and groundnuts (peanuts) are common commodities.

Tropical savannas receive their precipitation during less than six months of the year, when they are influenced by the ITCZ. The rest of the year they are under the drier influence of shifting subtropical high-pressure cells. Savanna shrubs and trees are frequently *xerophytic,* or drought resistant, with various adaptations to protect them from the dryness: small, thick leaves, rough bark, or waxy leaf surfaces.

Africa has the largest region of this biome, including the famous Serengeti Plains and the Sahel region. Sections of Australia, India, and South America also are part of the savanna biome. Some of the local names for these lands include the *Llanos* in Venezuela, stretching along the coast and inland east of Lake Maricaibo and the Andes; the *Campo Cerrado* of Brazil and Guiana; and the *Pantanal* of southwestern Brazil. Particularly in Africa, savannas are the home of large land mammals that graze on savanna grasses or feed upon the grazers themselves: lion, cheetah, zebra, giraffe, buffalo, gazelle,

wildebeest, antelope, rhinoceros, and elephant (Figure 16-27).

However, in our lifetime we may see the reduction of these animal herds to zoo stock only, because of poaching and habitat losses. The loss of both the black and white rhino is discussed in FYI Report 16-1. Establishment of large tracts of savanna as biosphere reserves is critical for the preservation of this biome and its associated fauna.

Midlatitude Broadleaf and Mixed Forest

Moist continental climates support mixed broadleaf and needleleaf forest in areas of warm-to-hot summers and cool-to-cold winters. This **midlatitude broadleaf and mixed forest** biome includes several distinct communities in North America, Europe, and Asia. Relatively lush evergreen broadleaf forests occur along the Gulf of Mexico. Northward are mixed deciduous and evergreen needleleaf stands associated with sandy soils and burning. When areas are given fire protection, broadleaf trees quickly take over.

Pines predominate in the southeastern and Atlantic coastal plains. Into New England and westward in a narrow belt to the Great Lakes, white and red pines and eastern hemlock are the principal evergreens, mixed with deciduous varieties of oak, beech, hickory, maple, elm, chestnut, and many others.

These mixed stands contain valuable timber, but their distribution has been greatly altered by human activity. Native stands of white pine in Michigan and Minnesota were removed before 1910; only later reforestation sustains their presence today. In northern China these forests have almost disappeared as a result of centuries of occupation. This deforestation, as elsewhere, was principally for agricultural purposes, as well as for construction materials and fuel.

This biome is quite consistent in appearance from continent to continent and at one time represented the principal vegetation of the *Cfa, Cwa humid subtropical hot summer climates, Cfb marine west coast,* and *Cwb cool summer, winter drought* climatic regions of North America, Europe, and Asia. To the north of this biome, poorer soils and colder climates favor stands of coniferous trees and a gradual transition to the northern needleleaf forests.

Northern Needleleaf Forest and Montane Forest

Stretching from the east coast of Canada and the Atlantic provinces westward to Alaska and continuing from Siberia across the entire extent of Russia to the European Plain is the **northern needleleaf forest** biome, also called the **boreal forest** (Figure 16-28). A more open form of boreal forest, transitional to arctic and subarctic regions, is termed the **taiga**. The Southern Hemisphere, lacking *D humid microthermal* climates except in mountainous locales, has no such biome. However, *montane forests* of needleleaf trees exist worldwide at high elevation.

Boreal forests of pine, spruce, and fir occupy most of the subarctic climates on Earth that are dominated by trees. Although these forests are similar in formation, individual species vary between North America and Eurasia. Certain regions of the northern needleleaf biome experience permafrost discussed in Chapter 14. When coupled with rocky and poorly developed soils, these conditions generally limit the existence of trees to those with shallow root systems.

The Sierra Nevada, Rocky Mountains, Alps, and Himalayas have similar forest communities occurring

FIGURE 16-28
Boreal forest of Canada. [Photo by author.]

at lower latitudes. Douglas and white fir grow in the western mountains in the United States and Canada. Economically, these forests are important for lumbering, with saw timber occurring in the southern margins of the biome and pulpwood throughout the middle and northern portions. Present logging practices and whether or not these yields are sustainable are issues of increasing controversy.

Temperate Rain Forest

The **temperate rain forest** biome is recognized by its lush forests at middle and high latitudes, occurring only along narrow margins of the Pacific Northwest in North America (Figure 16-29). This biome contrasts with the diversity of the equatorial and tropical rain forest in that only a few species

FIGURE 16-29
Temperate Hoh Rain Forest in Olympic National Park, Washington State. Club moss, big leaf maples, and sword ferns are pictured. [Photo by Steve Lambros/Lambros Photography.]

make up the bulk of the trees. The rain forest of the Olympic Peninsula in Washington State is a mixture of broadleaf and needleleaf trees, huge ferns, and thick undergrowth. Precipitation approaching 400 cm (160 in.) per year on the western slopes, moderate air temperatures, summer fog, and an overall maritime influence produce this moist, lush vegetation community.

The tallest trees in the world occur in this biome—the coastal redwoods *(Sequoia sempervirens)*. Their distribution is shown on the map in Figure 16-11a. These trees can exceed 1500 years of age and typically range in height from 60 to 90 m (200–300 ft), with some exceeding 100 m (330 ft). Virgin stands of other representative trees—Douglas fir, spruce, cedar, and hemlock—have been reduced to a few remaining valleys in Oregon and Washington, less than 10% of the original forest.

A study released in 1993 by the U.S. Forest Service noted the failing ecology of these forest ecosystems and suggested that timber management plans should include ecosystem preservation as a priority. The government held a "forestry summit" in Portland, Oregon in 1993, to bring together the opposing sides of the logging-versus-environment conflict. The ultimate solution must be one of economic-ecological synthesis rather than continuing conflict and forest losses.

Mediterranean Shrubland

The **Mediterranean shrubland** biome occupies those regions poleward of the shifting subtropical high-pressure cells. As those cells shift poleward with the high Sun, they cut off available storm systems and moisture. Their stable high-pressure presence produces the characteristic dry summer climate—Köppen's *Csa, Csb Mediterranean dry summer*—and establishes conditions conducive to fire. Plant ecologists think that this biome is well adapted to frequent fires, for many of its characteristically deep-rooted plants have the ability to resprout from their roots after a fire. Earlier stands of evergreen woodlands no longer dominate.

The dominant shrub formations that occupy these regions are stunted and tough in their ability to withstand hot-summer drought. The vegetation is called

FIGURE 16-30
Chaparral vegetation associated with the Mediterranean dry summer climate.
[Photo by author.]

sclerophyllous (from *sclero* or "hard" and *phyllos* for "leaf"); it averages a meter or two in height and has deep, well-developed roots, leathery leaves, and uneven low branches.

Typically, the vegetation varies between woody shrubs covering more than 50% of the ground and grassy woodlands with 25–60% coverage. In California, the Spanish word *chaparro* for "scrubby evergreen" gives us the word **chaparral** for this vegetation type (Figure 16-30). A counterpart to chaparral in the Mediterranean region, called *maquis,* includes live and cork oak trees (source of cork), as well as pine and olive trees. In Chile, such a region is called *mattoral;* in southwest Australia, *mallee scrub.* Of course, in Australia the bulk of the eucalyptus species is sclerophyllous in form and structure in whichever climate it occurs.

As described in Chapter 7, Mediterranean climates are important in commercial agriculture for subtropical fruits, vegetables, and nuts, with many food types produced only in this biome (e.g., artichokes, olives, almonds).

Midlatitude Grasslands

Of all the natural biomes, the **midlatitude grasslands** are most modified by human activity. Here are the world's "breadbaskets"—regions of grain and livestock production (Figure 16-31). In these regions, the only naturally occurring trees were deciduous broadleafs along streams and other limited sites. These regions are called grasslands because of the original predominance of grasslike plants:

> In this study of vegetation, attention has been devoted to grass because grass is the dominant feature of the Plains and is at the same time an index to their history. Grass is the visible feature which distinguishes the Plains from the desert. Grass grows, has its natural habitat, in the transition area between timber and desert. . . . The history of the Plains is the history of the grasslands.*

*From *The Great Plains* by Walter Prescott Webb, p. 32. © Copyright 1959 by Walter Prescott Webb, published by Ginn and Company, Needham Heights, MA.

FIGURE 16-31
Midlatitude grassland under cultivation. [Photo by author.]

In North America, tall grass prairies once rose to heights of 2 m (6.5 ft) and extended westward to about the 98th meridian, with short grass prairies farther west. The 98th meridian is roughly the location of the 51 cm (20 in.) isohyet, with wetter conditions to the east and drier to the west (see Figure 15-15).

The deep sod of those grasslands posed problems for the first settlers, as did the climate. The self-scouring steel plow, introduced in 1837 by John Deere, allowed the interlaced grass sod to be broken apart, freeing the soils for agriculture. Other inventions were critical to opening this region and solving its unique spatial problems: barbed wire (the fencing material for a treeless prairie), well-drilling techniques developed by Pennsylvania oil drillers but used for water wells, windmills for pumping, and railroads to conquer the distances.

Few patches of the original prairies (tall grasslands) or steppes (short grasslands) remain within this biome. For prairies alone, the reduction of natural vegetation went from 100 million hectares (250 million acres) down to a few areas of several hundred hectares each. A "prairie national park" was considered as a preservationist effort along the Kansas-Oklahoma border, but plans were put on hold. A state-designated 10 hectare (25 acre) remnant of the "prairie" is located 8 km north of Ames, Iowa. This is a patch of original grassland that never has felt the plow.

Outside North America, the Pampas of Argentina and Uruguay and the grasslands of Ukraine are char-acteristic midlatitude grassland biomes. In each region of the world where these grasslands have occurred, human development of them was critical to territorial expansion.

Desert Biomes

Earth's **desert biomes** cover more than one-third of its land area. In Chapter 12 we examined desert landscapes and in Chapter 7, desert climates. On a planet with such a rich biosphere, the deserts stand out as unique regions of fascinating adaptations for survival. Much as a group of humans in the desert might behave with short supplies, plant communities also compete for water and site advantage. Some desert plants, called *ephemerals,* wait years for a rainfall event, at which time their seeds quickly germinate, and the plants develop, flower, and produce new seeds, which then rest again until the next rainfall event. The seeds of some xerophytic species open only when fractured by the tumbling, churning action of flash floods cascading down a desert arroyo, and of course such an event produces the moisture that a germinating seed needs.

Perennial desert plants employ other adaptive features to cope with the desert, such as long, deep tap roots (e.g., the mesquite), succulence (i.e., thick, fleshy, water-holding tissue such as that of cacti), spreading root systems to maximize water availability, waxy coatings and fine hairs on leaves to retard water loss, leafless conditions during dry periods (e.g., palo verde and ocotillo), reflective surfaces to reduce leaf temperatures, and tissue that tastes bad to discourage herbivores.

The creosote bush *(Lorrea divaricata)* sends out a wide pattern of roots and contaminates the surrounding soil with toxins that prevent the germination of other creosote seeds, possible competitors for water. When a creosote bush dies and is removed, surrounding plants or germinating seeds work to occupy the abandoned site, but must rely on infrequent rains to remove the toxins.

We are witnessing an unwanted expansion of the desert biome. This is due principally to agricultural practices (overgrazing and inappropriate agricultural activities that abuse soil structure and fertility), improper soil-moisture management, erosion and salinization, deforestation, and the ongoing climatic change. A process known as *desertification* is now a worldwide phenomenon along the margins of semiarid and arid lands. The United Nations estimates that desertified lands have covered some 800 million hectares (2 billion acres) since 1930; many millions of additional hectares are added each year.

Earth's deserts are subdivided into desert and semidesert, to distinguish those with expanses of bare ground from those covered by xerophytic plants of various types. The two broad associations are further separated into warm deserts, principally tropical and subtropical, and cold deserts, principally midlatitude.

Warm Desert and Semidesert. Earth's warm desert and semidesert biomes are caused by the presence of dry air and low precipitation from subtropical high-pressure cells. These areas are very dry, as evidenced by the Atacama Desert of northern Chile, where only a minute amount of rain has ever been recorded—a 30-year annual average of only 0.05 cm (0.02 in.)! Like the Atacama, the true deserts of the Köppen *BW desert* classification are under the influence of the descending, drying, and stable air of high-pressure systems from 8 to 12 months of the year.

Vegetation ranges from almost none in the arid deserts to numerous xerophytic shrubs, succulents, and thorn tree forms. The lower Sonoran desert of southern Arizona is a warm desert (Figure 16-32). This desert landscape features the unique saguaro cactus *(Carnegiea gigantea)* which grows to many meters in height and up to 200 years in age if undisturbed. First blooms do not appear until it is 50 to 75 years old!

A few of the subtropical deserts—such as those in Chile, Western Sahara, and Namibia—are right on the sea coast and are influenced by cool offshore ocean currents. As a result, these true deserts expe-

FIGURE 16-32
Lower Sonoran desert west of Tucson, Arizona (32.3° north latitude; elevation 900 m or 2950 ft). A weather instrument shelter is pictured. [Photo by author.]

rience summer fog that mists the plant and animal populations with needed moisture.

The equatorward margin of the subtropical high-pressure cell is a region of transition to savanna, thorn tree, scrub woodland, and tropical seasonal forest. Poleward of the warm deserts, the subtropical cells shift to produce the Mediterranean dry summer regime along west coasts and may grade into cool deserts elsewhere.

Cold Desert and Semidesert. The cold desert and semidesert biomes tend to occur at higher latitudes where seasonal shifting of the subtropical high is of some influence less than six months of the year. Specifically, interior locations are dry because of their distance from moisture sources or their location in rain shadow areas on the lee side of mountain ranges such as the Sierra Nevada of the western United States, the Himalayas, and the Andes. The combination of interior location and rain shadow po-

sitioning produces the cold deserts of the Great Basin of western North America (Figure 16-33).

Winter snows, occur in the cold deserts, but are generally light. Summers are hot, with highs from 30° to 40°C (86° to 104°F). Nighttime lows, even in the summer, can cool 10° to 20C° (18° to 36F°) from the daytime high. The dryness, generally clear skies, and sparse vegetation lead to high radiative heat loss and cool evenings. Many areas of these cold deserts that are covered by sagebrush and scrub vegetation were actually dry shortgrass regions in the past, before extensive grazing altered their appearance. The look of the deserts in the upper Great Basin is the result of more than a century of such cultural practices.

Arctic and Alpine Tundra

The **arctic tundra** is found in the extreme northern area of North America and Russia, bordering on the Arctic Ocean and generally north of the 10°C (50°F)

FIGURE 16-33
Cold desert scene in the interior Great Basin of the western United States. [Photo by author.]

isotherm for the warmest month. Daylength varies greatly throughout the year, seasonally changing from almost continuous day to continuous night. The region, except a few portions of Alaska and Siberia, was covered by ice during all of the Pleistocene glaciations.

Winters in this biome, an *ET tundra* climate classification, are long and cold; cool summers are brief. Winter is governed by intensely cold continental polar air masses and stable high-pressure anticyclones. A growing season of sorts lasts only 60–80 days, and even then frosts can occur at any time. Vegetation is fragile in this flat, treeless world; soils are poorly developed periglacial surfaces, which are underlain by permafrost. In the summer months only the surface horizons thaw, thus producing a mucky surface of poor drainage. Roots can penetrate only to the depth of thawed ground, usually about a meter. The surface is shaped by freeze-thaw cycles that create the frozen ground phenomena discussed in Chapter 14.

Tundra vegetation is low, ground-level herbaceous plants such as sedges, mosses, arctic meadow grass, and snow lichen, and some woody species like dwarf willow (Figure 16-34). Owing to the short growing season, some perennials form flower buds one summer and open them for pollination the next.

Alpine tundra is similar to arctic tundra, but can occur at lower latitudes because it is associated with high elevation. This biome usually is described as above the timberline, that elevation above which trees cannot grow. Timberlines increase in elevation equatorward in both hemispheres. Alpine tundra communities occur in the Andes near the equator, the White Mountains of California, the Rockies, the Alps, and Mount Kilimanjaro of equatorial Africa, as well as mountains from the Middle East to Asia. Alpine meadows feature grasses and stunted shrubs, such as willows and heaths. Figure 16-35 shows a typical alpine setting.

FIGURE 16-34
Tundra on the Kamchatka Peninsula, Russia. The Uzon Caldera is in the center-background mountain range. [Photo by Wolfgang Kaehler.]

FIGURE 16-35
An alpine landscape near Logan Pass, Montana, in Waterton-Glacier International
Peace Park (48.8° north latitude; 2025 m or 6650 ft elevation). [Photo by Steve
Lambros/Lambros Photography.]

SUMMARY—Ecosystems and Biomes

Earth's biosphere is unique in the Solar System. The **diversity** of life on Earth is impressive. **Ecosystems** are self-regulating associations of living plants and animals and their nonliving physical environment. Earth's ecosystems are woven in a complex, interdependent web, with each life form mutually tied to the success of the others.

Biogeography, essentially a spatial **ecology**, is the study of the distribution of plants and animals and the diverse spatial patterns they create. Plants and animals have developed complex **community** relationships centered in specific **habitats** and **niches**. Plants perform **photosynthesis**, providing energy and oxygen to drive biological processes. **Net primary productivity** is an important aspect of a community being the net result of photosynthesis and **respiration** (opposite of photosynthesis) processes. Light, temperature, water, and the cycling of gases and nutrients constitute the supporting abiotic components

of ecosystems in **biogeochemical cycles**. **Biomass** and population pyramids characterize the flow of energy and the numbers of **producers**, **consumers**, and **decomposers** operating in an ecosystem.

Ecosystems appear to move toward stability and balance, with individual **terrestrial** and **aquatic ecosystems** undergoing a succession of stages. Human activity intercedes at almost every level of natural ecosystem operations, producing artificial conditions, ecosystems that lack diversity, and interrupted succession.

Biomes are Earth's major terrestrial ecosystems, each named for its dominant plant community. The 10 major biomes are generalized from **formation classes** that describe vegetation units on the basis of structure and appearance. A biome represents a climax community of natural vegetation, but few undisturbed biomes exist in the world. Many of Earth's plant and animal communities are experiencing accelerated change that could produce dramatic alterations within our lifetime.

525

KEY TERMS

aquatic ecosystem	life zone
biogeochemical cycle	limiting factor
biogeography	net primary productivity
biomass	niche
biome	omnivore
boreal forest	photoperiod
carnivore	photosynthesis
chaparral	producer
chlorophyll	respiration
community	stomata
consumer	terrestrial ecosystem
decomposer	arctic tundra
diversity	desert biome
ecological succession	equatorial and tropical rain forest
ecology	Mediterranean shrubland
ecosystem	midlatitude broadleaf and mixed
ecotone	forest
eutrophication	midlatitude grassland
fire ecology	northern needleleaf forest
food chain	taiga
food web	temperate rain forest
formation class	tropical savanna
habitat	tropical seasonal forest and scrub
herbivore	

REVIEW QUESTIONS

1. Briefly define an ecosystem and explain what its operation implies about the complexity of life.
2. Define a community within an ecosystem.
3. What do the concepts of habitat and niche involve? Relate them to some specific plant and animal communities.
4. Describe symbiotic and parasitic relationships in nature. Draw an analogy between these relationships and human societies on our planet. Explain.
5. What are the principal abiotic components in terrestrial ecosystems?
6. What are biogeochemical cycles? Describe several of the essential cycles.
7. What is a limiting factor? How does it function to control the spatial distribution of plant and animal species? Explain the concept of tolerance in this connection.
8. How do plants function to link the Sun's energy to living organisms?
9. What is net photosynthesis? How does the concept relate to photosynthesis, respiration, or net primary productivity?

10. Describe the relationship among producers, consumers, and decomposers in an ecosystem. What is the trophic nature of an ecosystem? What is the place of humans in a trophic system?

11. What is meant by ecosystem stability?

12. How does ecological succession proceed? What are the relationships that exist between existing communities and new, invading communities? Explain the principle of competition.

13. Define biome. What is the basis of the designation? What are formation classes?

14. Use the integrative chart in Table 16-2 and the world map in Figure 16-22 to correlate the description of each biome with its specific spatial distribution.

15. Describe the equatorial and tropical rain forests. Why is the rain forest floor somewhat clear of plant growth? Why are logging activities for specific species so difficult there?

16. What are the issues surrounding deforestation of the rain forest? What is the impact of these losses on the rest of the biosphere?

17. Describe the role of fire in the tropical savanna biome and the midlatitude broadleaf and mixed forest biome.

18. Why does the northern needleleaf forest biome not exist in the Southern Hemisphere? Where is this biome located in the Northern Hemisphere, and what is its relationship to climate type?

19. In which biome do we find Earth's tallest trees? Which biome is dominated by small, stunted plants, lichens, and mosses?

20. What type of vegetation predominates in the Mediterranean dry summer climates? Describe the adaptation necessary for these plants to survive.

21. Describe some of the unique adaptations found in a desert biome.

22. What is desertification? Explain its impact.

Traditional rural landscape in the Pamirs of Tajikistan [*Photo by Stephen Cunha.*]

17

EARTH, HUMANS, AND THE NEW MILLENNIUM

AN OILY BIRD
  The Larger Picture
GAIA HYPOTHESIS
THE NEED FOR INTERNATIONAL
  COOPERATION
  The Environmental Cost of Conflict
WHO SPEAKS FOR EARTH?
SUMMARY
FYI REPORT 17-1 EARTH SUMMIT 1992

During my space flight, I came to appreciate my profound connection to the home planet and the process of life evolving in our special corner of the Universe, and I grasped that I was part of a vast and mysterious dance whose outcome will be determined largely by human values and actions.*

—Rusty Schweickart

Earth is observed from many profound vantage points, as this astronaut experienced on the 1969 Apollo IX mission. Our vantage point in this book is physical geography. We have examined Earth through its energy, atmosphere, water, weather, climate, endogenic and exogenic systems, soils, ecosystems, and biomes—all of which lead to an examination of the planet's most abundant large animal, *Homo sapiens*.

We stand at the end of the Twentieth Century and at the brink of a new millennium. The Twenty-first Century will be an adventure for the global society, unparalleled in history in terms of experimenting with Earth's life-supporting systems. What preparations are we making for the advent of this new century?

In his 1980 book and PBS television series, astronomer Carl Sagan asked:

What account would we give of our stewardship of the planet Earth? We have heard the rationales offered by the nuclear superpowers. We know who speaks for the nations. But who speaks for the human species? Who speaks for Earth?*

We might attempt an answer: perhaps those who have studied Earth and know the operations of the global ecosystem should speak for Earth. Or perhaps those who have the greatest stake in the continued functioning of that ecosystem—the children—should address the importance of a workable future. Some might say that these questions of technology, poli-

tics, and future thinking belong outside of academia, that our job here is merely to learn how Earth's processes work. However, biologist-ecologist Marston Bates addressed this line of thought in 1960:

Then we came to humans and their place in this system of life. We could have left humans out, playing the ecological game of "let's pretend humans don't exist." But this seems as unfair as the corresponding game of the economists, "let's pretend nature doesn't exist." The economy of nature and the ecology of humans are inseparable and attempts to separate them are more than misleading, they are dangerous. Human destiny is tied to nature's destiny and the arrogance of the engineering mind does not change this. Humans may be a very peculiar animal, but they are still a part of the system of nature.*

Interestingly, most environmental problems have recognized solutions or strategies for positive resolution, if only this knowledge could be applied in constructive ways. Unfortunately, many of these solutions are not being acted upon by society. We do not seem to assess the true long-term costs of present economic thinking. A role for physical geography is to explain Earth systems through spatial analysis and through a synthesis of information from many disciplines—hopefully to assist an informed citizenry toward more considered judgments.

A trait of these times is that developed countries, perhaps unconsciously, speak for the billions living in developing nations. The fate of traditional modes of life may rest in some distant financial capital far from their homestead. The remote lands of Tajikistan (see chapter-opening photograph) are linked by Earth systems to the Pampas of Argentina and the Great Plains in North America.

There is an international environmental awareness and concern in people that is in turn driving new action by government. The George H. Gallup International Institute completed an extensive 22-nation, 22,000-person survey that exposed some false assumptions in the conventional wisdom. Survey re-

*Rusty Schweickart, "Our Backs Against the Bomb, Our Eyes on the Stars," *Discovery,* July 1987: 62.

*Carl Sagan, *Cosmos.* New York: Random House, 1980, p. 329.

*Marston Bates, *The Forest and the Sea.* New York: Random House, 1960, p. 247.

FIGURE 17-1
A western grebe contaminated with oil from the *Exxon Valdez* tanker accident in Prince William Sound, Alaska. [Photo by Geoffrey Orth/© Sipa Press.]

sults established that ". . . people in both poor and rich nations give priority to environmental protection over economic growth," ". . . accept responsibility for contributing to problems, and believe that citizen efforts can contribute significantly to a healthier planet," and ". . . environmental problems are at the very top of the list of problems in virtually all nations." Majorities in 20 nations are willing to "endorse environmental protection at the risk of slowing down economic growth."*

An Oily Bird

At first glance the chain of events that exposes wildlife to oil contamination seems to stem from a technological problem. An oil tanker splits open at sea and releases its petroleum cargo, which is moved by ocean currents toward shore, where it coats coastal waters, beaches, and animals (Figure 17-1). In response, concerned citizens mobilize and try to save as much of the spoiled environment as possi-

---

*R. E. Dunlap, G. H. Gallup, Jr., and A. M. Gallup, *The Health of the Planet Survey—A Preliminary Report on Attitudes Toward the Environment and Economic Growth Measured by Surveys of Citizens in 22 Nations to Date*. A George H. Gallup Memorial Survey. Princeton, NJ: George H. Gallup International Institute, July 1992.

ble, but the real problem goes far beyond the physical facts of the spill.

On March 24, 1989, in Prince William Sound off the southern coast of Alaska in clear weather and calm seas, a single-hulled supertanker operated by Exxon Corporation struck a reef. The tanker spilled 42 million liters (11 million gallons) of oil. It took only 12 hours for the *Exxon Valdez* to empty its contents, yet a reasonable cleanup is taking years and billions of dollars. (See the area of the spill noted on the map in Figure 17-2.)

Because contingency emergency plans were not in place, and promised equipment was unavailable, response by the oil industry took over 12 hours to activate, about the same time that it took the ship to drain. This initial response by Alyeska Pipeline Service Company and Exxon proved inadequate to the task. Badly needed equipment was unavailable in dry docks, in warehouses, and buried in snow. In fact initial efforts were directed at public relations. In the years following the spill, evidence of this inaction and confusion was documented. Total damage now is estimated between $15 and $20 billion, although Exxon settled with the government for $1.125 billion and several civil suits remain in the courts.

Eventually, almost 2415 km (1500 mi) of sensitive coastline was ruined. For perspective, had this spill occurred farther south, enough oil spilled to blacken every beach and bay along the Pacific Coast from southern Oregon to the Mexican border.

The death toll of animals was massive: at least 3000 sea otters killed (or about 20% of the resident otters), 300,000 birds, and uncounted fish, shellfish, plants, and aquatic microorganisms. Sublethal effects, namely mutations, now are appearing in fish. This latter side effect of the spill is serious because salmon fishing was the main economy in Prince William Sound, not oil. Conflicting scientific reports emerged in 1993 as to long-term damage in the region—scientific studies commissioned by Exxon disagreed with scientific studies prepared outside the industry.

The Larger Picture

The immediate effect of the oil spill on wildlife was contamination and death, but the issues involved are bigger than damaged ecosystems. Let's ask some fundamental questions: Why was the oil tanker there in the first place? Why is petroleum being imported into the continental United States from Alaska and abroad in such quantity? Is the demand for petroleum products based on real need, or does it reflect the business strategies of transnational oil corporations? The U.S. demand for oil is higher per capita than the demand of any other country; is this demand inflated by waste and inefficiency?

A great many factors influence our demand for oil. Well over half of our imported oil goes for transportation. Despite this enormous and expensive consumption, the 1980s saw a major rollback of auto efficiency standards, a reduction in gasoline prices, large reductions in funding for rapid transit development, and the continuing slow demise of America's passenger and freight rail system. Thus, waste, low prices, and a lack of alternatives all combined to spur demand for petroleum. In addition, land-use patterns continue to foster diffuse sprawl, thereby placing stress upon existing transportation systems.

The demand for fossil fuels in the 1980s also was affected by the slowing of domestic conservation programs, the elimination of research for energy alternatives such as solar and wind, and even delays of a law requiring small appliances to be more energy efficient. Increased fuel imports make the United States more vulnerable to foreign conflicts and war. By 1993, fuel demands reached 6.1 billion barrels a year in the United States!

Thus, the immediate problem of cleaning oil off a bird in Prince William Sound becomes a national and international concern with far-reaching significance. And while we search for answers, oil slicks continue their contamination (Figure 17-2).

In the Persian Gulf War of 1991, we saw over 1.1 billion liters (300 million gallons) flow into the Persian Gulf and additional millions of liters pour onto the land and burn into the atmosphere in purposefully set fires (Figure 17-3). Public opinion holds the second largest corporation in America responsible for what happened in Prince William Sound and a dictator in the Middle East as responsible for the tragedy in the Persian Gulf. Yet, there is apparent hypocrisy in our outrage over these events, since we continue to stand at the pump filling our automobiles with gas, thus creating the demand for imports. Physical geography is responsible for analyzing all the spatial aspects of these events and ironies that are symbolized by an oily bird.

Gaia Hypothesis

Just as the abiotic spheres affect the biosphere, so do living processes affect abiotic functions. All of these interactive effects in concert influence Earth's overall ecosystem. In essence, the planetary ecosystem sets the physical limits for life, which in turn evolves and helps to shape the planet. Thus, Earth can be viewed as one vast, self-regulating organism—a global symbiosis. This controversial concept is called the *Gaia hypothesis*. It was proposed in 1979 by James Lovelock, a British astronomer and inventor, and elaborated by American biologist Lynn Margulis.

Gaia is the ultimate *synergistic relationship*, in which the whole greatly exceeds the sum of the individual interacting components. The hypothesis contends that life processes control and shape inorganic physical and chemical processes, with the biosphere so interactive that a very small mass can affect a very large mass. Thus, Lovelock and Margulis think that

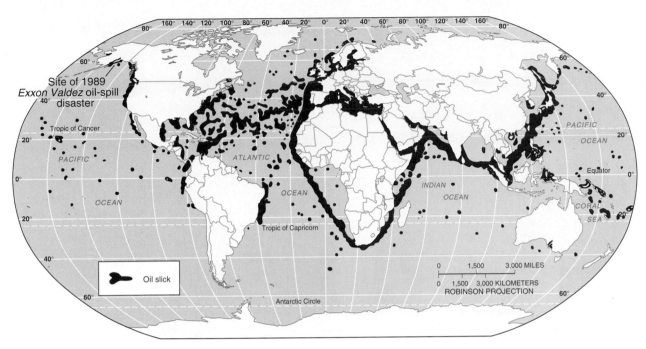

FIGURE 17-2
Location of recent visible oil slicks worldwide. The location of the 1989 *Exxon Valdez* oil spill is noted on the map. [Data from Organization for Economic Cooperation and Development, *The State of the Environment.* Paris: Organization for Economic Cooperation and Development, 1985.]

FIGURE 17-3
Kuwait oil well fires and smoke over thousands of square kilometers on July 1, 1991. [Satellite image courtesy of *EOSAT*, Earth Observation Satellite Company. Used by permission.]

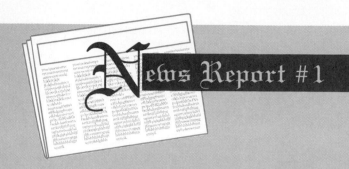

"Huge Oil Spills"

In just one year following the 1989 *Exxon Valdez* disaster in the waters of Prince William Sound, nearly 10,000 oil-spill accidents occurred worldwide. In addition, an even greater spill than that of the *Exxon Valdez* occurred off the coast of Morocco (in northwestern Africa) in the same year. Into those warm tropical waters, that vessel spilled 140.6 million liters (37 million gallons) of oil. A few examples of recent spills include: the *Aegean Sea*, off Spain in 1992, 83.6 million liters (22 million gallons); the *Maersk Navigator* in Indonesian waters in 1993, 30 million liters (7.8 million gallons); and the *Braer* off the Shetland Islands in 1993, 91 million liters (24 million gallons).

the material environment and the evolution of species are tightly joined; as the species evolve through natural selection, they in turn affect their environment. The present composition of the atmosphere is given as proof of this coevolution of living and nonliving systems.

From the perspective of physical geography, this hypothesis permits a view of all Earth and the spatial interrelationships among systems. Such a perspective is necessary for analyzing specific environmental issues. For instance, in deciding the fate of the Arctic National Wildlife Refuge (ANWR), we must weigh the supply, demand, and importation of oil, the public trust aspect of such wilderness, our will for conservation and efficiency, and our view of our place *in* nature or *outside* of nature (Figure 17-4). All these variables interact synergistically, producing both wanted and unwanted results.

One disturbing aspect of the Gaia hypothesis, is that any biotic threats to the operation of an ecosystem tend to move toward extinction *themselves,* thus preserving the system overall. Earth systems operation and feedback tend to eliminate offensive members. The degree to which humans represent a planetary threat, then, becomes a topic of great concern, for Earth (Gaia) will prevail, regardless of the outcome of the human experiment.

The maladies of Gaia do not last long in terms of her life span. Anything that makes the world uncomfortable to live in tends to induce the evolution of those species that can achieve a new and more comfortable environment. It follows that, if the world is made unfit by what we do, there is the probability of a change in regime to one that will be better for life but not necessarily better for us.[*]

The debate is vigorous regarding the true applicability of this hypothesis to nature. Regardless, it remains philosophically intriguing in its portrayal of the relationship between humans and Earth.

The Need for International Cooperation

We already have seen dramatic examples of international environmental problem solving: the Limited Test Ban Treaty of 1962, which bans atmospheric testing of nuclear weapons, and the 1987 and 1990 treaties proposing bans on ozone-destroying chlorofluorocarbons. All major problems need cooperative attention. For example, the escalating

---

[*]James Lovelock, *The Ages of Gaia—A Biography of Our Living Earth*. New York: W. W. Norton, 1988, p. 178.

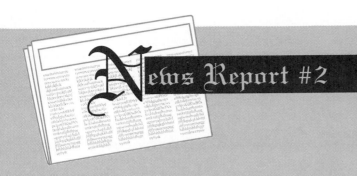

News Report #2

"ANWR Faces Threats"

Plans continue for development of the Arctic National Wildlife Refuge (ANWR) on Alaska's North Slope above the Arctic Circle, bordering on the Beaufort Sea, and adjoining the Yukon Territory. The refuge area sustains almost 200,000 caribou, polar and grizzly bears, musk-ox, and wolves. The ANWR remains the only portion of Alaska's Arctic coast that is not open to oil and gas exploration. Controversy over this refuge continues as political and corporate pressures mount to conduct oil exploration.

The resulting disruption would be devastating and the cost of development would price the oil at well over $30 per barrel, far above current market prices. The estimated oil reserve in the ANWR could be saved in only two years by a 2-mile-per-gallon increase in U. S. automobile efficiency. Economic activities always should be weighed against the nature of the ecosystem itself, its limitations, its uniqueness, and its fragility.

greenhouse effect that is discussed throughout this text and influences so many global systems certainly deserves international cooperation.

The largest gathering of nations and nongovernmental organizations ever held was about Earth—the United Nations Conference on Environment and Development (UNCED), the "Earth Summit of 1992." This important event and resulting agreements are surveyed in FYI Report 17-1. (See Appendix A for the numerous groups, organizations, and international action programs related to geography and environmental concerns.)

FIGURE 17-4
In Alaska's Arctic National Wildlife Refuge, Mount Chamberlin, the second highest peak of the Brooks Range, overlooks tundra in the foreground. Is this region destined for petroleum exploration and development, or for continuing preservation? [Photo by Scott T. Smith.]

Earth Summit 1992

The leaders of 100 nations, 10,000 delegates from over 160 countries, 9000 journalists, and the world's attention, all converged at Rio Centro, 40 minutes outside Rio de Janeiro, June 1–12, 1992 (Figure 1). An additional 1000 nongovernmental organizations (NGOs) with over 50,000 attendees assembled at Flamingo Park for a parallel, unofficial gathering dubbed the "Global Forum." The setting in Rio was ironic, for many of the problems discussed at the two conferences were evident on the streets of the city, where air pollution, water pollution, toxics, noise, disparity of income and wealth, and a struggle for health and education occur in abundance. Maurice F. Strong, a Canadian and Secretary-General of the UNCED, summarized in his conference address:

> The people of our planet, especially our youth and the generations which follow them, will hold us accountable for what we do or fail to do at the Earth Summit in Rio. Earth is the only home we have, its fate is literally in our hands. . . .The most important ground we must arrive at in Rio is the understanding that we are all in this together.

Setting the Stage

The idea for a global meeting on the environment was put forward at the 1972 U.N. Conference on the Human Environment held in Stockholm, also chaired by Maurice Strong. The U.N. General Assembly in 1987 achieved a landmark in global planning by agreeing to hold the Summit. *Our Common Future* set the tone for the 1992 Earth Summit:

> The Earth is one but the world is not. We all depend on one biosphere for sustaining our lives. Yet each community, each country, strives for survival and prosperity with little regard for its impact on others. Some consume the Earth's resources at a rate that would leave little for future generations. Others, many more in number, consume far too little and live with the prospect of hunger, squalor, disease, and early death.*

The Five Earth Summit Agreements

Five key agreements emerged from the conference, summarized here.

1. Climate Change Framework—This legally binding agreement is a first-ever attempt to evaluate and address global warming on an international scale. As of August 1992, 154 nations, including the United States and Canada, signed the Convention on Climate Change.

Initially the goal was to set specific timetables for cutting emissions of greenhouse gases. The United States objected to specific targets for controlling CO_2 emissions throughout the pre-summit sessions, citing economic uncertainties and unknown costs. The European Community, Canada, Japan, and the majority of attending nations favored stabilizing emissions at 1990 levels by the year 2000. The Office of Technology Assessment and National Academies of Science and Engineering, in separate assessments, concluded that the United States could hold to 1990 levels by the year 2015 at little or no additional cost.

2. Biological Diversity—This legally binding agreement is the first international attempt to protect Earth's biodiversity. It provides more equitable rights among nations in biotechnology and the genetic wealth of tropical ecosystems in particular.

Out of 161 signatories, the United States, Vietnam, Singapore, and Kiribati (a Pacific island nation) refused to sign the original treaty. This was a perplexing stand for the U.S. administration to take. Biodiversity is a divisive issue separating developed countries in the Northern Hemisphere from the predominantly developing nations of the equatorial and tropical regions and Southern Hemisphere. The developing countries want compensation for medicines derived from plants and animals

*World Commission on Environment and Development, *Our Common Future*. Oxford and New York: Oxford University Press, 1987, p. 27.

FIGURE 1
Leaders and delegates at the 1992 UNCED Earth Summit in Rio de Janeiro. (a) A portion of the delegates that attended from many nations. (b) Inset photo of Maurice Strong, Secretary-General of the UNCED (far left), a representative of indigenous peoples, and other delegates. [(a) Photo by Reuters/Bettmann; (b) photo by Ricardo Funavi/Imagens Da Terra/Impact Visuals.]

(b)

harvested from their indigenous genetic wealth. This return of some profit to them from transnational corporations and rich-nation enterprises creates incentive to further protect these critical biomes. The United States finally signed this treaty in 1993.

3. Management, Conservation, and Sustainable Development of All Types of Forests—This non-binding agreement guides world forestry practices toward a more sustainable future of forest yields and diversity.

The conflict between industrialized nations and developing nations is a classic confrontation of "north" and "south". How can rich northern nations continue to clear-cut their forests, yet turn to the developing countries, such as Brazil, and ask them to place their lands in a national preserve? How can the industrialized nations continue to produce excessive CO2, far beyond reasonable per capita limits, yet ask developing countries to cease destroying a principle sink for CO2, the tropical rain forest? The developing countries of the "south" insist on po-

litical and economic equality of forest practice. Along with sustainable timber practices, countries of the "north" need to begin government-sponsored paper recycling and packaging-reform programs. Such recycling now is part of the forestry debate.

4. The "Earth Charter"—This is a non-binding statement of 27 environmental and economic principles. They establish an ethical basis for a sustainable human-Earth relationship. An important emphasis is inclusion of *environmental costs* in economic assessments. Impoverishment of air, soil, water, and ecosystems sometimes is mistaken for progress. The environment is not an inexhaustible mine of resources to be tapped indefinitely. In terms of *natural capital*—air, water, timber, fisheries, petroleum—Earth is indeed a finite physical system.

5. Agenda 21 (Sustainable Development)—This non-binding action program is an 800-page guide for all nations into the 21st Century. The idea of "sustainable development," as opposed to a business proposal of "sustainable growth," is examined in Agenda 21.

Agenda 21 covers many key topics: *energy conservation and efficiency* to reduce consumption and related pollution; *climate change; stratospheric ozone depletion; transboundary air pollution; ocean and water resource protection; soil losses and increasing desertification; deforestation; regulations for safely handling radioactive waste and disposal; hazardous chemical exports for disposal in developing countries;* and *disparities of wealth* and *the plague of poverty.*

Agenda 21 also addresses the difficult question of financing sustainable development. Developing countries are asking the developed nations to spend 0.7% of their gross domestic product—approximately $125 billion per year—to assist them in implementing the Earth Charter and Agenda 21.

The Future

From the Earth Summit emerged a new organization—the *U.N. Commission on Sustainable Development*—to oversee the promises made in the five documents and agreements. Most of the participating countries completed State of the Environment Reports (SERs) and gathered environmental statistics for publication. These reports are an invaluable resource that will direct further research efforts in many countries. Considering these environmental problems and possible world actions, these are challenging times for humanity as we ponder our relationship to the home planet. Over the long term we no longer can sustain human activity through old patterns.

Asking whether the Earth Summit "succeeded" or "failed" is the wrong question. The occurrence of this largest-ever official gathering of Earthlings and the five years of preparatory effort and study that set the agenda are in themselves significant accomplishments. The challenge is one of education; the lesson is one of compromise and some sacrifice. Members of society should work to move the solutions for environmental and developmental problems off the bench and into play. The Earth Summit process continues as a good beginning.

A critical corollary to international efforts is the linkage of academic disciplines. A positive step in that direction is an *Earth systems science* approach, as illustrated by this text. Exciting progress toward an integrated understanding of the operation of Earth's physical systems is happening right now, driven by insights drawn from our remote-sensing capabilities. Never before has society been able to monitor Earth's physical geography so thoroughly.

The Environmental Cost of Conflict

Modern warfare has had a definite impact on the environment: scars from World Wars I and II still are visible on the landscapes of Europe and Asia, and more than 33 million bomb craters and massive defoliation provide environmental evidence of the more recent struggle in Vietnam. In fact, during the Vietnam War more than 2.3 times the equivalent tonnage of all munitions used in World War II were detonated. In terms of more modern weapons, just one of the planned MX missiles carries 3 megatons of explosive power, the equivalent of all the firepower of all the combatants in World War II! We must consider the environmental consequences of such weapons to air, water, land, and biotic systems, for modern technological warfare is not an environmentally sound activity.

In 1975, the National Academy of Sciences published its study on the potential impact of modern nuclear war on stratospheric ozone. This study states that a nuclear war could lead to the reduction of 40–70% of the stratospheric ozone, a catastrophic change for food chains, plants, animals, and humans.

In addition, from 1982 to the present, scientific publications have described the nuclear winter hypothesis. This hypothesis indicates that a "nuclear winter" might follow the detonation of only modest numbers of nuclear weapons. As you have learned, the atmosphere is very dynamic. The soot and ash from the many urban fires following a nuclear exchange would enter the atmosphere and raise Earth's albedo, absorb insolation in the stratosphere and upper troposphere, reradiate heat to space, and thus cool Earth's surface to below freezing even during summer months (as we saw to a much lesser degree following the eruption of Mount Pinatubo in 1991).

FIGURE 17-5

Time magazine cover for January 2, 1989, naming Earth "Planet of the Year." [Copyright © 1988 The Time Inc. Magazine Company. Reprinted by permission.]

As the nuclear winter hypothesis developed, others suggested a milder although still significant "nuclear autumn" as a consequence.

Refinement of the nuclear winter hypothesis continues, but the warnings are clear. One of the original nuclear winter study teams recently affirmed that "severe environmental anomalies—possibly leading to more human casualties globally than the direct effects of nuclear war—would be not just a remote possibility, but a likely outcome."*

These potential consequences demand that the methods of conflict resolution chosen by nations must acknowledge the limiting factors imposed by Earth's ecosystems. Modern technological conflicts no longer can be thought of only in a regional con-

text, for Earth's systems effectively spread the consequences around the globe. The real need for international cooperation should consider our *symbiotic relationships* with each other and with Earth's resilient, yet fragile, life-support systems.

Who Speaks For Earth?

Public awareness and education is a growing positive force on Earth. Geography awareness is improving each year. The National Geographic Society is conducting an annual National Geography Bee to promote geography education and global awareness to millions of 6th to 8th graders and their parents and teachers. In 1993 the first international Geography Olympiad was held in London. There are presently 60 geographic alliances in 47 states coordinating ge-

---

*R. P. Turco, O. B. Toon, T. P. Ackerman, J. B. Pollack, and C. Sagan, "Climate and Smoke: An Appraisal of Nuclear Winter," *Science* 247, no. 4939 January 22, 1990): 174.

ographic education among K–12, community college, college, and university teachers and students.

The global concern about environmental impacts prompted *Time* magazine in 1989 to deviate from its 60-year tradition of naming a prominent citizen as its person of the year, instead naming Earth the "Planet of the Year." The magazine featured 33 pages of Earth's physical and human geography, covering the endangered status of many ecosystems and cultures (Figure 17-5). We seem on the brink of an expanded awareness.

Carl Sagan asked, "Who speaks for Earth?" He answered with this perspective:

> We have begun to contemplate our origins: starstuff pondering the stars; organized assemblages of ten billion billion billion atoms considering the evolution

of atoms; tracing the long journey by which, here at least, consciousness arose. Our loyalties are to the species and the planet. We speak for Earth. Our obligation to survive is owed not just to ourselves but also to that Cosmos, ancient and vast, from which we spring.*

Each of us might consider where we will be in life's journey on Sunday night, at 11:59 P.M., December 31, 2000, with the new millennium just one minute away? May we all perceive our spatial importance within Earth's ecosystems and do our part to maintain a life-supporting and sustaining Earth for ourselves and countless generations in the future.

---

*Carl Sagan, *Cosmos*. New York: Random House, 1980, p. 345.

REVIEW QUESTIONS

1. What part do you think technology, politics, and future thinking should play in an academic physical geography course?
2. According to the discussion in the chapter, what global-scale factors led to the accident of the *Exxon Valdez*? Describe the complexity of that event.
3. What is meant by the Gaia hypothesis? Describe any concepts from this text that pertain to this hypothesis.
4. Relative to possible oil and gas exploration in the Arctic National Wildlife Refuge (ANWR), assume a geographical (spatial) perspective and analyze the pros and cons of such development. In your analysis, examine both supply-side and demand-side issues, as well as environmental and strategic factors.
5. This chapter states that we already know many of the solutions to the problems we face. Why do you think we are not implementing those solutions?
6. Who speaks for Earth?

APPENDIX A

INFORMATION SOURCES, ORGANIZATIONS, AND AGENCIES

This is a sampling of information sources for further geographic inquiry. Consult your campus library and instructor for related sources and research pathways.

General Sources

Academic Press Dictionary of Science and Technology edited by Christopher Morris. Published by Academic Press, San Diego, CA, 1992.

The Climates of Canada. Compiled by Canada's chief climatologist David Phillips. Available from Minister of Supply and Services Canada, Ottawa, Canada K1A 0S9, 1990.

The Complete Guide to Environmental Careers. The CEIP Fund. Island Press, Covela, CA 95428.

Congressional Directory. Published after every biennial congressional election. Available from U.S. Government Printing Office, Washington, DC 20402.

1994 Conservation Directory, 39th edition. National Wildlife Federation, Washington, DC, 1994.

Encyclopedia of Environmental Sources edited by Balachandran, Sarojini. Gale Research Inc., Washington, DC, 1993.

The Encyclopedia of Environmental Studies by William Ashworth. Published by Facts of File, New York, NY, 1991.

Environment. Heldref Publications, 4000 Albemarle St. NW, Washington, DC 20016.

1994 Environmental Almanac (Information Please). Compiled by the World Resources Institute and published by Houghton Mifflin Company, New York, NY, 1994.

Geographical Bibliography for American Libraries. Chauncy D. Harris, ed. A joint project of the Association of American Geographers and the National Geographic Society, Washington, 1985.

The Island Press Bibliography of Environmental Literature. Compiled by Joseph Miller, Sarah Friedman, and others at the Yale School of Forestry and Environmental Studies. Published by Island Press, Washington, DC, 1993.

National Reports Summaries–Nations of the Earth Report, in three volumes. United Nations Conference on Environment and Development, Geneva, Switzerland, 1992.

Source Book on the Environment—A Guide to the Literature. Kenneth A. Hammond, George Macinko, and Wilma B. Fairchild, University of Chicago Press, 1978.

State of the Environment—A View Toward the Nineties. A report from the Conservation Foundation, 1987, sponsored by the Charles Stewart Mott Foundation. Conservation Foundation, 1250 42nd Street NW, Washington, DC 20076.

State of the World 1995—Report on Progress Toward A Sustainable Society. Published annually by the Worldwatch Institute (Lester R. Brown, project director), 1776 Massachusetts Avenue NW, Washington, DC 20036.

Statistical Abstract of the United States. Published annually by the U.S. Bureau of the Census, Department of Commerce, Washington, DC. Also available from U.S. Government Printing Office, Washington, DC 20402.

U.S. Government Manual. Published by the Office of the Federal Register with annual updates. Available through the Superintendent of Documents, U.S. Government Printing Office, Washington, DC 20402.

U.S. Government Printing Office. North Capitol and H Streets NW, Washington, DC 20401. Superintendent of Documents provides monthly catalog of all U.S. government publications. Available at USGPO bookstores and offices.

The World Almanac. Published annually by World Almanac and Book of Facts, a Scripps Howard company, New York.

World Resources 1992–93. Published annually by the World Resources Institute, in collaboration with the United Nations Environment Programme, 1735 New York Avenue NW, Washington, DC. 20006.

Geography Organizations

American Geographic Society, 156 Fifth Avenue, Suite 600, New York, NY 10010–7002. Publishes *Focus* magazine quarterly.

Association of American Geographers, 1710 16th Street NW, Washington, DC 20009. Principal academic and professional geography organization—offers student memberships and career information. Publishes *Annals* and *The Professional Geographer*.

Canadian Association of Geographers, McGill University, 805 Sherbrooke Street West, Montréal, Québec H3A 2K6. Publishes *The Canadian Geographer* and *The Operational Geographer–The Canadian Journal for Practising Geographers*.

Geographic alliances, 60 state chapters. Information is available from college and university geography departments; generally coordinated by the National Geographic Society.

Institute of British Geographers, 1 Kingsington Gore, London, SW7 2AR, England. Publishes *Transactions of the Institute of British Geographers* and *Area*.

National Council for Geographic Education, Indiana University of Pennsylvania, Indiana, PA 15705. Publishes the *Journal of Geography*.

National Geographic Society, 17th and M Streets NW, Washington, DC 20036. Publishes maps, atlases, *National Geographic* magazine, and *Geography Education Update* through their Geographic Education Program.

Royal Canadian Geographical Society, 39 McArthur Ave., Vanier, Ontario, Canada K1L 8L7. Publishes *Canadian Geographic*.

Royal Geographical Society, 1 Kingsington Gore, London SW7 2AR, England.

Organizations of Interest

American Association for the Advancement of Science, 1333 H Street NW, Washington, DC 20005. Publishes *Science*.

American Chemical Society, 1155 16th Street NW, Washington, DC 20036.

American Geophysical Union, 2000 Florida Avenue NW, Washington, DC 20009.

American Meteorological Society, 45 Beacon Street, Boston, MA 02108. Publishes: *Journal of the Atmospheric Sciences, Journal of Climate, Monthly Weather Review, Weather and Forecasting,* and the *Bulletin of the AMS*.

Canadian Coalition on Acid Rain, 112 St. Clair Avenue West, Toronto, Ontario, Canada M4K 2Y3.

Center for Science in the Public Interest, 1501 16th Street NW, Washington, DC 20036

Climate Institute, 316 Pennsylvania Avenue, SE, Suite 403, Washington, DC 20003. Publishes *Climate Alert* and is an information clearing house.

Conservation Foundation, 1250 24th Street NW, Washington, DC 20037.

Conservation International (tropical ecosystems), 1015 18th Street NW, Suite 1000, Washington, DC 20036.

Cousteau Society, 930 West 21st Street, Norfolk, VA 23517.

Environmental Action Foundation, 1525 New Hampshire Avenue NW, Washington DC 20036.

Environmental Defense Fund, 257 Park Avenue South, New York, NY 10010.

Environmental Protection Agency, 401 M Street SW, Washington, DC 20460

Environment Canada, Ottawa, Ontario K1A OH3.

Friends of the Earth, Inc., 530 Seventh Avenue SE, Washington, DC 20003.

Greenpeace, U.S.A., Inc., 1436 U Street NW, Washington, DC 20009.

League of Conservation Voters, 1150 Connecticut Avenue NW, Suite 201, Washington, DC 20036.

National Parks and Conservation Association, 1015 31st Street NW, Washington, DC 20007. (Natural history associations are organized in conjunction with most national parks.)

National Recycling Coalition, 1101 30th Street NW, Washington, DC 20007.

National Weather Association, 4400 Stamp Road, Suite 404, Temple Hills, MD 20748.

National Wildlife Federation, 1400 16th Street NW, Washington, DC 20036. Publishes an annual updated list of all agencies and organizations dealing with natural resources—the *Conservation Directory*.

Natural Resources Defense Council, 122 East 42nd Street, New York, NY 10168. Publishes *The Amicus Journal*.

The Nature Conservancy, 1800 North Lynn Street, Arlington, VA 22209.

Rainforest Action Network, 301 Broadway, Suite A, San Francisco, CA 94133.

Rocky Mountain Institute, 1739 Snowmass Creek Road, Snowmass, CO 81654. Research in energy-related issues and conservation.

Sierra Club, 730 Polk Street, San Francisco, CA 94109.

Soil Conservation Society of America, 7515 Northeast Ankeny Road, Ankeny, IA 50021.

Solar Energy Research Institute, 6536 Cole Boulevard, Golden, CO 80401.

Union of Concerned Scientists, 26 Church Street, Cambridge, MA 02238.

United Nations Environment Programme, Office of Public Information, New York Liaison Office, Room DC 2–0803, United Nations, New York, NY 10017; or, UNEP, P.O. Box 30552, Nairobi, Kenya.

U.S. Geological Survey, 12201 Sunrise Valley Drive, Reston, VA 22092. Many regional offices, bookstores, and mapping centers.

The Wilderness Society, 1400 I Street NW, Washington, DC 20005.

Worldwatch Institute, 1776 Massachusetts Avenue NW, Washington, DC 20036.

World Wildlife Fund, 1250 24th Street NW, Washington, DC 20037.

APPENDIX B

TOPOGRAPHIC MAP SYMBOLS—USGS

Primary highway, hard surface	
Secondary highway, hard surface	
Light-duty road, hard or improved surface	
Unimproved road	
Road under construction, alinement known	
Proposed road	
Dual highway, dividing strip 25 feet or less	
Dual highway, dividing strip exceeding 25 feet	
Trail	

Railroad: single track and multiple track	
Railroads in juxtaposition	
Narrow gage: single track and multiple track	
Railroad in street and carline	
Bridge: road and railroad	
Drawbridge: road and railroad	
Footbridge	
Tunnel: road and railroad	
Overpass and underpass	
Small masonry or concrete dam	
Dam with lock	
Dam with road	
Canal with lock	

Buildings (dwelling, place of employment, etc.)	
School, church, and cemetery	Cem
Buildings (barn, warehouse, etc.)	
Power transmission line with located metal tower	
Telephone line, pipeline, etc. (labeled as to type)	
Wells other than water (labeled as to type)	Oil · Gas
Tanks: oil, water, etc. (labeled only if water)	Water
Located or landmark object; windmill	
Open pit, mine, or quarry; prospect	X
Shaft and tunnel entrance	Y

Horizontal and vertical control station:

Tablet, spirit level elevation	BM △ 5653
Other recoverable mark, spirit level elevation	△ 5455
Horizontal control station: tablet, vertical angle elevation	VABM △ 95I9
Any recoverable mark, vertical angle or checked elevation	△3775
Vertical control station: tablet, spirit level elevation	BM X957
Other recoverable mark, spirit level elevation	X 954
Spot elevation	X 7369 X 7369
Water elevation	670 670

Boundaries: National	
State	
County, parish, municipio	
Civil township, precinct, town, barrio	
Incorporated city, village, town, hamlet	
Reservation, National or State	
Small park, cemetery, airport, etc.	
Land grant	
Township or range line, United States land survey	
Township or range line, approximate location	
Section line, United States land survey	
Section line, approximate location	
Township line, not United States land survey	
Section line, not United States land survey	
Found corner: section and closing	
Boundary monument: land grant and other	
Fence or field line	

Index contour		Intermediate contour	
Supplementary contour		Depression contours	
Fill		Cut	
Levee		Levee with road	
Mine dump		Wash	
Tailings		Tailings pond	
Shifting sand or dunes		Intricate surface	
Sand area		Gravel beach	

Perennial streams		Intermittent streams	
Elevated aqueduct		Aqueduct tunnel	
Water well and spring		Glacier	
Small rapids		Small falls	
Large rapids		Large falls	
Intermittent lake		Dry lake bed	
Foreshore flat		Rock or coral reef	
Sounding, depth curve		Piling or dolphin	
Exposed wreck		Sunken wreck	
Rock, bare or awash; dangerous to navigation			

Marsh (swamp)		Submerged marsh	
Wooded marsh		Mangrove	
Woods or brushwood		Orchard	
Vineyard		Scrub	
Land subject to controlled inundation		Urban area	

SUGGESTED READINGS

Chapter 1 – Foundations of Geography

Avery, Thomas E. and Graydon Berlin. *Fundamentals of Remote Sensing and Airphoto Interpretation*, 5th ed. New York: Macmillan Publishing Company, 1992.

Dunn, Margery G., and Donald J. Crump. *Exploring Your World—The Adventure of Geography*. Washington, DC: National Geographic Society, 1989.

Gaile, Gary L., and Cort J. Willmott, eds. *Geography in America*. Columbus, OH: Macmillan Publishing Company, 1989.

Howse, Derek. *Greenwich Time and the Discovery of the Longitude*. London: Oxford University Press, 1980.

Miller, Victor C., and Mary E. Westerback. *Interpretation of Topographic Maps*. Columbus, OH: Macmillan Publishing Company, 1989.

Odum, Howard T. *Systems Ecology—An Introduction*. New York: John Wiley & Sons, 1983.

Rabenhorst, Thomas D., and Paul D. McDermott. *Applied Cartography—Introduction to Remote Sensing*. Columbus, OH: Macmillan Publishing Company, 1989.

Snyder, John P. *Map Projections—A Working Manual*. U.S. Geological Survey Professional Paper 1395. Washington, DC: U.S. Government Printing Office, 1987.

Star, Jeffrey and John Estes. *Geographic Information Systems An Introduction*. Englewood Cliffs, NJ: Prentice Hall, 1990.

Thompson, Morris M. *Maps for America*. 3d ed. U.S. Geological Survey. Washington, DC: U.S. Government Printing Office, 1987.

PART ONE : The Energy-Atmosphere System

Chapter 2 – Solar Energy, Seasons, and the Atmosphere

Albritton, D. L., R. Monastersky, J. A. Eddy, and others. "Our Ozone Shield," *Report to the Nation, Our Changing Planet*. Boulder, Colorado: University for Atmospheric Research, Fall 1992.

American Association for the Advancement od Science (AAAS). *Science*, 261 (August 27, 1993): 1128-58; features nine studies concerning stratospheric ozone.

Brodeur, Paul. "Annals of Chemistry—In the Face of Doubt," *The New Yorker* (9 June 1986): 70–87.

Gallant, Roy A. *National Geographic Picture Atlas of Our Universe*. Washington, DC: National Geographic Society, 1986.

Hall, Jane V., A. M. Winer, M. T. Kleinman, et al. "Valuing the Health Benefits of Clean Air," *Science* 255 (14 February 1992): 812–17.

Meinel, Aden, and Marjorie Meinel. *Sunsets, Twilights, and Evening Skies*. Cambridge: Cambridge University Press, 1983.

Miller, John M. "Acid Rain," *Weatherwise* 37, no. 5 (October 1984): 232–49.

Office of Technology Assessment. *Catching Our Breath: Next Steps for Reducing Urban Ozone*. OTA-0-412. Washington, DC: U.S. Government Printing Office, July 1989.

Stolarski, Richard, R. Bojkov, et al. "Measured Trends in Stratospheric Ozone," *Science* 256 (17 April 1992): 342–49.

United States Naval Observatory. *The Air Almanac.* Washington, DC: U.S. Government Printing Office, published annually.

Chapter 3 – Atmospheric Energy and Global Temperatures

Ardanuy, P. E., H. Lee Kyle, and D. Hoyt. "Global Relationships Among the Earth's Radiation Budget, Cloudiness, Volcanic Aerosols, and Surface Temperature," *Journal of Climate* 5, no. 10 (October 1992): 1120–39.

Budyko, M. I. *Climate and Life.* Edited by David H. Miller. International Geophysics Series, vol. 18. New York: Academic Press, 1974.

Budyko, M. I. *The Heat Balance of the Earth's Surface.* Translated by Nina A. Stepanova. Washington, DC: U.S. Department of Commerce, U.S. Weather Bureau, 1958.

Harrison, E. F., P. Minnis, B. R. Barkstrom, et. al. "Seasonal Variation of Cloud Radiative Forcing Derived from the Earth Radiation Budget Experiment," *Journal of Geophysical Research* 95, no. D11 (20 October 1990): 18,687–703.

Hubbard, H. M. "The Real Cost of Energy," *Scientific American* 264, no. 4 (April 1991): 36–42.

Landsberg, Helmut E. *The Urban Climate.* International Geophysics Series, vol. 28. New York: Academic Press, 1981.

Ramanathan, V., B. R. Barkstrom, and E. F. Harrison. "Climate and the Earth's Radiation Budget," *Physics Today* 42, no. 5 (May 1989): 22–32.

Sellers, William D. *Physical Climatology.* Chicago: University of Chicago Press, 1965.

Siddayao, Corazón M., ed. *Energy Investments and the Environment, Selected Topics,* the EDI Technical Materials series. Washington, DC: The World Bank, 1993.

Willmott, Cort J., John R. Mather, and Clinton M. Rowe. Average Monthly and Annual Surface Air Temperature and Precipitation Data for the World. Part 1, "The Eastern Hemisphere," and Part 2, "The Western Hemisphere." Elmer, NJ: C. W. Thornthwaite Associates and the University of Delaware, 1981.

Chapter 4 – Atmospheric and Oceanic Circulation

Brasseur, Guy, and Claire Granier. "Mount Pinatubo Aerosols, Chlorofluorocarbons, and Ozone Depletion." *Science* 257 (August 28, 1992): 1239–42.

Institute of Oceanographic Sciences, Deacon Laboratory. *The FRAM (Fine Resolution Antarctic Model) Atlas of the Southern Ocean.* Wormley, Surrey, England: Natural Environment Research Council, 1991.

McDonald, James E. "The Coriolis Effect." *Scientific American* 186 (May 1952): 72–76.

Neiburger, Morris, James G. Edinger, and William D. Bonner. *Understanding Our Atmospheric Environment.* San Francisco: W. H. Freeman, 1973.

Starr, Victor P. "The General Circulation of the Atmosphere." *Scientific American* 195 (December 1956): 40–45.

Stowe, L.L., R. M. Carey, and P. P. Pellegrino. "Monitoring the Mt. Pinatubo Aerosol Layer with NOAA -11 AVHRR Data," *Geophysical Research Letters* 19, no. 2 (January 24, 1992): 159–62.

Thurman, Harold V. *Introductory Oceanography.* 7th ed. New York: Macmillan Publishing Co., 1994.

Vesilind, Priit J. "Monsoon–Life Breath of Half the World," *National Geographic* 166, no. 6 (December 1984): 712–47.

Whipple, A. B. C., and the editors of Time-life Books. *Restless Oceans.* Alexandria, VA: Time-Life Books, 1983.

Wiebe, Peter H. "Rings of the Gulf Stream." *Scientific American* 246 (March 1982): 60–70.

PART TWO : Water, Weather, and Climate

Chapter 5 – Atmospheric Water and Weather

Allen, Oliver E., and the editors of Time-Life Books. *Atmosphere.* Planet Earth Series. Alexandria, VA: Time-Life Books, 1983.

Alper, Joe. "Everglades Rebound from Andrew," *Science* 257 (25 September 1992): 1852–54.

Hidore, John J., and John E. Oliver. *Climatology—An Atmospheric Science.* New York: Macmillan Publishing Company, 1993.

Moran, Joseph M., and Michael D. Morgan. *Meteorology—The Atmosphere and the Science of Weather,* 4th ed. New York: Macmillan Publishing Company, 1994.

Newcott, W. R. "Lightning, Nature's High-Voltage Spectacle," *National Geographic* 184, no. 1 (July 1993): 83–103.

Oliver, John E., and Rhodes W. Fairbridge, eds. *The Encyclopedia of Climatology.* New York: Van Nostrand Reinhold Co., 1987.

Schaeffer, Vincent J., and John A. Day. *A Field Guide to the Atmosphere.* Peterson Field Guide Series. Boston: Houghton Mifflin, 1981.

Whipple, A. B. C., and the editors of Time-Life Books. *Storm.* Planet Earth Series. Alexandria, VA: Time-Life Books, 1982.

Williams, Earle R. "The Electrification of Thunderstorms," *Scientific American* 259, no. 5 (November 1988): 88–99.

Williams, Jack, and Jack Faidley. "Hurricane Andrew in Florida," *Weatherwise* 45, no. 6 (December 1992/January 1993): 7–17.

Chapter 6 – Water Resources

Chahine, Moustafa T. "The Hydrologic Cycle and Its Influence on Climate," *Nature* 359 (October 1, 1992): 373–80.

Chelimsky, Eleanor. "Fighting Groundwater Contamination." Testimony given before Subcommittee on Water Resources, Transportation, and Infrastructure, Committee on Environment and Public Works, U.S. Senate, May 1988. Washington, DC: General Accounting Office.

Kliot, Nurit. *Water Resources and Conflict in the Middle East.* London: Routledge, 1994.

Postel, Sandra. *Last Oasis: Facing Water Scarcity.* New York: W. W. Norton and Company, 1992.

Pringle, Laurence, and the editors of Time-Life Books. *Rivers and Lakes.* Planet Earth Series. Alexandria, VA: Time-Life Books, 1985.

Rind, D., C. Rosenweig, and R. Goldberg. "Modelling the Hydrologic Cycle in Assessments of Climate Change," *Nature* 253 (July 9, 1992): 119–22.

Smith, Richard A., Richard B. Alexander, and M. Gordon Wolman. "Water Quality Trends in the Nation's Rivers," *Science* 235 (March 27, 1987): 1607–15.

Thornthwaite, Charles W., and John R. Mather. *The Water Balance.* Publications in Climatology, vol. 1. Centerton, NJ: Drexel Institute of Technology, Laboratory of Climatology, 1955.

U.S. Geological Survey. *National Water Summary 1984.* Water-Supply Paper No. 2275. (*N.W.S. 1985*, No. 2300; *N.W.S. 1986*, No. 2325; *N.W.S. 1987*, No. 2350; *N.W.S. 1988-1990*, No. 2375; *N.W.S. 1990-1991*, No. 2400). Washington, DC: Government Printing Office, 1985 (and each of the years listed).

van der Leeden, Frits, Fred L. Troise, and D. K. Todd. *The Water Encyclopedia*, 2nd ed. Chelsea, MI: Lewis Publishers, Inc., 1990.

Chapter 7 – Earth's Climates

Houghton, J. T., G. J. Jenkins, and J. J. Ephraums, eds. *Climate Change 1992—The Supplementary Report to the IPCC Scientific Assessment.* World Meteorological Organization/United Nations Environment Programme. New York: Cambridge University Press, 1992.

Houghton, J. T., G. J. Jenkins, and J. J. Ephraums, eds. *Climate Change—The IPCC Scientific Assessment.* Working Group I, World Meteorological Organization/United Nations Environment Programme. Port Chester, NY: Cambridge University Press, 1991.

Houghton, Richard A., and George M. Woodwell. "Global Climate Change," *Scientific American* 260, no. 4 (April 1989): 36–44.

Intergovernmental Panel on Climate Change. *Climate Change—The IPCC Response Strategies.* Working Group III. World Meteorological Organization/United Nations Environment Programme. Covelo, CA: Island Press, 1991.

Mather, John R. *Climatology—Fundamentals and Applications.* New York: McGraw-Hill, 1974.

Nordhaus, W. D. "An Optimal Transition Path for Controlling Greenhouse Gases," *Science* 258, no. 5086 (20 November 1992): 1315–19.

Oliver, John E., and Rhodes W. Fairbridge, eds. *The Encyclopedia of Climatology.* New York: Van Nostrand Reinhold Co., 1987.

Tegart, W. J. McG., G. W. Sheldon, and D. C. Griffiths. *Climate Change—The IPCC Impacts Assessment.* Working Group II. World Meteorological Organization/United Nations Environment Programme. Portland, OR: International Specialized Book Service/Australian Government, 1991.

Turco, R. P., et al. "Nuclear Winter: Global Consequences of Multiple Nuclear Explosions," *Science* 222, no. 4630 (23 December 1983): 1283–92.

Wilcox, Arthur A. "Köppen After Fifty Years," *Annals of the Association of American Geographers* 58, no. 1 (March 1968): 12–28.

PART THREE : Earth's Changing Landscapes

Chapter 8 – The Dynamic Planet

Anderson, Don L., Toshiro Tanimoto, and Yu-shen Zhang. "Plate Tectonics and Hotspots: The Third Dimension," *Science* 256 (June 19, 1992): 1645–51.

Bloxham, Jeremy, and David Gubbins. "The Evolution of the Earth's Magnetic Field," *Scientific American* 261, no. 6 (December 1989): 68–75.

Continents Adrift and Continents Aground—Readings from Scientific American. San Francisco: W. H. Freeman, 1976. "The Dynamic Earth"; "The Earth's Core"; "The Earth's Mantle"; "The Continental Crust." *Scientific American* 249, no. 3 (September 1983): entire issue.

Gore, Rick. "Our Restless Planet Earth," *National Geographic* 168 (August 1985): 142–81.

Hill, R. I., I. H. Campbell, G. F. Davies, and R. W. Griffiths. "Mantle Plumes and Continental Tectonics," *Science* 256 (April 10, 1992): 186–95.

Hurley, Patrick M. "The Confirmation of Continental Drift," *Scientific American* 218, no. 4 (April 1968): 52–64.

Jeanloz, Raymond and Thorne Lay. "The Core-Mantle Boundary," *Scientific American* 268, no. 5 (May 1993): 48-55.

Miller, Russel, and the editors of Time-Life Books. *Continents in Collision.* Planet Earth Series. Alexandria, VA: Time-Life Books, 1983.

Powell, Corey S. "Trends in Geophysics—Peering Inward," *Scientific American* 254, no. 6 (June 1991): 100–11.

Vink, Gregory E., W. Jason Morgan, and Peter R. Vogt. "The Earth's Hot Spots," *Scientific American* 252, no. 4 (April 1985): 50–57.

Chapter 9 – Earthquakes and Volcanoes

Decker, Robert W., Thomas L. Wright, and Peter H. Stauffer. *Volcanism in Hawaii.* 2 vols. U.S. Geological Survey Professional Paper 1350. Washington, DC: Government Printing Office, 1987.

Donnellan, Andrea, B. H. Hager, and R. W. King, "Discrepancy Between Geological and Geodetic Deformation Rates in the Ventura Basin," *Nature* 366, November 25, 1993: 299-301+.

Dvorak, J. J., Carl Johnson, and R. I. Tilling. "Dynamics of Kilauea Volcano," *Scientific American* 267, no. 2 (August 1992): 46–53.

Grove, Noel, and R. H. Ressmeyer. "Volcanoes—Crucibles of Creation," *National Geographic* 182, no. 6 (December 1992): 3–41.

Murphy, J. B., and R. D. Nance. "Mountain Belts and the Supercontinent Cycle," *Scientific American* 266, no. 4 (April 1992): 84–91.

Nur, Amos, Ron, H., and G. C. Beroza. "The Nature of the Landers-Mojave Earthquake Line," *Science* 261 (July 9, 1993): 201-03.

Rattanani, Lekha. "Earthquakes: High Risk Zones, Minimizing Damage, Coping with such Disasters," *India Today* (October 1993): 25-38.

Tilling, Robert I. and P. W. Lipman. "Lessons in Reducing Volcano Risk," *Nature* 364 (July 22, 1993): 277-80.

U.S. Geological Survey Staff. "The Loma Prieta, California, Earthquake: An Anticipated Event," *Science* 247 (January 19, 1990): 286–93.

Wallace, Robert E., ed. *The San Andreas Fault System, California.* U.S. Geological Survey Professional Paper 1515. Washington, DC: Government Printing Office, 1990.

Chapter 10 – Weathering, Karst Landscapes, and Mass Movement

Alden, William C. "Landslide and Flood at Gros Ventre, Wyoming," *Transactions of the American Institute of Mining Engineers* 76 (February 1928): 347–61.

Easterbrook, D. J. *Surface Processes and Landforms.* New York: Macmillan Publishing Co., 1993.

Fairbridge, Rhodes W., ed. *The Encyclopedia of Geomorphology.* Encyclopedia of Earth Sciences Series, vol. 3. New York: Reinhold Book Company, 1968.

Jackson, Donald Dale, and the editors of Time-Life Books. *Underground Worlds.* Planet Earth Series. Alexandria, VA: Time-Life Books, 1982.

McDowell, Bart. "Eruption in Colombia," *National Geographic* 165, no. 5 (May 1986): 640–53.

Peterson, Cass. "Killing Mountains for Coal—Federal Strip Mining Laws Have Not Healed Kentucky's Scars," *Washington Post National Weekly Edition*, June 29, 1987: 6–7.

Plafker, George and G. E. Ericksen. "Nevados Huascarán Avalanches, Peru." In Voight, B., ed., *Rockslides and Avalanches*, vol. 1 of Natural Phenomena: Developments in Geotechnical Engineering 14A, pp. 277–315. Amsterdam: Elsevier Scientific Publishing Co., 1978. (See maps and photo pp. 278–83.)

U.S. Geological Survey, National Park Service, Coast and Geodetic Survey, and Forest Service. *The Hebgen Lake, Montana, Earthquake of August 17, 1959.* U.S. Geological Survey Professional Paper 435. Washington, DC: Government Printing Office, 1964.

Chapter 11 – Rivers and Landforms

Chorley, Richard J., Stanley A. Schumm, and David E. Sugden. *Geomorphology.* New York: Methuen, 1985.

Clark, Champ, and the editors of Time-Life Books. *Flood.* Planet Earth Series. Alexandria, VA: Time-Life Books, 1982.

Doppelt, B., M. Scurlock, C. Frissell, J. Karr. *Entering the Watershed–A New Approach to Save America's River Ecosystems.* Washington, DC: Island Press, 1993.

Leopold, Luna B. *A View of the River.* Cambridge: Harvard University Press, 1994.

Mairson, Alan. "The Great Flood of '93," *National Geographic*, 185, no. 1 (January 1994): 42-81.

McCabe, G. J., Jr., and D. M. Wolock. "Effects of Climate Change and Climatic Variability on the Thornthwaite Moisture Index in the Delaware River Basin," *Climatic Change* 20 (1992): 143–153.

Myers, M. F., and Gilbert F. White. "The Challenge of the Mississippi Flood," *Environment* 35, no. 10 (December 1993): 2-9.

Pearse, P. H., F. Bertrand, and J. W. MacLaren. *Currents of Change—Final Report, Inquiry on Federal Water Policy.* Ottawa: Environment Canada, 1985.

Pringle, Laurence, and the editors of Time-Life Books. *Rivers and Lakes.* Planet Earth series. Alexandria, VA: Time-Life Books, 1985.

Rudloe, Jack, and Anne Rudloe. "Trouble in Bayou Country: Louisiana's Atchafalaya." *National Geographic*, September 1979, 377–97.

Chapter 12 – Wind Processes and Desert Landscapes

Abbey, Edward, and the editors of Time-Life Books. *Cactus Country.* American Wilderness Series. New York: Time-Life Books, 1973.

Cooke, R. U., and Andrew Warren. *Geomorphology in Deserts.* Los Angeles: University of California Press, 1973.

Hamblin, W. Kenneth. *The Earth's Dynamic Systems,* 7th ed. New York: Macmillan Publishing Company, 1994.

Hamilton, William J., III. "The Living Sands of the Namib," *National Geographic* 164, no. 3 (September 1983): 364–77.

Hundley, Norris. *Water and the West: The Colorado River Compact and the Politics of Water in the American West.* Berkeley: University of California Press, 1975.

Krutch, Joseph Wood. *The Voice of the Desert.* Wm. Sloane Associates, 1954.

McKee, Edwin D. *A Study of Global Sand Seas.* U.S. Geological Survey Professional Paper 1052. Washington, DC: Government Printing Office, 1979.

Page, Jake, and the editors of Time-Life Books. *Arid Lands.* Planet Earth Series. Alexandria, VA: Time-Life Books, 1984.

Péwé, T. L. *Desert Dust: Origin, Characteristics, and Effects on Man.* Geological Survey Special Paper 186. Washington, DC: Government Printing Office, 1981.

Stegner, Wallace. *Beyond the Hundredth Meridian.* New York: Houghton Mifflin, 1954.

Chapter 13 – Coastal Processes and Landforms

Blockson, Charles L. "Sea Change in the Sea Islands," *National Geographic* (December 1987): 735–63.

Brown, B. E., and J. C. Ogden. "Coral Bleaching," *Scientific American* 258, no. 1 (January 1993): 64–70.

Cameron, S. D. "The Living Beach," *Canadian Geographic* 113, no. 2 (March-April 1993): 66–79.

Dolan, Robert, Bruce Hayden, and Harry Lins. "Barrier Islands," *American Scientist* 68, no. 1 (January-February 1980): 16–25.

Kaufman, W., and O. H. Pilkey, Jr. *The Beaches are Moving: The Drowning of America's Shoreline.* Durham, NC: Duke University Press, 1983.

Platt, R. H., T. Bentley, and H. C. Miller. "The Failings of U.S. Coastal Erosion Policy," *Environment* 33, no. 9 (July 1992): 7–10.

Schwartz, Maurice L. *The Encyclopedia of Beaches and Coastal Environments.* Stroudsburg, PA: Hutchinson Ross, 1982.

Thurman, Harold V. *Introductory Oceanography,* 7th ed. New York: Macmillan Publishing Company, 1994.

Wanless, H. R. "The Inundation of Our Coastlines," *Sea Frontiers* 35, no. 5 (September-October 1989): 264–71.

Warrick, R. A., E. M. Barrow, and T. M. L. Wigley, eds. *Climate and Sea Level Change—Observations, Projections, and Implications.* Cambridge, England: Cambridge University Press, 1993.

Chapter 14 – Glacial and Periglacial Landscapes

"The Antarctic," *Scientific American* 207, no. 3 (September 1962): Entire issue.

Bailey, Ronald H., and the editors of Time-Life Books. *Glacier.* Planet Earth Series. Alexandria, VA: Time-Life Books, 1982.

Budyko, M. I. *The Earth's Climate: Past and Future.* International Geophysics Series, vol. 29. New York: Academic Press, 1982.

Chorlton, Windsor, and the editors of Time-Life Books. *Ice Ages.* Planet Earth Series. Alexandria, VA: Time-Life Books, 1983.

CLIMAP Project Members. "Seasonal Reconstructions of the Earth's Surface at the Last Glacial Maximum."

CLIMAP Project, compiled by Andrew McIntyre. Geological Society of America Map and Chart Series MC-36. Palisades, NY: 1981.

Kamb, Barclay, C. F. Raymond, W. D. Harrison, Herman Englehardt, K. A. Echelmeyer, N. Humphrey, M. M. Brugman, and T. Pfeffer. "Glacier Surge Mechanism: 1982–1983 Surge of Variegated Glacier, Alaska," *Science* 227, No. 4686 (February 1, 1985): 469–479.

Krimmel, Robert M. *Mass Balance, Meteorological, and Runoff Measurements at South Cascade Glacier, Washington, 1992 Balance Year*, Open-File Report 93-640, Tacoma, WA: U.S. Geological Survey, 1993.

Sharp, Robert P. *Living Ice—Understanding Glaciers and Glaciation*. Cambridge: Cambridge University Press, 1988.

Swithinbank, Charles. *Antarctica*. U.S. Geological Survey Professional Paper 1386B. Washington, DC: Government Printing Office, 1988.

Williams, Peter J., and Michael W. Smith. *The Frozen Earth—Fundamentals of Geocryology*. Cambridge: Cambridge University Press, 1989.

PART FOUR : Biogeography

Chapter 15 – The Geography of Soils

Brady, Nyle C. *The Nature and Properties of Soils*, 10th ed. New York: Macmillan Publishing Company, 1990.

Expert Committee on Soil Science, Agriculture Canada Research Branch. *The Canadian System of Soil Classification*, 2nd ed. Publication 1646. Ottawa K1A OS9: Supply and Services Canada, 1987. Replacing Agriculture Canada Publications 1455 (1974) and 1646 (1978).

Fairbridge, Rhodes W., and Charles W. Finkl, Jr. *The Encyclopedia of Soil Science*. Part 1, Physics, Chemistry, Biology, Fertility, and Technology. Encyclopedia of Earth Sciences, vol. 12. Stroudsburg, PA: Dowdon, Hutchinson, and Ross, 1979.

Finkl, Charles W., Jr., ed. *Soil Classification*, Benchmark Papers in Soil Science, vol. 1. Stroudsburg, PA: Hutchinson Ross, 1982.

Foth, Henry D. *Fundamentals of Soil Science*. 8th ed. New York: John Wiley & Sons, 1990.

Gersmehl, Philip J. "Soil Taxonomy and Mapping," *Annals of the Association of American Geographers* 67 (September 1977): 419–28.

Gibbons, Boyd. "Do We Treat Our Soil like Dirt?" *National Geographic*. September 1984: 350–89.

Singer, Michael J., and Donald N. Munns. *Soils—An Introduction*, 2nd ed. New York: Macmillan Publishing Company, 1987.

Soil Survey Staff and the Agronomy Department of Cornell University. *Keys to Soil Taxonomy*. Soil Management Support Services Technical Monograph 6. Ithaca, NY: Cornell University, 1987.

Soil Survey Staff. *Soil Taxonomy—A Basic System of Soil Classification for Making and Interpreting Soil Surveys*. Agricultural Handbook No. 436. Washington, DC: Government Printing Office, 1975.

Chapter 16 – Ecosystems and Biomes

Agee, J. K. *Fire Ecology of Pacific Northwest Forests*. Washington, DC: Island Press, 1993.

Aplet, G. H., Johnson, N., Olson, J. T., and V. A. Sample. *Defining Sustainable Forestry*. Washington, DC: Island Press, 1993.

Barbour, Michael G., and William D. Billings, eds. *North American Terrestrial Vegetation*. Cambridge: Cambridge University Press, 1988.

Berner, Robert A., and Antonio C. Lasaga. "Modeling the Geochemical Carbon Cycle," *Scientific American* 260 (March 1989): 74–84.

Botkin, D. B. *Discordant Harmonies—A New Ecology for the Twenty-first Century*. New York: Oxford University Press, 1990.

E. O. Wilson. *The Diversity of Life*. Cambridge, MA: The Belknap Press of Harvard University Press, 1992.

Millington, A. C., Critchley, R. W., Douglas, T. D., and P. Ryan. *Estimating Woody Biomass in Sub-Saharan Africa*. Washington, DC: The World Bank, 1994.

Mills, Susan M., and the Greater Yellowstone Coordinating Committee, eds. *The Greater Yellowstone Postfire Assessment*. Yellowstone National Park: National Park Service, 1989.

Peters, R. L. and T. E. Lovejoy, eds. *Global Warming and Biological Diversity*. New Haven: Yale University Press, 1992.

Smith, R. C., B. B. Prézelin, K. S. Baker, et al. "Ozone Depletion: Ultraviolet Radiation and Phytoplankton Biology in Antarctic Waters," *Science* 255 (February 12, 1992): 952–59.

Tucker, Compton J., John R. G. Townshend, and Thomas E. Goff. "African Land-Cover Classification Using Satellite Data," *Science* 227 (25 January 1985): 369–76.

World Resources Institute, The World Conservation Union, United Nations Environment Programme (UNEP), FAO, and UNESCO. *Global Biodiversity*

Strategy—Guidelines for Action to Save, Study, and Use Earth's Biotic Wealth Sustainably and Equitably. Washington, DC: World Resources Institute, 1992.

Chapter 17 – Earth, Humans, and the New Millennium

Association of American Geographers. "Statement of the Association of American Geographers on Nuclear War." Adopted by AAG, May 6, 1986, Minneapolis.

Barney, Gerald O. *The Global 2000 Report to the President.* New York: Penguin Books, 1982.

Brown, Lester R. *State of the World—1995.* Worldwatch Institute Series. New York: W. W. Norton, 1995. (Published every year.)

Canby, Thomas Y. "The Persian Gulf After the Storm," *National Geographic* 180, no. 2 (August 1991): 2–35.

Keeble, J. "A Parable of Oil and Water," and Sankovitch, N. "Lax Regulation of Oil Vessels and Processing Facilities Continues." *The Amicus Journal* 15, no. 1 (Spring 1993): 35–43.

Lovelock, James. *The Ages of Gaia—A Biography of Our Living Earth.* New York: W. W. Norton, 1988.

Parker, J. and C. Hope. "The State of the Environment— A Survey of Reports from Around the World," *Environment* 34, no. 1, (January/February 1992).

Repetto, Robert. "Accounting for Environmental Assets," *Scientific American* 266, no. 6 (June 1992): 94–100

Turco, R. P., et al. "Climate and Smoke: An Appraisal of Nuclear Winter," *Science* 247, no. 4939 (January 12, 1990): 166–74.

United Nations. *The Global Partnership for Environment and Development—A Guide to Agenda 21.* Geneva: United Nations Conference on Environment and Development, April 1992.

Westing, Arthur H., and E. W. Pfeiffer. "The Cratering of Indochina," *Scientific American* 226, no. 5 (May 1972): 21–29.

World Conservation Monitoring Centre and Brian Groombridge, ed. *Global Biodiversity: Status of the Earth's Living Resources.* Prepared in collaboration with The World Conservation Union (IUCN), United Nations Environment Programme (UNEP), World Wide Fund for Nature (WWF), and World Resources Institute. London: Chapman and Hall, 1992.

GLOSSARY

The chapter in which each term appears **boldfaced** is designated in parentheses, followed by a specific definition relevant to the key term's usage in the chapter.

Abiotic (1) Nonliving.

Ablation (14) Loss of glacial ice through melting, *sublimation*, wind removal by *deflation*, or the calving off of blocks of ice.

Abrasion (11, 12, 14) Mechanical wearing and *erosion* of bedrock accomplished by the rolling and grinding of particles and rocks carried in a stream, moved by wind in a "sandblasting" action, or imbedded in glacial ice.

Absorption (3) Assimilation and conversion of radiation from one form to another in a medium. In the process the temperature of the absorbing surface is raised, thereby affecting the rate and quality of radiation from that surface.

Actual evapotranspiration (6) Actual amount of *evaporation* and *transpiration* that occurs (ACTET); derived in the water-balance equation by subtracting the deficit from potential evapotranspiration.

Adiabatic (5) Pertaining to the heating and cooling of a descending or ascending parcel of air through compression and expansion, without any exchange of heat between the parcel and the surrounding environment.

Adsorption (15) The process whereby cations become attached to *soil colloids*, or the adhesion of gas molecules and ions to solid surfaces with which they come into contact.

Advection (3) Horizontal movement of air or water from one place to another.

Advection fog (5) Active *condensation* formed when warm, moist air moves laterally over cooler water or land surfaces, causing the lower layers of the overlying air to be chilled to the *dew-point temperature*.

Aggradation (11) The general building up of land surface because of deposition of material; opposite of *degradation*.

When the sediment load of a stream exceeds the stream's capacity, the stream channel is filled through this process.

Air mass (5) A distinctive, homogeneous body of air in terms of temperature and humidity, that takes on the moisture and temperature characteristics of its source region.

Air pressure (4) Pressure produced by the motion, size, and number of gas molecules and exerted on surfaces in contact with the air. Normal sea level pressure, as measured by the height of a column of mercury (Hg), is expressed as 1013.2 millibars, 760 mm of Hg, or 29.92 inches of Hg. Air pressure can be measured with mercury or aneroid barometers.

Albedo (3) The reflective quality of a surface, expressed as the relationship of incoming to reflected insolation and stated as a percentage; a function of surface color, angle of incidence, and surface texture.

Alfisols (15) Moderately weathered forest soils that are moist versions of *Mollisols*, with productivity dependent on specific patterns of moisture and temperature; rich in organics; most wide-ranging of the 11 soil orders in the *Soil Taxonomy* classification.

Alluvial fan (12) Fan-shaped *fluvial* landform at the mouth of a canyon, generally occurs in arid landscapes where streams are intermittent.

Alluvial terraces (11) Level areas that appear as topographic steps above the river, created by a stream as it scours with renewed downcutting into its floodplain; composed of unconsolidated alluvium.

Alluvium (11) General descriptive term for clay, silt, and sand, transported by running water and deposited in sorted or semisorted sediment on a floodplain, delta, or stream bed.

Alpine glacier (14) A glacier confined in a mountain valley or walled basin, consisting of three subtypes: *valley glacier* (within a valley), *piedmont glacier* (coalesced at the base of a mountain, spreading freely over nearby lowlands), and outlet glacier (flowing outward from a continental glacier).

Altitude (2) The angular distance between the horizon (a horizontal plane) and the Sun (or any point).

Andisols (15) A soil order in the Soil Taxonomy derived from volcanic parent materials in areas of volcanic activity. A new order created in 1990 of soils previously considered under *Inceptisols* and *Entisols*.

Anemometer (4) A device to measure wind velocity.

Aneroid barometer (4) A device to measure air pressure using a partially emptied, sealed cell (see air pressure).

Anticline (9) Upfolded strata in which layers slope away from the axis of the fold, or central ridge.

Anticyclone (4) A dynamically or thermally caused area of high atmospheric pressure with descending and diverging air flows.

Aphelion (2) The most distant point in Earth's elliptical orbit about the Sun; reached on July 4 at a distance of 152,083,000 km (94.5 million mi); variable over a 100,000-year cycle.

Apparent temperature (3) The temperature subjectively perceived by each individual, also known as *sensible temperature*.

Aquatic ecosystem (16) An association of plants and animals and their nonliving environment in a water setting.

Aquiclude (6) A body of rock that does not conduct water in usable amounts; an impermeable layer.

Aquifer (6) Rock strata permeable to groundwater flow.

Aquifer recharge area (6) The surface area where water enters an aquifer to recharge the water-bearing strata in a groundwater system.

Arctic tundra (16) A biome in the northernmost portions of North America, Europe, and Russia, featuring low ground-level herbaceous plants as well as some woody plants.

Arête (14) A sharp ridge that divides two cirque basins. Means "fish bone" in French. *Arêtes* form sawtooth and serrated ridges in glaciated mountains.

Aridisols (15) Largest single soil order in the *Soil Taxonomy* classification and typical of dry climates; low in organic matter and dominated by *calcification* and *salinization*.

Artesian water (6) Pressurized groundwater that rises in a well or a rock structure above the local water table; may flow out onto the ground.

Asthenosphere (8) Region of the upper mantle just below the lithosphere known as the plastic layer; the least rigid portion of Earth's interior; shatters if struck yet flows under extreme heat and pressure.

Atmosphere (1) The thin veil of gases surrounding Earth, that forms a protective boundary between outer space and the *biosphere*; generally considered to be below 480 km (300 mi).

Aurora (2) A spectacular glowing light display in the *ionosphere*, stimulated by the interaction of the *solar wind* with oxygen and nitrogen gases; at high latitudes: aurora borealis (Northern Hemisphere) and aurora australis (Southern Hemisphere).

Autumnal (September) equinox (2) The time around September 22–23 when the Sun's *declination* crosses the equatorial parallel; all places on Earth experience days and nights of equal length. The Sun rises at the South Pole and sets at the North Pole.

Available water (6) The portion of *capillary water* that is accessible to plant roots; usable water held in soil moisture storage.

Axial parallelism (2) Earth's axis is parallel to itself throughout the year; the North Pole points to near Polaris.

Axis (2) An imaginary line, extending through Earth from the geographic North Pole to the geographic South Pole, around which Earth rotates.

Badland (12) Rugged topography, usually of relatively low, varied relief and barren of vegetation; associated with arid and semiarid regions and rocks with low resistance to weathering.

Bajada (12) A continuous apron of coalesced alluvial fans, formed along the base of mountains in arid climates; presents a gently rolling surface from fan to fan.

Barrier beach (13) Narrow, long depositional feature, generally composed of sand, that forms offshore and is roughly parallel to the coast; may appear as *barrier islands* and long chains of barrier beaches.

Barrier island (13) Generally, a broadened *barrier beach*.

Barrier spit (13) A depositional form that develops when transported sand in a barrier beach or island is deposited in long ridges that are attached at one end to the mainland and partially cross the mouth of a bay.

Basalt (8) A common extrusive *igneous* rock; its mafic composition is fine-grained, comprising the bulk of the ocean floor crust, lava flows, and volcanic forms.

Base level (10) A hypothetical level below which a stream cannot erode its valley, and thus the lowest operative level for denudation processes; in an absolute sense represented by sea level extending back under the landscape.

Basin and Range Province (12) A region of dry climates, few permanent streams, and interior drainage patterns in the western United States; composed of a sequence of *horsts* and *grabens*.

Batholith (8) The largest plutonic form exposed at the surface; an irregular intrusive mass (>100 km^2; >40 mi^2) that invades crustal rocks, cooling slowly so that crystals develop.

Bay barrier (13) An extensive sand spit that encloses a bay, cutting it off completely from the ocean and forming a lagoon; produced by littoral drift and wave action; sometimes referred to as a baymouth bar.

Beach (13) The portion of the coastline where an accumulation of sediment is in motion.

Beach drift (13) Material, sand, gravel, and shells, that are moved by the *longshore current* in the effective direction of the waves.

Bed load (11) Coarse materials that are dragged along the bed of a stream by *traction* or by the rolling and bouncing motion of *saltation*; involves particles too large to remain in suspension.

Bedrock (10) The rock of Earth's crust that is below the soil and basically unweathered; such solid crust sometimes is exposed as an outcrop.

Biogeochemical cycles (16) The various circuits of flowing elements and materials (carbon, oxygen, nitrogen, phosphorus, water) that combine Earth's *biotic* and *abiotic* systems; the cycling of materials is continuous and renewed through the *biosphere* and the life processes.

Biogeography (16) The study of the distribution of plants and animals and related *ecosystems*; the geographical relationships with related environments over time.

Biomass (16) The total mass of living organisms on Earth or per unit area of a landscape; also, the weight of the living organisms in an ecosystem.

Biome (16) A large *terrestrial ecosystem* characterized by specific plant communities and formations; usually named after the predominant vegetation in the region.

Biosphere (1) That area where the *atmosphere*, *lithosphere*, and *hydrosphere* function together to form the context within which life exists; an intricate web that connects all organisms with their physical environment.

Biotic (1) Living.

Blowout depression (12) Eolian erosion whereby deflation forms a basin in areas of loose sediment. Size may range up to hundreds of meters.

Bolson (12) The slope and basin area between the crests of two adjacent ridges in a dry region.

Boreal forest (16) See northern needleleaf forest.

Braided stream (11) A stream that becomes a maze of interconnected channels laced with excess sediments. Braiding often occurs with a reduction of discharge that affects a stream's transportation ability or an increase in sediment load.

Breaker (13) The point where a wave's height exceeds its vertical stability and the wave breaks as it approaches the shore.

Calcification (15) The illuviated accumulation of calcium carbonate or magnesium carbonate in the B and C soil horizons.

Caldera (9) An interior sunken portion of a composite volcanic crater; usually steep-sided and circular, sometimes containing a lake; also found in conjunction with *shield volcanoes*.

Caliche (15) A cemented or hardened subsurface diagnostic soil horizon; found in the southwestern United States, usually in arid and semiarid climates.

Capillary water (6) Soil moisture, most of which is accessible to plant roots; held in the soil by surface tension and cohesive forces between water and soil (see also *available water*, *field capacity*, and *wilting point*).

Carbonation (10) A process of chemical weathering by a weak carbonic acid (water and carbon dioxide) that reacts with many minerals, especially limestone, containing calcium, magnesium, potassium, and sodium transforming them into carbonates.

Carbon dioxide (2) A natural by-product of life processes and complete combustion; the principal radiatively active gas in the greenhouse effect; CO_2.

Carbon monoxide (2) An odorless, colorless, tasteless combination of carbon and oxygen produced by the incomplete combustion of fossil fuels or other carbon-containing substance; CO.

Carnivore (16) A secondary *consumer* that principally eats meat for sustenance. The top carnivore in a food chain is considered a tertiary consumer.

Cartography (1) The making of maps and charts; a specialized science and art that blends aspects of geography, engineering, mathematics, graphics, computer science, and artistic specialties.

Cation-exchange capacity (CEC) (15) The ability of soil colloids to exchange cations between their surfaces and the soil solution; a measured potential.

Chaparral (16) Dominant shrub formations of Mediterranean dry summer climates; characterized by (sclerophyllous) scrub and short, stunted, and tough forests; derived from the Spanish *chapparo*; specific to California.

Chemical weathering (10) Decomposition and decay of the constituent minerals in rock through chemical alteration of those minerals. Water is essential, with rates keyed to temperature and precipitation values. Processes include *hydrolysis*, *oxidation*, *carbonation*, and *solution*.

Chlorofluorocarbon compounds (CFCs) (2) Large manufactured molecules (polymers) containing chlorine, fluorine, and carbon; inert and possessing remarkable heat properties; also known as hologens. After slow transport to the stratospheric ozone layer CFCs react with ultraviolet radiation freeing chlorine atoms that act as a catalyst to produce reactions that destroy ozone.

Chlorophyll (16) A light-sensitive pigment that resides within the chloroplast bodies of plants in leaf cells; the basis of *photosynthesis*.

Cinder cone (9) A landform of *tephra* and scoria, usually small and cone-shaped and generally not more than 450 m (1500 ft) in height; with a truncated top.

Circle of illumination (2) The division between lightness and darkness on Earth; a day-night *great circle*.

Circum-Pacific belt (9) A tectonically and volcanically active region encircling the Pacific Ocean; also known as the "ring of fire."

Cirque (14) A scooped-out, amphitheater-shaped basin at the head of an alpine glacier valley; an erosional landform.

Cirrus (5) Wispy filaments of ice-crystal clouds that occur above 6000 m (20,000 ft); appear in a variety of forms, from feathery hairlike fibers to veils of fused sheets.

Classification (7) The process of ordering or grouping data or phenomena in related classes; results in a regular distribution of information; a taxonomy.

Climate (7) The consistent, long-term, behavior of *weather* over time, including its variability; in contrast to weather, which is the condition of the atmosphere at any given place and time.

Climatic regions (7) Areas of similar *climate*, which contain characteristic regional weather and air mass patterns.

Climatology (7) A scientific study of climate and climatic patterns and the consistent behavior of weather and weather variability and extremes over time in one place or region; including the effects of climate change on human society and culture.

Climographs (7) Graphs that plot daily, monthly, or annual temperature and precipitation values for a selected station; may also include additional weather information.

Closed system (1) A system that is shut off from the surrounding environment so that it is entirely self-contained in terms of energy and materials; Earth is a closed material system (see open system).

Cloud (5) An aggradation of moisture droplets and ice crystals that are suspended in air and are great enough in volume and density to be visible; basic forms include *stratiform*, *cumuliform*, and *cirroform*.

Coal (8) A biochemical rock, rich in carbon; including lignite, bituminous, and anthracite forms.

Col (14) Formed by two headward eroding *cirques* that reduce an *arête* (ridge crest) to form a high pass or saddlelike narrow depression.

Cold front (5) The leading edge of a cold *air mass*; identified on a weather map as a line marked with a series of triangular spikes, pointing in the direction of frontal movement.

Composite volcano (9) A volcano formed by a sequence of explosive volcanic eruptions; steep-sided, conical in shape; sometimes referred to as a stratovolcano, although composite is the preferred term.

Condensation nuclei (5) Necessary microscopic particles on which water vapor condenses to form moisture droplets; can be sea salts, dust, soot, or ash.

Conduction (3) The slow molecule-to-molecule transfer of heat through a medium, from warmer to cooler portions.

Cone of depression (6) The depressed shape of the water table around a well after active pumping from an aquifer. The *water table* adjacent to the well is drawn down during the process of water removal.

Confined aquifer (6) An *aquifer* that is bounded above and below by impermeable layers of rock or sediment.

Consumer (16) Organisms in an *ecosystem* that depend on *producers* (autotrophs), organisms capable of using carbon dioxide as their sole source of carbon) for their source of nutrients.

Continental divide (11) A ridge or elevated area that determines the drainage pattern of drainage basins; specifically, that ridge in North America that separates drainage to the Pacific in the west from drainage to the Atlantic and Gulf in the east and to Hudson Bay and the Arctic Ocean in the north.

Continental glacier (14) A continuous mass of unconfined ice, covering at least 50,000 km² (19,500 mi²); most extensive as ice sheets covering Greenland and Antarctica.

Continentality (3) A qualitative designation applied to regions that lack the temperature-moderating effects of the sea and that exhibit a greater range between minimum and maximum temperatures, both daily and annually.

Continental platforms (9) The broadest category of landforms, including those masses of crust that reside above or near sea level and the adjoining undersea continental shelves along the coastline.

Continental shield (9) Generally old, low-elevation heartland regions of continental crust; various cratons (granitic cores) and ancient mountains exposed at the surface.

Contour lines (1) Isolines on a topographic map that connect all points at the same elevation relative to a reference elevation called the vertical datum.

Convection (3) Vertical transfer of heat from one place to another through the actual physical movement of air; involves a strong vertical motion.

Coordinated Universal Time (UTC) (1) The official reference time in all countries, formerly known as Greenwich Mean Time; now measured by six primary standard atomic clocks whose time calculations are collected in Paris, France, by the Bureau International de l'Heure.

Coral (13) A simple, cylindrical marine animal with a saclike body that secretes calcium carbonate to form a hard external skeleton; lives symbiotically with nutrient-producing algae.

Cordilleran system (9) One of two large mountain systems on Earth; refers to the relatively young mountains along the western margins of North and South America, from the tip of Tierra del Fuego to the massive peaks of Alaska.

Core (8) The deepest inner portion of Earth, representing one-third of its entire mass; differentiated into two zones—a solid iron *inner core* surrounded by a dense, molten, fluid metallic-iron *outer core*.

Coriolis force (4) The apparent deflection of moving objects on Earth from a straight path, in relationship to the differential speed of rotation at varying latitudes. Deflection is to the right in the Northern Hemisphere and to the left in the Southern Hemisphere; it produces a maximum effect at the poles and zero effect along the equator.

Crater (9) A circular surface depression formed by volcanism (accumulation, collapse, or explosion); usually located at a volcanic vent or pipe; can be at the summit or on the flank of a volcano.

Crevasses (14) Vertical cracks that develop in a *glacier* as a result of friction between valley walls, or tension forces of extension on convex slopes, or compression forces on concave slopes.

Crust (8) Earth's outer shell of crystalline surface rock, ranging from 5 to 60 km (3 to 38 mi) in thickness from oceanic crust to mountain ranges.

Cumulonimbus (5) A towering, precipitation-producing *cumulus* cloud that is vertically developed across altitudes associated with other clouds; frequently associated with lightning and thunder and thus sometimes termed a thunderhead.

Cumulus (5) Bright and puffy cumuliform clouds up to 2000 m in altitude (6500 ft).

Cut bank (11) A steep bank formed along the outer portion of a meandering stream; produced by lateral, erosive, under-cutting action of a stream.

Cyclone (4) A dynamically or thermally caused low-pressure area of converging and ascending air flows (see wave cyclone and tropical cyclone).

Daylength (2) Duration of exposure to *insolation*, varying during the year depending on *latitude*; an important aspect of seasonality.

Daylight saving time (1) Time is set ahead one hour in the spring and set back one hour in the fall in the Northern Hemisphere. Time is set ahead on the first Sunday in April and set back on the last Sunday in October—only Hawaii, Arizona, portions of Indiana, and Saskatchewan exempt themselves.

Debris avalanche (10) A mass of falling and tumbling rock, debris, and soil; can be dangerous because of the tremendous velocities achieved by the onrushing materials.

Declination (2) The latitude that receives direct overhead (perpendicular) insolation on a particular day; migrates annually through 47° of latitude between the tropics.

Decomposer (16) Microorganisms that digest and recycle organic debris and waste in the environment: includes bacteria, fungi, insects, and worms.

Deficit (6) In a water balance, the amount of unmet, or unsatisfied, potential evapotranspiration (DEFIC).

Deflation (12) A process of wind erosion that removes and lifts individual particles, literally blowing away unconsolidated, dry, or noncohesive sediments.

Delta (11) A depositional plain formed where a river enters a lake or an ocean; named after the triangular shape of the Greek letter *delta*.

Denudation (10) A general term that refers to all processes that cause degradation of the landscape: *weathering, mass movement, erosion*, and *transport*.

Deposition (11) The process whereby weathered, wasted, and transported sediments are laid down; deposited by air, water, or ice.

Desert biome (16) Arid landscapes of uniquely adapted dry-climate plants and animals.

Desert pavement (12) In arid landscape, a surface formed when wind *deflation* and *sheet flow* remove smaller particles, leaving residual pebbles and gravels to concentrate at the surface; resembles a cobblestone street.

Dew-point temperature (5) The temperature at which a given mass of air becomes *saturated*, holding all the water it can hold. Any further cooling or addition of water vapor results in active condensation.

Differential weathering (10) The effect of different resistance in rock, coupled with variations in the intensity of physical and chemical weathering.

Diffuse radiation (3) The downward component of scattered incoming insolation from clouds and the atmosphere.

Discharge (6) The measured volume of flow in a river that passes by a given cross section of a stream in a given unit of time; in cubic meters per second or cubic feet per second.

Dissolved load (11) Materials carried in chemical solution in a stream derived from minerals such as limestone and dolomite, or from soluble salts.

Diversity (16) A principle of ecology: the more diverse the species population (both in number of species and quantity of members in each species), the more risks are spread over the entire community, which results in greater overall stability.

Downwelling current (4) An area of the sea where a convergence or accumulation of water thrusts excess water downward; occurs, for example, at the western end of the equatorial current or along the margins of Antarctica.

Drainage basin (11) The basic spatial geomorphic unit of a river system; distinguished from a neighboring basin by ridges and highlands that form divides.

Drainage pattern (11) A geometric arrangement of streams in a region; determined by slope, differing rock resistance to weathering and erosion, climatic and hydrologic variability, and structural controls of the landscape.

Drawdown (6) See cone of depression.

Drumlin (14) A depositional landform related to glaciation that is composed of *till* and is streamlined in the direction of continental ice movement; blunt end upstream and tapered end downstream with a rounded summit.

Dry adiabatic rate (DAR) (5) The rate at which a parcel of air that is less than saturated cools (if ascending) or heats (if descending); a rate of 10C° per 1000 m (5.5F° per 1000 ft) (see also *adiabatic*).

Dune (12) A depositional feature of sand grains deposited in transient mounds, ridges, and hills; extensive areas of sand dunes are called sand seas.

Dust dome (3) A dome of airborne pollution associated with every major city; may be blown by winds into elongated plumes downwind from the city.

Dynamic equilibrium model (10) The balancing act between tectonic uplift and reduction rates of erosion, between the resistance of crust materials and the work of denudation processes. Landscapes evidence ongoing adaptation to rock structure, climate, local relief, and elevation.

Earthquake (9) A sharp release of energy that produces shaking in Earth's crust at the moment of rupture along a fault or in association with volcanic activity. Earthquake magnitude is estimated by the *Richter scale*; intensity is described by the *Mercalli scale*.

Ecological succession (16) The process whereby different and usually more complex assemblages of plants and animals replace older and usually simpler communities. Changes apparently move toward a more stable and mature condition.

Ecology (16) The science that studies the interrelationships among organisms and their environment and among various ecosystems.

Ecosphere (1) Another name for the *biosphere*.

Ecosystem (16) A self-regulating association of living plants, animals, and their nonliving physical and chemical environment.

Ecotone (16) A boundary transition zone between adjoining ecosystems that may vary in width and represent areas of tension as similar species of plants and animals compete for the resources.

Effusive eruption (9) An eruption characterized by low viscosity, basaltic magma, with low-gas content readily escaping. *Lava* pours forth onto the surface with relatively small explosions and little *tephra*; tends to form *shield volcanoes*.

Elastic-rebound theory (9) A concept describing the faulting process, in which the two sides of a fault appear locked despite the motion of adjoining pieces of crust, but with accumulating strain they rupture suddenly, snapping to new positions relative to each other.

Electromagnetic spectrum (2) All the radiant energy produced by the Sun placed in an ordered range, divided according to wavelengths.

Eluviation (15) The downward removal of finer particles and minerals from the upper horizons of soil.

Endogenic system (8) The system internal to Earth, driven by radioactive heat derived from sources within the planet.

In response, the surface is fractured, mountain building occurs, and earthquakes and volcanoes are activated.

Entisols (15) A soil order in the Soil Taxonomy classification that specifically lacks vertical development of horizons; usually young or undeveloped; found in active slopes, alluvial-filled floodplains, poorly drained tundra.

Entrenched meander (11) Incised river meanders excavated into the landscape; thought to be evidence of stream rejuvenation.

Environmental lapse rate (2, 5) The actual lapse rate in the lower atmosphere at any particular time under local weather conditions; may deviate above or below the average *normal lapse rate* of 6.4C° per 1000 m (3.5F° per 1000 ft).

Eolian (12) Caused by wind; refers to the erosion, transportation, and deposition of materials; spelled aeolian in some countries.

Epicenter (9) The surface area directly above the subsurface *focus* where movement along the fault plane was initiated. Shock waves radiate outward from the epicenter area.

Epipedon (15) The diagnostic soil horizon that forms at the surface; not to be confused with the A horizon; may include all or part of the illuviated B horizon.

Equal area (1) A trait of a *map projection*; indicates the equivalence of all areas on the surface of the map, although shape is distorted.

Equatorial and tropical rain forest (16) A lush biome of tall broadleaf evergreen trees and diverse plants and animals. The dense canopy of leaves is usually arranged in three levels.

Equatorial countercurrent (4) A strong countercurrent that travels east along the full extent of the Pacific, Atlantic, and Indian oceans; usually flows alongside or just beneath the westward-flowing surface current.

Equatorial low-pressure trough (4) A thermally caused low-pressure area that almost girdles Earth, with air converging and ascending all along its extent; also called the *intertropical convergence zone* (*ITCZ*).

Equilibrium (1) The status of a balanced energy and material system; maintains the system's general structure and character; a balance between form and process.

Equilibrium line (14) The area of a *glacier* where accumulation (gain) and *ablation* (loss) are balanced.

Erg desert (12) Sandy deserts, or areas where sand is so extensive that it constitutes a *sand sea*.

Erosion (11) *Denudation* by wind, water, and ice, which dislodges, dissolves, or removes surface material.

Esker (14) A sinuously curving, narrow deposit of coarse gravel that forms along a meltwater stream channel, developing in a tunnel beneath the glacier.

Estuary (11) The point at which the mouth of a river enters the sea and freshwater and seawater are mixed; a place where tides ebb and flow.

Eurasian-Himalayan system (9) One of two major mountain chains on Earth, stretching from southern Asia, China, and northern India and continuing in a belt through the upper Middle East to Europe and the European Alps.

Eustasy (5) Refers to worldwide changes in sea level that are not related to movements of land but rather to a rise and fall in the volume of water in the oceans.

Eutrophication (16) A natural process in which lakes receive nutrients and sediment and become enriched; the gradual filling and natural aging of water bodies.

Evaporation (6) The movement of free water molecules away from a wet surface into air that is less than saturated; the phase change of water to water vapor.

Evaporation fog (5) A fog formed when cold air flows over the warm surface of a lake, ocean, or other body of water; forms as the water molecules evaporate from the water surface into the cold, overlying air; also known as a steam fog or sea smoke.

Evaporite (8) Chemical sediments formed from inorganic sources when water evaporates and leaves behind a residue of salts previously in solution.

Evapotranspiration (6) The merging of *evaporation* and *transpiration* water loss into one term (see potential and actual evapotranspiration).

Exfoliation dome (10) A dome-shaped feature of weathering, produced by the response of *granite* to the overburden removal process, which relieves pressure from the rock. Layers of rock sluff off in slabs or shells in a *sheeting* process.

Exogenic system (8) The external surface system, powered by *insolation*, that energizes air, water, and ice and sets them in motion, under the influence of gravity. Includes all processes of landmass *denudation*.

Exosphere (2) An extremely rarefied outer atmospheric halo beyond the *thermopause* (at an altitude of 480 km (300 mi); probably composed of hydrogen and helium atoms, with some oxygen atoms and nitrogen molecules present near the thermopause.

Exotic stream (6) A river that rises in a humid region and flows through an arid region, with discharge decreasing toward the mouth; for example, the Nile River and the Colorado River.

Explosive eruption (9) A violent and unpredictable eruption, the result of magma that is thicker, stickier (more viscous), and higher in gas content and silica, than that of an *effusive eruption*; tends to form blockages within a volcano; produces composite volcanic landforms.

Faulting (9) The process whereby displacement and fracturing occurs between two portions of Earth's crust; usually associated with earthquake activity.

Feedback loop (1) When a portion of the system output cycles back as an information input, causing changes that guide further system operations (see negative and positive feedback).

Field capacity (6) Water held in the soil by hydrogen bonding against the pull of gravity, remaining after water drains from the larger pore spaces; the available water for plants.

Fire ecology (16) Recognizes fire as a dynamic ingredient in community succession. Controlled fires secure plant reproduction and prevent the accumulation of forest litter and brush; widely regarded as a wise forest management practice.

Firn (14) Snow of a granular texture that is transitional in the slow transformation from snow to *glacial ice;* snow that has persisted through a summer season in the zone of accumulation.

Firn line (14) The snow line that is visible on the surface of a glacier, where winter snows survive the summer *ablation* season; analogous to a snowline on land.

Fjord (14) A drowned glaciated valley, or glacial trough, along a coast which is filled by the sea.

Flash flood (12) A sudden and short-lived torrent of water that exceeds the capacity of a stream channel; associated with desert and semiarid washes.

Flood (11) A high water level that overflows the natural (or artificial) banks along any portion of a stream.

Floodplain (11) A low-lying area near a stream channel, subject to recurrent flooding; alluvial deposits generally mask underlying rock.

Fluvial (11) Stream-related processes; from the Latin *fluvius* for "river" or "running water."

Fog (5) A cloud, generally stratiform, in contact with the ground, with visibility usually restricted to less than 1 km (3300 ft).

Folding (9) A process that bends and deforms beds of various rocks subjected to compressional forces.

Food chain (16) The circuit along which energy flows from producers, who manufacture their own food, to consumers; a one-directional flow of chemical energy, ending with decomposers.

Food web (16) A complex network of interconnected food chains.

Formation class (16) That portion of a *biome* that concerns the plant communities only, subdivided by size, shape, and structure of the dominant vegetation present.

Friction force (4) The effect of drag by the wind as it moves across a surface; may be operative through 500 m (1640 ft) of altitude. Surface friction slows the velocity of the wind and therefore reduces the effectiveness of the Coriolis force.

Front (5) The leading edge of an advancing air mass; a line of contrasting weather conditions.

Frost action (10) A powerful mechanical force produced as water expands 9% of its volume as it freezes and can exceed the tensional strength of rock.

Fusion (2) The process of forcibly joining, under extreme temperature and pressure, positively charged hydrogen and helium nuclei; occurs naturally in thermonuclear reactions within stars, such as our Sun.

General circulation model (GCM) (7) Complex computer-based climate models that produce generalizations of reality and produce predictive forecasts of future weather and climate conditions.

Genetic classification (7) A type of classification that uses causative factors to determine climatic regions; for example, an analysis of the effect of interacting air masses.

Geodesy (1) The science that determines Earth's shape and size through surveys, mathematical means, and remote sensing.

Geographic information system (GIS) (1) A computer-based data processing tool or methodology used for gathering, manipulating, and analyzing geographic information to produce a holistic, interactive analysis.

Geography (1) The science that studies the interdependence among geographic areas, natural systems, processes, society, and cultural activities over space—a *spatial science*. The five themes of geographic education include: location, place, movement, regions, and human-Earth relationships.

Geoid (1) A word that describes Earth's shape, literally, "the shape of Earth is Earth-shaped." A theoretical surface at sea level that extends through the continents; deviates from a perfect sphere.

Geologic cycle (8) A general term characterizing the vast cycling (hydrologic, tectonic, and rock) in and on the *lithosphere*.

Geologic time scale (8) A listing of eras, periods, and epochs that span Earth's history; reflects the relative relationship of various layers of rock strata and the absolute dates as determined by scientific methods such as radioactive isotopic dating.

Geomorphic cycle model (10) A conceptual model proposed by William Morris Davis to characterize landscape development as a cyclic pattern from initial uplift to final erosional surface, or peneplain.

Geomorphic threshold (10) The threshold up to which landforms change before lurching to a new set of relationships, with rapid realignments of landscape materials and slopes.

Geomorphology (10) The science that analyzes and describes the origin, evolution, form, classification, and spatial distribution of landforms.

Geostrophic wind (4) Winds moving between pressure areas along paths that are parallel to the *isobars*. In the upper *troposphere* the *pressure gradient force* equals the *Coriolis force* so that the amount of deflection is proportional to air movement.

Geothermal energy (9) The energy potential of boiling steam produced by subsurface magma in near contact with groundwater. Active examples include Iceland, New Zealand, Italy, and northern California.

Glacial drift (14) The general term for all glacial deposits, both unsorted and sorted.

Glacial ice (14) A hardened form of ice, very dense in comparison to normal snow or firn; under pressure within a glacier is capable of downhill movement.

Glacial surge (14) The rapid, lurching, unexpected movement of a glacier.

Glacier (14) A large mass of perennial ice resting on land or floating shelflike in the sea adjacent to the land; formed from the accumulation and recrystallization of snow.

Goode's homolosine projection (1) An equal-area projection developed in 1925 by Dr. Paul Goode; combining two oval projections: the holographic and the sinusoidal; presented interrupted in the text.

Graben (9) Pairs or groups of faults that produce downward faulted blocks; characteristic of the basins of the interior western United States (see horst and Basin and Range Province).

Graded stream (11) A condition in a stream of mutual adjustment between the load carried by a stream and the related landscape through which the stream flows, forming a state of dynamic equilibrium among erosion, transported load, deposition, and the stream's capacity.

Gradient (11) The drop in elevation from a stream's headwaters to its mouth, ideally forming a concave slope.

Granite (8) A coarse-grained (slow-cooling) intrusive igneous rock of 25% quartz and more than 50% potassium and sodium feldspars; characteristic of the continental crust.

Gravitational water (6) That portion of surplus water that percolates downward from the capillary zone, pulled by gravity to the groundwater zone.

Great circle (1) Any circle of circumference drawn on a globe with its center coinciding with the center of the globe. An infinite number of great circles can be drawn, but only one parallel is a great circle—the equator.

Greenhouse effect (3) The process whereby radiatively active gases absorb and delay the loss of heat to space, thus keeping the lower troposphere moderately warmed through the radiation and reradiation of infrared wavelengths.

Greenwich Mean Time (GMT) (1) Former world standard time, now known as *Coordinated Universal Time* (*UTC*) (see Coordinated Universal Time).

Ground ice (14) Subsurface water that is frozen in regions of *permafrost*. The moisture content of areas with ground ice may vary from nearly absent in regions of drier permafrost to almost 100% in saturated soils.

Gulf Stream (3) A strong northward-moving warm current off the east coast of North America, which carries its water far into the North Atlantic.

Gyre (4) The dominant circular ocean current beneath subtropical high-pressure cells in both hemispheres; offset to the western margin of each ocean basin.

Habitat (16) That physical location in which an organism is biologically suited to live. Most species have specific habitat parameters, or limits.

Hadley cell (4) The vertical convection cell in each hemisphere that is generated along the low-pressure system of converging and ascending air along the equator, which then subsides and diverges at subtropical latitudes.

Hail (5) A type of precipitation formed when a raindrop is repeatedly circulated above and below the freezing level in a cloud, with each cycle adding more ice to the hailstone until it becomes too heavy to stay aloft.

Headlands (13) Protruding landforms that are extensions of the coast; generally composed of more resistant rocks.

Herbivore (16) The primary *consumer* in a food chain, which eats plant material formed by a *producer* that has synthesized organic molecules.

Heterosphere (2) A zone of the atmosphere above the mesopause, 80 km (50 mi) in altitude; composed of rarified layers of oxygen atoms and nitrogen molecules; includes the ionosphere.

Histosols (15) A soil order in the Soil Taxonomy classification that is formed from thick accumulations of organic matter, such as beds of former lakes, bogs, and layers of peat.

Homosphere (2) A zone of the atmosphere from the surface up to 80 km (50 mi), composed of an even mixture of gases including nitrogen, oxygen, argon, carbon dioxide, and trace gases.

Horn (14) A pyramidal, sharp-pointed peak that results when several cirque glaciers gouge an individual mountain summit from all sides.

Horst (9) Upward-faulted blocks produced by pairs or groups of faults; characterized by the mountain ranges of the interior of the western United States (see graben and Basin and Range Province).

Hot spot (8) An individual point of upwelling material originating in the asthenosphere; tend to remain fixed relative to migrating plates; some 100 are identified worldwide; exemplified by Yellowstone National Park, Hawaii, and Iceland.

Human-Earth relationships (1) One of the oldest themes of geography (the human-land tradition); includes the spatial analysis of settlement patterns, resource utilization and exploitation, hazard perception and planning, and the impact of environmental modification and artificial landscape creation.

Humidity (5) Water vapor content of the air. The capacity of the air to hold water vapor is mostly a function of the water vapor temperature and air temperature.

Humus (15) A mixture of organic debris in the soil, worked by consumers and decomposers in the humification process; characteristically formed from plant and animal litter laid down at the surface.

Hurricane (5) A tropical cyclone that is fully organized and intensified in inward-spiraling rainbands; ranges from 160 to 960 km (100 to 600 mi) in diameter, with wind speeds in excess of 119 kmph (65 knots, or 74 mph); a name used specifically in the Atlantic and eastern Pacific.

Hydration (10) A process involving water, although not involving any chemical change; water is added to a mineral, which initiates swelling and stress within the rock, mechanically forcing grains apart as the constituents expand; this is a physical weathering process.

Hydraulic action (11) The erosive work accomplished by the turbulence of water; causes a squeezing and releasing action in joints in bedrock; capable of prying and lifting rocks.

Hydrograph (11) A graph of stream discharge over a period of time (minutes, hours, days, years) at a specific place on a stream. The relationship between stream discharge and precipitation input is illustrated on the graph.

Hydrologic cycle (6) A simplified model of the flow of water and water vapor from place to place as energy powers system operations. Water flows through the atmosphere, across the land where it is also stored as ice, and within groundwater.

Hydrolysis (10) When minerals chemically combine with water; a decomposition process that causes silicate minerals in rocks to break down and become altered. This is a chemical weathering process.

Hydrosphere (1) An abiotic open system that includes all of Earth's water.

Ice age (14) A cold episode, with accompanying alpine and continental ice accumulations, that has repeated roughly every 200 to 300 million years since the late Precambrian era (1.25 billion years ago); includes the most recent episode of during the Pleistocene Ice Age, which began 1.65 million years ago.

Ice cap (14) A dome-shaped glacier, less extensive than an ice sheet (<50,000 km²), although it buries mountain peaks and the local landscape.

Ice field (14) An extensive form of land ice, with mountain ridges and peaks visible above the ice.

Ice sheet (14) An enormous continuous *continental glacier*. The bulk of glacial ice on Earth covers Antarctica and Greenland in two ice sheets.

Ice wedge (14) Formed when water enters a thermal contraction crack in permafrost and freezes. Repeated seasonal freezing and melting progressively expand the wedge.

Iceberg (14) Floating ice created by calving ice or large pieces breaking off and floating adrift; a hazard to shipping because the ice is 91% submerged.

Igneous rock (8) Rocks that solidify and crystallize from a hot molten state.

Illuviation (15) A depositional soil process as differentiated from *eluviation*; usually in the B horizon, where accumulations of clays, aluminum, iron, and some humus occur.

Impermeable (6) Subsurface structures that obstruct water flows in the groundwater system.

Inceptisols (15) An order in the Soil Taxonomy classification; weakly developed soils that are inherently infertile; usually young soils that are weakly developed, although they are more developed than Entisols.

Industrial smog (2) Air pollution associated with coal-burning industries; may contain sulfur oxides, particulates, carbon dioxide, and exotics.

Infiltration (6) Water access to subsurface regions of *soil moisture storage* through penetration of the soil surface.

Insolation (2) Solar radiation that is intercepted by Earth.

Internal drainage (6) In regions where rivers do not flow into the ocean, outflow is through evaporation or subsurface gravitational flow. Portions of Africa, Asia, Australia, and the western United States have such drainage.

International Date Line (1) The 180° meridian; an important corollary to the *prime meridian* at 0°, Greenwich, London, England; established by the treaty of 1884.

Intertropical convergence zone (ITCZ) (4) See equatorial low-pressure trough.

Ionosphere (2) A layer in the *atmosphere* above 80 km (50 mi) where gamma, X-ray, and some ultraviolet radiation is absorbed and converted into infrared, or heat energy, and where the *solar wind* stimulates the *auroras*.

Isobar (4) An isoline connecting all points of equal pressure.

Isostasy (8) A state of equilibrium formed by the interplay between portions of the *lithosphere* and the *asthenosphere*; the crust depresses with weight and recovers with the melting of the ice or removal of the load, in isostatic rebound.

Isotherm (3) An isoline connecting all points of equal temperature.

Jet stream (4) The most prominent movement in upper-level westerly wind flows; irregular, concentrated, sinuous bands of geostrophic wind, traveling at 300 kmph (190 mph).

Joints (10) Fractures or separations in rock without displacement of the sides; increases the surface area of rock exposed to a weathering processes.

Kame (14) A depositional feature of glaciation; a small hill of poorly sorted sand and gravel that accumulates in crevasses or in ice-caused indentations in the surface.

Karst topography (10) Distinctive topography formed in a region of chemically weathered limestone with poorly developed surface drainage and solution features that appear pitted and bumpy; originally named after the Krš Plateau of Yugoslavia.

Katabatic wind (4) Air drainage from elevated regions, flowing as gravity winds.

Kettle (14) Forms when an isolated block of ice persists in a ground moraine, an outwash plain, or valley floor after a glacier retreats; as the block finally melts, it leaves behind a steep-sided hole that frequently fills with water.

Kinetic energy (2) The energy of motion in a body; derived from the vibration of the body's own movement and stated as temperature.

Köppen-Geiger climate classification (7) An empirical climate classification system based on average monthly temperature, average monthly precipitation, and total annual precipitation. Capital letters A, B, C, D, E, and H designate climate categories.

Lacustrine deposit (14) Deposit associated with lake-level fluctuations; for example, benches or terraces marking former shorelines.

Lagoon (13) A portion of coastal seawater that is virtually cut off from the ocean by a *bay barrier* or *barrier beach*; also, the water surrounded and enclosed by an atoll.

Landslide (10) A sudden, rapid, downslope movement of a cohesive mass of regolith and/or bedrock in a variety of mass-movement forms under the influence of gravity.

Land-water heating differences (3) The differences in the way land and water heat, as a result of contrasts in *transmission*, *evaporation*, mixing, and specific heat capacities. Land surfaces heat and cool faster than water and are characterized as having aspects of *continentality*, whereas water is regarded as producing a *marine* influence.

Latent heat (5) Heat energy that is absorbed or released in the phase change of water and is stored in one of the three states—ice, water, or water vapor; includes the latent heat of melting, freezing, vaporization, evaporation, and condensation.

Latent heat of condensation (5) The heat energy released in a phase change from water vapor to liquid; 585 calories is released from 1 gram of water vapor that condenses at 20°C (68°F).

Latent heat of evaporation (3) The heat energy required to change phase from liquid to water vapor under normal sea-level pressure, 585 calories must be added to 1 gram of water (at 20°C) to achieve a phase change to water vapor.

Lateral moraine (14) Debris transported by a glacier that accumulates along the sides of the glacier and is deposited along these margins.

Laterization (15) A pedogenic process operating in well-drained soils that are found in warm and humid regions; typical of Oxisols. High precipitation values leach soluble minerals and soil constituents usually reddish or yellowish colors.

Latitude (1) The angular distance measured north or south of the equator from a point at the center of Earth. A line connecting all points of the same latitudinal angle is called a *parallel*.

Lava (8, 9) *Magma* that issues from volcanic activity onto the surface; the extrusive rock that results when magma solidifies.

Life zone (16) An altitudinal zonation of plants and animals that form distinctive communities. Each life zone possesses its own temperature and precipitation relationships.

Lightning (5) Flashes of light, caused by tens of millions of volts of electrical charge igniting the air to temperatures of 15,000°C to 30,000°C.

Limestone (8) The most common chemical sedimentary rock (nonclastic); lithified calcium carbonate ($CaCO_3$), which is very susceptible to chemical weathering.

Limiting factor (16) The physical or chemical factor that most inhibits (either through lack or excess) biotic processes.

Lithification (8) The compaction, cementation, and hardening of sediments into *sedimentary rock*.

Lithosphere (1) Earth's *crust* and that portion of the uppermost mantle directly below the crust that extends down to 70 km (45 mi). Some use this term to refer to the entire Earth.

Littoral zone (13) A specific coastal environment; that region between the high water line during a storm and a depth at which storm waves are unable to move sea-floor sediments.

Longshore current (13) A current that forms parallel to a *beach* as waves arrive at an angle to the shore; generated in the surf zone by wave action, transporting large amounts of sand and sediment (also known as littoral current).

Loam (15) A mixture of sand, silt, and clay in almost equal proportions, with no one texture dominant.

Location (1) A basic theme of *geography* dealing with the absolute and relative position of people, places, and things on Earth's surface.

Loess (12) Large quantities of fine-grained clays and silts left as glacial outwash deposits; subsequently blown by the wind great distances and redeposited as a generally unstratified, homogeneous blanket of material covering existing landscapes; in China loess originated from desert lands.

Longitude (1) The angular distance measured east or west of a *prime meridian* from a point at the center of Earth. A line connecting all points of the same longitude is called a *meridian*.

Magma (8) Molten rock from beneath the surface of Earth; fluid, gaseous, under tremendous pressure, and either intruded into country rock or extruded onto the surface as *lava*.

Magnetic reversal (8) With an uneven regularity the magnetic field fades to zero, then phases back to full strength, but with the magnetic poles reversed. Reversals have occurred nine times during the past 4 million years.

Magnetosphere (2) Earth's magnetic force field, which is generated by dynamolike motions within the planet's outer core; deflects the solar wind toward each pole.

Mangrove swamp (13) Wetland ecosystems between 30° N or S and the equator; tend to form a distinctive *community* of mangrove plants.

Mantle (8) An area within the planet representing about 80% of Earth's total volume, with densities increasing with depth; occurs above the core and below the crust; is rich in iron and magnesium oxides and silicates.

Map projection (1) The reduction of a spherical globe onto a flat surface in some orderly and systematic realignment of the latitude and longitude grid.

Marine (3) Regions that are dominated by the moderating effects of the ocean and that exhibit a smaller minimum and maximum temperature range than continental stations (see land-water heating differences).

Mass movement (10) All unit movements of materials propelled by gravity; can range from dry to wet, slow to fast, small to large, and free-falling to gradual or intermittent; sometimes used interchangeably with mass wasting.

Meandering stream (11) The sinuous, curving pattern common to *graded streams*, with the outer portion of each curve subjected to the greatest erosive action and the inner portion receiving sediment deposits

Mean sea level (13) The average of tidal levels recorded hourly at a given site over a long period of time, which must be at least a full lunar tidal cycle.

Medial moraine (14) Debris transported by a glacier that accumulates down the middle of the glacier when two glaciers merge and their lateral moraines combine; forms a depositional feature following glacial retreat.

Mediterranean shrubland (16) A major *biome* dominated by Mediterranean dry summer climates and is characterized by scrub and short, stunted, tough (sclerophyllous) forests.

Mercator projection (1) A cylindrical, true-shape, projection developed by Gerardus Mercator in 1569. All straight lines represent true constant directions, or rhumb lines.

Mercury barometer (4) A device that measures *air pressure* with a column of mercury in a tube that is inserted in a vessel of mercury.

Meridian (1) See *longitude*.

Mesocyclone (5) A large rotating circulation initiated within a parent cumulonimbus cloud at the mid-troposphere level; generally produces heavy rain, large hail, blustery winds, and lightning; may lead to tornado activity.

Mesosphere (2) The upper region of the *homosphere* from 50 to 80 km (30 to 50 mi) above the ground; designated by temperature criteria and very low pressures.

Metamorphic rock (8) Existing rock, both *igneous* and *sedimentary*, that goes through profound physical and chemical changes under increased pressure and temperature. Constituent mineral structures may exhibit foliated or nonfoliated textures.

Meteorology (5) The scientific study of the *atmosphere* that includes a study of the atmosphere's physical characteristics and motions, related chemical, physical, and geological processes, the complex linkages of atmospheric systems, and weather forecasting.

Methane (7) A radiatively active gas that participates in the *greenhouse effect*; derived from the organic processes of burning, digesting, and rotting in the presence of oxygen; CH_4.

Microclimatology (3) The study of climates at or near Earth's surface.

Midlatitude broadleaf and mixed forest (16) A biome in moist continental climates in areas of warm-to-hot summers and cool-to-cold winters; relatively lush stands of broadleaf forests trend northward into needleleaf evergreen stands.

Midlatitude grassland (16) The major *biome* most modified by human activity; so named because of the predominance of grasslike plants, although deciduous broadleafs appear along streams and other limited sites; location of the world's breadbaskets of grain and livestock production.

Mid-ocean ridge (8) Submarine mountain ranges that extend more than 65,000 km (40,000 mi) worldwide and average more than 1000 km (620 mi) in width; centered along sea-floor spreading centers.

Milky Way Galaxy (2) A flattened, disk-shaped mass estimated to contain up to 400 billion stars; includes our Solar System.

Mineral (8) An element or combination of elements that form an inorganic natural compound; described by a specific formula and qualities of a specific nature.

Model (1) A simplified version of a *system*, representing an idealized part of the real world.

Mohorovičić discontinuity (8) The boundary between the *crust* and the rest of the lithospheric upper mantle; named for the Yugoslavian seismologist Mohorovicic; a zone of sharp material and density contrasts, also called the *Moho*.

Moist adiabatic rate (MAR) (5) The rate at which a parcel of saturated air cools in ascent; a rate of 6C° per 1000 m (3.3F° per 1000 ft). This rate may vary, with moisture con-

tent and temperature, from 4C° to 10C° per 1000 m (2F° to 6F° per 1000 ft) (see adiabatic).

Moisture droplets (5) Initial composition of clouds. Each droplet measures approximately 0.002 cm (0.0008 in.) in diameter and is invisible to the human eye.

Mollisols (15) A soil order in the Soil Taxonomy classification that has a humus-rich organic content high in alkalinity; some of the world's most significant agricultural soils.

Monsoon (4) From the Arabic word *mausim*, meaning "season"; refers to an annual cycle of dryness and wetness, with seasonally shifting winds produced by changing atmospheric pressure systems; affects India, Southeast Asia, Indonesia, northern Australia, and portions of Africa.

Moraine (14) Marginal glacial deposits (lateral, medial, terminal, ground) of unsorted and unstratified material.

Movement (1) A major theme in *geography* involving migration, communication, and the interaction of people and processes across space.

Natural levees (11) Long, low ridges that occur on either side of a river in a developed floodplain; depositional by-products (coarse gravels and sand) of river-flooding episodes.

Nebula (2) A cloud of dust and gas in space.

Negative feedback (1) A *feedback loop* that tends to slow or dampen response in a *system*; promotes self-regulation in a system; far more common than *positive feedback* in living systems.

Net primary productivity (16) The net photosynthesis (photosynthesis minus respiration) for a given community; considers all growth and all reduction factors that affect the amount of useful chemical energy (biomass) fixed (chemically bound) in an *ecosystem*.

Net radiation (NET R) (3) The net all-wave radiation available; the final outcome of the radiation balance process between incoming and outgoing shortwave and longwave energy.

Niche (16) The basic function, or occupation, of a lifeform within a given *community*; the way an organism obtains its food, air, and water.

Nickpoint (knickpoint) (11) The point at which the longitudinal profile of a stream is abruptly broken by a change in gradient; for example, a waterfall, rapids, or cascade.

Nimbostratus (5) A rain-producing, dark, grayish, stratiform cloud characterized by gentle drizzles.

Nitrogen dioxide (2) A reddish-brown choking gas produced in high-temperature combustion engines; can be damaging to human respiratory tracts and to plants; participates in photochemical reactions and acid deposition.

Normal fault (9) Or *tension fault*, occurs when rocks are pulled apart; vertical motion along an inclined fault plane with one block of land ending up lower than the other.

Normal lapse rate (2, 5) The average rate of temperature decrease with increasing altitude in the lower atmosphere; an average value of 6.4C° per km, or 1000 m (3.5F° per 1000 ft).

Northern needleleaf forest (16) Forests of pine, spruce, fir, and larch, stretching from the east coast of Canada westward to Alaska and continuing from Siberia westward across the entire extent of Russia to the European Plain; called the *taiga* (a Russian word) or the *boreal forest*; principally in the D climates. Includes montane forests.

Nuclear winter hypothesis (7, 17) An increase in Earth's *albedo* and upper atmospheric absorption of insolation resulting in surface cooling; associated with the detonation of a relatively small number of nuclear warheads within the biosphere. Now encompasses a whole range of ecological, biological, and climatic impacts.

Occluded front (5) In a cyclonic circulation, the overrunning of a surface *warm front* by a *cold front* and the subsequent lifting of the warm air wedge off the ground; initial precipitation is moderate to heavy.

Ocean basins (9) The physical containers for Earth's oceans.

Oceanic trench (8) The deepest single features of Earth's *crust*; associated with *subduction zones*. The deepest is the Mariana Trench near Guam, which descends to 11,033 m (36,198 ft).

Omnivore (16) A *consumer* that feeds on both *producers* (plants) and consumers (meat)—a role occupied by humans, among other animals.

Open system (1) A system with inputs and outputs crossing back and forth between the system and the surrounding environment. Earth is an open system in terms of energy.

Orogenesis (9) The process of mountain building that occurs when large-scale compression leads to deformation and uplift of the crust; literally the birth of mountains.

Orographic lifting (5) The uplift of migrating air masses in response to the physical presence of a mountain, a topographic barrier. The lifted air cools adiabatically as it moves upslope; may form clouds and produce increased precipitation.

Outgassing (5) The release of trapped gases from rocks, forced out through cracks, fissures, and volcanoes from within Earth; the terrestrial source of Earth's water.

Outwash plain (14) Glaciofluvial deposits of stratified drift from meltwater-fed, braided, and overloaded streams; beyond a glacier's morainal deposits.

Oxbow lake (11) A lake that was formerly part of the channel of a *meandering stream*; isolated when a stream eroded its outer bank forming a cutoff through the neck of a looping meander.

Oxidation (10) A chemical weathering process whereby oxygen oxidizes (combines with) certain metallic elements to form oxides; most familiar as the "rusting" of iron in a rock or soil that produces a reddish-brown stain of iron oxide (Fe_2O_3).

Oxisols (15) In the Soil Taxonomy classification a soil order of tropical soils that are old, deeply developed, and lacking in horizons wherever well-drained; heavily weathered, low in cation exchange capacity, and low in fertility.

Ozone layer (2) See ozonosphere.

Ozonosphere (2) A layer of ozone (O_3) occupying the full extent of the *stratosphere* (20 to 50 km or 12 to 30 mi) above the surface; the region of the *atmosphere* where ultraviolet wavelengths are principally absorbed and converted into heat.

Paleoclimatology (7) The science that studies the climates of past ages.

Paleolakes (14) See pluvial.

PAN (2) See peroxyacetyl nitrates.

Pangaea (8) The supercontinent formed by the collision of all continental masses approximately 225 million years ago; named by Wegener (1912) in his continental drift theory.

Parallel (1) See latitude.

Parent material (10) The unconsolidated material, from both organic and mineral sources, that is the basis of soil development.

Patterned ground (14) Areas in the periglacial environment where freezing and thawing of the ground create polygonal forms of arranged rocks at the surface.

Pedon (15) A soil profile extending from the surface to the lowest extent of plant roots or to the depth where regolith or bedrock is encountered; imagined as a hexagonal column; the basic soil sampling unit.

Percolation (6) The process by which water permeates through to the subsurface environment; vertical water movement through soil or porous rock.

Periglacial (14) Cold-climate processes, landforms, and topographic features along the margins of glaciers, past and present, that occupy over 20% of Earth's land surface; including *permafrost, frost action,* and *ground ice.*

Perihelion (2) That point in Earth's elliptical orbit about the Sun where Earth is closest to the Sun; occurs on January 3 at 147,255,000 km (91,500,000 mi); variable over a 100,000-year cycle.

Permafrost (14) Forms when soil or rock temperatures remain below 0°C (32°F) for at least two years in areas considered periglacial; criterion is based on temperature and not on whether water is present.

Permeable (6) The ability of water to flow through soil or rock; a function of the texture and structure of the medium.

Peroxyacetyl nitrates (PAN) (2) A pollutant formed from photochemical reactions involving nitric oxide (NO) and hydrocarbons (HC). They produce no known human health effects but are particularly damaging to plants.

Phase change (5) The change in phase, or state, between ice, water, and water vapor; involves the absorption or release of *latent heat.*

Photochemical smog (2) Air pollution produced by the interaction of ultraviolet light, nitrogen dioxide, and hydrocarbons; produces *ozone* and *PAN* through a series of complex photochemical reactions. Automobiles are the major source of the contributive gases.

Photoperiod (16) The duration of daylight experienced at a given location.

Photosynthesis (16) The joining of carbon dioxide and oxygen in plants, under the influence of certain wavelengths of visible light; releases oxygen and produces energy-rich organic material (sugars and starches).

Physical geography (1) A science that studies the spatial aspects of the physical elements and processes that make up the environment: energy, air, water, weather, climate, landforms, soils, animals, plants, and Earth.

Physical weathering (10) The breaking and disintegrating of rock without any chemical alteration; sometimes referred to as mechanical or fragmentation weathering.

Place (1) A major theme in geography focused on the tangible and intangible characteristics that make each location unique.

Plane of the ecliptic (2) A plane intersecting all the points of Earth's orbit.

Planetesimal hypothesis (2) Early protoplanets formed from the condensing masses of a nebular cloud of dust, gas, and icy comets; a formation process now being observed in other parts of the galaxy.

Planimetric map (1) A basic map showing the horizontal position of boundaries, land-use activities, and political, economic, and social outlines.

Plate tectonics (8) The conceptual model that encompasses continental drift, *sea-floor spreading*, and related aspects of crustal movement; accepted as the foundation of crustal tectonic processes.

Plateau basalts (9) An accumulation of horizontal flows formed when lava spreads out from elongated fissures onto the surface in extensive sheets; associated with effusive eruptions; also known as flood basalts.

Playa (12) An area of salt crust left behind by evaporation on a desert floor usually in the middle of a *bolson* or valley; intermittently wet and dry.

Plinthite (15) An ironstone hardpan structure formed in *Oxisol* subsurfaces or surface horizons that are exposed to repeated wetting and drying sequences; quarried in some areas for building materials.

Pluton (8) A mass of intrusive igneous rock that has cooled slowly in the crust; forms in any size or shape. The largest partially exposed pluton is a *batholith.*

Pluvial (14) Referring to a period in the past, principally during the Pleistocene, when there was greater moisture availability or efficiency; resulted in increased lake levels in arid and seasonally arid regions—*paleolakes.*

Podzolization (15) A pedogenic process in cool, moist climates; forms a highly leached soil with strong surface acidity because of humus from acid-rich trees.

Point bar (11) The inner portion of a *meander* which receives sediment fill.

Polar easterlies (4) Variable weak, cold, and dry winds moving away from the polar region; an anticyclonic circulation.

Polar front (4) A significant zone of contrast between cold and warm air masses; roughly situated between 50° and 60° latitude.

Polar high-pressure cells (4) A weak, anticyclonic, thermally produced pressure system positioned roughly above each pole.

Polypedon (15) The identifiable soil in an area, with distinctive characteristics differentiating it from surrounding polypedons forming the basic soil mapping unit; composed of many pedons.

Porosity (6) The total volume of available pore space in soil; a result of the texture and structure of the soil.

Positive feedback (1) Feedback that amplifies or encourages responses in a system.

Potential evapotranspiration (6) The amount of moisture that would evaporate and transpire if adequate moisture were available; the amount lost under optimum moisture conditions—the moisture demand (POTET).

Precipitation (6) Rain, snow, sleet, and hail—the moisture supply (PRECIP).

Pressure gradient force (4) Causes air to move from areas of higher barometric pressure to areas of lower barometric pressure due to pressure differences.

Prime meridian (1) An arbitrary *meridian* designated as 0° *longitude*; the point from which longitude is measured east or west; agreed to in an 1884 treaty.

Process (1) A set of actions and changes that occur in some special order; analysis of processes is central to modern geographic study.

Producer (16) Organisms that are capable of using carbon dioxide as their sole source of carbon, which they chemically fix through *photosynthesis* to provide their own nourishment; also called an *autotroph.*

Radiation fog (5) Formed by radiative cooling of the surface, especially on clear nights in areas of moist ground; occurs when the air layer directly above the surface is chilled to the dew-point temperature, thereby producing saturated conditions.

Rain gauge (6) A standardized device that catches and measures rainfall.

Rain shadow (5) The area on the leeward slopes of a mountain range; in the shadow of the mountains, where precipitation receipt is greatly reduced compared to windward slopes.

Reflection (3) The portion of arriving energy that returns directly back to space without being converted into heat or performing any work (see *albedo*).

Refraction (3) The bending effect that occurs when *insolation* enters the atmosphere or another medium; the same process by which a crystal, or prism, disperses the component colors of the light passing through it.

Region (1) A geographic theme that focuses on areas that display unity and internal homogeneity of traits; includes the study of how a region forms, evolves, and interrelates with other regions.

Regolith (10) Partially weathered rock overlying bedrock, whether residual or transported.

Relative humidity (5) The ratio of water vapor actually in the air (content) compared to the maximum water vapor the air is able to be hold (capacity) by at that temperature; expressed as a percentage.

Relief (9) Elevation differences in a local landscape; an expression of the unevenness, height, and slope variation.

Remote sensing (1) Information acquired from a distance, without physical contact with the subject; for example, photography or orbital imagery.

Respiration (16) The process by which plants derive energy for their operations; essentially, the reverse of the photosynthetic process; releases carbon dioxide, water, and heat into the environment.

Reverse fault (9) Or *compression fault*, one side of the fault moves upward vertically in comparison to the other side.

Revolution (2) The annual orbital movement of Earth about the Sun; determines year length and the length of seasons.

Rhumb line (1) A line of constant compass direction, or constant bearing, which crosses all meridians at the same angle. A portion of a *great circle*.

Richter scale (9) An open-ended, logarithmic scale that estimates earthquake magnitude; designed by Charles Richter in 1935.

Robinson projection (1) An oval projection developed by A. Robinson 1963.

Rock (8) An assemblage of minerals bound together (like granite), or may be a mass of a single mineral (like rock salt).

Rock cycle (8) A model representing the interrelationships among the three rock-forming processes: *igneous, sedimentary*, and *metamorphic*.

Rockfall (10) Free-falling movement of debris from a cliff or steep slope, generally falling straight down or bounding downslope.

Rossby waves (4) Undulating horizontal motions in the upper-air westerly circulation at middle and high latitudes.

Rotation (2) The turning of Earth on its axis; averages 24 hours in duration; determines day-night relationships.

Salinity (6) The concentration of natural elements and compounds dissolved in solution, as solutes; measured by weight in parts per thousand (‰) in seawater.

Salinization (15) A pedogenic regime that results from high potential evapotranspiration rates in deserts and semiarid regions. Soil water is drawn to surface horizons, and dissolved salts are deposited as the water evaporates.

Saltation (11) The transport of sand grains (usually larger than 0.2 mm, or 0.008 in.) by stream or wind, bouncing the grains along the ground in asymmetrical paths.

Salt marsh (13) Wetland ecosystems characteristic of latitudes poleward of the 30th parallel.

Sand sea (12) An extensive area of sand and dunes; characteristic of Earth's *erg deserts*.

Saturated (5) Air that is holding all the water vapor that it can hold at a given temperature.

Scale (1) The ratio of the distance on a map to that in the real world; expressed as a representative fraction, graphic scale, or written scale.

Scarification (10) Human-induced mass movements of Earth materials, such as large-scale open-pit mining and strip mining.

Scattering (3) Deflection and redirection of insolation by atmospheric gases, dust, ice, and water vapor; the shorter the wavelength, the greater the scattering.

Scientific method (2) An approach that uses applied common sense in an organized and objective manner; based on observation, generalization, formulation of a hypothesis, and ultimately the development of a theory.

Sea-floor spreading (8) As proposed by Hess and Dietz, the mechanism driving the movement of the continents; associated with upwelling flows of *magma* along the worldwide system of *mid-ocean ridges*.

Sediment (10) Fine-grained mineral matter that is transported and deposited by air, water, or ice.

Sedimentary rock (8) One of three basic rock types; formed from the compaction, cementation, and hardening of sediments derived from former rocks.

Seismic waves (8) The shock waves sent through the planet by an earthquake or underground nuclear test. Transmission varies according to temperature and the density of various layers within the planet.

Seismograph (9) A device that measures seismic waves of energy transmitted throughout Earth's interior.

Sensible heat (2) Heat that can be measured with a thermometer; a measure of the concentration of *kinetic energy* from molecular motion.

Sheet flow (11) Water that moves downslope in a thin film as overland flow, not concentrated in channels larger than rills.

Sheeting (10) A form of weathering associated with fracturing or fragmentation of rock by pressure release; often related to exfoliation processes.

Shield volcano (9) A symmetrical mountain landform built from *effusive eruptions*; gently sloped, gradually rising from the surrounding landscape to a summit crater; typical of the Hawaiian Islands.

Sinkholes (10) Nearly circular depressions created by the weathering of karst landscapes; also known as dolines; may collapse through the roof of an underground space.

Slipface (12) Formed as dune height increases above 30 cm (12 in.) on the leeward side at an angle (30° to 34°) at which loose material is stable—its angle of repose.

Slopes (10) Curved, inclined surfaces that bound landforms.

Small circle (1) Circles on a globe's surface that do not share Earth's center; for example, all parallels other than the equator.

Soil (15) A dynamic natural body made up of fine materials covering Earth's surface in which plants grow; composed of both mineral and organic matter.

Soil colloid (15) Tiny clay and organic particles in soil; provide chemically active sites for mineral ion *adsorption*.

Soil creep (10) A persistent mass movement of surface soil where individual soil particles are lifted and disturbed by the expansion of soil moisture as it freezes, or even by grazing livestock or digging animals.

Soil fertility (15) The ability of soil to support plant productivity when it contains organic substances and clay minerals that absorb water and certain elemental ions needed by plants.

Soil horizon (15) The various layers exposed in a *pedon*; roughly parallel to the surface and identified as O, A, E, B, and C.

Soil moisture storage (6) The retention of moisture within soil; represents a savings account that can accept deposits (**soil moisture recharge**) or experiences withdrawals (**soil moisture utilization**) as conditions change.

Soil Taxonomy (15) A soil classification system based on observable soil properties actually seen in the field; published in 1975 by the U.S. Soil Conservation Service.

Solar constant (2) The amount of insolation intercepted by Earth on a surface perpendicular to the Sun's rays when Earth is at its average distance from the Sun; a value of 1370 watts per m²; averaged over the entire globe at the thermopause; or, 1.968 calories per cm² per minute.

Solar wind (2) Clouds of ionized (charged) gases emitted by the Sun and traveling in all directions from the Sun's surface. Effects on Earth include auroras, disturbance of radio signals, and possible influences on weather.

Solifluction (14) Gentle downslope movement of a saturated surface material (soil and regolith), in various climatic regimes, where temperatures are above freezing.

Solum (15) A true soil profile in the *pedon*; ideally, a combination of A and B horizons.

Solution (10) The dissolved load of a stream.

Spatial analysis (1) The examination of spatial interactions, patterns, and variations over area and/or space; a key integrative approach of geography.

Specific heat (3) The increase of temperature in a material when energy is absorbed; water is said to have a higher specific heat.

Specific humidity (5) The mass of water vapor (in grams) per unit mass of air (in kilograms) at any specified temperature. The maximum mass of water vapor that a kilogram of air can hold at any specified temperature is termed its *maximum specific humidity*.

Speed of light (2) Specifically, 299,792 km per second (186,282 mi per second), covering more than 9.4 trillion km per year (5.9 trillion mi per year)—a distance known as a light-year.

Spheroidal weathering (10) A *chemical weathering* process in which the sharp edges and corners of boulders and rocks are weathered in thin plates that create a rounded, spheroidal form.

Spodosols (15) A soil order in the *Soil Taxonomy* classification that occurs in northern coniferous forests; best developed in cold, moist, forested climates of Dfb, Dfc, Dwc, Dwd; lacks humus and clay in the A horizons, with high acidity associated with *podzolization* processes.

Stability (5) The condition of a parcel, whether it remains where it is or changes its initial position. The parcel is stable if it resists displacement upwards, unstable if it continues to rise.

Steady-state equilibrium (1) The condition that occurs in a system when rates of input and output are equal and the amounts of energy and stored matter are nearly constant around a stable average.

Stomata (16) Small openings on the undersides of leaves, through which water and gases pass.

Storm surge (5) Large quantities of seawater pushed inland by the strong winds associated with a *tropical cyclone*.

Stratified drift (14) Sediments deposited by glacial meltwater that appear sorted; a specific form of *glacial drift*.

Stratigraphy (8) An analysis of the sequence, spacing, and spatial distribution of rock strata.

Stratocumulus (5) A lumpy, grayish, low-level cloud, patchy with sky visible, sometimes present at the end of the day.

Stratosphere (2) That portion of the *homosphere* that ranges from 20 to 50 km (12.5 to 30 mi) above Earth's surface; temperatures ranging from −57°C (−70°F) at the *tropopause* to 0°C (32°F) at the stratopause. The functional *ozonosphere* is within the stratosphere.

Stratus (5) A stratiform (flat, horizontal) cloud generally below 2000 m (6500 ft).

Strike-slip fault (9) Horizontal movement along a faultline, that is, movement in the same direction as the fault; also known as a transcurrent fault. Such movement is described as right-lateral or left-lateral, depending on the relative motion observed.

Subduction zone (8) An area where two plates of crust collide and the denser oceanic crust dives beneath the less dense continental plate, forming deep oceanic trenches and seismically active regions.

Sublimation (5) A process in which ice evaporates directly to water vapor or water vapor freezes to ice (deposition).

Subpolar low-pressure cell (4) A region of low pressure centered approximately at 60° latitude in the North Atlantic near Iceland and in the North Pacific near the Aleutians, as well as in the Southern Hemisphere. Air flow is cyclonic, that weakens in summer and strengthens in winter.

Subsolar point (2) The only point receiving perpendicular insolation at a given moment—the Sun directly overhead.

Subsurface diagnostic horizon (15) A soil horizon that originates below the *epipedon* at varying depths; may be part of the A and B horizons; important in soil description as part of the *Soil Taxonomy.*

Subtropical high-pressure cells (4) Dynamic high-pressure areas covering roughly the region from 20° to 35° N and S latitudes; responsible for the hot, dry areas of Earth's arid and semiarid deserts.

Sulfur dioxide (2) A colorless gas detected by its pungent odor; produced by the combustion of fossil fuels that contain sulfur as an impurity; can react in the atmosphere to form sulfuric acid, a component of acid deposition.

Summer (June) solstice (2) The time when the Sun's *declination* is at the Tropic of Cancer, at 23.5° N latitude; June 20–21 each year.

Sunspots (2) Magnetic disturbances on the surface of the Sun; occurring in an average 11-year cycle; related flares, prominences, and outbreaks produce surges in solar wind.

Surface creep (12) A form of eolian transport that involves particles too large for *saltation;* a process whereby individual grains are impacted by moving grains and slide and roll.

Surplus (6) The amount of moisture that exceeds *potential evapotranspiration;* moisture oversupply when *soil moisture storage* is at *field capacity.*

Suspended load (11) Fine particles held in suspension in a stream. The finest particles are not deposited until the stream velocity nears zero.

Swells (13) Regular patterns of smooth, rounded waves in open water; can range from small ripples to very large waves.

Syncline (9) A trough in folded strata, with beds that slope toward the axis of the downfold.

System (1) Any ordered, interrelated set of materials or items existing separate from the environment, or within a boundary; energy transformations and energy and matter storage and retrieval occur within a system.

Taiga (16) See northern needleleaf forest.

Tectonic processes (8) Driven by internal energy from within Earth; refers to large-scale movement and deformation of the crust.

Temperate rain forest (16) A major *biome* of lush forests at middle and high latitudes; occurs along narrow margins of the Pacific Northwest in North America, among other locations; includes the tallest trees in the world.

Temperature (3) A measure of *sensible heat* energy present in the atmosphere and other media, indicates the average *kinetic energy* of individual molecules within the atmosphere.

Temperature inversion (2) A reversal of the normal decrease of temperature with increasing altitude; can occur anywhere from ground level up to several thousand meters.

Tephra (9) Pulverized rock and clastic materials ejected violently during a volcanic eruption—all pyroclastics from a volcano.

Terranes (9) Migrating crustal pieces, dragged about by processes of mantle convection and *plate tectonics.* Displaced terranes are distinct in their history, composition, and structure from the continents that accept them.

Terrestrial ecosystem (16) A self-regulating association characterized by specific plant formations; usually named for the predominant vegetation and known as a *biome* when large and stable.

Thermal equator (3) A line on an isothermal map that connects all points of highest mean temperature.

Thermopause (2) A zone approximately 480 km (300 mi) in altitude that serves conceptually as the top of the atmosphere; an altitude used for the determination of the *solar constant.*

Thermosphere (2) A region of the *heterosphere* extending from 80 to 480 km (50 to 300 mi) in altitude; contains the functional *ionosphere* layer.

Thrust fault (9) A low-angle fault plane, relative to the horizontal, where an overlying block shifts (thrusts) over an underlying block.

Thunder (5) The violent expansion of suddenly heated air, created by *lightning* discharges, sending out shock waves as an audible sonic bang.

Tide (13) Patterns of daily oscillations in sea level produced by astronomical relationships among the Sun, the Moon, and Earth; experienced in varying degrees around the world.

Till (14) Direct ice deposits that appear unstratified and unsorted; a specific form of *glacial drift.*

Till plains (14) Large, relatively flat plains composed of unsorted glacial deposits behind a terminal or end moraine. Low-rolling relief and unclear drainage patterns are characteristic.

Tilted fault block (9) A tilted landscape produced by a *normal fault* on one side of a range; for example, the Sierra Nevada and the Grand Tetons.

Topographic map (1) A map that portrays physical relief through the use of elevation *contour lines* that connect all points at the same elevation above or below a vertical datum, such as *mean sea level.*

Topography (9) The undulations and configurations that give Earth's surface its texture; the heights and depths of local relief including both natural and human-made features.

Tornado (5) An intense, destructive cyclonic rotation, developed in response to extremely low pressure; associated with *mesocyclone* formation.

Total runoff (6) Surplus water that flows across a surface toward stream channels; formed by *sheet flow*, combined with precipitation and subsurface flows into those channels.

Traction (11) A type of sediment transport that drags coarser materials along the bed of a stream.

Trade winds (4) Northeast and southeast winds that converge in the *equatorial low pressure trough*, forming the *intertropical convergence zone.*

Transparency (3) The quality of light able to shine through a medium.

Transpiration (6) The movement, as *water vapor*, out through the pores in leaves of water drawn by plant roots from the *soil moisture storage.*

Transport (11) The actual movement of weathered and eroded materials by air, water, and ice.

Tropical cyclone (5) A cyclonic circulation originating in the tropics, with winds between 30 and 64 knots (39 to 73 mph); characterized by closed *isobars*, circular organization, and heavy rains (see *hurricane* and *typhoon*).

Tropical savanna (16) A major *biome* containing large expanses of grassland interrupted by trees and shrubs; a transitional area between the humid rain forests and tropical seasonal forests and the drier, semiarid tropical steppes and deserts.

Tropical seasonal forest and scrub (16) A variable *biome* on the margins of the rain forests, occupying regions of lesser and more erratic rainfall; the site of transitional communities between the rain forests and tropical grasslands.

Tropic of Cancer (2) The northernmost point of the Sun's *declination* during the year; 23.5° N latitude.

Tropic of Capricorn (2) The southernmost point of the Sun's *declination* during the year; 23.5° S latitude.

Troposphere (2) The home of the *biosphere*; the lowest layer of the *homosphere*, containing approximately 90% of the total mass of the *atmosphere*; extends up to the tropopause, defined by a temperature of –57°C (–70°F); occurring at an altitude of 18 km (11 mi) at the equator, 13 km (8 mi) in the middle latitudes, and at lower altitudes near the poles.

True shape (1) A map property showing the correct configuration of coastlines; a useful trait of conformality for navigational and aeronautical maps, although areal relationships are distorted.

Tsunami (13) A seismic sea wave, traveling at high speeds across the ocean, formed by sudden and sharp motions in the seafloor, such as sea-floor earthquakes, submarine landslides, or eruptions from undersea volcanoes.

Typhoon (5) A *tropical cyclone* in excess of 65 knots (74 mph) that occurs in the western Pacific.

Ultisols (15) A soil order in the *Soil Taxonomy* classification, featuring highly weathered forest soils, principally in the Cfa climatic classification. Increased weathering and exposure can degenerate an *Alfisol* into the reddish color and texture of these more humid to tropical soils. Fertility is quickly exhausted when Ultisols are cultivated.

Unconfined aquifer (6) The zone of saturation in water-bearing rock strata, with no impermeable overburden and recharge generally accomplished by water percolating down from above.

Uniformitarianism (8) An assumption that physical processes active in the environment today are operating at the same pace and intensity that has characterized them throughout geologic time; proposed by Hutton and Lyell.

Upslope fog (5) Forms when moist air is forced to higher elevations along a hill or mountain and is thus cooled.

Upwelling current (4) Ocean currents that cause surface waters to be swept away from the coast by surface divergence or offshore winds. Cool, deep waters, which are generally nutrient rich, rise to replace the vacating water.

Urban heat islands (3) Urban microclimates, which are warmer on the average than areas in the surrounding countryside because of various surface characteristics.

Valley fog (5) The settling of cooler, more dense air in low-lying areas, produces saturated conditions and fog.

Valley glacier (14) A type of *alpine*, or mountain, *glacier* within the confines of a valley; can range from 100 m (328 ft) to 100 km (62 mi) in length.

Vapor pressure (5) That portion of total *air pressure* that results from water vapor molecules; expressed in millibars (mb). At a given temperature the maximum capacity of the air is termed its *saturation vapor pressure.*

Ventifacts (12) The individual pieces and pebbles etched and smoothed by eolian erosion—abrasion by wind-blown particles.

Vernal (March) equinox (2) The time when the Sun's *declination* crosses the equatorial parallel and all places on Earth experience days and nights of equal length; around March 20–21 each year. The Sun rises at the North Pole and sets at the South Pole.

Vertisols (15) A soil order in the *Soil Taxonomy* classification that features expandable clay soils; composed of more than 30% swelling clays; occurs in regions that experience highly variable soil moisture balances through the seasons.

Volcano (9) A landform at the end of a conduit or pipe which rises from below the crust and vents to the surface. *Magma* rises and collects in a magma chamber deep below, resulting in eruptions that are effusive or explosive forming the mountain landform.

Warm front (5) The leading edge of an advancing warm air mass, which is unable to push cooler, passive air out of the way; tends to push the cooler, underlying air into a wedge shape; noted on weather maps with a series of rounded knobs placed along the front in the direction of the frontal movement.

Wash (12) An intermittently dry streambed that fills with torrents of water after rare precipitation events in arid lands.

Watershed (11) The catchment area of a *drainage basin*; delimited by divides.

Waterspout (5) An elongated, funnel-shaped circulation formed when a *tornado* takes place over water.

Water table (6) The upper limit of *groundwater*; that contact point between zones of saturation and aeration in an unconfined aquifer.

Wave (13) Undulations of ocean water produced by the conversion of solar energy to wave energy; produced in a generating region or a stormy area of the sea.

Wave-cut platform (13) A flat, or gently sloping, tablelike bedrock surface that develops in the tidal zone, where wave action cuts a bench that extends from the cliff base out into the sea.

Wave cyclone (5) An organized area of low pressure, with converging and ascending air flow producing an interaction of air masses; migrates along storm tracks. Such lows or depressions form the dominant weather pattern in the middle and higher latitudes of both hemispheres.

Wavelength (2) Measurement of waveform; the actual distance between the crests of successive waves. The number of waves passing a fixed point in one second is called the frequency of the wavelength.

Wave refraction (13) A process that concentrates energy on *headlands* and disperses it in coves and bays; a coastal straightening process.

Weather (5) The short-term condition of the *atmosphere*, as compared to *climate*, which reflects long-term atmospheric conditions and extremes. *Temperature*, *air pressure*, *relative humidity*, wind speed and direction, *daylength*, and Sun angle are important measurable elements that contribute to the weather.

Weathering (10) The related processes by which surface and subsurface rock disintegrate, or dissolve, or are otherwise broken down. Rocks at or near Earth's surface are exposed to physical, organic, and chemical weathering processes.

Westerlies (4) The predominant wind flow pattern from the subtropics to high latitudes in both hemispheres.

Western intensification (4) The piling up of ocean water along the western margin of each ocean basin, to a height of about 15 cm (6 in.); produced by the *trade winds* that drive the oceans westward in a concentrated channel.

Wilting point (6) That point in the soil moisture balance when only *hygroscopic water* and some bound *capillary water* remains. Plants wilt and eventually die after prolonged stress from a lack of available water.

Wind (4) The horizontal movement of air relative to Earth's surface; produced essentially by *air pressure* differences from place to place; also influenced by the *Coriolis force* and surface *friction*.

Wind chill factor (3) An indication of the enhanced rate at which body heat is lost to the air. As wind speed increases, heat loss from the skin also increases.

Wind vane (4) A device to determine wind direction.

Winter (December) solstice (2) That time when the Sun's *declination* is at the Tropic of Capricorn, at 23.5° S latitude, December 21–22 each year.

Zone of aeration (6) A groundwater zone above the *water table*, which may or may not hold water in pore spaces.

Zone of saturation (6) A groundwater zone below the *water table*, in which all pore spaces are filled with water.

INDEX

A

Abbey, Edward, 384-389
Abiotic spheres, 9, 9*f*
Ablation, 427, 447
Abrasion
    eolian, 373-374
    fluvial, 348
    glacial, 428
Absolute age, 252
Absorption, 80
Acid deposition, 205-206
Acidity, in soils, 458-459
Actual evapotranspiration (ACTET), 190
Adaptation, to climate change, 498
Adiabatic processes, 151
Adiabatic rate
    dry (DAR), 152, 153*f*
    moist (MAR), 152, 153*f*
Adsorption, 458
Advanced very high resolution
    radiometer (AVHRR), 29
Advection, 80
Advection fog, 159
Aeration, 196
Aerosol optical thickness (AOT), 108
Agassiz, Louis, 434
Agenda 21 (of Earth Summit), 537-538
Aggradation, 350
Agricultural extension service, 454
Air masses, 161-163, 162*f*
    classification of, 161-162
Air parcel, 151, 152*f*
Air pollution, 62-70
    anthropogenic, 64-70, 65*t*
    industrial smog, 68-70
    natural factors affecting, 63-64
    natural sources of, 62*t*, 62-63
    ozone, 67-68
    photochemical smog, 65-67
Air pressure, 111-114, 115*f*, 120*f*

gravity and, 115
    primary high-pressure and low-
      pressure areas, 119-125
Alabama Hills, 326
Albedo, 76-78, 78*f*
Albers conic projection, 23*f*, 25
Aleutian low, 124
Alfisols, 471, 472*f*-473*f*
Alkalinity, in soils, 458-459
Alleghany orogeny, 293
Alluvial fan, 383-384, 385*f*
Alluvial terraces, 356-357, 357*f*
Alluvium, 342
Alpine glaciers, 422-424, 426*f*
Alpine permafrost, 438
Alpine tundra, 506*f*-507*f*, 524, 525*f*
    characteristics of, 508*t*-509*f*
Altitude
    of Sun, 47
    and temperature, 86-89, 89*f*
Amazon River, 202*f*, 361
Andisols, 476, 477*f*
Anemometer, 110
Aneroid barometer, 114
Angle of equilibrium, 318
Angle of repose, 376
Annapolis Tidal Generating Station,
    399, 400*f*
Annular drainage pattern, 347, 347*f*
Antarctica, ozone loss over, 59
Antarctic circle, 13-14
Antarctic high, 125
Antarctic ice sheet, 446
Antarctic region, 445-446, 447*f*
Antarctic zones, 13, 13*f*
Anthracite coal, 265
Anticline, 287
Anticyclone, 118, 124*f*
Aphelion, 38
Apollo XI, 41, 42*f*
Appalachian mountains, 293

Apparent temperature, 98
Aquatic ecosystems, 502
Aquatic succession, 501-502
Aquiclude, 197
Aquifer, 197
Aquifer recharge area, 197
Arboreal, definition of, 511
Arctic circle, 13-14
Arctic National Wildlife Refuge
    (ANWR), 535, 535*f*
Arctic region, 445, 447*f*
Arctic tundra, 506*f*-507*f*, 523-524, 524*f*
    characteristics of, 508*t*-509*f*
Arctic zones, 13, 13*f*
Arcuate delta, 358, 359*f*
Arecibo Observatory, 328*f*
Arête, 431
Aridisols, 468-471, 469*f*, 472*f*
Ar Rub' al Khali Erg, 375, 377*f*
Artesian water, 197
Association of American Geographers
    (AAG), 4
Asthenosphere, 257
Atmosphere, 9, 9*f*, 53-70, 55*f*
    density of, 112*f*
    driving forces in, 115-118
    energy in, 76-80
    fluctuation in gases of, 444, 444*f*
    lifting mechanisms in, 163-168
    origin of, 41
    patterns of motion of, 118-134
    pollution of. *See* Air pollution
    stability of, 151-154
    stable and unstable conditions of,
      152-154
    upper, circulation in, 125-126
    variable components of, 62-70
Atmospheric circulation. *See* Wind(s)
Atmospheric conditions, 152-154
Atoll, 412, 413*f*
Aurora australis, 42

Aurora borealis, 42, 43*f*
Autotrophs, 493
Autumnal equinox, 52
Available water, 191-192
Avalanche, debris, 324
Avalanche slope, 376
Avoided damage, 362
Axial parallelism, 50
Axis, 49
Azimuthal projection, 22, 23*f*
Azores high, 124

B

Backshore, 397*f*, 406-408
Backswamp, 355*f*, 356
Bacon, Sir Francis, 267
Badland, 389
Bagnold, Ralph, 372
Bajada, 384, 392*f*
Balanced Rock, 389, 390*f*
Bank storage, 388
Barchan dune, 377, 378*f*
Barchanoid ridge, 377, 378*f*
Barometer, 113, 113*f*
Barrier beaches, 408-409
Barrier islands, 408-409, 411*f*
Barrier reef, 412, 413*f*
Barrier spit, 405, 407*f*
Basalt(s), 257, 305*f*
    plateau, 306
Basaltic lava flows, 263*f*
Base flow, 365
Base level, 316, 317*f*
Basin and Range Province, 389-390, 391*f*
Basins, 288
Bates, Marston, 530
Batholith, 262, 263*f*
Bay barrier, 405, 407*f*
Baymouth bar, 405
Bay of Fundy, 399, 400*f*
Beach(es), 406-408, 407*f*
Beach drift, 402, 403*f*
Beach Management Act, 418
Beach nourishment, 408, 410
Bed load, 350
Bedrock, 320
Bergschrund crevasse, 427
Berm, 397*f*, 406-408
Bermuda high, 124
Biodiversity. *See also* Diversity, of
        ecosystems
    Earth Summit agreement on, 536-
        537
    efforts to preserve, 513-515
Biogeochemical cycles, 490-491
Biogeographical communities, 515
Biogeography, 482
    island, 513-515

Biomass, 486, 487*t*
Biome(s), 214, 503, 506*f*-507*f*
    characteristics of, 508*t*-509*t*
    desert, 521-523
    equatorial and tropical rain forest,
        505-511, 510*f*
    Mediterranean shrubland, 519-520
    midlatitude broadleaf and mixed
        forest, 517-518
    midlatitude grasslands, 520-521,
        521*f*
    northern needleleaf forest and
        montane forest, 518, 518*f*
    temperate rain forest, 518-519,
        519*f*
    tropical savanna, 515-517, 517*f*
    tropical seasonal forest and scrub,
        511-515, 516*f*
Biosphere, 9*f*, 10
Biosphere reserves, 513-515
Biotic sphere, 9, 9*f*
Bird-foot delta, 358, 360*f*
Bituminous coal, 265
Blowout depressions, 373
Bolson, 390, 392*f*
Bora, 130
Boreal forest, 518, 518*f*
Boulder Canyon Act, 386
Bow Glacier, 420*f*
Braided stream pattern, 350, 350*f*
Breaker, 400
Breakwater, 408, 409*f*
Bridalveil Falls, 432*f*
Brine, definition of, 209
Bryophytes, 505
Bypass channels, 362

C

Cairo Compact (1989), 244
Calcification, 471, 473*f*
Caldera, 303
Caliche, 471
California, earthquakes in, 295-302
Calving, definition of, 424
Capillary water, 191
Capture, 347, 351
    elbows of, 347
Carbon, in atmosphere,
        anthropogenic, 493
Carbonation, 325
Carbon cycle, 490-491, 491*f*
Carbon dioxide
    in atmosphere, 57
    and global warming, 238*f*, 239-240
Carbon monoxide, 65
Carlsbad Caverns, 329*f*
Carnivore, 494
Carson, Rachel, 396

Carter Lake, Nebraska, 351, 352*f*
Cartography, 19-25
Cascade Glacier avalanche, 334, 335*f*
Catastrophism, 252
Cation-exchange capacity (CEC), 458
Caverns, 328, 329*f*
Caves, 328, 330*f*
    exploration of, 331
Celsius, Anders, 87
Celsius scale, 87-88
Channel patterns, 350-351
Chaparral, 228, 520, 520*f*
Chemical sediments, 264-265
Chemical weathering, 323-325
Chlorofluorocarbon compounds
        (CFCs), 60
    and global warming, 241
Chlorophyll, 485
Cinder cone, 303
Circle of illumination, 49, 51*f*
    definition of, 49
Circum-Pacific belt, 292
Cirque, 422-424
Cirque glacier, 424
Cirrus clouds, 157*f*, 158
Clastic sediments, 264-265
Climate(s), 142, 214. *See also* Climatic
        regions
    changes in, 237-244
    Earth Summit agreement on, 536
    components of, 214-218
    definition of, 214
    of deserts, 382
    effects on ecosystem, 489, 497-498,
        498*f*
    global patterns, 222
    models for, 237-238
    warming of. *See* Global warming
    and weathering, 320
Climatic designations, 219-222
Climatic regions
    classification of, 218-237
    criteria for, 219
    empirical, 219
    genetic, 218-219
    Köppen, 219-224
    moisture, 221*f*
    thermal, 220*f*-221*f*
    cold midlatitude desert, 236, 237*f*
    cold midlatitude steppe (BSk), 237,
        238*f*
    dry arid and semiarid (B),
        235-237
    and ecosystems, 214
    hot low-latitude desert (BWh), 236,
        236*f*
    hot low-latitude steppe (BSh), 236-
        237

humid continental hot summer (Dfa, Dwa), 230-231, 231*f*
humid continental mild summer (Dfb, Dwb), 231, 232*f*
humid subtropical hot summer (Cfa), 225-226, 226*f*
humid subtropical winter drought (Cwa), 226-227
marine west coast (Cfb, Cfc), 227-228, 229*f*
Mediterranean dry summer (Csa, Csb), 228-229, 230*f*
mesothermal (C), 225-229
microthermal (D), 229-234
polar (E), 234-235
subarctic (Dfc, Dwc, Dwd), 231-234, 233*f*
tropical (A), 222-224
tropical monsoon (Am), 223-224
tropical rain forest (Af), 222-223, 223*f*
tropical savanna (Aw), 224
Climatic sensitivity, 238
Climatic warming. *See* Global warming
Climatology, 214
Climax community, 498
Climographs, 222
Closed system, 5, 6*f*
definition of, 5
Earth's matter as example of, 8-9
Cloud cover, 89-90
Clouds, classification of, 154-158
Coal, 265
Coast, 396, 397*f*
Coastal environment(s), 396-397
human impact on, 418
Coastal processes
depositional, 405-409
erosional, 404-405, 416-418, 417*f*
Coastal straightening, 403*f*
Coastal system
actions of, 397-405
components of, 396-397
outputs of, 405-414
Coastline processes, organic, 410-412
Coastlines
emergent, 409
environmental planning for, 416-418
submergent, 409-410
Cockpit karst, 326, 328*f*
Col, 431
Cold front, 166-167, 167*f*
Cold midlatitude desert climates, 236, 237*f*
Cold midlatitude steppe climates, 237, 238*f*

Colorado River, 317*f*, 386-389, 387*f*
water budget for, 388*t*
Colorado River Compact, 386
Column, 328, 329*f*
Community, 482-484
Competence, definition of, 349
Competition exclusion principle, 484
Composite cone, 306-307
Composite overlay, 31
Composite volcano, 306-307, 307*f*
Compound valley glacier, 427
Compression, 287-291
Condensation, 145
Condensation nuclei, 154
Conduction, 80
Cone of depression, 198
Confined aquifer, 197
Conic projection, 22, 23*f*
Consistence, 456
Consumers, 493
primary, 494
secondary, 494
tertiary, 494
Contact metamorphism, 266
Continental divides, 342, 343*f*
Continental drift, 267-271
Continental glaciation, erosional and depositional features of, 433-436
Continental glacier, 447
Continental glaciers, 424
Continentality, definition of, 92
Continental platforms, 283
Continental polar (cP) air masses, 161
Continental shields, 285-286
Continents, 271, 272*f*
Contour interval, definition of, 26
Contour lines, 26, 26*f*
Convection, definition of, 80
Convectional lifting, 163*f*, 163-164, 164*f*
Convergence, 292-293, 294*f*
Coordinated Universal Time (UTC), 18-19
Coral, 410
bleaching of, 414
distribution of, 411*f*
reefs, 412, 413*f*
Cordilleran system, 295
Core of Earth, 255
Coriolis, Gaspard, 121
Coriolis force, 115-118, 117*f*-118*f*, 121
Coupled general circulation models (CGCMs), 238
Coupled Ocean-Atmosphere Response Experiment (COARE), 92
Crater, 303
Craton, 285

Crescentic dunes, 377, 378*f*
Crevasse(s), 427, 429*f*
bergschrund, 427
headwall, 427
Crust, 257-258
deformation of, 287-291
formation of, 285-287
new, 269-270
subduction of, 270
Crystallization, 321-322
Cumulonimbus clouds, 157*f*, 158, 158*f*
Cumulus clouds, 155, 157*f*
Cut bank, 350, 351*f*
Cutoff, 351
Cyanobacteria, 485
Cyclogenesis, 168
Cyclone(s), 118, 126*f*
midlatitude, 168-172, 169*f*
tropical, 175-179, 177*f*
classification of, 176*t*
Cylindrical projection, 22, 23*f*

D
Daily water budget, 204-206
Dams
functions of, 362
geologic assessment of site for, 365
DAR (Dry adiabiatic rate), 152, 153*f*
Darwin, Charles, 412
Davis, Margaret, 498
Davis, William Morris, 429-430
Davis Dam, 386, 387*f*
Dawn, 53
Daylength, 47
Daylight saving time, 19
Death Valley, 392*f*
Debris avalanche, 324
Debris slope, 318
Declination, of Sun, 47
Decomposers, 495
Deficit (DEFIC), 190
Deflation, eolian, 373
Deforestation, 511, 512*f*, 516
Delaware River basin, 345
Delta, 357-361
arcuate, 358, 359*f*
bird-foot, 358
estuarian, 358
Dendritic drainage pattern, 345, 346*f*-347*f*
Dentrifying bacteria, 491
Denudation, 316-319
Deposition, 144-145, 342
of streams, 354-361
Depositional landforms
coastal, 405-409, 407*f*
glacial, 432-433
Deranged drainage pattern, 347, 347*f*

Desert(s)
  climates of, 382
  cold, 523, 523*f. See also* Desert
    biomes
  distribution of, 383*f*
  effects of off-road vehicles on, 373
  effects of rainfall on, 384*f*
  effects of warfare on, 373
  erg, 375, 377*f*
  fluvial processes in, 382-384
  landscapes of, 382-390
  warm, 522*f*, 522-523. *See also*
    Desert biomes
Desert biomes, 506*f*-507*f*, 521-523
  characteristics of, 508*t*-509*f*
Desert pavement, 372*f*, 373
Design flood, 362
Dew-point temperature, 149
Differential weathering, 321
Diffuse radiation, 79
Diffuse reflection, 78-79
Discharge, 201, 202*f*, 342
Discontinuity, 257
Dissolved load, 349
Distributaries, 357
Diversity, of ecosystems, 496-498
Doline karst, 326
Dome dune, 379*f*
Domes, 288
Downwelling current, 135
Drainage basin, 342, 344*f*
Drainage basin system, 342-347
Drainage divides, 342, 343*f*
Drainage patterns, 345-347, 346*f*-347*f*
Drainage water, problems in disposal
  of, 470
Drawdown, 198
Drip curtain, 330*f*
Dripstones, 328
Drip stones, 330*f*
Drumlin, 434-436, 435*f*-436*f*
Dry adiabatic rate (DAR), 152, 153*f*
Dry arid and semiarid climates, 235-
  237
Dry soil, 457
Dune(s), 375
  forms of, 376-379, 378*f*-379*f*
Dust Bowl, 374, 382
Dust dome, 101
Dust storm, 374
Dynamic equilibrium, 7, 353, 499
Dynamic equilibrium model, of
  landforms, 318-319

E
Earth
  advocacy for, 530, 539-540
  axis of, 49, 49*t*

core of, 255
in cross section, 254-258, 255*f*-256*f*
deformation of crust of, 287-291
formation of crust of, 285-287
human relationships to, 3*f*, 4
internal energy of, 252-258
location on, 12-17
magnetism of, 255
mantle of, 257
measurement of, 10*f*, 10-12
orbit of, 38
physical matter of, 8-9
Planet of the Year, 539*f*
relief features of surface of, 282-
  285
revolution about Sun, 48, 49*f*
rotation about axis, 49, 49*f*, 116-118
spheres of, 9*f*, 9-10
sphericity of, 10*f*, 10-12, 49*t*
structural regions of, 293-295, 296*f*
structure of, 252-258
as system, 8-9
temperature patterns of, 93-101
topographic regions of, 284*f*, 284-
  285
Earth charter, 537
Earth Observation System (EOS), 8
Earthquakes, 295-303, 298*t*
  damage caused by, 299
  forecasting of, 302-303
  frequency of, 297*t*
  measurement of, 297-298
  and plate boundaries, 275
  tectonic, 297
Earth Summit (1992), 536-538, 537*f*
Earth systems, concepts of, 5-10
Easterlies, polar, 125
Ebb tide, 398
Ecliptic, 38, 40*f*, 49
Ecological climax, 498
Ecological pyramids, 495-496, 496*f*
Ecological succession, 498-502
Ecology, 482
Ecosphere, 9*f*, 10
Ecosystems, 214, 482, 502-505
  aquatic, 502
  components of, 483*f*
  abiotic, 486-493
  biotic, 485-486
  diversity of, 496-498
  effects of humans on, 497
  large marine (LME), 503
  management of, 503
  operations of, biotic, 493-496
  primary production and plant
    biomass of, 487*t*
  stability of, 496-498, 497*f*
  terrestrial, 503-505

Ecotone, 505
Effusive eruptions, 304-306
Elastic-rebound theory, 298
Elbows of capture, 347
El Chichón, 244
Electromagnetic spectrum, 42-44, 43*f*
El Niño/Southern Oscillation (ENSO),
  215-216, 216*f*, 414
Eluviation, 456
Embayments, 409-410
Emergence, 397
Emergent coastlines, 409
Empirical classification, 219
End moraine, 433, 433*f*, 435*f*
Endogenic system, 249
Energy
  of Earth, 8, 252-258
  at surface, 81-83
  geothermal, 303
  intercepted, 44-46
  kinetic, 56
  radiant, 42-44
  solar, 39-44, 48*f*, 84-85, 489
Energy balance, of Earth's surface, 82-
  83
Entisols, 474-475, 476*f*
Entrenched meanders, 353, 354*f*
Environmental lapse rate, 58, 151
Environmental problems
  international cooperation on, 534-
    539
  popular concern with, 530-531
Eolian, definition of, 372
Eolian processes, 372-380
  deposition, 375-379
  erosion, 373-374
  transport, 374-375
  prevention of, 375, 376*f*
Ephemeral lake, 383
Ephemerals, 521
Epicenter, 298
Epipedon, 461
Epiphytes, 505
Equal area, 22
Equatorial and tropical rain forest,
  505-511, 506*f*-507*f*, 510*f*
  characteristics of, 508*t*-509*f*
Equatorial countercurrent, 135
Equatorial low-pressure trough, 119,
  121-122
Equatorial zones, 13, 13*f*
Equilibrium
  definition of, 7
  dynamic, 7
  of Earth's energy, 8
  steady state, 7
Equilibrium line, 427
Equinox

autumnal, 52
vernal, 50
Eratosthenes, 11-12
Erg desert, 375, 377*f*
Erosion, 342
  fluvial, 342
  glacial, 428, 430*f*
  sand, prevention of, 376*f*
Erosional landforms
  coastal, 404-405
  glacial, 429-432, 431*f*
Eruptions
  effusive, 304-306
  explosive, 306-308
Escarpment, 290
Esker, 434
Estuarian delta, 358
Estuary, 358, 410
Eurasian-Himalayan system, 295
European Space Agency, 30
Eustasy, 143
Eutrophication, 501
Evaporation, 90, 145, 188
Evaporation fog, 159
Evaporation pan, 189, 189*f*
Evaporimeter, 189, 189*f*
Evaporites, 265*f*, 265-266
Evapotranspiration, 188-189
Exfoliation, 323
Exfoliation dome, 323, 324*f*
Exogenic system, 249
Exosphere, 55, 55*f*
Exotic stream, 201, 383, 386-389
Explosive eruptions, 306-308
Extinction, 513
Exxon Valdez oil spill, 531*f*, 531-532, 533*f*
Eyewall swirls, 177

F

Fahrenheit scale, 87-88
Fault block, 292
Faulting, 290-291, 298-302
Fault plane, 290
Fault scarp, 290
Feedback, 6-7
Feedback loops, 6
Field capacity, 192, 457
Fire ecology, 500-501
Firn, 424, 447
Firn line, 427
Fissures, 304
Fjord(s), 410, 432
Flash flood, 382
Flood(s), 356, 357*f*, 361
  control on Colorado River, 389
  definition of, 361
  management of, 361-366
  probable maximum (PMF), 362

Floodplains, 354-356, 355*f*
  agricultural use of, 356
  management of, 362-365
  strategies for, 362
Flood tide, 398
Flow, 333, 336
Fluvial, definition of, 342
Fluvial erosion, 342
Fluvial processes, 342-361
  in deserts, 382-384
Fluvial transport, 348-350, 349*f*
Fog, 154, 159-161, 160*f*
  and desert organisms, 161
  types of, 159
Folding, 287-290, 288*f*-289*f*
Food, world resources of, 494, 495*f*
Food chain, 493
Food web, 493-494, 495*f*
Foreshore, 397*f*, 406-408
Forests. *See also* Rain forest,
    Midlatitude broadleaf and mixed
    forest, Needleleaf and Montane
    forest, Temperate forest
  Earth Summit agreement on care
    of, 537
Formation classes, 503
Fractional scale, 20
Freedunes, 376-377
Free face, 318
Freeze-thaw action. *See* Frost action
Frequency, definition of, 42
Friable, 457
Friction force, 115, 117*f*, 118
Fringing reef, 412, 413*f*
Front(s), 165. *See also specific fronts*
Frontal lifting, 165-166
Frost action, 322-323, 323*f*
  processes of, 438
Frost-wedging, 323
Frozen groud phenomena, 438
Funnel clouds, 173
Fusion, definition of, 40

G

Gaia hypothesis, 532-534
Galileo, 15, 112
Ganges River, delta of, 357, 359*f*
Gaseous cycles, 490-491
Geiger, Rudolph, 219
Gelifluction, 438
General circulation models (GCMs),
    238, 241, 243*f*
Generating region, 400
Genetic classification, 218-219
Geodesy, definition of, 10
Geographic continuum, 4*f*, 4-5
Geographic information system (GIS),
    30-31, 31*f*

Geographic zones, latitudinal, 13*f*, 13-14
Geography
  definition
    of, 2-3
  foundations of, 1-36
  physical versus human/cultural, 4
  science of, 2-5
Geoid, definition of, 10-12
Geologic cycle, 259-267, 260*f*
Geologic time scale, 252, 253*f*
Geomorphic threshold, 318
Geomorphology, 316
Geostationary Operational
    Environmental Satellite (GOES),
    30, 171*f*, 172
Geostrophic winds, 116, 119*f*
Geothermal energy, 303
Glacial, 441
Glacial drift, 432, 447
Glacial erosion, 428, 430*f*
Glacial erratic, 434
Glacial ice, 424-427, 447
  formation of, 424-427
Glacial landforms
  depositional, 432-433
  erosional, 429-437, 431*f*
Glacial movement, 427-428, 429*f*
Glacial plucking, 428
Glacial polish, 428, 430*f*
Glacial processes, 424-428
Glacial surge, 428
Glacial system, mass balance of, 426*f*,
    427
Glaciation
  astronomical factors in, 443, 444*f*
  geographical-geological factors in,
    444
  geophysical factors in, 443-444
Glacier(s), 243-244, 422
  alpine, 422-424, 426*f*
  cirque, 424
  compound valley, 427
  continental, 424, 447
  loss of mass, 428
  piedmont, 424
  tidal, 424
  types of, 422-424
  valley, 422, 423*f*, 447
Glen Canyon dam, 389
Global Absolute Sea Level Monitoring
    System, 397
Global Change Program, 244
Global cooling, 244
Global Energy and Water Cycle
    Experiment (GEWEX), 187
Global Positioning System (GPS), 397
Global warming, 239-241, 242*f*
  consequences of, 241-244
  effects on biosphere, 242-243

Global warming, *(cont.)*
   effects on world food supply, 242-243
   solutions to, 244
Global winds, 110-111, 111*f*
GMS weather satellite, 30
Goddard Institute, 241
GOES, 171*f,* 172
Goode, J. Paul, 25
Goode's homolosine projection, 24*f,* 25
Graben, 291, 390, 392*f*
Graded stream, 353
Gradient, 353, 353*f*
Grand Erg Oriental, 375
Granite, 257
Granular disintegration, 325
Graphic scale, 21
Gravitational force, and air pressure, 115
Gravitational water, 192
Great circle, 16-17
   definition of, 16
Great Salt Lake, rising level of, 446
Greenhouse effect, 80, 239-241, 493
Greenwich Mean Time (GMT), 18
Groin, 408, 409*f*
Gros Ventre landslide, 336
Ground ice, 438, 439*f*
Ground moraine, 433, 435*f*
Groundwater, 196*f*-197*f,* 196-200
   mining of, 198
   pollution of, 198-200
   utilization of, 198
Groundwater mining, 199-200
Groundwater resource potential, 195*f*
Gulf Stream, 91, 91*f*
Gully, 345
Guttenberg discontinuity, 257
Gyre, 134

H
Habitat, 484
Habitat niche, 484
Hadley cell, 122, 123*f*
Hail, 172
Halophytic, definition of, 413
Hanging valleys, 432, 432*f*
Harrison, John, 15-16
Hawaiian-Emperor islands chain, 276, 276*f*
Headlands, 401
Headwall crevasse, 427
Heat, 56. *See also* Latent heat
Heat index, 99, 99*f*
Heat lightning, 172
Herbivore, 494
Herbs, 505

Heterosphere, 54-56, 55*f*
Heterotrophs, 493
Hills, 285
Histosols, 478
Homosphere, 55*f,* 56-62
Hoover Dam, 386, 387*f*
Horn, 431
Horst, 291, 390, 392*f*
Hot low-latitude desert climates, 236, 236*f*
Hot low-latitude steppe climates, 236-237
Hot spots, 275*f,* 275-277
Hubbard Glacier, 428, 430*f*
Human-Earth relationships, 3*f,* 4
Humid continental hot summer climates, 230-231, 231*f*
Humid continental mild summer climates, 231, 232*f*
Humidity, 148-151
   relative, 148-150, 149*f*
   specific, 150-151
Humid subtropical hot summer climates, 225-226, 226*f*
Humid subtropical winter drought climates, 226-227
Humus, 455
Hurricane Andrew, 2, 29, 106*f,* 180-181
Hurricane Emily, 411*f*
Hurricane Gilbert, 177-178, 178*f*
Hurricane Hazel, 412
Hurricane Hugo, 412
Hurricanes, 177-179
   benefits of, 179
Hutton, James, 252
Hydration, 322
Hydraulic action, 348
Hydrogen bonding, 144
Hydrograph, 365, 366*f*
   effects of urbanization on, 365-366, 366*f*
Hydrologic cycle, 186-187, 187*f,* 259-267
   model of, 186-187, 187*f*
Hydrolysis, 324-325
Hydrosphere, 9*f,* 9-10
Hydrothermal vents, 209
Hygroscopic nuclei, 154
Hygroscopic water, 191
Hypocenter, 298
Hypsographic curve, 283, 283*f*
Hypsometry, 283-284

I
Ice, 145
Ice age, 441, 448. *See also* Glaciation
Icebergs, 424
Ice cap, 234, 424, 425*f*

Ice field, 424, 425*f*
Icelandic low, 124
Ice sheet(s), 243, 424
Ice shelves, 446, 447*f*
Ice wedge, 438, 439*f*
Igneous rocks, 261, 263*f*
Illuviation, 456
Impermeable, 197
Inceptisols, 475476
Industrial smog, 68-70
Infiltration, 186
Inselberg, 389
Insolation, 44-45, 46*f,* 76
   in city environments, 100*f,* 100-101
Intercepted energy, 44-46
Interglacial, 441
Internal differentiation, 254
Internal drainage, 203
International Date Line, 18
Intertropical convergence zone (ITCZ), 120*f,* 121, 122*f,* 125, 224
Ion, definition of, 56
Ionosphere, 55*f,* 56
Ions, 458
Intergovernmental Panel on Climate Change (IPCC), 241
Isobar, 115
Isostasy, 257, 258*f*
Isostatic rebound, 258
Isotherm, 94

J
Jakobshavn Glacier, 428
James Bay Project, 363
Japan Current, 91-92
Japan Weather Association, 30
Jet stream, 125-126, 127*f*-128*f*
Jetty, 408, 409*f*
John Hopkins Glacier, 426*f,* 428
Johnson, David, 309
Joint-block separation, 323, 323*f*
Joints, 320

K
Kame, 434, 435*f*
Karst
   cockpit, 326, 328*f*
   doline, 326
   tower, 328, 329*f*
Karst topography, 326-328, 327*f*
Karst valley, 326, 327*f*
Katabatic winds, 129
Kelvin scale, 87-88
Kesterton National Wildlife Refuge, 470
Kettle, 434, 435*f*
Kinetic energy, 56
Köppen, Wladimir, 219

Köppen-Geiger climate classification, 219-224, 220*f*-221*f*
Krakatoa, 307
Kramer International solar-thermal energy installation, 84, 85*f*
Kuroshio, 91-92

L
Lacustrine deposits, 445
Lagoon, 405, 407*f*
Lahar, 331
Lake Bonneville, 445, 445*f*
Lake Lahontan, 445, 445*f*
Lakes, paternoster, 432
Landsat satellite, 28*f*, 29
Landslide, 324-325
Land-water heating differences, 90-93
Lapse rate, 58, 62*f*, 151-152
Large marine ecosystem (LME), 503
Latent heat, 146, 147*f*
    of condensation, 146
    of evaporation, 83, 147
Lateral moraine, 427, 432, 433*f*
Laterization, 466, 467*f*
Latitude, 13-14
    definition of, 13
    methods for determination of, 13
    and temperature, 86-89, 89*f*
Latitudinal geographic zones, 13*f*, 13-14
Lava, 261, 303
Leeward side, 376
Levees, natural, 355-356
Lianas, 505
Lichen, 484
Life forms, 504*f*, 505
Life zones, 489*f*, 489-490
Light
    effects on ecosystem, 489
    speed of, 38
Lightning, 172
Lignite, 265
Limestone, 265
    dissolution of, 325
Limited Test Ban Treaty (1962), 534
Limiting factor, 493, 494*f*
Linear dunes, 377-378, 379*f*
Lithification, 264
Lithosphere, 9*f*, 10, 257-258, 305*f*
Littoral current, 401-402
Littoral drift, 403*f*, 405
Littoral zone, 396, 397*f*
Lives of a Cell, The, 54
Local base level, 316, 317*f*
Location, 3*f*, 4
    definition of, 3*f*
    on Earth, 12-17
Loess deposits, 379-380, 381*f*

Longitude, 14*f*, 14-16
    definition of, 14
Longitudinal dune(s), 378, 379*f*
Longitudinal profile, of stream, 353, 353*f*
Longshore current, 401-402, 403*f*
Longwave radiation, 43
Lovelock, James, 532-534
Lyell, Charles, 252

M
MacArthur, R. H., 513
Madison River Canyon landslide, 332, 332*f*
Magma, 261
Magnetic reversal, 256, 269, 269*f*
Magnetism, of Earth, 255-257
Magnetosphere, 41
Mangrove swamps, 412-413, 415*f*
Mantle, 257
    attempts to drill to, 259
Map(s), 19-27
    definition of, 19
    planimetric, 25
    scale, 20-21
    temperature, 94*f*, 94-97, 96*f*-97*f*
    topographic, 25-27
Mapmaking, 12-13
Map projection, 21-25
    classes of, 22-24, 23*f*
    definition of, 21-25
    properties of, 22
Maquis, 520
Margulis, Lynn, 532-534
Marine, definition of, 92
Marine polar (mP) air masses, 161-162
Marine west coast climates, 227-228, 229*f*
Maritime tropical (mT) air masses, 163
MAR (Moist adiabiatic rate), 152, 153*f*
Mass movement(s), 330-337
    classes of, 333*f*, 333-336
    definition of, 331
    human-induced, 336-337
    mechanics of, 331-333
McHarg, Ian, 416-418
McPhee, John, 390
Meandering stream, 350-351, 351*f*
Meanders, 350-351, 351*f*-352*f*
    entrenched, 353, 354*f*
Mean sea level, 397
Medial moraine, 427, 432, 433*f*
Mediterranean dry summer climates, 228-229, 230*f*
Mediterranean shrubland, 506*f*-507*f*, 519-520
    characteristics of, 508*t*-509*f*
Mercalli scale, 297

Mercator, Gerardus, 24
Mercator projection, 23, 23*f*
Mercury barometer, 113-114
Meridian
    definition of, 14
    standard, 22-23, 23*f*
Meridional flows, 110
Mesocylone, 173
Mesosphere, 55*f*, 57
Mesothermal climates, 225-229
Metamorphic rock, 266*f*, 266-268
    hardness of, 267
Meteorology, 142
METEOSAT weather satellite, 30
Methane, and global warming, 240-241
Miami silt loam, 457*f*
Michener, James, 286
Microclimate, 490
Microclimatology, 82
Microecosystems, 490, 490*f*
Microthermal climates, 229-234
Midlatitude broadleaf and mixed forest, 506*f*-507*f*, 517-518
    characteristics of, 508*t*-509*f*
Midlatitude cyclone, 168-172, 169*f*
    daily weather map of, 170-172, 171*f*
Midlatitude grasslands, 506*f*-507*f*, 520-521, 521*f*
    characteristics of, 508*t*-509*f*
Midlatitude zones, 13, 13*f*
Mid-ocean ridges, 269, 269*f*
Milankovitch, Milutin, 443
Milky Way Galaxy, 38-39
Mineral, 260-261
Mirage, 79-80
Mississippi River, 358, 361
    delta of, 358, 360*f*
    floods on (1993), 356, 357*f*
Mistral, 129-130
Model
    definition of, 7
    of system, 7-8
Mohorovičić; discontinuity (Moho), 257
Moist adiabatic rate (MAR), 152, 153*f*
Moist soil, 456-457
Moisture droplets, 154
Mollisols, 470-471, 471*f*-472*f*
Monsoon, 132-134, 223-224
Montane forest, 518. *See also* Northern needleleaf and montane forest
Mont Pelée, 308
Moon, influence on tides, 398, 399*f*
Moraine, 427, 432
    end, 433, 433*f*, 435*f*
    ground, 433, 435*f*

Moraine, *(cont.)*
lateral, 427, 432, 433*f*
medial, 427, 432, 433*f*
recessional, 433
terminal, 432-433, 433*f*
Morelos Dam, 386
Mountains, 285
Appalachian, 293
Cordilleran system, 295
Eurasian-Himalayan system, 295
formation of, 291-295
Mount Pinatubo, 2, 7, 7*f*, 308, 538
and atmospheric circulation, 109*f*
and global cooling, 244
and global temperatures, 241
Mount Saint Helens, 309-311, 310*f*
succession on land surrounding,
499, 499*f*
Mount Vesuvius, 307
Movement, 3*f*, 4
definition of, 3*f*
Mutualism, 484

N
National Council for Geographic
Education (NCGE), 4
National Flood Insurance Act, 418
National Oceanic and Atmospheric
Administration (NOAA), 29
National Ocean Service, 24
National Weather Service, 142
Natural levees, 355-356
NAVSTAR satellites, 397
Neap tide, 398, 399*f*
Nebula, 38
Needleleaf forest and montane forest,
506*f*-507*f*
Negative feedback, 6
Net primary productivity, 486, 487*f*
Net radiation (NET R), 45, 83
Nevado del Ruiz, 330
Nevado Huascarán avalanche, 324
New Jersey shore, 416-418
Newton, Sir Isaac, 10
Next Generation Water Level
Measurement System, 397
Niagara Falls, 354, 355*f*, 356
Niche, 484
Nickpoints, 353-354, 354*f*-355*f*
Nile River, 356
delta of, 358, 359*f*
Nimbostratus clouds, 155, 156*f*
Nimbus-7, 45, 77
Nitrogen cycle, 491, 492*f*
Nitrogen dioxide, and air pollution,
66
Nitrogen-fixing bacteria, 491
Nonradiative transfer, 81

Normal fault, 290, 291*f*
Normal lapse rate, 58, 151
Northern needleleaf and montane
forest, characteristics of, 508*t*-
509*f*
Northern needleleaf forest and
montane forest, 518, 518*f*
Nuclear winter, 538-539
Nuclear winter hypothesis, 244
Nullarbor National Park, 394*f*

O
Occluded front, 170
Ocean basins, 283
Ocean currents, 124*f*
deep, 135, 135*f*
and sea-surface temperatures, 91-92
surface, 134-135
Ocean floor, 282
Oceanic trenches, 270
Oceans, 207*f*, 207-209, 208*f*
depths of, 209*t*
Ogallala aquifer, 199-200
Ohio River, 346*f*
Oil spills, 531*f*, 531-532, 533*f*, 534
Omnivore, 494
Open system, 5, 6*f*
definition of, 5
Earth's energy equilibrium as, 8
Orogenesis, 291-295
types of, 292-293
Orographic lifting, 164, 165*f*
Ortelius, Abraham, 267
Outgassing, 142
Outwash plain, 434
Oval projection, 22, 23*f*
Overland flow, 190
Overlay analysis, 31
Oversteepened slope, 332, 332*f*
Oxbow lake, 351, 352*f*
Oxidation, 325
Oxisols, 466-468, 467*f*
Oxygen cycle, 490, 491*f*
Ozone layer, 57
loss of, 59-61
Ozone pollution, 67-68
Ozonosphere, 57-58

P
Pacific high, 124
Paleoclimatology, 239, 443-444
definition of, 442
Paleolakes, 445, 445*f*
Pangaea, 268, 270-271, 272*f*
Parabolic dune(s), 377, 378*f*
Parallel
definition of, 13
standard, 22-23, 23*f*

Parallel drainage pattern, 347, 347*f*
Parasitic relationships, 484
Parcel, 151, 152*f*
Parent material, 320, 321*f*
of soil, 454
Parker Dam, 386
Paternoster lakes, 432
Patterned ground, 438, 439*f*
Peak flow, 365
Ped, 456
Pediment, 318, 392*f*
Pedon, 454, 455*f*
Percolation, 186
Periglacial, definition of, 437
Periglacial landscapes, 437-441
humans and, 438-441
location of, 437
Perihelion, 38
Permafrost, 232, 437-438
alpine, 438
and construction, 438-441, 440*f*
continuous, 438
discontinuous, 438
distribution in Northern
Hemisphere, 437*f*
Permeable, 197
Peroxyacetyl nitrates (PAN), and air
pollution, 66
Persian Gulf War, 532, 533*f*
effects on desert pavement, 373
Peyto Glacier, 433, 435*f*
Phase change, 144, 145*f*
Photic layer, 90-91, 502
Photochemical smog, 65-67
Photoperiod, 489
Photosynthesis, 485
Phreatophytes, 235
Physical geography, 4
Physical weathering, 321-323, 324*f*
Phytoplankton, 502
Piedmont glacier, 424
Piezometric surface, 197
Pioneer community, 499
Place, 3*f*, 4
Plains, 284
Planar projection, 22, 23*f*
Plane of the ecliptic, 38
Planetesimal hypothesis, 38
Planimetric map, 25
Plants
leaf activity in, 485
photosynthesis and respiration in,
485, 486*f*
primary productivity in, 485-486
Plate(s), 274*f*
collisions of, 292-293, 294*f*
convergence of, 292-293, 294*f*
convergent boundaries of, 273*f*, 274

divergent boundaries of, 271-274, 273f
transform boundaries of, 273f, 274
Plateau basalts, 306
Plate tectonics, 267-277, 273f
Platform, wave-cut, 405, 407f
Playa, 383, 392f
Pleistocene ice age epoch, 441-445
    extent of glaciation in Northern Hemisphere, 442f
    temperatures in, 441
Plinthite, 466, 468f
Pluton, 262
Pluvial periods, 445
Podzolization, 474, 475f
Point bar, 350, 351f
Polar climates, 234-235
Polar easterlies, 125
Polar front, 124
Polar high-pressure cells, 119, 125
Pollution
    of air. See Air pollution
    of groundwater, 198-200
Polypedon, 454, 455f
Porosity, 196-197
Positive feedback, 6
Potential evapotranspiration (POTET), 188-191, 191f
    definition of, 189
    determination of, 189
POTET. See Potential evapotranspiration (POTET)
Powell, John Wesley, 316, 386
Precession, 443, 444f
Precipitation (PRECIP), 188-190, 190f
    probable maximum (PMP), 362
    and temperature, 217-218, 218f-219f
Pressure gradient force, 115f, 115-116, 117f
Pressure-release jointing, 323
Primary productivity, 485
    net, 486, 487f
Prime meridian, 17-19
    definition of, 14-16
Principles of Geology (Charles Lyell), 252
Probable maximum flood (PMF), 362
Probable maximum precipitation (PMP), 362
Process, 4
Producers, 493
Projection. See Map projection
Province, 390
Ptolemy, 12-13
Public Lands Survey System, 25
Pyroclastics, 307
Pythagoras, 10

Q
Qattara Depression, 373
Quadrangle system (United States Geological Survey), 27f
Quantity equilibrium, of water on Earth, 142-143

R
Radial drainage pattern, 347, 347f
Radiant energy, 42-44
Radiation
    diffuse, 79
    global net, 45
    longwave, 43
    net (NET R), 83
    shortwave, 43
    solar, 77f
Radiation balance, Earth-atmosphere, 81, 81f
Radiation curves, 82, 82f
Radiation fog, 159
Radiatively active gasses, 239
Radiative transfer, 81
Radiosondes, 151
Rain bands, 177
Raindrops, formation of, 155f
Rain forest, 220-224
    canopy, 505, 510f, 511
    deforestation of, 511, 512f, 516
    Earth Summit agreement on care of, 557
    equatorial and tropical, 505-511, 510f
    temperate, 518-519, 519f
Rain gauge, 188, 189f
Rain shadow, 164, 165f
Rayleigh, Lord, 79
Rayleigh scattering, 79
Recessional moraine, 433
Rectangular drainage pattern, 347, 347f
Reflection, 76-78
Refraction, 79-80
Reg desert, 373
Region, 3f, 4
    definition of, 3f
Regolith, 320, 321f
Rejuvenation, 353, 354f
Relative humidity, 148-151, 149f
Relative time, 252
Relief, 282
    crustal orders of, 283-284
Remote sensing, 27-30, 28f
    active systems, 29
    passive systems, 29-30
Representative fraction (RF), definition of, 20
Reproductive niche, 484
Reservoirs, 362-363, 363f

Resilience, 496
Reverse fault, 290, 291f
Reversing dune, 379f
Revolution, 48, 49t
Rhinoceroses, risk of extinction, 513, 514f
Rhumb lines, 24, 24f
Ria, 410
Richter scale, 297
Rift valleys, 290
Rill, 345
Ring of fire, 292
Ripples, 375, 376f
River basin, 345
Rivers. See also Streams
    largest, 203t
    management of, 361-366
Robinson, Arthur H., 25
Robinson projection, 25, 25f
Roche moutonée, 434, 436f
Rock(s)
    definition of, 261
    igneous, 261-262
Rock cycle, 260-267, 262f
Rockfall, 333-334, 334f
Rossby, G. G., 125
Rossby waves, 125, 127f
Rotation, 49, 49t
Rotational slide, 333f, 335-336
Runoff, 190, 202f

S
Sagan, Carl, 530, 540
Salinity, 208-209
Salinization, 468
Saltation, 350
    eolian, 375, 375f
Salt-crystal growth, 321-322
Salt marshes, 412-413, 415f
San Andreas fault, 292f, 299, 300f
Sand erosion, prevention of, 376f
Sand sea, 375, 377f
Sandstone, 389
    cross bedding in, 391f
Sand storm, 374
San Francisco earthquakes, 295-299
San Joaquin Valley, 470
San Juan River channel, changes in, 348, 348f
Santa Ana winds, 130
Satellite(s). See also specific satellite
    polar-orbiting, 8
Saturated, definition of, 149
Saturation, 196
Savanna, 224
Scale
    definition of, 20
    in mapmaking, 20-21

Scarification, 337

Scattering, 78-79

Schweickart, Rusty, 530

Scientific Assessment of Climate Change (1990), 241

Scientific method, 38-39

Sclerophyllus vegetation, 519-520

Scripps Institution of Oceanography, 241

Sea cliffs, 404

Sea-floor spreading, 268f, 269-270, 305f

Sea level, 243-244, 396-397, 397f
  changes in, 404
  mean, 397
  variation along U.S. coastline, 398

Seamount, 277

Seas, 207-209
  list of, 208f

Seasat satellite, 29, 118
  and surface winds, 110-111, 111f

Seasonality, 47-48
  definition of, 47

Seasons, 46-53, 50t, 51f
  reasons for, 48-50

Sea-surface temperature (SST), 91-92, 92f

Seawater intrusion, into groundwater, 198

Sediment(s), 264-265, 320

Sedimentary cycles, 490-491

Sedimentary rocks, 264-266

Seif, 378

Seif dune, 379f

Seismic gaps, 302

Seismic sea waves, 402-404

Seismic waves, 255

Seismograph, 297

Selva, 505

Semideciduous, definition of, 515

Semidesert
  cold, 523, 523f. *See also* Desert biomes
  warm, 522f, 522-523. *See also* Desert biomes

Sensible heat, 56
  turbulent transfer of, 83

Sensible temperature, 98

Sheet erosion, 460

Sheet flow, 345

Sheeting, 323, 324f

Shields, 285-286

Shield volcano, 306

Shoreline, 396, 397f

Shortwave radiation, 43

Shrubs, 505

Sima, 257

Singh, Khushwant, 133-134

Sinkhole(s), 326, 327f, 330f

Slash-and-burn shifting cultivation, 466-468

Slide, 333
  rotational, 333f, 335-336
  translational, 333f, 335

Slipface, 376, 377f

Slope(s), 318
  debris, 318
  in disequilibrium, 318, 320f
  forces acting on, 319f
  oversteepened, 332, 332f
  role in mass movement, 331-333
  waning, 318
  waxing, 318

Small circle, 16-17
  definition of, 17

Smog
  industrial, 68-70
  photochemical, 65-67

Snow dunes, 379

Snowfield, 422, 424

Snowflakes, formation of, 155f

Snowline, 422

Soil, 454
  acid deposition, 205-206
  characteristics of, 454-460
  chemistry of, 457-459
  color of, 456
  consistence of, 456-457
  dry, 457
  moist, 456-457
  moisture of, 457-458
  parent materials of, 454
  pH of, 458, 459f
  porosity of, 457
  properties of, 456-458
  structure of, 456
  texture of, 456, 457f
  wet, 456
  wilting point of, 191

Soil classification, 460-478
  history of, 460

Soil colloids, 458, 459f

Soil creep, 336, 337f

Soil fertility, 454, 458

Soil formation, 459-460
  human factors in, 460, 461f
  natural factors in, 459

Soil horizons, 454-456
  diagnostic, 460-461

Soil loss, 466
  worldwide distribution of, 461f

Soil moisture recharge, 193

Soil moisture storage (STRGE), 191-193, 192f

Soil moisture utilization, 193

Soil orders
  Alfisols, 471, 472f-473f

Andisols, 476, 477f

Aridisols, 468-471, 469f, 472f

Entisols, 474-475, 476f

Inceptisols, 475476

Mollisols, 471f-472f

Oxisols, 466-468, 467f

Spodosols, 474, 475f

taxonomy of, 464t-465t

Ultisols, 473, 474f

Vertisols, 476, 477f

worldwide distribution of, 462f-463f

Soil profiles, 454

Soil science, 454

Soil separates, 456

Soil series, 460

Soil solution, 458

Soil Taxonomy, 460

Solar constant, 44-45
  definition of, 44

Solar energy, 39-44, 48f, 84-85
  collection and concentration of, 84-85
  in ecosystem, 489

Solar radiation, 77f

Solar system, 38-39, 40f

Solar wind, 41-42
  effect on Earth's weather, 44

Solifluction, 438

Solstice, 50-51
  summer, 51
  winter, 50

Solum, 456

Solution, 325, 349

South Cascade Glacier, 428

Spatial, definition of, 3

Spatial analysis, 4

Specific heat, 91

Specific humidity, 150-151, 151f

Speed of light, 38

Speleology, 331

Sphagnum peat, 478

Spheroidal weathering, 324, 325f

Spodosols, 474, 475f

Spring tide, 398, 399f

Squall line, 167, 167f

Stability, 151-154
  of ecosystems, 496-498, 497f
  inertial, 496

Staff gauge, 365

Stalactite(s), 328, 329f-330f

Stalagmite(s), 328, 330f

Standard line or point, in map projections, 22-23, 23f

Standard meridian, 22-23, 23f

Standard parallel, 22-23, 23f

Standard time, 17-19

Star dune(s), 377-378, 379f-380f

Statistical construction, 218
Steady-state equilibrium, 7
Steppes, 235-237
Stilling well, 365
Stomata, 485
Storms, 172
Storm surges, 179
Stovepipe Wells dune field, 384f
Strain, 287, 287f
Stratified drift, 432, 434, 448
Stratigraphy, 264
Stratocumulus clouds, 157f, 158
Stratosphere, 57
Stratovolcanoes, 306
Stratus clouds, 155, 156f
Stream(s), 200-203
    channels, changes in, 348, 348f
    deposition of, 354-361
    erosion of, 348
    exotic, 201
    graded, 353
    gradient of, 353f, 353-354
    as political boundaries, 351
    terraces of, 356-357
    transport by, 348-350, 349f
    underground, 330f
Stream course, 345
Streamflow
    characteristics of, 347-351
    measurement of, 361-366, 366f
    volume/velocity relationship, 347-
        348, 349f
Stream piracy, 347
Stress, 287, 287f
STRGE. See Soil moisture storage
Strike-slip fault, 290, 291f
Strong, Maurice F., 536
Subantarctic zones, 13, 13f
Subarctic climates, 231-234, 233f
Subarctic zones, 13, 13f
Subduction, 268f, 270
Subduction zone, 268f, 270, 305f
Sublimation, 144-145
Submergence, 397
Submergent coastlines, 409-410
Subpolar low-pressure cells, 119, 124-
    125
Subsolar point, 45
Subsurface diagnostic horizon, 461
Subtropical high-pressure cells, 119,
    122-124
Subtropical zones, 13, 13f
Succession, 499f
    aquatic, 501-502, 502f
    ecological, 498-502
    primary, 499
    secondary, 500, 500f
    terrestrial, 499-500

Sulfur dioxide, 69
Sulfur oxides, and air pollution, 68-70
Summer solstice, 51
Sun
    altitude of, 47
    declination of, 47
    influence on tides, 398, 399f
Sunspots, 41, 42f
    definition of, 41
Superposition, 252
Surface creep, 375
Surface tension, 144
Surface water resource, 190
Surplus (SURPL), 190-191
Suspended load, 349
Sustainable development, Earth
        Summit agreement on, 537-538
Swelling clays, 476
Swells, 400
Symbiotic, definition of, 410
Symbiotic relationships, 484
Syncline, 287
Synclinical ridge, 288, 288f
System
    closed, 5, 8-9
    definition of, 5
    model of, 7-8
    open, 5, 8
System Probatoire d'Observation de la
    Terre (SPOT), 29-30
Systems theory, 5-8

T
Tablelands, 284-285
Taiga, 518
Tajikistan, 528f, 530
Taku, 130
Talus cone(s), 334, 334f, 430
Tambora, 108
Tarn, 431-432
Tectonic cycle, 260
Tectonic processes, 267
Tectonics, plate, 267-277, 273f
Temperate rain forest, 506f-507f, 518-
    519, 519f
    characteristics of, 508t-509f
Temperature, 56, 83
    annual range of, 97-98
    apparent, 98
    dew-point, 149
    effects on ecosystem, 488f, 489
    global warming, 239-241
    heat index for, 99, 99f
    and human body, 98-99
    patterns, 93-101, 94f, 96f-97f
    and precipitation, 217-218, 218f-
        219f
    principal controls of, 83-93

    scales of, 87-88
    sensible, 98
    wind chill factor in, 98f, 98-99
Temperature inversion, 63-64
Tephra, 303
Terminal moraine, 432-433, 433f
Terminus, 432
Terrace(s)
    alluvial, 356-357, 357f
    wave-cut, 405, 407f
Terranes, 286-287
Terrestrial ecosystems, 503-505
Terrestrial succession, 499-500
Teton Dam, collapse of, 364f, 365
Thallophytes, 505
Thermal equator, 94
Thermopause, 56
    definition of, 44
Thermosphere, 55f, 56
Thomas, Lewis, 54
Thomson, William (Lord Kelvin), 87
Thoreau, Henry David, 142
Thornthwaite, Charles Warren, 187,
    189, 219
Thrust fault, 290
Thunder, 172
Thunderstorms, 172, 173f
Tidal bulge, 398
Tidal glacier, 424
Tidal power, 399
Tidal range, 398, 400f
Tide(s), 398-399
    ebb, 398
    factors affecting, 398-399
    flood, 398
    neap, 398, 399f
    spring, 398, 399f
Tied dune, 377
Till, 432, 447-448
Till plain, 433-434
Tilted fault block, 292, 293f
Time
    Coordinated Universal Time (UTC),
        18-19
    Daylight saving time, 19
    geologic, 252, 253f
    Greenwich Mean Time (GMT), 18
    relative, 252
    standard, 17-19
    zones, 20f
Tombolo, 407f, 408, 408f
TOPEX/Poseidon satellite, 397
Topographic maps, 25-27, 26f
Topography, 282
Tornadoes, 172-175, 174f
    damage caused by, 174f, 176
    frequency of, 175f
Torricelli, Evangelista, 112-114

Total runoff, 190
Tower karst, 328, 329*f*
Traction, 350
Trade winds, 122
Transcurrent fault, 290
Transform fault(s), 274, 290, 305*f*
Translational slide, 333*f*, 335
Transparency, and distribution of heat
    energy, 90-91
Transpiration, 188
Transport, 342
    fluvial, 348-350, 349*f*
Transverse dune(s), 377, 378*f*
Trees, 505
Trellis drainage pattern, 347, 347*f*
Trophic niche, 484
Tropical climates, 222-224
Tropical cyclones, 175-179, 177*f*
    classification of, 176*t*
Tropical monsoon climates, 223-224
Tropical Ocean Global Atmosphere
    (TOGA), 92
Tropical rain forest climates, 222-223,
    223*f*
Tropical savanna, 506*f*-507*f*, 515-517,
    517*f*
    characteristics of, 508*t*-509*f*
Tropical savanna climates, 224
Tropical seasonal forest and scrub,
    506*f*-507*f*, 511-515, 516*f*
    characteristics of, 508*t*-509*f*
Tropical zones, 13, 13*f*
Tropic of Cancer, 13, 47, 51
Tropic of Capricorn, 13, 47, 50
Troposphere, 58-62
    energy balance in, 76-81
True shape, 22
Tsunami, 402-404
Tundra, 234
    alpine. *See* Alpine tundra
    arctic. *See* Arctic tundra
Turbulence, 348
Twilight, 53
Typhoons, 177-179

U
Ultimate base level, 316, 317*f*
Ultisols, 473, 474*f*
Ultraviolet index, 58
Unconfined aquifer, 197
Uniformitarianism, 252
United Nations Conference on
    Environment and Development
    (UNCED), 244
United Nations Environment
    Programme, 241

United Nations Environment
    Programme (UNEP), 244
United Nations Framework
    Convention on Climate Change,
    244
United States Geological Survey
    (USGS), 25-27
    National Mapping Program, 27
    quadrangle system, 27*f*
Upslope fog, 159
Upwelling current(s), 135, 268*f*
Urban environments, and
    temperature, 100-101, 102*t*
Urban heat islands, 100-101
Urbanization, effects on streamflow,
    365-366, 366*f*
Utilidors, 438, 440*f*
Uvala, 326, 327*f*

V
Valley fog, 159
Valley glacier(s), 422, 423*f*, 447
    compound, 427
Valleys
    hanging, 432, 432*f*
    rift, 290
Valley train deposit, 433, 435*f*
Vaporization, 145
Vapor pressure, 150
Vegetation, and weathering, 320-321,
    321*f*
Ventifacts, 374
Vernal equinox, 50
Vertical datum, definition of, 26
Vertisols, 476, 477*f*
Vespucci, Amerigo, 15
Volcanic activity
    locations of, 303-304
    types of, 304*f*, 304-308, 305*f*
Volcanoes, 303-308
    composite, 306-307
    list of, 308*t*
    shield, 306
von Humboldt, Alexander, 489

W
Walden (Henry David Thoreau), 142
Waning slope, 318
Warfare
    effects on desert pavement, 373
    environmental cost of, 538-539
Warm front, 167-168
Wash, 382
Water, 142
    artesian, 197
    available, 191-192

brackish, 209
capillary, 191
composition of, 208-209
consumptive use, 204-206
daily budget, 204-206
distribution of, on Earth, 143*f*, 143-
    144
effects of freezing, 148
effects on ecosystem, 488*f*, 489
gas phase of, 146-147
gravitational, 192
groundwater resources of, 196*f*-
    197*f*, 196-200
hygroscopic, 191
liquid phase of, 145-146
properties of
    heat, 144-148
    in nature, 147-148
salinity of, 208-209
solid phase of, 145
as universal solvent, 325
withdrawal of, 206
world economy of, 206-207
Water balance, 187-196, 193*f*-194*f*
    examples of, 193
Water-balance equation, 187-193,
    188*f*
Water budget, daily, 204-206
Waterfalls, 354, 355*f*
Water gaps, 293
Water resource management, 186
Watershed, 342
Waterspout, 175
Water supply, 203-207, 204*t*
Water table, 197
Water vapor, 146
Wave-cut platform, 405, 407*f*
Wave-cut terrace, 405, 407*f*
Wave cyclone, 168
Wavelength
    definition of, 42
    and Rayleigh scattering, 79
Wave refraction, 401-402, 403*f*
Waves, 400-404
    killer, 402
    seismic sea, 402-404
    of transition, 400, 401*f*
    of translation, 400-401, 401*f*
Wave trains, 400
Waxing slope, 318
Weather, 142, 214
    violent, 172-179
Weathering, 320
    chemical, 323-325
    differential, 321
    physical, 321-323, 324*f*

processes, 320-325
   spheroidal, 324, 325*f*
Wegener, Alfred, 267-268
Weir, 362
Westerlies, 122-123
Western intensification, 134
Western Pacific Warm Pool, 92
Wet soil, 456
Wilson, E. O., 513
Wilting point, 191
Wind, 372-380
Wind(s)
   classification of, 108-109
   definition of, 110
   as energy resource, 131-132
   geostrophic, 119*f*
   global, 110-111
   katabatic, 129
   land-sea breezes, 126-128, 129*f*

measurement of, 110
monsoonal, 132-134, 133*f*
mountain-valley breezes, 129,
   130*f*
Santa Ana, 130
surface, 110-111, 111*f*
upper atmospheric circulation of,
   125-126
Wind chill factor, 98*f*, 98-99
Wind compass, 110*f*
Wind Energy Association, 132
Wind vane, 110
Windward side, 376
Winter solstice, 50
Woodwell, George, 239
World Climate Research Program, 187
World structural regions, 293-295,
   296*f*
World water economy, 206-207

World Weather Watch System, 244
Written scale, definition of, 20

X
Xerophytic, definition of, 517
Xerophytic vegetation, 235

Y
Yardangs, 374
Yazoo tributary, 355*f*, 356
Yellostone National Park, fire ecology
   in, 501, 501*f*
Yellowstone National Park, 277

Z
Zonal flows, 110
Zone of aeration, 196
Zone of saturation, 196
Zones, latitudinal geographic, 13*f*, 13-14

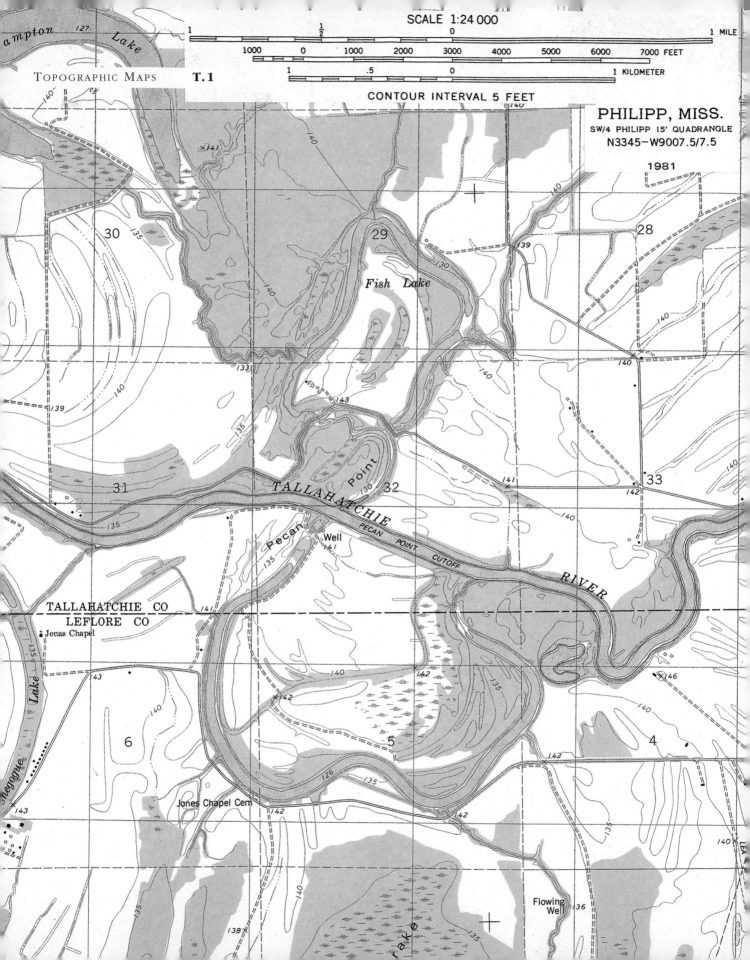

SCALE 1:24 000

CONTOUR INTERVAL 5 FEET

TOPOGRAPHIC MAPS **T.1**

PHILIPP, MISS.
SW/4 PHILIPP 15' QUADRANGLE
N3345—W9007.5/7.5

1981

Fish Lake

Point

TALLAHATCHIE

Pecan   PECAN   POINT   CUTOFF   RIVER

Well

TALLAHATCHIE CO
LEFLORE CO

Jonas Chapel

Jones Chapel Cem

Flowing Well

OMAHA NORTH, NEBR.—IOW.

41095-C8-TF-024

1956

PHOTOREVISED 1984

SCALE 1:24 000

CONTOUR INTERVAL 10 FEET

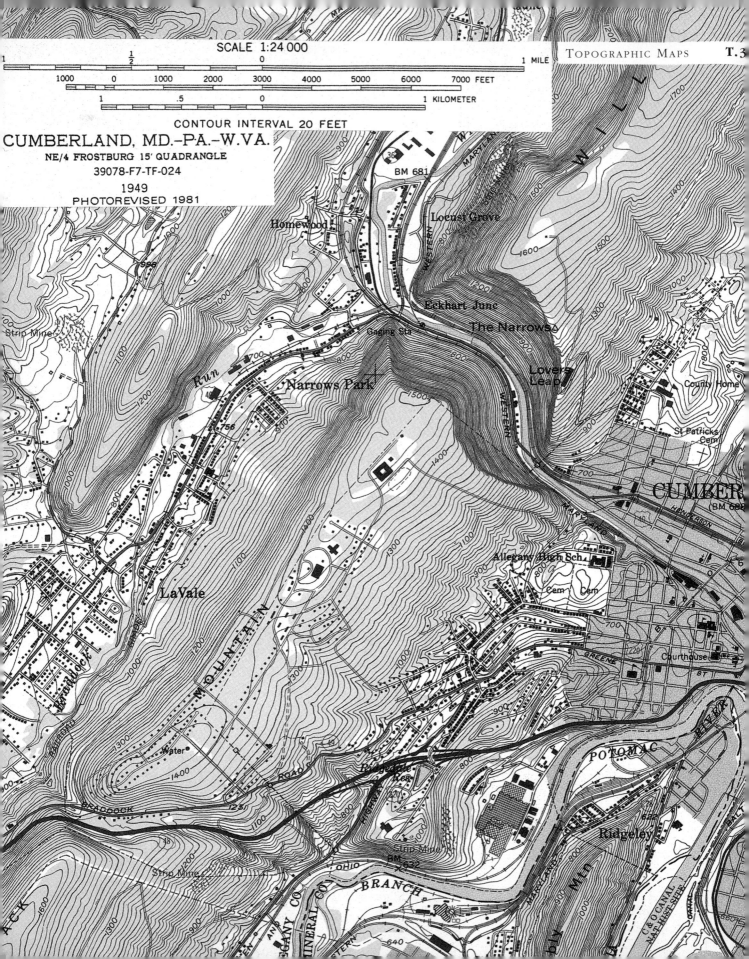

SCALE 1:24 000

1 | ½ | 0 | 1 MILE

1000 | 0 | 1000 | 2000 | 3000 | 4000 | 5000 | 6000 | 7000 FEET

1 | .5 | 0 | 1 KILOMETER

CONTOUR INTERVAL 20 FEET

CUMBERLAND, MD.–PA.–W.VA.
NE/4 FROSTBURG 15' QUADRANGLE
39078-F7-TF-024
1949
PHOTOREVISED 1981

Homewood

Locust Grove

BM 681

Eckhart Junc

Gaging Sta

The Narrows

Run

Narrows Park

Lovers Leap

County Home

St Patricks Cem

CUMBER

(BM 695)

Allegany High Sch.

Cem   Cem

LaVale

MOUNTAIN

Courthouse

Water

GREENE

POTOMAC RIVER

Ridgeley

Res.

ROAD

BRADDOCK

Strip Mine

Strip Mine

BM 632

BRANCH

BM 640

OHIO

MINERAL CO

ALLEGANY CO

C&O CANAL NAT HIST SIDE

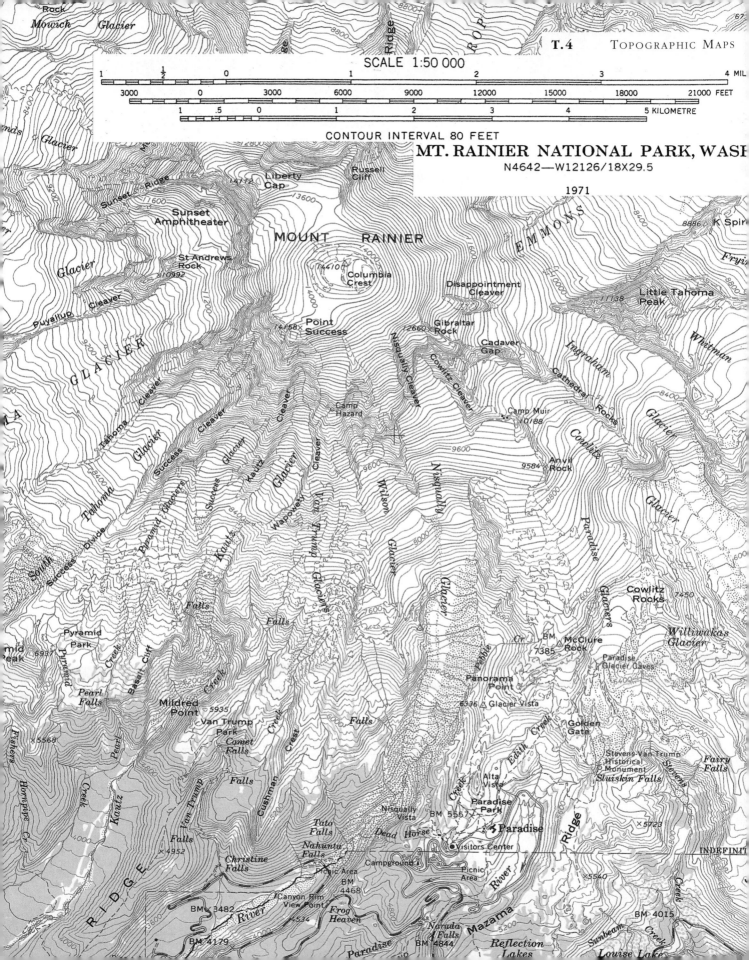

Topographic Maps   T.5

LIBERTY

Prospect Hill
Crouch Sch.
Kam Sch.
BM 1047
Skiff Lake
East Liberty Sch.
Clarklake
Clark
Mud Lake
Crispell Lake
Liberty
River
Grand Lake
Infant Sch.
Putney Mill Pond
Grand Lakes
Sutton Sch.
JACKSON CO
HILLSDALE CO
Braxee Lake
Crystal Lake
Bunday Hill
Irene Lake
Perch Lake
Somerset Center
Somerset
Goose Lake
Briggs Sch.
Gravel Sch.

MERIDIAN
MICHIGAN
NORTHERN
BLUE RIDGE

ACKSON, MICH.
N4200—W8415/15

1935

SCALE 1:62500

4 MILES

3000   0   3000   6000   9000   12000   15000   18000   21000 FEET

1 .5 0   1   2   3   4   5 KILOMETERS

CONTOUR INTERVAL 10 FEET